Phylogeography and Population Genetics in Crustacea

CRUSTACEAN ISSUES

General editor
Prof. Dr. Stefan Koenemann
Department of Biology and Didactics
Science and Technology
University of Siegen
Germany

1   Crustacean Phylogeny
     Schram, F.R. (ed.)
     1983   ISBN 90 6191 231 8 Sold out

2   Crustacean Growth: Larval Growth
     Wenner, A. (ed.)
     1985   ISBN 90 6191 294 6

3   Crustacean Growth: Factors in Adult Growth
     Wenner, A. (ed.)
     1985   ISBN 90 6191 535 X

4   Crustacean Biogeography
     Gore, R.H. & Heck, K.L. (eds.)
     1986   ISBN 90 6191 593 7

5   Barnacle Biology
     Southward, A.J. (ed.)
     1987   ISBN 90 6191 628 3

6   Functional Morphology of Feeding and Grooming in Crustacea
     Felgenhauer, B.E., Thistle, A.B. & Watling, L. (eds.)
     1989   ISBN 90 6191 777 8

7   Crustacean Egg Production
     Wenner, A. & Kuris, A. (eds.)
     1991   ISBN 90 6191 098 6

8   History of Carcinology
     Truesdale, F.M. (ed.)
     1993   ISBN 90 5410 137 7

9   Terrestrial Isopod Biology
     Alikhan, A.M.
     1995   ISBN 90 5410 193 8

10   New Frontiers in Barnacle Evolution
     Schram, F.R. & Hoeg, J.T. (eds.)
     1995   ISBN 90 5410 626 3

11   Crayfish in Europe as Alien Species - How to Make the Best of a Bad Situation?
      Gherardi, F. & Holdich, D.M. (eds.)
      1999   ISBN 90 5410 469 4

12   The Biodiversity Crisis and Crustacea - Proceedings of the Fourth International
      Crustacean Congress, Amsterdam, Netherlands
      Vaupel Klein, J.C. von & Schram, F.R. (eds.)
      2000   ISBN 90 5410 478 3

13   Isopod Systematics and Evolution
      Kensley, B. & Brusca, R.C. (eds.)
      2001   ISBN 90 5809 327 1

14   The Biology of Decapod Crustacean Larvae
      Anger, K.
      2001   ISBN 90 2651 828 5

15   Evolutionary Developmental Biology of Crustacea
      Scholtz, G. (ed.)
      2004   ISBN 90 5809 637 8

16   Crustacea and Arthropod Relationships
      Koenemann, S. & Jenner, R.A. (eds.)
      2005   ISBN 0 8493 3498 5

17   The Biology and Fisheries of the Slipper Lobster
      Lavalli, K.L. & Spanier, E. (eds.)
      2007   ISBN 0 8493 3398 9

18   Decapod Crustacean Phylogenetics
      Martin, J.W., Crandall, K.A. & Felder, D.L. (eds.)
      2009   ISBN 1 4200 9258 8

19   Phylogeography and Population Genetics in Crustacea
      Held, C., Koenemann, S. & Schubart, C.D. (eds.)
      2011   ISBN 978 1 4398 4074 3

Phylogeography and Population Genetics in Crustacea

Edited by

Christoph Held

Alfred Wegener Institute for Polar and Marine Research
Bremerhaven, Germany

Stefan Koenemann

University of Siegen
Siegen, Germany

Christoph D. Schubart

University of Regensburg
Regensburg, Germany

CRC Press
Taylor & Francis Group
Boca Raton London New York

CRC Press is an imprint of the
Taylor & Francis Group, an **informa** business

CRC Press
Taylor & Francis Group
6000 Broken Sound Parkway NW, Suite 300
Boca Raton, FL 33487-2742

First issued in paperback 2019

© 2011 by Taylor & Francis Group, LLC
CRC Press is an imprint of Taylor & Francis Group, an Informa business

No claim to original U.S. Government works

ISBN-13: 978-1-4398-4073-3 (hbk)
ISBN-13: 978-0-367-38199-8 (pbk)

Visit the Taylor & Francis Web site at
http://www.taylorandfrancis.com

and the CRC Press Web site at
http://www.crcpress.com

Contents

Preface

CHRISTOPH HELD[1], STEFAN KOENEMANN[2], & CHRISTOPH D. SCHUBART[3]

[1] Division of Biosciences, Alfred Wegener Institute for Polar and Marine Research, Bremerhaven, Germany
[2] Department of Biology and Didactics, University of Siegen, Siegen, Germany
[3] Biology 1, University of Regensburg, Regensburg, Germany

Crustacean Issues 19, "Phylogeography and Population Genetics in Crustacea," deals with a subject, the analysis and interpretation of intraspecific genetic variation, that has become again the focus of attention after a long period of neglect by the scientific community. The theoretical foundation for this field was laid in the first half of the last century; the Hardy-Weinberg principle was published more than 100 years ago, and in the decades that followed it was mainly the works of R. A. Fisher, S. Wright, and J. B. S. Haldane that formed the basis of population genetics as we know it today. At that time its practical use was severely limited by the fact that the underlying variants in the genotype had to be visible in the phenotype in order to be amenable to analysis. Now, however, the use of molecular tools has become more widespread in application and effectiveness, so much so that putting the predictions of population genetic theory to the test became feasible in real-world scenarios.

The advance of phylogenetic analysis during the 1990s has permanently added sophisticated molecular techniques to the zoologist's toolbox. However, it was almost exclusively the genetic variation to be found among species that was under investigation whereas the intraspecific variance was routinely down-weighted or entirely excluded in earlier phylogenetic studies. Therefore new modes of analyzing have had to be adopted to adequately handle and appreciate the part of the genetic variation to be found within species. In addition to population genetics, which describes patterns of differentiation at any given moment, phylogeography emphasizes the element of time and quickly gained a foothold as an alternative method of analysis. Phylogeography makes use of spatial information mapped onto haplotype networks or gene trees and is perhaps more intuitively understood than population genetics statistics, especially for an audience trained in phylogeny. Nevertheless, more recently both methods have evolved in ways that have bridged the gap between phylogenetic (macroevolutionary) studies and population genetic (microevolutionary) studies.

We all have seen how the increased efforts focused on phylogenetic research and international barcoding campaigns of the past ten years have increased the availability of data useful for phylogeography on a massive scale. Phylogeography, population genetics, and the ever-accelerating progress in technology collectively have put larger and larger molecular datasets into play, rekindling the interest in the study of intraspecific variation.

As crustacean zoologists with very diverse backgrounds and research interests set out to investigate patterns on the scale of populations using molecular tools, they are exposed to a bewildering variety of genetic markers, analytical methods, and computer programs from which to choose. The feedback during organization of two symposia at The Crustacean Society Summer Meeting 2009 in Tokyo ("Phylogeography and Population Genetics in Decapod Crustacea" and "Speciation and Biogeography in Non-Decapod Crustaceans" organized by Christoph D. Schubart and Christoph Held, respectively), and the response these programs generated confirmed our feeling that the crustacean community on the whole was watching the field of intraspecific genetic variability attentively and would welcome an up-to-date summary.

Crustacean Issues 19 "Phylogeography and Population Genetics in Crustacea" intends to open the various uses of intraspecific genetic variability to a wider readership in the crustacean community and to provide guidance at this pivotal moment to workers interested in applying these techniques. To do this, we present examples of current practice, pointing out shortcomings and existing alternatives in the choice of molecular markers and methods of analysis, and provide an overview into the possible future of this field of endeavor.

The book is organized in three sections—theoretical aspects (section 1), and case studies in marine (section 2) and limnic (section 3) crustaceans—comprising 17 chapters in total. The first chapter, a practical guide to the analysis of intraspecific genetic variation by Leese and Held, illustrates how the most relevant shortcomings of mtDNA-only studies can be avoided by adding microsatellites as an additional source of information. The second chapter by Bird et al. on detecting and measuring genetic variation demonstrates how the choice of a statistical estimator of population differentiation can greatly influence the outcome of an analysis and suggests what statistic to use under which circumstances. The next two chapters by Yednock and Neigel, and Toonen and Grosberg, ask to what degree two often-ignored parameters in population genetic studies—temporal variation and selection—could bias the answers to commonly asked questions in phylogeography and population genetics.

The second section of the volume comprises five case studies involving marine crustaceans, the first three of which, by Tsang et al., Barber et al., and Froufe et al. make use of the cytochrome c oxidase subunit I barcoding fragment (COI) to estimate population subdivision, demography, and cryptic diversity of decapod and nondecapod crustaceans. The next two chapters, one by Wanna et al. on penaeid shrimps in the Thai Peninsula and the other by Fratini et al. dealing with the population genetics of *Pachygrapsus marmoratus*, employ nuclear markers (nuclear introns and microsatellites, respectively) and critically compare their results with findings derived from mitochondrial studies.

The third section comprises eight studies on nonmarine crustaceans. We begin with a review by Dufresne of a rarely considered manifestation of genetic variation, namely the ploidy level in *Daphnia*. This is followed by a contribution by Kappas et al. that reviews worldwide phylogeographic patterns in hypersaline brine shrimps. Chapters 12 written by von Rintelen and 13 by Cook et al. use mitochondrial genes to define bioregions and their degrees of affinity in freshwater shrimps in ancient lakes in Sulawesi and rivers in northern Australia. The next two chapters authored by Pérez-Losada et al. and Bracken-Grissom et al. compare the phylogeography and molecular diversity of South American Aeglidae across the Andes, and between river and lake habitats, respectively. Chapter 16 presented by Breinholt et al. explores how the different physiological constraints of two species of crayfish from the same study area have resulted in different colonization histories inferred from mitochondrial DNA. The last chapter written by Schubart et al. employs morphometric, nuclear, and mitochondrial datasets, and in the process the authors examine the underlying reasons for the striking differences in diversity of freshwater brachyuran crabs from Caribbean islands.

We hope that this collection of papers, written by experienced experts in the various uses and aspects of intraspecific genetic variation, may provide a stimulating and reliable resource for established researchers and students alike.

Finally, we would like to extend our sincere thanks to all contributing authors and all our reviewers, both of whom had to endure tight deadlines. John Sulzycki, Pat Roberson, John Edwards, and Joselyn Banks-Kyle provided feedback and instructions from the CRC Press throughout the preparation of the manuscript. Marco Thomas Neiber afforded invaluable support during the proof-reading stage and the production of the LaTeX manuscript.

Christoph Held, Bremerhaven, Germany
Stefan Koenemann, Siegen, Germany
Christoph D. Schubart, Regensburg, Germany
June 2011

I

Analyses of population genetics: guidelines and developments

Analyzing intraspecific genetic variation: a practical guide using mitochondrial DNA and microsatellites

FLORIAN LEESE[1,3] & CHRISTOPH HELD[2]

[1] *Ruhr University Bochum, Department of Animal Ecology, Evolution and Biodiversity, Universitätsstraße 150, D-44801 Bochum, Germany*
[2] *Alfred Wegener Institute for Polar and Marine Research, Functional Ecology, Am alten Hafen 26, D-27568 Bremerhaven, Germany*
[3] *British Antarctic Survey, High Cross, Madingley Road, Cambridge CB3 0ET, United Kingdom*
Both authors contributed equally to the manuscript.

ABSTRACT

Population geneticists and phylogeographers are interested in understanding the processes that have shaped the present distribution and diversity of organisms and the genetic variation they contain. In this context, the analysis of spatial patterns of molecular variation within a species has a long-standing tradition. Recently, as a by-product of the rise of DNA barcoding efforts and technological progress, the availability of datasets on intraspecific variability has increased significantly. Researchers with many different backgrounds are therefore increasingly being faced with the need to carry out a meaningful analysis in the context of population genetics and phylogeography. A central limitation present in many studies today is that either a single or few mitochondrial gene fragments are being studied as a basis for reconstructing the genetic structure of species and inferring their evolutionary history, even though evidence is mounting that analyses based on a single marker in general, and on a single mitochondrial marker in particular, may be severely biased and may even lead to false conclusions.

In this chapter, we emphasize that an independent class of markers should be chosen to complement a mitochondrial dataset. Accordingly, we outline a workflow based on several independent microsatellites in addition to mitochondrial DNA. This guideline builds upon earlier synopses and adds new perspectives afforded by novel laboratory and bioinformatic microsatellite screening procedures. Finally, we summarize basic analytic concepts and common problems associated with their application and highlight the utility of different programs in a workflow towards a hypothesis-driven analysis of intraspecific variation. Where possible, we chose studies on crustaceans as examples.

1 INTRODUCTION

The main intention of phylogeographic and population genetic projects is to study patterns of intraspecific genetic variation in order to infer the underlying microevolutionary processes in a species' evolutionary history that may have generated the observed structure. The major microevolutionary forces shaping the genetic variation within a species are random genetic drift, mutation, diversifying and stabilizing (homogenizing) selection, and gene flow (Figure 1). Whereas genetic drift, mutation, and diversifying selection have the potential to disrupt the gene pool of a species over time, stabilizing selection and gene flow among populations contribute to preserving genetic homogeneity, and ultimately species integrity. Hence, from understanding the patterns of genetic

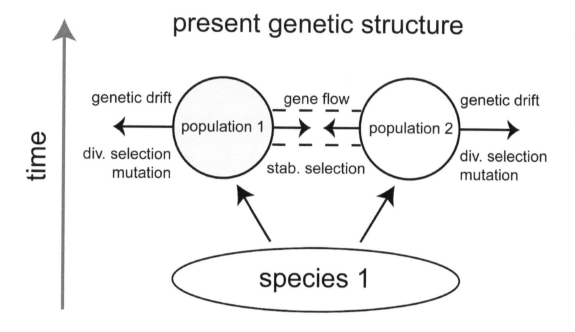

Figure 1. Disrupting and unifying microevolutionary forces acting on a species' gene pool. Genetic structure and species integrity over time depends on the predominance of either the unifying or disrupting microevolutionary forces.

variation, one can deduce both past and present genetic composition of a species, and possibly even forecast future trends.

Both fields, population genetics and phylogeography, have grown dramatically over the past two decades due to novel laboratory, computational and conceptual approaches (Hickerson et al. 2010). Much of this progress has been driven by studies on genetic model organisms such as *Homo sapiens* and *Drosophila* or commercially important species (e.g., Cann et al. 1987; Edwards et al. 1992; Garnery et al. 1992; Rogers & Harpending 1992; Bowcock et al. 1994). In these cases, a wealth of background knowledge about the target genomes makes screening of hundreds of thousands of different markers possible (Henn et al. 2011), leading to statistically solid inferences regarding population structure and past evolutionary processes. Although we generally welcome these contributions to the field, studies on model organisms are not representative of typical phylogeographic studies at large. With the notable exception of *Daphnia*, which has attained the role of an ecological model organism (Colbourne et al. 2011), almost all crustaceans fall into the category of being understudied taxa to date. In such nonmodel species, high-quality markers are hard to come by, and hence the majority of published analyses still rely on only one or a few mitochondrial DNA (mtDNA) markers (e.g., Milligan et al. 2011; Teixeira et al. 2011; Yebra et al. 2011; see figures 6 and 7 in Beheregaray 2008 and Figure 2 in this study). However, numerous studies have rendered such inferences as oversimplified or even unrepresentative of an organisms' evolutionary history (e.g., Shaw 2002; Bensch et al. 2006; Wahlberg et al. 2009; see Ballard & Whitlock 2004 for a review). It is therefore obvious that there is a striking discrepancy between what we know in theory about the optimal design of population genetic and phylogeographic analyses and the way they are usually performed in studies of nonmodel organisms.

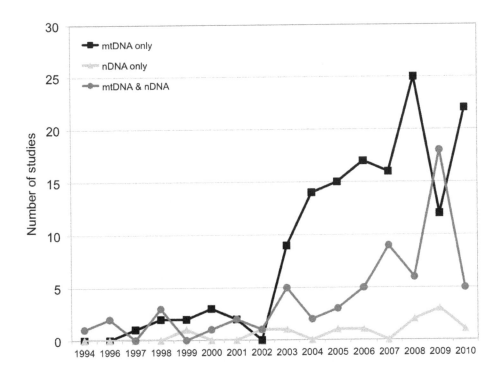

Figure 2. Studies using either mitochondrial or nuclear DNA markers only or a combination of both marker types based on 216 original crustacean phylogeographic studies found in the Zoological Record (access date: Feb. 23rd 2011, not all data for 2010 available yet). The graph illustrates that mtDNA is still increasingly popular in phylogeographic studies, while nuclear DNA (nDNA) as a marker system is much less frequently used. With the exception of the year 2009, studies using only mtDNA predominate.

1.1   The boon and bane of using mitochondrial DNA as a marker

The wide application of mtDNA markers in animal phylogenetic, phylogeographic and also population studies has a long-standing tradition and relies on a number of inherent characteristics of the mtDNA (e.g., see Avise et al. 1987). The mitochondrial genome is haploid and is generally considered to lack recombination in most taxa. Moreover, it has elevated mutation rates compared to coding regions in the nuclear genome, and it is easy to isolate due to its high copy number per cell. These features have made mtDNA, in particular the cytochrome c oxidase subunit I (COI) gene, the most popular molecular marker for barcoding animal species (Hebert et al. 2003). The presence of COI in almost all extant taxa facilitates the amplification and sequencing of this gene region using a small set of universal primers and allows for comparable studies across a broad taxonomic range (see Ratnasingham & Hebert 2007 and references therein).

Some shortcomings of mtDNA as a marker notwithstanding, there is no doubt that the use of mtDNA as a marker in phylogeographic and population genetics studies has been hugely successful. The limitations of mtDNA are due to its unique evolution thus making it rather unrepresentative compared with other genomic regions (Ballard & Kreitman 1995; Rokas et al. 2003; Ballard & Whitlock 2004; Hurst & Jiggins 2005). Although mtDNA has originally been regarded to evolve in agreement with neutral expectations and in a clock-like fashion, several examples now indicate that these criteria may be violated in many cases. Consequently, selective neutrality needs to be explicitly tested for prior to analyses, as opposed to simply assuming it. If indeed the mitochondrial genome in general did not evolve in a neutral fashion, the practical consequences for the interpretation of data are potentially far-ranging as several widely used tests designed to infer population demography tacitly assume the neutrality of mitochondrial substitution patterns—a condition that may not be met in a majority of taxa (Wares 2010). Furthermore, the mitochondrial mutation rate is not necessarily high in all animals (Shearer et al. 2002) making mtDNA a too slowly evolving marker for analyses of intraspecific variation in some cases.

Due to its high sensitivity to random genetic drift, an analysis based on mtDNA may lead to an overestimation of differentiation, i.e., indicate complete lineage sorting and reciprocal monophyly of subpopulations even in the event of ongoing but infrequent nuclear gene flow. On the other hand, even in the absence of gene flow, mitochondrial haplotypes may remain similar in two distinct populations or even species due to introgression or the adoption of entire foreign mitochondrial genomes (Shaw 2002; Funk & Omland 2003). Other misleading patterns may be generated by nuclear dysfunctional copies of mitochondrial genes; these so-called pseudogenes or numts have been reported in many taxa (Sunnucks & Hales 1996; Bensasson et al. 2001) including crustaceans (e.g., Williams & Knowlton 2001; Buhay 2009). If the translocation to the nuclear genome occurred recently enough (so that the primer regions have not accumulated sufficient mutations to prohibit amplification), the universal primers may co-amplify these "molecular poltergeists" (Hazkani-Covo et al. 2010), sometimes even preferentially. Other biases caused by recombination, biparental inheritance, heteroplasmy, and *Wolbachia*-infections have also been reported (see Ballard & Whitlock 2004; Rokas et al. 2003; Hurst & Jiggins 2005 for reviews).

A great number of contemporary population genetic and phylogeographic studies still rely on a single mitochondrial gene, often simply for reasons of convenience and tradition. However, new marker isolation protocols and next-generation sequencing technologies are bringing novel genetic tools within reach of scientists working on nonmodel species (e.g., Santana et al. 2009).

The main objective of this chapter is, therefore, to outline a practical hands-on strategy for using microsatellites on nonmodel organisms as an additional marker system complementing the mitochondrial genes that are seeing widespread use. Although there are a number of excellent reviews on both of these topics (e.g., Zane et al. 2002; Excoffier & Heckel 2006; Selkoe & Toonen 2006), we feel there is a need for an update in the light of the rapidly developing advances in the field of sequencing technologies and bioinformatic approaches. However, rather than evaluating every software program in the field, we cover the most common tasks and outline an exemplary workflow using one out of many possible combinations of software, highlighting potential pitfalls along the way. We are aware that the rapid development of new program versions may considerably shorten the practical half-life of the information contained in this section. Nonetheless, we hope that the examples presented here may help raise awareness for the types of problems that need to be solved, and that our present workflow can easily be updated to evaluate forthcoming markers and new software tools with different capabilities.

In a concluding section we briefly discuss the prospects of emerging sequencing technologies for population genetic and phylogeographic research on nonmodel organisms.

2   CHOOSING THE RIGHT MARKER TO OVERCOME THE LIMITATIONS OF mtDNA-ONLY STUDIES

The shortcomings of studies limited to mtDNA are to be taken seriously, but we wish to emphasize that something can and should be done to reduce both bias and error caused by relying solely on a single source of information. Several classes of molecular markers with different properties can be chosen to complement mtDNA sequences in a population genetic and phylogeographic context (Sunnucks 2000). Which marker to choose determines a balance between the precision of the marker on the one hand and the convenience of its application on the other.

RAPDs (Randomly Amplified Polymorphic DNA) and AFLPs (Amplified Fragment Length Polymorphisms) are markers that are quick to establish and can be a good first foray into evaluating the amount of intraspecific genetic variation among populations. RAPDs, however, are notoriously difficult to interpret, since nonspecific amplification conditions are a characteristic of this method. AFLPs are not much harder to establish and should be preferred over RAPDs due to their greater consistency. Both are multilocus markers, i.e., the band pattern is caused by an unknown number of loci in the genome, hence more powerful statistics, requiring information about the allelic composition of an individual, cannot be applied. Furthermore, both marker systems are dominant and therefore have less information per locus. This limitation is at least partly offset by the larger number of loci that can be established and studied in a given amount of time (Sunnucks 2000).

Single Nucleotide Polymorphisms (SNPs; Brumfield et al. 2003; Seeb et al. 2011), anonymous markers (Jennings & Edwards 2005), Restriction-Site-Associated DNA genotyping (RAD; Miller et al. 2007), Exon-Primed Intron Crossing Sequencing (EPIC; e.g., Chenuil et al. 2010), shotgun next-generation sequencing (Gompert et al. 2010), and microsatellites are all potentially powerful marker systems to study intraspecific variation in nonmodel organisms. These marker systems can complement data derived from an mtDNA marker. The decision of which marker system makes the most sense for a given biological question will depend on individual experience with the markers and future technical developments.

In this chapter, we have chosen microsatellites as additional markers complementing mtDNA, because a) they are well-researched, powerful markers, b) tried and true statistics have been developed for them, and c) user-friendly programs and plugins are readily available to aid in their analysis.

3   MICROSATELLITES

Microsatellites are short DNA motifs between two and six nucleotides long, which occur tandemly repeated, often many times, and in high frequency in the genomes of every higher organism investigated to date (Tóth et al. 2000; Mayer et al. 2010). Microsatellites are appealing because they are single-locus, co-dominant, multiallelic markers (for reviews see Jarne & Lagoda 1996; Goldstein & Pollock 1997; Goldstein & Schlötterer 1999) and therefore offer more analytical power than dominant marker systems. In the 1990s when the implementation of microsatellites in population genetics was becoming increasingly popular (Schlötterer 2004), one of their main advantages was that studying the fragment length circumvented the need to carry out mass sequencing. This aspect has, however, now taken a back seat, putting more emphasis on the fact that multiple unlinked loci can be studied.

The mutation rates of microsatellites generally exceed those of coding regions in the nuclear genome by one or several orders of magnitude (Weber & Wong 1993; Buschiazzo & Gemmel 2006; Kelkar et al. 2010). Consequently, the amount of allelic diversity in microsatellites is advantageous for studying microevolutionary processes that generally act on short timescales typically encountered in population genetics and phylogeography (Figure 1). At the same time, the fast mutation rates may make extrapolating the evolution of single microsatellites to the entire genome difficult (Väli et al. 2008; Ljungqvist et al. 2010), and increase the risk of homoplasious (i.e., similar in state, not by descent) characters due to back mutations (Balloux et al. 2000; Balloux & Lugon-Moulin 2002). Furthermore, the high mutation rates of microsatellites come at the cost of their limited applicability outside the taxon for which they have been established. In particular, the flanking regions that serve as priming sites for PCR amplification tend to have slightly elevated mutation rates compared to genomic regions unconnected to the repeats, and may thus be less conserved across species boundaries (Bailie et al. 2010). Although some microsatellite loci amplify successfully in more than one species (Rico et al. 1996), this is often true for the least variable and thus least informative loci. If the mutation rate of flanking regions was positively correlated with that of the microsatellite itself, placing primers in the flanking regions would lead to an increasing number of mismatches with the DNA when studying fast-evolving microsatellites and/or distantly related populations or species. However, even if the amplification succeeds in a species other than the one a microsatellite locus was designed for, one should be prepared (and test) for amplification problems that affect only a subset of the alleles of the respective locus (nonamplifying alleles or "null alleles" that cannot be amplified due to mutations in the flanking regions).

Developing microsatellites de novo for species without extensive information about their genomes is therefore the norm. It usually requires a well-equipped molecular laboratory and considerable amounts of time and money. One of the main reasons microsatellites did not become more popular since their discovery was down to their high unpredictability. The success of establishing microsatellites becomes obvious only after investing considerable time and effort to amplify and reliably genotype them (Zhang 2004; see Sands et al. 2009 for a discussion). Indeed, for a long time, microsatellites as markers in nonmodel species seemed a poor choice for anyone but the most determined and experienced population geneticists. Early microsatellite extraction protocols tended to yield a small number of candidate loci only, which forced researchers to include loci with unsuitable properties into their studies.

More recently, however, a number of microsatellite extraction protocols have streamlined the process of marker development considerably (Zane et al. 2002; Glenn & Schable 2005; Nolte et al. 2005; Leese et al. 2008; Santana et al. 2009; Malausa et al. 2011) and microsatellites are now frequently reported also from nonmodel organisms (see Molecular Ecology Resources: Permanent Genetic Resources). Their increased efficiency leads to a much larger number of initial candidate loci to choose from. Furthermore, much of the difficulties with low amplification success may result from unknowingly placing primers in flanking regions with cryptic simple repeat sequences (Bailie et al. 2010). This approach inevitably leads to competition among equally suitable, nonunique priming sites and therefore unspecific amplification products (see below).

In the following, we outline an updated strategy for the isolation of microsatellites from nonmodel species that has proven successful for a wide range of taxa. It features greatly reduced marker attrition by helping identify loci that will amplify more reliably, thus establishing microsatellites more efficiently.

3.1   Starting points

Since the appearance of the guideline paper by Selkoe and Toonen (2006), the greatly reduced cost of mass sequencing has provided alternative starting points for a microsatellite extraction strategy in nonmodel organisms (Figure 3).

Next-gen sequencing data, in particular 454 sequence reads due to their greater lengths compared to competing technologies (SOLiD, Illumina), are an alternative starting point (Santana et al. 2009; Castoe et al. 2010; Csencsics et al. 2010; Malausa et al. 2011) that requires little amount of work in the laboratory. After extraction of genomic DNA, size selection, and enrichment, mass sequencing can be outsourced to a company making cloning obsolete. It is even possible to outsource size selection and skip the enrichment, reducing laboratory work to the extraction of genomic DNA. Finding microsatellite candidate loci is achieved in silico by evaluating sequences using a repeat detecting program (see below and Figure 3).

Another valuable source of microsatellites without the need to dedicate time to the extraction of loci in the lab are EST libraries (for example, see Maneeruttanarungroj et al. 2006; Hoffman & Nichols 2011). Microsatellites rarely occur in coding sequences. However, in the 5' leader se-

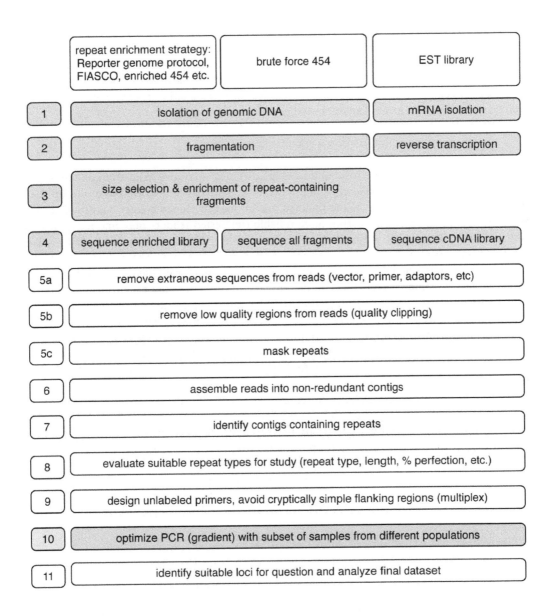

Figure 3. In addition to traditional enrichment protocols, other starting points for establishing microsatellite markers in unknown genomes are possible. Next-generation sequencing is becoming more affordable and read lengths of 300 bp or more make the assembly of nonredundant contigs feasible. EST libraries can also be valuable sources of microsatellites. Microsatellites derived from EST libraries are more likely to be subject to natural selection, but depending on the type of study, this may be acceptable or even desirable. In all cases, picking the suitable repeat loci takes place in silico (step 7). The protocols differ in how much preselection of repeat-containing fragments takes place in the lab (step 3). Decreasing costs for sequencing render this lab-based enrichment obsolete.

quences, which are also present in expressed sequence tag (EST) libraries, microsatellites can be found. The effort of producing such a library solely for the purpose of microsatellite detection is unreasonable. However, if an EST library for a species under study or a close relative already exists for other purposes, it may be a convenient source for microsatellites that can also be explored at very little cost. However, it should be kept in mind that the physical proximity to expressed genes also increases the probability that the microsatellites associated with them have not evolved completely neutrally. Nevertheless, selective neutrality can be explicitly tested for (e.g., Antao et al. 2008) and exploring already existing EST libraries may still be worthwhile if the alternative is to establish microsatellites de novo. In some instances, the study of microsatellites responding to selection may even be desirable (e.g., local adaptation studies).

3.2   An ounce of prevention is worth a pound of cure

Much of the reputation of microsatellites being "difficult" may originate from attempts to cope with intricate results rather than spending more time earlier in the workflow to ensure the data quality. Here, we report some of the findings that in our experience greatly contribute to making microsatellite generation and scoring more reliable (interested readers may find useful additional information in Hoffman & Amos 2005; Pompanon et al. 2005; Dewoody et al. 2006; Selkoe & Toonen 2006).

If at all possible, the DNA for the initial production of microsatellite candidate loci should be of the highest possible quality. Later in the process, once the loci are established and optimized, even material in poor condition often works well due to the comparatively small size of amplicons (typically 100–350 bp). Size selection of the DNA before library generation and sequencing ensures that neither insufficient space for primers (fragments too small) nor incompletely sequenced fragments (fragments too large) reduce the efficiency of the screening process. In practice, 400 to 800 bp fragments provide a good balance for Sanger sequencing, and not more than 500 bp for 454 sequencing (steps 1–4 in Figure 3).

For initial screening, the DNA from representatives of physically isolated populations should be pooled. Pooling what may turn out to be distinct genotypes reduces the probability that the candidate markers produced work fine only for one outlier population but not for others and helps to identify polymorphic loci early in the process (step 6 in Figure 3; see below).

Every microsatellite isolation strategy has the potential to pick up homologous fragments more than once. The identity of the fragments is not easily detectable by eye because slightly different starting and end points and reverse sequencing directions obscure the identify of the homologous fragments. An assembly step is required to merge the partly redundant reads into nonredundant contigs.

Three basic tasks need to be performed (steps 5a–c in Figure 3):

a) Extraneous reads that do not originate from the target organism (vector, primer, 454 tag, adaptor, etc., depending on the protocol used) must be eliminated.

b) Low quality regions that typically occur at the ends of sequence reads should be discarded or ignored (trimmed) for the assembly.

c) Repeat sequences should be masked, i.e., bioinformatically tagged so that they can be ignored during assembly.

BLAST searches against databases allow the identification of problematic sequences (step 5a). Repeat masking prevents the assembly program from mistakenly joining reads based on the repeats (e.g., a long AC repeat stretch) rather than on the homology of the flanking regions. The purpose of the repeat detection for masking is quite different from the identification of repeats for the purpose of isolating microsatellites further down in the workflow (Kraemer et al. 2009), and hence,

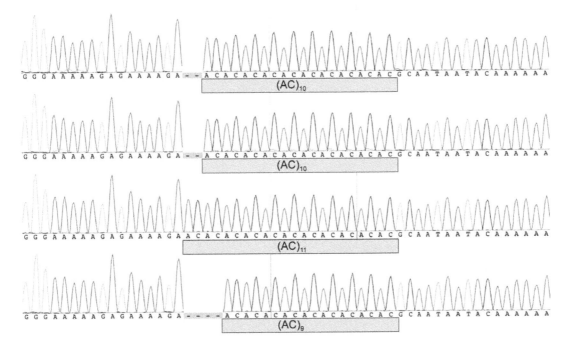

Figure 4. Using a DNA mix of different individuals allows testing for variation at individual loci after shotgun 454 sequencing. In this case, three different alleles for the microsatellite locus, i.e., $(CA)_9$, $(CA)_{10}$ and $(CA)_{11}$, are clearly distinguishable (own data from an enrichment according to Leese et al. 2008) (see Figure 1 in Color insert).

even though often the same program is used for these two steps, different settings should be applied. During the assembly, the main objective is to ensure that the remaining unmasked regions are unique, so that relaxed, i.e., more inclusive and computationally less demanding, settings are acceptable. The flexible and fast repeat identifying program Phobos (Mayer 2010) is available either in a stand-alone version or integrated into molecular workflow pipelines (Staden and Geneious), the latter features two different presets of parameters for masking and analysis.

All three steps can be carried out automatically in many programs used for sequence assembly such as Geneious, Staden.

3.3   Which microsatellite to choose?

Traditional protocols based on hybridization with synthetic probes tended to yield a limited number of candidate loci hence the necessity to choose among them rarely became an issue. Modern protocols afford a much greater number of candidate loci (e.g., Leese et al. 2008, Castoe et al. 2010), of which typically only a small percentage get analyzed in depth. What properties characterize a "good" microsatellite and how can one be recognized? Although the exact properties of a locus are only known a posteriori, there are a couple of things that can help guide the choice which loci to choose and which to ignore.

Dinucleotide repeats as the most common repeat type consist of four different units (AC, AG, AT, CG), all other types are variations (Chambers & MacAvoy 2000). Dinucleotide repeats also tend to be the most variable microsatellite loci.

If enough potential candidate loci are left after the previous steps of removing redundancies and candidates with cryptically repetitive flanking regions or insufficient room for primer placement

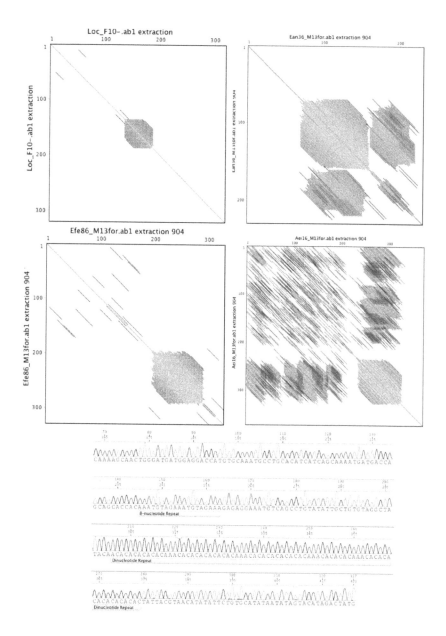

Figure 5. Dotplots of four microsatellite-containing contigs against themselves (Geneious Pro version 5.3). The main diagonal indicates the perfect identity of the sequence (X axis) against itself (Y axis). Off-diagonal parallels to the main diagonal indicate that sequence motifs occur elsewhere in the sequence, albeit slightly modified and thus not easily detected using bioinformatic approaches or inspection by eye. Candidate sequences with enough room to place primers in unique parts of the flanking sequence (indicated by lack of parallels, upper left panel) are to be preferred over candidates with cryptically repetitive flanking regions. Note that the Phobos search for repeats with commonly used parameters only identifies one additional 8 bp repeat (annotations in lower panel; Geneious Pro). The remaining repeat structures are camouflaged by their lower degree of sequence conservation in the sequence view (lower panel), but clearly visible in the dotplot of the same sequence (lower left panel) (see Figure 2 in Color insert).

on either side of the repeat, several rough guidelines can help identify loci with desirable properties: If pooled DNA of several specimens from different populations is the starting material for the screening (as suggested), sequence reads of similar allelic variants based on flanking regions will be detected in the assembly. This step provides a good overview over polymorphic markers even prior to genotyping (see Figure 4). If the nature of the study demands highest variability (small geographic scale, paternity studies), dinucleotide microsatellites with many repeats (> 20) are a common choice (see Selkoe & Toonen 2006; Hoffman & Nichols 2011), but some loci with longer repeat units (tri- or tetranucleotide repeats) should be included as well. Although longer repeat types tend to be less polymorphic compared to loci with shorter types, they do not suffer as much from size homoplasy and in vitro artifacts, thus reducing the effect of allele scoring errors (Wang 2010) and suffer less from size homoplasy. Particularly when dealing with species with large population sizes and larger geographic and temporal scales, less variable loci are suitable, e.g., tetra- or pentamer motifs.

The identification of candidate microsatellite loci can be carried out, for example, using the software program Phobos. During this operation the contigs, i.e., the nonredundant consensus sequences from the assembly (not the individual reads; see above), serve as input for a further round of repeat detection. Now more stringent search criteria (higher mismatch penalties, longer microsatellites such as di-, tri-, tetranucleotide repeats, $> 95\%$ perfection, > 6 repeat units in tandem) should be used than during the repeat masking step in the assembly. If the number of loci reported is too small, the criteria can be relaxed in a second Phobos run (step 8 in Figure 3).

3.4   Primer design considerations

In the next step, consensus sequences of contigs with promising microsatellites serve as input for a primer design program, e.g., Primer 3 (Rozen & Skaletsky 2000), which is available online or integrated into Staden and Geneious. The targeted repeat is specified as the insert, i.e., Primer 3 is instructed to search for primers on either side of the repeat but not within (step 9 in Figure 3). Candidates with insufficient space for placing upstream and/or downstream primers should be discarded right away. It is also advisable to place primers in such a way that the resulting amplicon spans one microsatellite and not a complex of different microsatellites.

The size of the amplicon should be in the range of 80 to 350 bp, longer fragments are in general more likely to be misscored (due to under- or overestimating its true size) as well as to suffer mutations in the flanking regions.

One often overlooked characteristic of microsatellites is that their flanking sequences can frequently contain cryptic simple repeat-like elements (Bailie et al. 2010). Moreover, the whole microsatellite may just be a subelement of a higher-order repeat (see dotplots in Leese et al. 2008). The repeat-like structure of microsatellite flanking regions may be the result of partial degradation of larger microsatellites, a process that has been proposed to be responsible for size constraints in microsatellites (Buschiazzo & Gemmell 2006). Due to the low degree of repeat motif preservation, these repetitive elements are not easily detected either by eye or by using repeat finder programs. Dotplots of a candidate sequence against itself, using a sliding window and an adjustable sequence conservation, are a quick and useful tool to visualize these cryptic repeats and help to avoid primer placement in these areas (see Figure 5). Dotplot tools are integrated in molecular workflow pipelines (Kraemer et al. 2009), standard packages (Staden, Geneious), or can be used as stand-alone tools or web applications (Junier & Pagni 2000). Unknowingly placing primers for the amplification of a microsatellite locus within these flanking cryptic repeats, however, can cause a primer to find several suitable annealing sites for a single locus. This effect can occur, in particular, under conditions of somewhat relaxed PCR stringency (which are advisable to avoid an excessive occurrence of null alleles; see below). The co-amplification of homologous fragments with varying amounts of flanking regions results in uninterpretable banding patterns, for example, more than two alleles for a diploid organism. Such patterns may be the main reason for the unpredictable success and high marker attrition of microsatellites reported in the literature (Bailie et al. 2010; Buschiazzo & Gemmell 2010).

3.5   Reaction set-up

The cost of fragment analysis can be reduced by optimizing loci in such a way that they can be analyzed simultaneously in a single reaction (multiplexing), each locus being labeled by a different fluorescent dye. Loci can only be multiplexed if their amplification conditions are similar (annealing temperatures, cycling times), and it is also advisable to combine loci possessing non-overlapping allele size spectra (see Li et al. 2007 for an example). Although in theory the wavelengths of the dyes should be clearly discernible, some dyes have partly overlapping wavelength ranges, which make interpretation difficult, especially when background noise is complex (pull-up peaks; see Selkoe & Toonen 2006).

The use of 5'-tailed primers allows using a small number of fluorescently labeled primers among all markers under analysis (Boutin-Ganache et al. 2001).

The amplification of each locus needs to be optimized subsequently. Several specimens should be tested simultaneously at this step (individually or pooled DNA) to reduce the danger of mistakenly excluding a locus due to bad DNA quality or particular sensitivity to non-amplifying alleles (null alleles). Minimizing the number of PCR cycles (usually 28–32) can help reduce stutter bands. (A collection of other possible ways to decrease stutter bands has been posted on the website for evolutionary scientists, EvolDir, at http://evol.mcmaster.ca/~brian/evoldir/Answers/Micro.stutter. answers). However, we do not recommend investing too much time on stuttering microsatellite loci, but rather on selecting other locus candidates, if available. The absence of bands on an agarose gel does not necessarily mean that the PCR was unsuccessful. A subset of samples should be checked on the sequencer, which is more sensitive.

During the preparation of fragment analyses a prolonged final extension of at least 30 minutes should be used as this significantly decreases the percentage of prematurely terminated amplicons lacking a terminal adenine (Figure 6). In fragment analysis, this effect can make interpretation of some loci very difficult and lead to inconsistent results, which may depend on different laboratory conditions and protocols (Figure 6, see the 5-minute chromatogram vs. 45-minute chromatogram).

Contrary to common practice in PCR, one should avoid choosing the most stringent amplification conditions for subsequent fragment generation as this can also increase the probability of low amplification success or even failure for some alleles (allelic drop-out or null alleles; see above).

Many errors can be introduced during genotyping. Hence, independent PCR and genotyping runs must be conducted for some samples. Furthermore, chromatograms or gel-pictures should be scored independently by two researchers. Reference samples should be used for every gel to avoid shifts in the size standards. For a good guide on how to genotype different microsatellite loci we refer to the Appendix S2 in Selkoe and Toonen (2006).

3.6   Marker quantity and quality

As stated above, empirical and simulation studies have shown that the number of independent loci used to infer population genetic structure is a crucial aspect concerning a maximized robustness of inferences (Koskinen et al. 2004; Felsenstein 2006). For pragmatic reasons, namely laboratory time and costs, the number of microsatellites used for exploratory studies on nonmodel organisms is usually in the range of 6–10 (Selkoe & Toonen 2006). However, besides marker quantity the quality of the markers is of great importance. Outlier loci with a very obvious pattern of selection or extreme

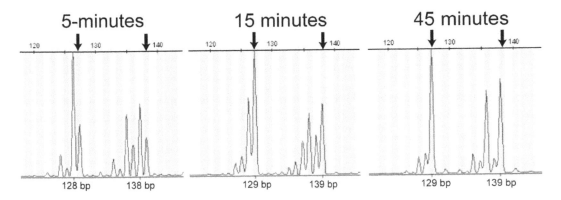

Figure 6. Chromatograms of the same heterozygous microsatellite with 5-minute, 15-minute, and 45-minute final elongation time. Note that a final extension of 5 minutes not only generates a more complex stutter pattern but also distorts the amplitude ratio between the peak representing the correct allele (arrow) and some pronounced in vitro artifacts.

interpopulation differences in allele lengths can be detected and excluded from population genetic analyses (Landry et al. 2002; Beaumont 2005; Antao et al. 2008). Furthermore, in vitro artifacts such as null alleles, stutter bands, allelic dropout or the amplification of paralogous loci may bias the analyses (Hoffman & Amos 2005; Pompanon et al. 2005; Dewoody et al. 2006). An additional problem is that two microsatellite alleles can be identical in length, but be independently derived from different ancestors (homoplasy). Some cases of homoplasy can be resolved by sequencing (different sequences of the same length), whereas others are fundamentally undetectable (different evolutionary history but now identical in state). Several microsatellite studies have revealed the presence of detectable homoplasies (Estoup et al. 1995; Estoup et al. 2002 for review). Homoplasy can become a great concern, in particular, in analyses based on highly mutable loci that have a length constraint and in species with large effective population sizes (Estoup et al. 2002). Resequencing of alleles identical in state and further techniques such as SSCP (Single Strand Conformation Polymorphisms), DGGE (Denaturating Gradient Gel Electrophoresis), and HRM (High Resolution Melting) can better differentiate homoplasious alleles. However, this would raise time and costs, rendering microsatellites much less attractive, and should therefore be tested using a subset of the markers. Studying a number of microsatellite loci in addition to a mitochondrial gene may not completely eliminate biases, but will nevertheless reduce their influence because not all loci will be biased in the same way.

4   GENETIC MARKER ANALYSIS

A review of population genetic programs and their underlying assumptions by Excoffier and Heckel (2006) may still serve as an indispensable "survival guide" for those interested in starting or extending population genetic analyses. Over the past years, several software updates and novel programs have appeared, some of which are even capable of analyzing microsatellites and sequence data simultaneously and test complex evolutionary scenarios (see the data analysis section below). For inferences of past processes, the development of coalescent samplers that allow the incorporation of stochasticity in the data by estimating multiple genealogies has considerably changed approaches to intraspecific data analyses (see Kuhner 2008).

In the following section, we focus on a number of programs that are useful for detecting problems in raw data, and converting datasets. Most importantly, however, we highlight programs that can assess the genetic structure of populations and test competing hypotheses dealing with microevolutionary processes.

4.1   File format conversion

Due to the lack of a standardized file format, many programs use proprietary and mutually incompatible formats. A number of software packages exist that are capable of converting data matrices (see Table 1).

4.2   Data validation

4.2.1   *Analyses of mtDNA*

For the analysis of sequence data it is highly important to initially perform BLAST database searches to verify the origin of the sequences under study. For queries involving COI sequences, the BOLD database often provides a more reliable tentative identification than GenBank, because the BOLD metadata are more trustworthy in general and searches include hits from datasets, which have not yet been made publicly available.

In order to identify potentially mislabeled sequences in the databases, it is important to explore the sequence representing the best hit as well as its neighborhood in a tree-based approach. Furthermore, for protein-coding genes of nonmodel organisms such as COI, translation into protein sequences with the correct translation table and in the correct orientation can help identify pseudogenes, which may have been amplified unknowingly (Bensasson et al. 2001; Buhay 2009). Dotplots are useful tools to detect similarities either to the sense or antisense strands even in very distantly related sequences.

A particular concern when studying intraspecific genetic variation is cross-contamination among samples, i.e., contamination by a specimen of the target species other than the specimen under scrutiny. Unlike contaminations by foreign DNA, cross-contaminations among conspecifics cannot be detected a posteriori by running a BLAST search against a suitable database. Hence, precautions against carryover of tissue and DNA are mandatory.

4.2.2   *Microsatellites*

Several reviews cover the various steps that need to be performed to verify data and minimize bias due to in vitro artifacts (Dewoody et al. 2006; Selkoe & Toonen 2006). For microsatellites, precautions mainly focus on testing for genotyping errors, allelic dropout, presence of null alleles, evidence of selection or unequal contribution to the variation (Hoffman & Amos 2005; Pompanon et al. 2005). To this end, software tools such as MICROCHECKER (van Oosterhout et al. 2004; see Table 1) are mandatory. Loci that suffer from artifacts need to be reanalyzed or excluded from the analysis (Leese et al. 2008). A second step in the process of marker validation is to detect outlier loci that may be under selection (Beaumont 2005) or may bias the analyses for other reasons (Landry et al. 2002). At this stage in the workflow an initial investment into obtaining a sufficiently high number of quality candidate loci pays off.

4.3   Assessing the genetic structure

The main interest in most studies is to analyze the genetic structure of populations and interpret it in the light of present and past geographic and demographic scenarios. Analysis of the genetic structure permits inferences on microevolutionary processes in the species' evolutionary history (see Bilodeau et al. 2005 for an exemplary workflow in decapod crustaceans). Phylogeographers and population geneticists traditionally follow two different "schools of thought," the differences of which are summarized in Hey and Machado (2003). Phylogeographers traditionally start estimating gene trees or networks using software packages such as MEGA (Kumar et al. 2008), TCS (Clement et al. 2000), or Splitstree (Huson & Bryant 2006). Subsequently, they base their inferences on underlying processes on the topology of this single gene tree, neglecting stochasticity in the data. In contrast, population geneticists use statistical models for the analysis of intraspecific variation. We will outline both different strategies and show how these can potentially be merged using recently developed approaches and programs.

4.3.1 *Population genetic approaches*

The most commonly used method for assessing intraspecific genetic structure among geographically defined populations is by F–statistics (Wright 1943). In the F–statistic framework, several hierarchical estimators that describe the partitioning of genetic variation within and among populations are usually assessed (see Holsinger & Weir 2009 for overview). F_{ST} is still one of the most widely used statistics in the context of phylogeography and population genetics, since it abstracts from differences among individuals and concentrates on how the composition of groups of individuals differ on a more inclusive geographic level. An inherent assumption of F–statistics is that any patterns observed are the result of the distribution of existing variation in space rather than the generation of new variation, i.e., the rate of mutation is assumed to be much smaller than the rate of migration.

Several variants of F_{ST} have been developed which incorporate intrinsic information about the data source such as R_{ST} (incorporating a microsatellite stepwise mutation model), Φ_{ST} (for sequence data), Q_{ST} (quantitative analysis of continuously varying traits; see Holsinger & Weir 2009; Meirmans & Hedrick 2011 for review; see Bird et al. in this volume for a review and comparison for these and other metrics of genetic differentiation). Today, estimation of F_{ST} is usually performed by comparing allele frequencies among a priori defined populations in a variance framework after Weir and Cockerham (1984) and assuming an infinite allele model. If F_{ST} is low this means that allele frequencies among the populations are very similar. If F_{ST} is large it means that populations are differentiated, i.e., either they are genetically independent or the locus is subject to diversifying selection (Beaumont 2005). F_{ST} values among populations and their significance can be assessed using the programs GENEPOP (Rousset 2008), ARLEQUIN (Excoffier et al. 2005), or GenoDive (Meirmans & Van Tienderen 2004). A problem that needs to be kept in mind when comparing F_{ST} values is that in a variance framework the theoretical maximum of 1 can actually be achieved only for biallelic loci. If more than two alleles are present at any one locus, the maximum F_{ST} value may be significantly smaller than 1 even if the populations under study do not share a single allele. This makes corrections necessary to allow comparisons among loci, otherwise the magnitude of F_{ST} becomes rather arbitrary. Following Hedrick (2005) and Meirmans (2006) one can recode the dataset using the program recodedata (Meirmans 2006) or GenoDive (Meirmans & Van Tienderen 2004). With this recoded dataset it is possible to estimate the maximum possible F_{STmax} for this dataset using, e.g., GENEPOP, ARLEQUIN or GenoDive. In a subsequent step the corrected F'_{ST} can be calculated by dividing the observed measure of population differentiation estimates (F_{ST}) by the value of their maximal expected differentiation (F_{STmax}). This approach increases the data spread for each locus between 0 and 1 again. Another strong and novel measure is Jost's D (Jost 2008). Although this correction aims at the symptoms and is subject to sampling error, it is an improvement that aids in comparing different loci. Most importantly, it minimizes the possibility of wrongly rejecting the hypothesis of population differentiation on the grounds of low uncorrected F_{ST} values (e.g., $F_{ST} < 0.05$) when in fact the populations are strongly differentiated (see Heller &

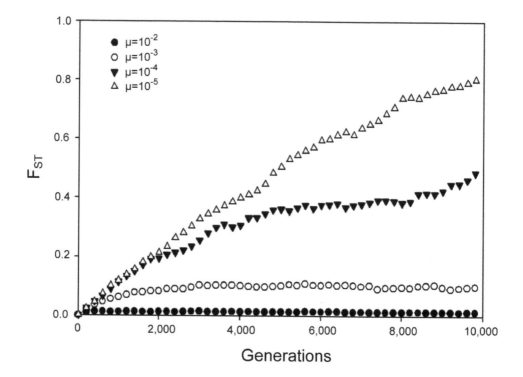

Figure 7. Mutation-rate dependency of microsatellite-based F_{ST}–estimates under a two-population scenario (effective size $N_e = 2000$) without migration, and mutation rates ranging from $\mu = 10^{-2}$ to 10^{-5}. The dataset with the lowest mutation rate (10^{-5}) reaches equilibrium latest ($> 10,000$ generations) and shows the highest differentiation values. The locus with the fastest mutation rate of 10^{-2} reaches equilibrium quickly (400 generations), and the maximum differentiation value for F_{ST} is low (0.013). The dataset was simulated using EASYPOP version 2.01 (Balloux 2001). F_{ST}–estimates were calculated according to Weir and Cockerham (1984). Data points represent average F_{ST} values calculated from five independent simulation runs for a microsatellite dataset of $n = 10$ loci with a maximum of 30 allelic states per locus. Mutations at microsatellite loci followed a two-phase mutation model (90% according to the stepwise mutation model, 10% according to the infinite allele model).

Siegismund 2009; Meirmans & Hedrick 2011, Bird et al. this volume). One should keep in mind, however, that microsatellites with their generally high mutation rates may lead to systematically underestimated differentiation among populations due to high homoplasy if mutation becomes the dominant microevolutionary force over migration (see above). As a consequence, mutations leading to identity in state can deflate the observed differentiation pattern severely even in the complete absence of migration (see Figure 7 for a theoretical example). A pattern indicating differentiation can take longer to manifest even in the complete absence of migration when the effective population size is large (see section "Gene Flow" below).

The most common estimator of genetic differentiation between populations based on microsatellite allele size data is R_{ST} (Slatkin 1995). In contrast to F_{ST}, R_{ST}–estimates assume an underlying stepwise mutation model of microsatellite evolution and take the magnitude of differences in allele lengths into account. The stepwise mutation model is regarded to represent better the mutational processes of microsatellites, which often tend to occur in a somewhat stepwise fashion (but see Ellegren 2004). According to this model, alleles with very different lengths are more distantly related than two alleles of rather similar length. While R_{ST}–estimates provide a welcome addition for assessing genetic differentiation it is important to test whether extreme patterns of differentiation may rely on very few outlier loci only (Landry et al. 2002).

Due to the hierarchical framework of F–statistics one can easily add additional layers and perform tests for geographical superorder partitioning of intraspecific genetic variation (AMOVA; Excoffier et al. 1992). For both microsatellites and sequence data, subpopulations are combined into regional clusters assigned in an a priori fashion. A hierarchical analysis with ARLEQUIN (Excoffier et al. 2005) tests how the variance is partitioned within populations, among populations within regions, and among regions.

As stated above, this F_{ST} approach can be done with microsatellite allele length data as well as with sequence data (Excoffier et al. 1992). However, when comparing population data, it cannot always be defined a priori what a population actually is (Waples & Gaggiotti 2006) since the geographically defined origin of the samples might be a poor predictor of their genetic relatedness. Even cryptic species may occur in sympatry (e.g. Held 2003; Leese & Held 2008). As an approach to more objectively analyzing the real population boundaries, genetic clustering methods have recently been implemented in several programs. Good clustering programs are STRUCTURE (Pritchard et al. 2000; Falush et al. 2003) and TESS (Chen et al. 2007). They rely on probabilistic solutions to complex population genetic problems using Bayesian approaches (Beaumont & Ranalla 2004). With Bayesian population clustering algorithms it is possible to estimate the number of distinct subpopulations (clusters) from the data in an a posteriori fashion and assign individuals to them. To this end the true, or rather the most probable number of populations K are estimated from the data by a model-based algorithm. It is important to run the programs without prior population information and let the program calculate the probability of the data for a given K in the range of $K = 1$ to $K = n$ (n being greater than the assumed number of populations). The highest likelihood compared over all K gives an indication of the number of genetic clusters in the data. When the true number of K is high, the likelihoods of K may constantly increase with K. However, a potential solution has been suggested (Evanno et al. 2005; Chen et al. 2007). With cluster algorithms it is furthermore possible to detect migrants in the populations or test for clinal variation in geographically defined populations (Chen et al. 2007; Durand et al. 2009). Different runs of the software can be compared, and a consensus membership (Q) matrix with individual assignment proportions to the different genotypes be calculated in CLUMPP (Jakobsson & Rosenberg 2007). Individual or population-spanning representations of the probability proportions can be visualized using Distruct (Rosenberg 2004).

4.3.2 *Phylogeographic approach*

A number of phylogeographic methods explicitly rely on gene trees and not allele frequency data. A method that has received much attention in the last two decades is nested clade phylogeographic analysis (NCPA; Templeton 1998). This method directly uses a single maximum parsimony network and geographic data of the samples to make a posteriori inferences on a number of different demographic processes that may have generated the pattern of the gene network. After a period of intensive use, NCPA became criticized for lacking a statistical framework and thus putting too much confidence into the scenario that supposedly explains the data best without explicitly evaluating alternative hypotheses (Knowles 2008; Nielsen & Beaumont 2009). However, in model-based software packages solutions to this problem are available. Testing phylogeographic hypotheses using sequence data can be carried out using the software programs such as Migrate (Beerli 2006), IMa2 (Hey & Nielsen 2007), or PhyloMapper (Lemmon & Lemmon 2008) and BEAST (Drummond & Rambaut 2007).

4.4   Gene flow

One very important measure determining the genetic structure of a species is the level of gene flow (synonymous to migration in a population genetic context) among populations. Some approaches deliberately simplify the problem, e.g., by neglecting the contribution of mutation to the molecular variance. However, these approaches are error-prone when the real population violates the assumptions of this model (Whitlock & McCauley 1999; see also Neigel 2002 for discussion). Powerful programs for estimating population genetic parameters such as present migration rates or present and past effective population sizes in a much more realistic way are Migrate (Beerli 2006) and IMa2 (Hey & Nielsen 2007). Both programs are complex but the respective authors provide excellent introductions to the underlying models and tutorials guiding through the practical use of the programs.

A particular strength of Migrate is that a priori hypotheses regarding gene flow can explicitly be tested based on the marginal likelihoods calculated for the different models tested in a log Bayes factor framework (Beerli & Palczewski 2010). A case study from the Southern Ocean showed that whilst the information from mtDNA and the microsatellites differed somewhat (Leese et al. 2010), testing different models in a log Bayes factor framework nevertheless demonstrated that asymmetric gene flow consistent with one of the major ocean current regimes explains the data significantly better than other models (Leese et al. 2010).

Since the number of parameters to infer increases dramatically with increasing number of populations and loci, estimates with Migrate can take a long time to complete (weeks to several months) but tasks can be parallelized. Novel approximate Bayesian approaches as implemented in popABC and DIYABC (Cornuet et al. 2010) provide new and promising alternatives that can yield results in a comparatively short time. A further advantage of both of these programs is that they are capable of a combined analysis of microsatellite and sequence data. However, although a recent split without gene flow and very low level of gene flow over a long time represent quite different biological scenarios, distinguishing between them is not a trivial problem for any method.

5   SAMPLING BIAS

Most population genetic and phylogeographic studies are based on a fraction of the true diversity of populations/specimens since it is generally impossible to sample all members of a species. Therefore, a typical population genetic or phylogeographic study suffers from a twofold reduction in information. First, a single or a few molecular markers are chosen out of hundreds of thousands of potential loci of the species' genome and, second, this marker is then analyzed for a small subset of the specimens representing the species. Furthermore, from this locus generally only single genetrees are drawn (e.g., the most parsimonious) which are then regarded to truly reflect the species' evolutionary history. While it is obvious that this twofold reduction of information content cannot be avoided fully, it is important to be aware that possibilities exist to reduce this bias by a) increasing the sample size, b) selecting several unlinked genomic loci, and c) using statistical approaches that can incorporate this statistical uncertainty, e.g., coalescent-based approaches (see Table 1).

Table 1. Overview of selected programs for sequence (S) and microsatellite (M) data conversion, validation, and analysis. These programs are examples that have proven very useful in our studies, although numerous other programs might be equally suitable.

Purpose & Program	Data type	Tasks	Weblink	Citation
Marker development				
Geneious*	S (M)	Sequence editing, repeat search, masking, assembly, primer design	http://www.geneious.com	Drummond et al. 2009
Phobos	S	Repeat search (either within Geneious or stand-alone)	http://www.ruhr-uni-bochum.de/spezzoo/cm/cm-phobos.htm (also plugin for Geneious)	Mayer 2010
Dotlet	S	Search for cryptic simplicity, higher-order repeat structure, duplications	http://myhits.isb-sib.ch/cgi-bin/dotlet	Junier & Pagni 2000
STAMP	S	Microsatellite development using the Staden Package and Phobos	http://www.ruhr-uni-bochum.de/spezzoo/cm/	Kraemer et al. 2009
Data conversion				
Convert	M	Convert microsatellite tables into different formats	http://www.agriculture.purdue.edu/fnr/html/faculty/rhodes/students%20and%20staff/glaubitz/software.htm	Glaubitz 2004
Create	M	Convert microsatellite tables into different formats	https://bcrc.bio.umass.edu/pedigreesoftware/node/2	Coombs et al. 2008
Formatomatic	M	Convert microsatellite tables into different formats	http://taylor0.biology.ucla.edu/~manoukis/Pub_programs/Formatomatic/	Manoukis 2007
PGDSpider	M, S	Convert microsatellite and sequence data into different formats	http://www.cmpg.unibe.ch/software/PGDSpider/	Lischer 2009
ALTER	S	Convert sequences into different formats	http://sing.ei.uvigo.es/ALTER/	Glez-Peña et al. 2010
Fabox	S	Convert sequences into different formats	http://www.birc.au.dk/~biopv/php/fabox/	Villesen 2007
Data verification				
MICROCHECKER	M	Test for genotyping errors, null alleles, dropout	http://www.microchecker.hull.ac.uk/	van Oosterhout et al. 2004
Dropout	M	Test for allelic dropout. Good overview over allele data	http://www.rmrs.nau.edu/wildlife/genetics/software.php	McKelvey & Schwartz 2005
Animalfarm	M	Test for unequal contributions of individual loci to variance	http://users.utu.fi/primmer/publications/31/AnimalFarm.html	Landry et al. 2002
LOSITAN	M	Test for loci under selection (balancing or purifying)	http://popgen.eu/soft/lositan/	Antao et al. 2008
Fdist2	M	Test for loci under selection (balancing or purifying)	http://www.rubic.rdg.ac.uk/~mab/software.html	Beaumont & Nichols 1996
BOLD BLAST	S	Verify sequence identity	http://www.boldsystems.org/views/idrequest.php	Ratnasingham & Hebert 2007

Table 1. Continuation.

Data analysis

Purpose & Program	Data type	Tasks	Weblink	Citation
Recodedata	M	Recode dataset to calculate maximum possible F_{ST}	http://www.bentleydrummer.nl/software/software/Other%20Software.html	Meirmans 2006
GENEPOP	M	Population genetic software for various tests	http://kimura.univ-montp2.fr/%7Erousset/Genepop.htm	Rousset 2008
ARLEQUIN	M, S	Population genetic software for various tests	http://cmpg.unibe.ch/software/arlequin35/	Excoffier et al. 2005
GenoDive	M	Population genetic software for various tests including standardized differentiation	http://www.bentleydrummer.nl/software/software	Meirmans & Van Tienderen 2004
Splitstree	S	Visualization of genetic structure from sequence data (various algorithms)	http://www.splitstree.org/	Huson & Bryant 2006
TCS	S	Parsimony Network construction from sequence data	http://darwin.uvigo.es/software/tcs.html	Clement et al. 2000
MEGA	S	Calculation of phylogenetic trees and distance matrices	http://www.megasoftware.net/	Tamura et al. 2011
Phylomapper	S	Tests different phylogeographic scenarios based on sequence data	http://www.evotutor.org/LemmonLab/Software.html	Lemmon & Lemmon 2008
Structure	M	Tests for population structure in the data with and without prior information on population sampling sites using a clustering method	http://pritch.bsd.uchicago.edu/structure.html	Pritchard et al. 2000
TESS	M	Tests for population structure using a clustering method as in with Structure but explicitly incorporates geographical information	http://membres-timc.imag.fr/Olivier.Francois/tess.html	Chen et al. 2007
Migrate-n	M, S	Bayesian and likelihood estimation of demographic parameters (gene flow, effective population size)	http://popgen.sc.fsu.edu/Migrate/Migrate-n.html	Beerli 2006
DIYABC	M, S	Estimation of demographic parameters using approximate Bayesian computation; can combine sequence and microsatellite data	http://www1.montpellier.inra.fr/CBGP/diyabc/	Cornuet et al. 2010
popABC	M, S	Estimation of demographic parameters using approximate Bayesian computation; can combine sequence and microsatellite data	http://code.google.com/p/popabc/	Lopes et al. 2009
BEAST	S	Tree-based analysis of sequence data, hypothesis testing	http://beast.bio.ed.ac.uk/Main_Page	Drummond & Rambaut 2007

* There are various commercial and freeware packages available for sequence editing and processing, such as BioEdit, Staden Package, Sequencher, and DNAStar.

6   A PEEK INTO THE FUTURE

Mitochondrial DNA has proven extremely useful for studying phylogeography and population genetics of nonmodel species and will continue to contribute valuable information in the future. However, it is clear that the evolution of a marker and of the species carrying it, are not the same thing. If the task is not the reconstruction of the evolution of the marker itself, then complementing mtDNA with data from another independent and informative marker system is the only way to overcome the limitation and potential bias inherent in any single locus.

For nonmodel species whose genomes are typically poorly known, the decision which new class of markers to select has to balance the effort needed to develop the markers on the one hand and the power of these markers to answer the scientific question on the other hand. This balance is dynamic because the molecular genetic tools are in constant development. New technologies, in particular the advent of next generation sequencing technologies will continue to push the boundaries for marker development but some problems remain to be solved (e.g., the number of usable individual sequence tags and the fidelity of the sequencing) before they can be routinely applied to phylogeographic or population genetics studies (Holsinger 2010).

While many exciting new developments are on the horizon, meanwhile a significant improvement over the typical single mtDNA fragment study can be made using today's technology by combining tried and true markers (mtDNA and, e.g., microsatellites) and existing tools for their analysis in a statistical framework that accommodates the inherent stochasticity in intraspecific genetic datasets and tests competing hypotheses.

ACKNOWLEDGEMENTS

We gratefully acknowledge the detailed comments from Stefan Koenemann, Kathrin Lampert, Chester Sands and our anonymous reviewers, which helped to improve the manuscript considerably. Markus Gronwald helped collecting and analyzing the literature from databases for Figure 2. We thank Mark Harrison for proofreading. FL and CH were supported by DFG grants HE 3391/5 and LE 2323/2, respectively.

REFERENCES

Antao, T., Lopes, A., Lopes, R.J. Beja-Pereira, A. & Luikart, G. 2008. LOSITAN: a workbench to detect molecular adaptation based on a Fst-outlier method. *BMC Bioinform.* 9: 323.
Avise, J.C., Arnold J., Ball, R.M., Bermingham, E., Lamb, T., Neigel, J.E., Reeb, C.A. & Saunders, N.C. 1987. Intraspecific phylogeography: the mitochondrial DNA bridge between population genetics and systematics. *Annu. Rev. Ecol. Syst.* 18: 489–522.
Bailie D.A., Fletcher, H. & Prodöhl, P.A. 2010. High incidence of cryptic repeated elements in microsatellite flanking regions of galatheid genomes and its practical implications for molecular marker development. *J. Crust. Biol.* 30: 664–672.
Ballard J.W.O. & Kreitman M. 1995. Is mitochondrial DNA a strictly neutral marker? *Trends Ecol. Evol.* 10: 485–488.
Ballard, J.W. & Whitlock, M.C. 2004. The incomplete natural history of mitochondria. *Mol. Ecol.* 13: 729–744.
Balloux, F. 2001. EASYPOP (version 1.7): a computer program for population genetics simulations. *J. Hered.* 92: 301–202.
Balloux, F., Brünner, H., Lugon-Moulin, N., Hausser, J. & Goudet, J. 2000. Microsatellites can be misleading: an empirical and simulation study. *Evolution* 54: 1414–1422.

Balloux, F. & Lugon-Moulin, N. 2002. The estimation of population differentiation with microsatellite markers. *Mol. Ecol.* 11: 155–165.

Beaumont, M.A. 2005. Adaptation and speciation: what can F_{ST} tell us? *Trends Ecol. Evol.* 20: 435–440.

Beaumont, M.A. & Nichols, R.A. 1996. Evaluating loci for use in the genetic analysis of population structure. *Proc. Biol. Sci.* 263: 1619–1626.

Beaumont, M.A. & Rannala, B. 2004. The Bayesian revolution in genetics. *Nat. Rev. Genet.* 5: 251–261.

Beerli, P. 2006. Comparison of Bayesian and maximum-likelihood inference of population genetic parameters. *Bioinformatics* 22: 341–345.

Beerli, P. & Palczewski, M. 2010. Unified framework to evaluate panmixia and migration direction among multiple sampling locations. *Genetics* 185: 313–326.

Beheregaray, L.B. 2008. Twenty years of phylogeography: the state of the field and the challenges for the Southern Hemisphere. *Mol. Ecol.* 17: 3754–3774.

Bensasson, D., Zhang, D.-X., Hartl, D.L. & Hewitt, G.M. 2001. Mitochondrial pseudogenes: evolution's misplaced witnesses. *Trends Ecol. Evol.* 16: 314–321.

Bensch, S., Irwin, D.E., Irwin, J.H., Kvist, L. & Akesson, S. 2006. Conflicting patterns of mitochondrial and nuclear DNA diversity in *Phylloscopus* warblers. *Mol. Ecol.* 15: 161–171.

Bilodeau, A.L., Felder, D.L. & Neigel, J.E. 2005. Population structure at two geographic scales in the burrowing crustacean *Callichirus islagrande* (Decapoda, Thalassinidea): historical and contemporary barriers to planktonic dispersal. *Evolution* 59: 2125–2138.

Boutin-Ganache, I., Raposo, M., Raymond, M. & Deschepper, C.F. 2001. M13-tailed primers improve the readability and usability of microsatellite analyses performed with two different allele-sizing methods. *Biotechniques* 31: 24–26, 28.

Bowcock, A.M., Ruiz-Linares, A., Tomfohrde, J., Minch, E., Kidd, J.R. & Cavalli-Sforza, L.L. 1994. High resolution of human evolutionary trees with polymorphic microsatellites. *Nature* 368: 455–457.

Brumfield, R.T., Beerli, P., Nickerson, D.A. & Edwards, S.V. 2003. The utility of single nucleotide polymorphisms in inferences of population history. *Trends Ecol. Evol.* 18: 249–256.

Buhay, J.E. 2009. "COI-like" sequences are becoming problematic in molecular systematic and DNA barcoding studies. *J. Crust. Biol.* 29: 96–110.

Buschiazzo, E. & Gemmell, N.J. 2006. The rise, fall and renaissance of microsatellites in eukaryotic genomes. *BioEssays* 28: 1040–1050.

Buschiazzo, E. & Gemmell, N.J. 2010. Conservation of human microsatellites across 450 million years of evolution. *Genome Biol. Evol.* 2: 153–165.

Cann, R.L., Stoneking, M. & Wilson, A.C. 1987. Mitochondrial DNA and human evolution. *Nature* 325: 31–36.

Castoe, T.A., Poole, A.W., Gu, W., De Koning, A.P.J., Daza, J.M., Smith, E.N. & Pollock, D.D. 2010. Rapid identification of thousands of copperhead snake (*Agkistrodon contortrix*) microsatellite loci from modest amounts of 454 shotgun genome sequence. *Mol. Ecol. Resour.* 10: 341–347.

Chambers, G.K. & MacAvoy, E.S. 2000. Microsatellites: consensus and controversy. *Comp. Biochem. Physiol. B Biochem. Mol. Biol.* 126: 455–476.

Chen, C., Durand, E., Forbes, O. & François, O. 2007. Bayesian clustering algorithms ascertaining spatial population structure: a new computer program and a comparison study. *Mol. Ecol. Notes* 7: 747–756.

Chenuil, A., Hoareau, T.B., Egea, E., Penant, G., Rocher, C., Aurelle, D., Mokhtar-Jamai, K., Bishop, J.D.D, Boissin, E., Diaz, A., Krakau, M., Luttikhuizen, P.C., Patti, F.P., Blavet, N. & Mousset, S. 2010. An efficient method to find potentially universal population genetic markers, applied to metazoans. *BMC Evol. Biol.* 10: 276.

Clement, M., Posada, D. & Crandall, K.A. 2000. TCS: a computer program to estimate gene genealogies. *Mol. Ecol.* 9: 1657–1659.

Colbourne, J.K., Pfrender, M.E., Gilbert, D., Thomas, W.K., Tucker, A., Oakley, T.H., Tokishita, S., Aerts, A., Arnold, G.J., Basu, M.K., Bauer, D.J., Cáceres, C.E., Carmel, L., Casola, C., Choi, J.-H., Detter, J.C., Dong, Q., Dusheyko, S., Eads, B.D., Fröhlich, T., Geiler-Samerotte, K.A., Gerlach, D., Hatcher, P., Jogdeo, S., Krijgsveld, J., Kriventseva, E.V., Kültz, D., Laforsch, C., Lindquist, E., Lopez, J., Manak, J.R., Muller, J., Pangilinan, J., Patwardhan, R.P., Pitluck, S., Pritham, E.J., Rechtsteiner, A., Rho, M., Rogozin, I.B., Sakarya, O., Salamov, A., Schaack, S., Shapiro, H., Shiga, Y., Skalitzky, C., Smith, Z., Souvorov, A., Sung, W., Tang, Z., Tsuchiya, D., Tu, H., Vos, H., Wang, M., Wolf, Y.I., Yamagata, H., Yamada, T., Ye, Y., Shaw, J.R., Andrews, J., Crease, T.J., Tang, H., Lucas, S.M., Robertson, H.M., Bork, P., Koonin, E.V., Zdobnov, E.M., Grigoriev, I.V., Lynch, M. & Boore, J.L. 2011. The ecoresponsive genome of *Daphnia pulex. Science* 331: 555–561.

Coombs, J.A., Letcher, B.H. & Nislow, K.H. 2008. CREATE: a software to create input files from diploid genotypic data for 52 genetic software programs. *Mol. Ecol. Resour.* 8: 578–580.

Cornuet, J.M., Ravigné, V. & Estoup, A. 2010. Inference on population history and model checking using DNA sequence and microsatellite data with the software DIYABC (v1.0). *BMC Bioinform.* 11: 401.

Csencsics, D., Brodbeck, S. & Holderegger, R. 2010. Cost-effective, species-specific microsatellite development for the endangered dwarf bulrush (*Typha minima*) using next-generation sequencing technology. *J. Hered.* 101: 789–793.

Dewoody, J., Nason, J.D. & Hipkins, V.D. 2006. Mitigating scoring errors in microsatellite data from wild populations. *Mol. Ecol. Notes* 6: 951–957.

Drummond, A.J., Ashton, B., Cheung M., Heled, J., Kearse, M., Moir, R., Stones-Havas, S., Thierer, T. & Wilson, A. 2009. *Geneious v4. 7.* Biomatters, Ltd., Auckland: New Zealand.

Drummond, A.J. & Rambaut, A. 2007. BEAST: Bayesian evolutionary analysis by sampling trees. *BMC Evol. Biol.* 7: 214.

Durand, E., Jay, F., Gaggiotti, O.E. & François, O. 2009. Spatial inference of admixture proportions and secondary contact zones. *Mol. Biol. Evol.* 26: 1963–1973.

Edwards, A.L., Hammond, H.A., Jin, L., Caskey, C.T. & Chakraborty, R. 1992. Genetic variation at five trimeric and tetrameric tandem repeat loci in four human population groups. *Genomics* 12: 241–253.

Ellegren, H. 2004. Microsatellites: simple sequences with complex evolution. *Nat. Rev. Genet.* 5: 435–445.

Estoup, A., Jarne, P. & Cornuet, J.-M. 2002. Homoplasy and mutation model at microsatellite loci and their consequences for population genetic analysis. *Mol. Ecol.* 11: 1591–1604.

Estoup, A., Tailliez, C., Cornuet, J.-M. & Solignac, M. 1995. Size homoplasy and mutational processes of interrupted micorsatellites in two bee species, *Apis mellifera* and *Bombus terrestris* (Apidae). *Mol. Biol. Evol.* 12: 1074–1084.

Evanno, G., Regnaut, S. & Goudet, J. 2005. Detecting the number of clusters of individuals using the software STRUCTURE: a simulation study. *Mol. Ecol.* 14: 2611–2620.

Excoffier, L. & Heckel, G. 2006. Computer programs for population genetics data analysis: a survival guide. *Nat. Rev. Genet.* 7: 745–758.

Excoffier, L., Laval, G. & Schneider, S. 2005. Arlequin (version 3.0): an integrated software package for population genetics data analysis. *Evol. Bioinform. Online* 1: 47–50.

Excoffier, L., Smouse, P.E. & Quattro, J.M. 1992. Analysis of molecular variance inferred from metric distnaces among DNA haplotypes: application to human mitochondrial DNA restriction data. *Genetics* 131: 479–491.

Falush, D., Stephens, M. & Pritchard, J.K. 2003. Inference of population structure using multilocus genotype data: linked loci and correlated allele frequencies. *Genetics* 164: 1567–1587.

Felsenstein, J. 2006. Accuracy of coalescent likelihood estimates: do we need more sites, more sequences, or more loci? *Mol. Biol. Evol.* 23: 691–700.

Funk, D.J. & Omland, K.E. 2003. Species-level paraphyly and polyphyly: frequency, causes, and consequences, with insights from animal mitochondrial DNA. *Annu. Rev. Ecol. Evol. Syst.* 34: 397–423.

Garnery, L., Cornuet, J.-M. & Solignac, M. 1992. Evolutionary history of the honey bee *Apis mellifera* inferred from mitochondrial DNA analysis. *Mol. Ecol.* 1: 145–154.

Glaubitz, J.C. 2004. CONVERT: a user-friendly program to reformat diploid genotypc data for commonly used population genetic software packages. *Mol. Ecol. Notes* 4: 309–310.

Glenn, T.C. & Schable, N.A. 2005. Isolating microsatellite DNA loci. *Methods Enzymol.* 395: 202–222.

Glez-Peña, D., Gómez-Blanco, D., Reboiro-Jato, M., Fdez-Riverola, F. & Posada, D. 2010. ALTER: program-oriented conversion of DNA and protein alignments. *Nucleic Acids Res.* 38: W14–W18.

Goldstein, D.B. & Pollock, D.D. 1997. Launching microsatellites: a review of mutation processes and methods of phylogenetic interference. *J. Hered.* 88: 335–342.

Goldstein, D.B. & Schlötterer, C. 1999. *Microsatellites: Evolution and Applications.* New York, NY: Oxford University Press.

Gompert, Z., Forister, M.L., Fordyce, J.A., Nice, C.C., Willamson, R.J. & Buerkle, C.A. 2010. Bayesian analysis of molecular variance in pyrosequences quantifies population genetic structure across the genome of *Lycaeides* butterflies. *Mol. Ecol.* 19: 2455–2473.

Hazkani-Covo, E., Zeller, R.M. & Martin, W. 2010. Molecular poltergeists: mitochondrial DNA copies (numts) in sequenced nuclear genomes. *PLoS Genet.* 6: e1000834.

Hebert, P.D., Ratnasingham, S. & deWaard, J.R. 2003. Barcoding animal life: cytochrome c oxidase subunit 1 divergences among closely related species. *Proc. Biol. Sci.* 270 *Suppl.* 1: S96–S99.

Hedrick, P.W. 2005. A standardized measure of genetic differentiation. *Evolution* 59: 1633–1638.

Held, C. 2003. Molecular evidence for cryptic speciation within the widespread Antarctic crustacean *Ceratoserolis trilobitoides* (Crustacea, Isopoda). In: Huiskes, A.H., Gieskes, W.W., Rozema, J., Schorno, R.M., van der Vies, S.M. & Wolff, W.J. (eds.) *Antarctic Biology in a Global Context*: 135–139. Leiden: Backhuys Publishers.

Heller, R. & Siegismund, H.R. 2009. Relationship between three measures of genetic differentiation G_{ST}, D_{EST} and G'_{ST}: how wrong have we been? *Mol. Ecol.* 18: 2080–2083.

Henn, B.M., Gignoux, C.R., Jobin, M., Granka, J.M., Macpherson, J.M., Kidd, J.M., Rodríguez-Botigué, L., Ramachandran, S., Hon, L., Brisbin, A., Lin, A.A., Underhill, P.A., Comas, D., Kidd, K.K., Norman, P.J., Parham, P., Bustamante, C.D., Mountain, J.L. & Feldman, M.W. 2011. Hunter-gatherer genomic diversity suggests a southern African origin for modern humans. *Proc. Natl. Acad. Sci. USA* 108: 5154–5162.

Hey, J. & Machado, C.A. 2003. The study of structured populations—new hope for a difficult and divided science. *Nat. Rev. Genet.* 4: 535–543.

Hey, J. & Nielsen, R. 2007. Integration within the Felsenstein equation for improved Markov chain Monte Carlo methods in population genetics. *Proc. Natl. Acad. Sci. USA* 104: 2785–2790.

Hickerson, M.J., Carstens, B.C., Cavender-Bares, J., Crandall, K.A., Graham, C.H., Johnson, J.B., Victoriano, P.F. & Yoder, A.D. 2010. Phylogeography's past, present, and future: 10 years after Avise, 2000. *Mol. Phylogent. Evol.* 54: 291–301.

Hoffman, J.I. & Amos, W. 2005. Microsatellite genotyping errors: detection approaches, common sources and consequences for paternal exclusion. *Mol. Ecol.* 14: 599–612.

Hoffman, J.I. & Nichols, H.J. 2011. A novel approach for mining polymorphic microsatellite markers *in silico*, *PLoS One* 6: e23283.

Holsinger, K.E. 2010. Next generation population genetics and phylogeography. *Mol. Ecol.* 19: 2361–2363.

Holsinger, K.E. & Weir, B.S. 2009. Genetics in geographically structured populations: defining, estimating and interpreting F_{ST}. *Nat. Rev. Genet.* 10: 639–650.

Hurst, G.D.D. & Jiggins, F.M. 2005. Problems with mitochondrial DNA as a marker in population, phylogeographic and phylogenetic studies: the effects of inherited symbionts. *Proc. R. Soc. Lond. B* 272: 1525–1534.

Huson, D.H. & Bryant, D. 2006. Application of phylogenetic networks in evolutionary studies. *Mol. Biol. Evol.* 23: 254–267.

Jakobsson, M. & Rosenberg, N.A. 2007. CLUMPP: a cluster matching and permutation program for dealing with label switching and multimodality in analysis of population structure. *Bioinformatics* 23: 1801–1806.

Jarne, P. & Lagoda, P.J.L. 1996. Microsatellites, from molecules to populations and back. *Trends Ecol. Evol.* 11: 424–429.

Jennings, W.B. & Edwards, S.V. 2005. Speciational history of Australian grass finches (*Poephila*) inferred from thirty gene trees. *Evolution* 59: 2033–2047.

Jost, L. 2008. G_{ST} and its relatives do not measure differentiation. *Mol. Ecol.* 17: 4015–4026.

Junier, T. & Pagni, M. 2000. Dotlet: diagonal plots in a web browser. *Bioinformatics* 16: 178–179.

Kelkar, Y.D., Strubczewski, N., Hile, S.E., Chiaromonte, F., Eckert, K.A. & Makova, K.D. 2010. What is a microsatellite: a computational and experimental definition based upon repeat mutational behavior at A/T and GT/AC Repeats. *Genome Biol. Evol.* 2: 620–635.

Knowles, L.L. 2008. Why does a method that fails continue to be used? *Evolution* 62: 2713–2717.

Koskinen, M.T., Hirvonen, H., Landry, P.A. & Primmer, C.R. 2004. The benefits of increasing the number of microsatellites utilized in genetic population studies: an empirical perspective. *Hereditas* 141: 61–67.

Kraemer, L., Beszteri, B., Gäbler-Schwarz, S., Held, C., Leese, F., Mayer, C., Pohlmann, K. & Frickenhaus, S. 2009. STAMP: Extensions to the STADEN sequence analysis package for high throughput interactive microsatellite marker design. *BMC Bioinform.* 10: 41.

Kuhner, M.K. 2008. Coalescent genealogy samplers: windows into population history. *Trends Ecol. Evol.* 24: 86–93.

Kumar, S., Nei, M., Dudley, J. & Tamura, K. 2008. MEGA: a biologist-centric software for evolutionary analysis of DNA and protein sequences. *Brief. Bioinform.* 9: 299–306.

Landry, P.-A., Koskinen, M.T. & Primmer, C.R. 2002. Deriving evolutionary relationships among populations using microsatellites and $(\delta\mu)^2$: all loci are equal, but some are more equal than others. *Genetics* 161: 1339–1347.

Leese, F., Agrawal, S. & Held, C. 2010. Long-distance island hopping without dispersal stages: transportation across major zoogeographic barriers in a Southern Ocean isopod. *Naturwissenschaften* 97: 583–594.

Leese, F. & Held, C. 2008. Identification and characterization of microsatellites from the Antarctic isopod *Ceratoserolis trilobitoides*—nuclear evidence for cryptic species. *Conserv. Genet.* 9: 1369–1372.

Leese, F., Mayer, C. & Held, C. 2008. Isolation of microsatellites from unknown genomes using known genomes as enrichment templates. *Limnol. Oceanogr. Methods* 6: 412–426.

Lemmon, A.R. & Lemmon, E.M. 2008. A likelihood framework for estimating phylogeographic history on a continuous landscape. *Syst. Biol.* 57: 544–561.

Li, Y., Wongprasert, K., Shekhar, M., Ryan, J., Dierens, L., Meadows, J., Preston, N., Coman, G. & Lyons, R.E. 2007. Development of two microsatellite multiplex systems for black tiger shrimp *Penaeus monodon* and its application in genetic diversity study for two populations.

Aquaculture 266: 279–288.

Lischer, H. 2009. *PGDSpider: a program for converting data between population genetics programs.* Berne: University of Berne.

Ljungqvist, M., Akesson, M. & Hansson, B. 2010. Do microsatellites reflect genome-wide genetic diversity in natural populations? A comment on Väli et al. (2008). *Mol. Ecol.* 19: 851–855.

Lopes, J.S., Balding, D. & Beaumont, M.A. 2009. PopABC: a program to infer historical demographic parameters. *Bioinformatics* 25: 2747–2749.

Malausa, T., Gilles, A., Meglécz, E., Blanquart, H., Duthoy, S., Costedoat, C., Dubut, V., Pech, N., Castagnone-Sereno, P., Délye, C., Feau, N., Frey, P., Gauthier, P., Guillemaud, T., Hazard, L., Le Corre, V., Lung-Escarmant, B., Malé, P.-J. G., Ferreira, S. & Martin, J.-F. 2011. High-throughput microsatellite isolation through 454 GS-FLX Titanium pyrosequencing of enriched DNA libraries. *Mol. Ecol. Resour.* doi: 10.1111/j.1755-0998.2011.02992.x

Maneeruttanarungroj, C., Pongsomboon, S., Wuthisuthimethavee, S., Klinbunga, S., Wilson, K.J., Swan, J., Li, Y., Whan, V., Chu, K.-H., Li, C.P., Tong, J., Glenn, K., Rothschild, M., Jerry, D. & Tassanakajon, A. 2006. Development of polymorphic expressed sequence tag-derived microsatellites for the extension of the genetic linkage map of the black tiger shrimp (*Penaeus monodon*). *Anim. Genet.* 37: 363–368.

Manoukis, N.C. 2007. FORMATOMATIC: a program for converting diploid allelic data between common formats for population genetic analysis. *Mol. Ecol. Notes* 7: 592.

Mayer, C. 2010. Phobos version 3.3.12. A tandem repeat search program. http://www.ruhr-uni-bochum.de/spezzoo/cm/cm_phobos.htm.

Mayer, C., Leese, F. & Tollrian, R. 2010. Genome-wide analysis of tandem repeats in *Daphnia pulex*—a comparative approach. *BMC Genomics* 11: 277.

McKelvey, K.S. & Schwartz, M.S. 2005. DROPOUT: a program to identify problem loci and samples for noninvasive genetic samples in a capture-mark-recapture framework. *Mol. Ecol. Notes* 5: 716–718.

Meirmans, P.G. 2006. Using the AMOVA framework to estimate a standardized genetic differentiation measure. *Evolution* 60: 2399–2402.

Meirmans, P.G. & Hedrick, P.W. 2011. Assessing population structure: F_{ST} and related measures. *Mol. Ecol. Resour.* 11: 5–18.

Meirmans, P.G. & Van Tienderen, P.H. 2004. GENOTYPE and GENODIVE: two programs for the analysis of genetic diversity of asexual organisms. *Mol. Ecol. Notes* 4: 792–794.

Miller, M.R., Dunham, J.P., Amores, A., Cresko, W.A. & Johnson, E.A. 2007. Rapid and cost-effective polymorphism identification and genotyping using restriction site associated DNA (RAD) markers. *Genome Res.* 17: 240–248.

Milligan, P.J., Stahl, E.A., Schizas, N.V. & Turner, J.T. 2011. Phylogeography of the copepod *Acartia hudsonica* in estuaries of the northeastern United States. *Hydrobiologia* 666: 155–165.

Neigel, J.E. 2002. Is F_{ST} obsolete? *Conserv. Genet.* 3: 167–173.

Nielsen, R. & Beaumont, M.A. 2009. Statistical inferences in phylogeography. *Mol. Ecol.* 18: 1034–1047.

Nolte, A.W., Stemshorn, K.C. & Tautz, D. 2005. Direct cloning of microsatellite loci from *Cottus gobio* through a simplified enrichment procedure. *Mol. Ecol. Notes* 5: 628–636.

Pompanon, F., Bonin, A., Bellemain, E. & Taberlet, P. 2005. Genotyping errors: causes, consequences and solutions. *Nat. Rev. Genet* 6: 847–859.

Pritchard, J.K., Stephens, M. & Donnelly, P. 2000. Inference of population structure using multilocus genotype data. *Genetics* 155: 945–959.

Ratnasingham, S. & Hebert, P.D.N. 2007. BOLD: the Barcode of Life Data System (www.barcodinglife.org). *Mol. Ecol. Notes* 7: 355–364.

Rico, C. Rico, I. & Hewitt, G. 1996. 470 million year of consevation of microsatellite loci among fish species. *Proc. R. Soc. Lond. B* 263: 549–557.

Rogers, A.R. & Harpending, H. 1992. Population growth makes waves in the distribution of pairwise genetic differences. *Mol. Biol. Evol.* 9: 552–569.

Rokas, A., Ladoukakis, E. & Zouros, E. 2003. Animal mitochondrial DNA recombination revisited. *Trends Ecol. Evol.* 18: 411–417.

Rosenberg, N.A. 2004. Distruct: a program for the graphical display of population structure. *Mol. Ecol. Notes* 4: 137–138.

Rousset, F. 2008. GENEPOP'007: a complete re-implementation of the GENEPOP software for Windows and Linux. *Mol. Ecol. Resour.* 8: 103–106.

Rozen, S. & Skaletsky, H.J. 2000. Primer3 on the WWW for general users and for biologist programmers. In: Krawetz, S. & Misener, S. (eds.), *Bioinformatics Methods and Protocols: Methods in Molecular Biology*: 365–386. Totowa, NJ: Humana Press.

Sands, C.J., Lancaster, M.L., Austin, J.J. & Sunnucks, P. 2009. Single copy nuclear DNA markers for the onychophoran *Phallocephale tallagandensis. Conserv. Genet. Resour.* 1: 17–19.

Santana, Q.C., Coetzee, M.P.A., Steenkamp, E.T., Mlonyeni, O.X., Hammond, G.N.A., Wingfield, M.J. & Wingfield, B.D. 2009. Microsatellite discovery by deep sequencing of enriched genomic libraries. *BioTechniques* 46: 217–223.

Schlötterer, C. 2004. The evolution of molecular markers—just a matter of fashion? *Nat. Rev. Genet.* 5: 63–69.

Seeb, J.E., Carvalho, G., Hauser, L., Naish, K., Roberts, S. & Seeb, L.W. 2011. Single-nucleotide polymorphism (SNP) discovery and applications of SNP genotyping in nonmodel organisms. *Mol. Ecol. Resour.* 11: 1–8.

Selkoe, K.A. & Toonen, R.J. 2006. Microsatellites for ecologists: a practical guide to using and evaluating microsatellite markers. *Ecol. Lett.* 9: 615–629.

Shaw, K.L. 2002. Conflict between nuclear and mitochondrial DNA phylogenies of a recent species radiation: what mtDNA reveals and conceals about modes of speciation in Hawaiian crickets. *Proc. Natl. Acad. Sci. USA* 99: 16122–16127.

Shearer, T.L., Van Oppen, M.J.H., Romano, S.L. & Wörhcide, G. 2002. Slow mitochondrial DNA sequence evolution in the Anthozoa (Cnidaria). *Mol. Ecol.* 11: 2475–2487.

Slatkin, M. 1995. A measure of population subdivision based on microsatellite allele frequencies. *Genetics* 139: 457–462.

Sunnucks, P. 2000. Efficient genetic markers for population biology. *Trends Ecol. Evol.* 15: 199–203.

Sunnucks, P. & Hales, D.F. 1996. Numerous transposed sequences of mitochondrial cytochrome oxidase I–II in aphids of the genus *Sitobion* (Hemiptera: Aphididae). *Mol. Biol. Evol.* 13: 510–524.

Tamura, K., Peterson, D., Peterson, N., Stecher, G., Nei, M. & Kumar, S. 2011. MEGA5: molecular evolutionary genetics analysis using maximum likelihood, evolutionary distance, and maximum parsimony methods. *Mol. Biol. Evol.* (online early view).

Teixeira, S., Cambon-Bonavita, M.A., Serrão, E.A., Desbruyéres, D. & Arnaud-Haond, S. 2011. Recent population expansion and connectivity in the hydrothermal shrimp *Rimicaris exoculata* along the Mid-Atlantic Ridge. *J. Biogeogr.* 38: 564–574.

Templeton, A.R. 1998. Nested clade analyses of phylogeographic data: testing hypotheses about gene flow and population history. *Mol. Ecol.* 7: 381–397.

Tóth, G., Gáspári, Z. & Jurka, J. 2000. Microsatellites in different eukaryotic genomes: survey and analysis. *Genome Res.* 10: 967–981.

Väli, U., Einarsson, A., Waits, L. & Ellegren, H. 2008. To what extent do microsatellite markers reflect genome-wide genetic diversity in natural populations? *Mol. Ecol.* 17: 3808–3817.

van Oosterhout, C., Hutchinson, W.F., Wills, D.P.M. & Shipley, P. 2004. MICRO-CHECKER: software for identifying and correcting genotyping errors in microsatellite data. *Mol. Ecol. Notes* 4: 535–538.

Villesen, P. 2007. FaBox: an online toolbox for fasta sequences. *Mol. Ecol. Notes* 7: 965–968.

Wahlberg, N., Weingartner, E., Warren, A.D. & Nylin, S. 2009. Timing major conflict between mitochondrial and nuclear genes in species relationships of *Polygonia* butterflies (Nymphalidae: Nymphalini). *BMC Evol. Biol.* 9: 92.

Wang, J. 2010. Effects of genotyping errors on parentage exclusion analysis. *Mol. Ecol.* 19: 5061–5078.

Waples, R.S. & Gaggiotti, O. 2006. What is a population? An empirical evaluation of some genetic methods for identifying the number of gene pools and their degree of connectivity. *Mol. Ecol.* 15: 1419–1439.

Wares, J.P. 2010. Natural distributions of mitochondrial sequence diversity support new null hypotheses. *Evolution* 64: 1136–1142.

Weber, J.L. & Wong, C. 1993. Mutation of human short tandem repeats. *Hum. Mol. Genet.* 2: 1123–1128.

Weir, B.S. & Cockerham, C.C. 1984. Estimating F–statistics for the analysis of population structure. *Evolution* 38: 1358–1370.

Whitlock, M.C. & McCauley, D.E. 1999. Indirect measures of gene flow and migration: $F_{ST} \neq 1/(4Nm+1)$. *Heredity* 82: 117–125.

Williams, S.T. & Knowlton, N. 2001. Mitochondrial pseudogenes are pervasive and often insidious in the snapping shrimp genus *Alpheus*. *Mol. Biol. Evol.* 18: 1484–1493.

Wright, S. 1943. Isolation by distance. *Genetics* 28: 114–138.

Yebra, L., Bonnet, D., Harris, R.P., Lindeque, P.K. & Peijnenburg, T.C.A. 2011. Barriers in the pelagic: population structuring of *Calanus helgolandicus* and *C. euxinus* in European waters. *Mar. Ecol. Progr. Ser.* 428: 135–149.

Zane, L., Bargelloni, L. & Patarnello, T. 2002. Strategies for microsatellite isolation: a review. *Mol. Ecol.* 11: 1–16.

Zhang, D.-X. 2004. Lepidopteran microsatellite DNA: redundant but promising. *Trends Ecol. Evol.* 19: 507–509.

Detecting and measuring genetic differentiation

CHRISTOPHER E. BIRD[1], STEPHEN A. KARL[1], PETER E. SMOUSE[2] & ROBERT J. TOONEN[1]

[1] *Hawai'i Institute of Marine Biology, School of Ocean and Earth Sciences and Technology, University of Hawai'i at Mānoa, Kāne'ohe, HI 96744, U. S. A.*

[2] *Ecology, Evolution and Natural Resources, School of Environmental and Biological Sciences, Rutgers, The State University of New Jersey, New Brunswick, NJ 08901, U. S. A.*

ABSTRACT

F_{ST}, F'_{ST}, Φ_{ST}, Φ'_{ST}, and D_{est} are the primary metrics utilized for empirically estimating and testing the magnitude of genetic divergence among populations. There is currently active discussion in the literature about which of these metrics are most appropriate for empirical surveys of genetic differentiation. Here we compare the performance of each metric in 80 simulated population comparisons with an a priori known level of genetic differentiation that ranges from 0 to 100%. In these simulations, we manipulate population characteristics such as the genetic distance between haplotypes and the diversity of haplotypes that are shared among populations, as well as those that are unique to specific populations, illustrating key features and differences among the metrics, with an eye toward separating biological signal from statistical noise. F_{ST} is the best choice for datasets consisting of neutral unlinked single nucleotide polymorphism (SNP) datasets involving two alleles per locus. D_{est} and F'_{ST} tend to be the best metrics for analyses with more than two alleles, at least where the genetic distance among the alleles is not important, but the use of F'_{ST} and, to a greater extent, D_{est} is currently limited to relatively simple datasets, due to a dearth of computer software. If genetic distance among alleles is an important consideration, then Φ_{ST} is the better metric, but we demonstrate that Φ_{ST} and Φ'_{ST} can accentuate either noise or signal, depending upon the characteristics of the populations and the hypotheses being tested. In many cases, it can be informative to apply both distance-based and allele/haplotype-based metrics, or both fixation and genetic differentiation indices. All of these measures are highly sensitive to the diversity of alleles shared between populations, with common alleles dominating the behavior of all of these metrics. Φ_{ST} is shown to be relatively unaffected by the phenomenon of high allelic diversity driving down estimates of genetic differentiation that plague F_{ST}. Overall, there is no single metric that best captures population genetic differentiation, and we recommend that researchers report both a fixation index (F_{ST} or Φ_{ST}) and an index of genetic differentiation (F'_{ST} or D_{est}) for their datasets because they represent different properties of population partitioning. When indices of fixation and genetic differentiation are in agreement, one can be sure of the conclusion. When the two methods yield differing results, the pattern and direction of discord can be diagnostic of a particular phenomenon, and we provide a range of simulations across parameter space to illustrate both points.

1 INTRODUCTION

A primary goal of empirical population genetic studies is the identification, quantification, and comparison of genetic differentiation among loci, individuals, populations, species, and studies. As the field has pushed to find ever more polymorphic markers, with the expectation that a higher mutation rate would provide finer temporal resolution, there has been an unanticipated consequence for studies of genetic differentiation. Several authors have pointed out that the widely used metrics for describing genetic differentiation, Sewall Wright's F–statistics (1943), originally introduced for bi-allelic loci, decrease with increasing allelic diversity (Wright 1978; Charlesworth 1998; Hedrick 1999, 2005; Jost 2008; Meirmans & Hedrick 2011), potentially affecting the conclusions of a number of published studies (cf., Neigel 2002; Heller & Siegismund 2009). Recent realization of the problem in relation to the estimation of population partitioning has resulted in a proliferation of standardization procedures for F-statistics (Meirmans 2006), G-statistics (Hedrick 2005), a novel method (D_{est}) of assessing differentiation (Jost 2008), and some understandable confusion about which metric to use for any particular analysis. As often happens, there is no "silver bullet"; the optimal choice depends on the specific situation (Meirmans & Hedrick 2011). Here, we review several of these methods of measuring genetic differentiation, outline the strengths and weaknesses of each, illustrate their behavior across a range of hypothetical datasets, and provide some guidelines for selecting the appropriate analysis in different situations.

1.1   Fixation indices versus differentiation indices

1.1.1   *Fixation indices (F_{ST}, θ, Φ_{ST}, G_{ST})*

The most widely used metric of genetic differentiation has been Wright's (1943, 1951, 1965) fixation index, F_{ST}, which was developed as part of a set of hierarchical parameters (F_{ST}, F_{IS}, and F_{IT}) to assess the way in which genetic variation is hierarchically partitioned in natural populations. Modern analogues to F_{ST} include G_{ST} (Nei 1973, 1987), θ (Cockerham 1969, 1973; Weir & Cockerham 1984), and Φ_{ST} (Excoffier et al. 1992). As formulated, Wright's F_{ST} ranges from 0.0 to 1.0, where 0.0 indicates identical allele frequencies in a pair of populations (no differentiation) and 1.0 indicates alternate fixation for a single unique allele in each population. Extending beyond Wright's (1951, 1965) original intent of estimating the degree of fixation, F_{ST} and its analogues (fixation indices) are often employed in modern empirical studies to identify, quantify, and compare the magnitude of genetic differentiation among population samples. In this context, a minimum value of 0.0 is interpreted as identical allelic composition, but a maximum value of 1.0 is interpreted as the absence of any shared alleles, and intermediate values as the magnitude of genetic differentiation among samples. If any population sample has more than one allele, however, even when the samples share no alleles in common, the maximum value of 1.0 is never observed for fixation indices (Wright 1978; Hedrick 1999; Jost 2008). Further, as the internal genetic diversity of population samples increases, the maximum possible value decreases. Consequently, using fixation indices will systematically underestimate genetic differentiation, especially when using highly polymorphic markers such as microsatellites (Hedrick 1999).

1.1.2   *Standardized fixation indices (G'_{ST}, F'_{ST}, Φ'_{ST})*

Hedrick (2005) proposed that standardizing G_{ST} relative to mean within population heterozygosity (i.e., genetic diversity) yields an improved (0,1)–scaled estimate of genetic differentiation (G'_{ST}), calculated by scaling the observed (G_{ST}) relative to its maximum achievable value (G_{STmax}), given the observed within-population heterozygosity. Following Hedrick, Meirmans (2006) applied the same standardization concept to analysis of molecular variance (AMOVA; Excoffier et al. 1992), resulting in a standardized index of genetic differentiation, ϕ'_{ST}, relative to their maximum achievable values. Note: Excoffier et al. (1992) used Φ_{ST} as an index-label for both "degree of difference" (distance-based) and "simply different" (allele/haplotype-based) usage, but we will here denote indices based on the distance-based fixation indices as Φ and those that do not take distance into account as F, to make a useful distinction. With that same rationale $\phi'_{ST} = F'_{ST}$, because variable genetic distances between alleles were not addressed by Meirmans (2006). When population samples are completely differentiated (i.e., have no shared alleles), both Hedrick's and Meirmans' standardized fixation indices reach the maximum of 1.0, regardless of allelic variation within populations, providing useful (scaled) measures of genetic differentiation.

1.1.3   *Pure indices of genetic differentiation* (D, D_{est})

Taking a different tack, Jost (2008) uses true allelic diversity (Δ) to derive a measure of genetic differentiation, D_{est}, an estimate of divergence (D), corrected for sampling bias, that is not rooted in F–statistic-based metrics. Jost (2008) argues that D is superior, because although they are related, (1) heterozygosity (H) or allele frequency variance (nonlinear measures) are not equal to true diversity (a linear measure). Instead, true diversity can be defined as $\Delta = 1/(1 - H)$. Equally important is that (2) the relationship among the mean within-populations (Δ_{WP}), among-populations (Δ_{AP}), and the total diversity components (Δ_T) is multiplicative ($\Delta_T = \Delta_{WP}\Delta_{AP}$), rather than additive, as is the case for the AMOVA variance components ($\sigma_T^2 = \sigma_{WP}^2 + \sigma_{AP}^2$, where σ_T^2, σ_{WP}^2, and σ_{AP}^2 are the total, within population, and among population variance components, respectively). Jost (2008) demonstrates that $D = ((\Delta_{WP}/\Delta_T) - 1)/((1/n) - 1)$. No standardization is required to mitigate the effects of within-population allelic diversity on the estimate of genetic differentiation among populations, as the diversity components within and among populations are independent by construction (unlike the variance components used to estimate F_{ST} or Φ_{ST}). Thus, D is not affected by within-population allelic variation, and when two population samples are completely differentiated (i.e., share no alleles), $D = 1$. The calculation of D can also be extended to more complex hierarchical sampling schemes (Tuomisto 2010b, a). D, however, does not incorporate the degree of genetic distance between alleles/haplotypes into the estimation of genetic differentiation among population samples (unlike Φ_{ST}). While no test of the statistical significance of D was offered by Jost (2008), incorporating resampling procedures such as those employed by Excoffier et al. (1992) for AMOVA is straightforward.

1.2   Comparisons of genetic differentiation

The majority of published studies that seek to evaluate the performance of one estimate of genetic differentiation versus another use such complex datasets that there is no clear expectation of the actual level of genetic differentiation. For example, Meirmans (2006) simulates complex real-world datasets to test his F'_{ST} against F_{ST}. The simulation consists of two groups of five populations, with 100 individuals each and a migration rate of 0.01 among populations within groups and 0.001 between populations in different groups. Each individual has 25 linked loci, with a K–allele mutation model having 100 possible allelic states and two mutation rates, 0.01 and 0.001. F'_{ST} yields fairly consistent results between simulations with different mutation rates, but F_{ST} does not. Such studies rely on precision to assess which technique yields the best results. No a priori assessments of genetic differentiation are made for these complex simulations, so the reader can only assume that F'_{ST} was accurate as well as precise. We also need to know whether the estimates of genetic differentiation are unbiased. Typically, few datasets are simulated, and they are rarely designed to explore the extremes of the models in question. In general, we need to evaluate procedures under a wide variety of circumstances.

1.3   Effects of allelic diversity and genetic distance

Jost (2008) argues that by using D_{est}, we avoid any impact of within-population diversity on estimates of genetic differentiation among populations. It is worth noting, however, that both D_{est} and F–statistics and analogues are based on the Gini-Simpson index of diversity (Gini 1918; Simpson 1949), a second order diversity measure. Second order diversity measures deliberately weight common alleles more heavily than rare alleles (Jost 2007, 2008). How well any of these metrics perform when the alleles shared among populations are numerous but at low frequency (rare and diverse), versus few and at higher frequency (common and not diverse) is unknown.

With the exception of Φ_{ST} and Φ'_{ST}, all fixation and genetic differentiation indices discussed here assume that the alleles being compared are equally and maximally distant from one another. As might be expected, however, distance-based AMOVA (yielding Φ_{ST}), when applicable, is currently the most commonly reported analysis of genetic differentiation in the literature, because it incorporates more genetic information into the analysis and is viewed as superior to allele/haplotype-based methods. (Note: We refer to differentiation as differences among groups of individuals and to distance as an estimate of the number of differences between two alleles at a particular locus). Thus far, however, the sensitivity of Φ_{ST} to the effects of increasing allelic diversity within populations has not been evaluated (Hedrick 1999, 2005; Meirmans 2006; Meirmans & Hedrick 2010). Moreover, it is generally unknown whether there are situations in which the additional information of genetic distances between alleles, which is used to effectively bin similar alleles together, could decrease the accuracy of an analysis.

Here we seek to elucidate the selection of a genetic differentiation measure by comparing the performance of the commonly reported indices of genetic differentiation, specifically F_{ST}, F'_{ST}, Φ_{ST}, Φ'_{ST}, and D_{est}, using simulated datasets constructed with varying levels of known differentiation, overall diversity, shared allelic diversity, and unshared diversity among populations. Our goal is to show how these metrics perform in a variety of situations, with the immediate object being to distinguish biologically relevant phenomena from measurement artifacts.

2   MATERIAL AND METHODS

2.1   AMOVA calculations

2.1.1 F_{ST} and F'_{ST}

To fully understand the results of an AMOVA, it is important to have some familiarity with the methodology. The finer calculation details can be found in a variety of sources (Weir & Cockerham 1984; Excoffier et al. 1992; Weir 1996; Weir & Hill 2002; Excoffier et al. 2006). Such publications often contain complex mathematical equations and proofs, making them somewhat inaccessible to nonspecialists. Here, we review the AMOVA in a mathematically simplified and visually intuitive manner. Basically, AMOVA is analysis of variance for molecular markers and one uses a matrix of genetic distances among sampled individuals as a starting point. The widely familiar output is an AMOVA table that details the degrees of freedom (df), sum of squared differences (SS), mean square differences (MS), and resulting F–statistics for comparisons within and among populations. Here, we outline the calculation of each of these parameters, starting with the matrix of genetic distances among individuals (Figure 1).

AMOVA is based on the matrix of genetic distances (Figure 1A). The genetic distance matrix has one row and one column for each individual sampled—N^2 cells—where N is the total number of individuals sampled. The matrix is subdivided by the population samples to which the individuals belong. In Figure 1A, there are three individuals sampled from population I ($n_1 = 3$) and three from

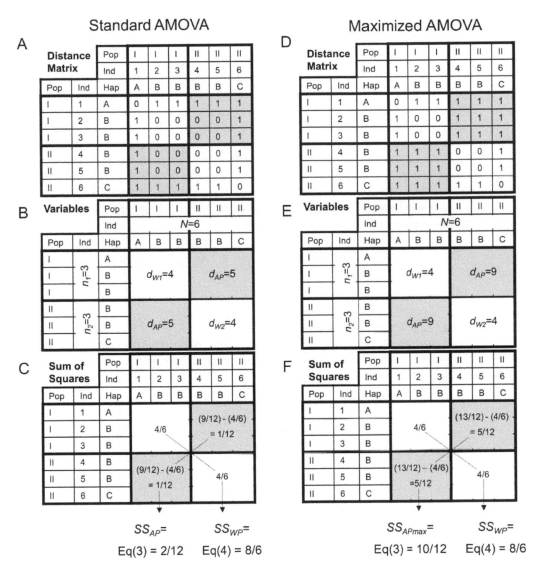

Figure 1. Illustrated diagram, demonstrating the calculation of the within-population (white-filled cells) and among-population (gray-filled cells) sums of squared differences used in the calculation of analysis of molecular variance (AMOVA). Panel A is the distance matrix. Panel B shows the calculation of sample size and summing the four regions of the distance matrix. Panel C demonstrates the specific calculation of the sum of squared differences. Panels D–F show the same procedure for the case where the among populations differences are maximized (Meirmans 2006). N is the total sample size; n_1 and n_2 are sample sizes for populations I and II respectively; d_{W1} and d_{W2} are the sum of differences within populations I and II, respectively; d_{AP} is the sum of differences among populations I and II; SS_{AP} and SS_{WP} are the sum of squared differences among and within populations I and II; and SS_{APmax} is the maximum possible sum of squared differences among populations I and II.

population II ($n_2 = 3$), for a total of six individuals ($N = 6$). Three different haplotypes (A, B, and C) are sampled, and they are treated as being equally different; we are thus calculating F_{ST}.

Each cell in the matrix is filled with either a 1 if the haplotype for that pair of individuals (row and column) are different or a 0 if they are the same. The degrees of freedom for the among-populations (df_{AP}) and within-populations (df_{WP}) components are calculated as follows:

$$df_{AP} = P - 1, \tag{1}$$

$$df_{WP} = (n_1 - 1) + (n_2 - 1), \tag{2}$$

where P is the number of populations sampled (here $P = 2$). For the example in Figure 1, $df_{AP} = 2 - 1 = 1$ and $df_{WP} = (3 - 1) + (3 - 1) = 4$ (Table 1A).

The calculation of the sum of squared differences (SS) is somewhat more complicated. The matrices in Figure 1 are divided into two gray blocks for comparisons among populations and two white blocks for comparisons within populations. The first step in the AMOVA is to sum the square of each distance in each of the four blocks (Figure 1C), where the sum of differences within population I is d_{W1}, that within population II is d_{W2}, and that among populations is d_{AP}. The SS values among and within populations are calculated as follows:

$$SS_{AP} = \left(\frac{d_{AP} + d_{W1}}{2N} - \frac{d_{W1}}{2n_1} \right) + \left(\frac{d_{AP} + d_{W2}}{2N} - \frac{d_{W2}}{2n_2} \right), \tag{3}$$

$$SS_{WP} = \frac{d_{W1}}{2n_1} + \frac{d_{W2}}{2n_2} \tag{4}$$

(Figure 1C). The total sum of squared differences, $SS_{Total} = SS_{AP} + SS_{WP}$. The mean squared differences (MS) are simply the sum of squared differences, divided by their degrees of freedom. Both the within- and among-population variance components are required to calculate F_{ST}, and we estimate them as functions of the mean squares:

$$\sigma^2_{WP} = MS_{WP}, \tag{5}$$

$$\sigma^2_{AP} = \frac{MS_{AP} - MS_{WP}}{n_c}, \tag{6}$$

$$n_c = \frac{N - \left\lfloor \frac{n_1^2 + n_2^2}{N} \right\rfloor}{P - 1}, \tag{7}$$

from which we estimate the fraction of total variation that is explained by differentiation among populations:

$$F_{ST} = \frac{\sigma^2_{AP}}{\sigma^2_{WP} + \sigma^2_{AP}}. \tag{8}$$

The next step is to standardize F_{ST}, so that it is 1.0 when no alleles are shared between populations. Meirmans' (2006) F'_{ST} relies upon determining the maximum F_{ST} that can be obtained for the observed within-population variance. To do this, the genetic distances in the among-populations portions of the distance matrix (Figure 1D, gray fill) are maximized (set to 1) and the AMOVA calculations are repeated (Figures 1D–F, Table 1). Meirmans' F'_{ST} represents the proportion of total variance explained by genetic differentiation among populations, relative to the maximum proportion attainable, given the observed variation within populations, calculated as:

$$F'_{ST} = \frac{F_{ST}}{F_{STmax}}. \tag{9}$$

F'_{ST} is thus independent of the level of within-population variation. Meirmans' F'_{ST} equation works well in most situations, but assumes the minimum possible F_{ST} is 0.0 and it can yield large negative numbers when sample sizes are small (negative F_{ST} values are a function of a sampling bias correction in the AMOVA calculations).

Table 1. AMOVA tables for the calculation of (A) F_{ST} and (B) F_{STmax} data are drawn from Figure 1. Because AMOVA accounts for sampling bias, it is possible to obtain negative values of F_{ST}.

A. Observed divergence

Source	df	SS	MS	σ^2	F_{ST}
Among populations	1	0.17	0.17	−0.06	−0.2
Within populations	4	1.33	0.33	0.33	
Total	5	1.5			

B. Maximum divergence

Source	df	SS	MS	σ^2	F_{STmax}
Among populations	1	0.83	0.87	0.17	0.33
Within populations	4	1.33	0.33	0.33	
Total	5	2.17			

C. Minimum or zero divergence

Source	df	SS	MS	σ^2	F_{STmin}
Among populations	1	0	0	−0.11	−0.5
Within populations	4	1.33	0.33	0.33	
Total	5	1.33			

2.1.2 Φ_{ST} and Φ'_{ST}

The methods above are the same for calculating Φ_{ST}, but the genetic distance matrix can be populated by any range of genetic distances among alleles, rather than a binary code of zeros and ones (Figure 2). In Figure 2, each individual sampled has a unique haplotype, the genetic distance among haplotypes ranges from one to five, and $\Phi_{ST} = 0.520$ (Table 2). For the maximized AMOVA (Figure 2D), the selection of the maximum genetic distance should be carefully considered. The reason for standardizing Φ is to make results broadly comparable; therefore, the maximum genetic distance should not vary with sample size. For example, when employing the simple pairwise genetic distance among DNA sequences, such as mitochondrial gene fragments, it is most conservative to use the fragment length, in basepairs, as the maximum genetic distance among alleles. In certain situations, however, the maximum genetic distance could be set to some other value, such as the maximum observed distance. In Figures 2D–F, we use five as the maximum genetic distance among haplotypes for simplicity of illustration and $\Phi'_{ST} = 0.709$. All other calculations of Φ'_{ST} herein utilize a maximum genetic distance of 600 basepairs (bp) because we simulate 600 bp mitochondrial DNA (mtDNA) fragments.

2.1.3 D_{est}

For D_{est}, we used a formulation corrected for sampling bias, using the method of Chao et al. (2006, 2008), as described in Jost (2008). Unlike AMOVA, D_{est} is not based upon a matrix of genetic distances. Rather, the computation is based on the formalization of the Gini-Simpson diversity index (Gini 1918; Simpson 1949), which is also the method commonly used to calculate genetic

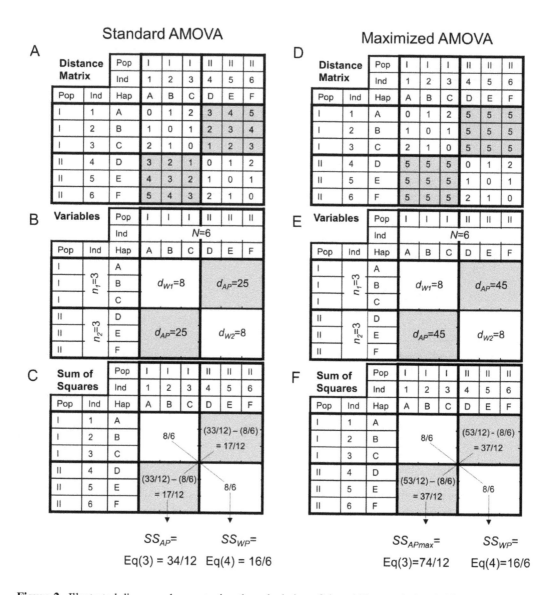

Figure 2. Illustrated diagram, demonstrating the calculation of the within-population (white-filled cells) and among-population (gray-filled cells) sums of squared differences used in the calculation of distance-based analysis of molecular variance (AMOVA). Panel A is the distance matrix. Panel B shows the calculation of sample size and summing the four regions of the distance matrix. Panel C demonstrates the specific calculation of the sum of squared differences. Panels D–F show the same procedure for the case where the among populations differences are the maximum (Meirmans 2006).

diversity or heterozygosity by population geneticists. Following Jost (2008), we calculate D_{est} as follows:

$$D_{est} = 1 - \frac{x}{y}, \tag{10}$$

Table 2. AMOVA tables for the calculation of (A) Φ_{ST} and (B) Φ_{STmax} data are drawn from Figure 2.

A. Observed divergence

Source	df	SS	MS	σ^2	Φ_{ST}
Among populations	1	2.833	2.833	0.721	0.520
Within populations	4	2.667	0.668	0.668	
Total	5	5.167			

B. Maximum divergence

Source	df	SS	MS	σ^2	Φ_{STmax}
Among populations	1	6.167	6.167	1.833	0.733
Within populations	4	2.667	0.668	0.668	
Total	5	8.834			

$$x = \sum_{i=1}^{A} \left\{ \left[\left(\sum_{j=1}^{P} p_{ij} \right)^2 - \sum_{j=1}^{P} p_{ij}^2 \right] / (P-1) \right\}, \tag{11}$$

$$y = \sum_{i=1}^{A} \sum_{j=1}^{P} \frac{N_{ij} (N_{ij} - 1)}{N_j (N_j - 1)}, \tag{12}$$

where A is the total number of alleles sampled, P is the total number of population samples, p_{ij} is the proportion of allele i in population sample j, N_{ij} is the number of allele i found in population sample j, and N_j is the number of copies of all alleles found in population sample j. As with F_{ST}, D_{est} yields a negative value (-0.33) due to very low sample size (Table 3).

2.2 Simulations

We simulate pairwise population comparisons of 600 bp mtDNA fragments, where (1) the level of genetic differentiation is known a priori and genetic diversity is equal between the populations (Section 2.2.1), and (2) the level of genetic differentiation is known and the diversity of alleles that are shared between the populations is manipulated independently from the diversity of private

Table 3. Terms used in calculation of D_{est} (Equations 10–12) for the example dataset in Figure 1.

		$j = 1$	$j = 2$	$j = 1$	$j = 2$		$j = 1$	$j = 2$			
	Haplotype	N_{ij}	N_{ij}	N_j	N_j	P	p_{ij}	p_{ij}	x	y	D_{est}
$i = 1$	A	1	0				0.33	0.00			
$i = 2$	B	2	2	3	3	2	0.67	0.67	0.89	0.67	-0.33
$i = 3$	C	0	1				0.00	0.33			

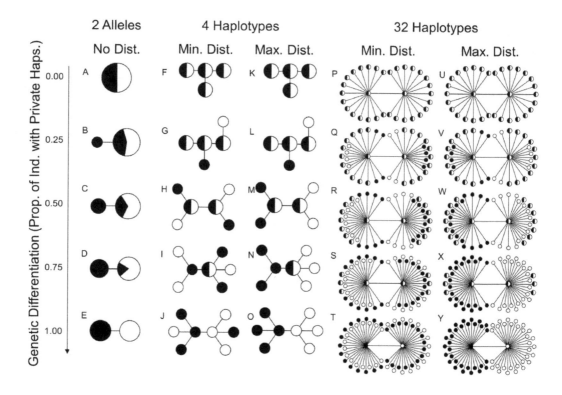

Figure 3. Haplotype networks depicting simulated population comparisons where genetic differentiation is 0% (top row), 25% (second row), 50% (third row), 75% (fourth row), or 100% (bottom row). Simulations were conducted for populations with two alleles (A–E, SNP simulation), four haplotypes (F–O), and 32 haplotypes (P–Y). In the cases where populations have four or 32 haplotypes per population, haplotypes were arranged to maximize and minimize genetic distance among haplotypes unique to different populations, compatible with the level of shared overlap. Each circle represents a haplotype, and the area of each circle is proportional to its abundance in the two populations. The color for each haplotype is indicative of population affiliation. Each solid line connection between haplotypes is a single nucleotide difference.

alleles (Section 2.2.2). No two alleles differ by more than three nucleotides in these simulations, and Φ_{ST} and Φ'_{ST} are nearly identical, when employing a maximum possible genetic distance of 600 bp, hence we also (3) explore the circumstances required for Φ_{ST} and Φ'_{ST} to yield different values (Section 2.2.3). Finally we (4) demonstrate the circumstances under which distance-based AMOVA (yielding Φ_{ST} and Φ'_{ST}) can be misleading, relative to allele-based analysis (F_{ST}, F'_{ST}, D_{est}; Section 2.2.4). In all cases, the populations of interest were assumed to be of infinite size, and we bootstrapped 10,000 replicate samples of 50 individuals from each population for each assessment of genetic differentiation. All bootstrap resampling simulations were run in Visual Basic 6.0 (Microsoft Inc., Redman, WA). Φ_{ST}, Φ'_{ST}, and D_{est} values were calculated in Visual Basic, following Excoffier et al. (1992), Meirmans (2006), and Jost (2008), respectively. The estimated parameters were plotted against the known levels of genetic differentiation with SPSS 13 (SPSS Inc., Chicago, IL).

Pairwise Population Comparison

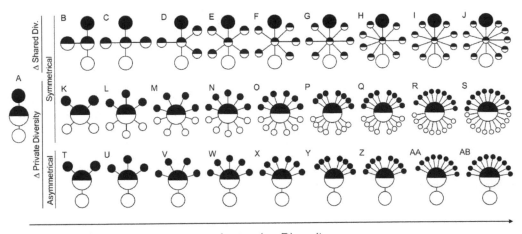

Increasing Diversity

Figure 4. Haplotype networks depicting simulated population comparisons, where the diversity of shared haplotypes ranges from low to high (A–J), the diversity of unique haplotypes (unshared diversity) ranges from low to high (A, K–S), and the diversity of unique haplotypes (unshared diversity, asymmetrical network) in one population ranges from low to high (A, T–AB). Each circle represents a haplotype, and the area of each circle is proportional to its proportional representation in the two populations. The color of the fill for each haplotype is indicative of population affiliation. In each solid line connection between haplotypes is a single nucleotide difference.

2.2.1 *Simple pairwise comparisons*

In order to compare the performance of F_{ST}, F'_{ST}, Φ_{ST}, Φ'_{ST}, and D_{est}, five comparisons of population pairs were conducted at low (two SNP alleles), medium (four mtDNA haplotypes), and high diversity (32 mtDNA haplotypes) for the minimum and maximum possible haplotypic distance between populations, given the structure of haplotype networks (Figure 3A–Y). Each population pair has an even distribution and equal number of haplotypes, thus the actual level of genetic differentiation can be accurately calculated as the parametric proportions of haplotypes shared between the populations sampled (0.00, 0.25, 0.50, 0.75, and 1.00).

2.2.2 *Pairwise comparisons with unequal shared versus unshared allelic diversity*

We construct a specific set of paired population comparisons to assess the impact of the diversity of shared and private alleles on F_{ST}, F'_{ST}, Φ_{ST}, Φ'_{ST}, and D_{est} (Figure 4A–AB). Here, all comparisons consist of pairs of populations where 50% of the individuals in each population exhibit a private haplotype (i.e., 50% population genetic differentiation). We manipulate the number of haplotypes that are shared between population pairs in single haplotype steps from one to ten, while holding the number of haplotypes that are private in each population to one (Figure 4B–J), and vice versa (Figure 4K–S). In a third set of comparisons, we hold one population at two haplotypes, while allowing the private haplotypes in the other population to increase in diversity (Figure 4T–AB) to assess the impact of unequal allele frequencies among populations.

Pairwise Population Comparison

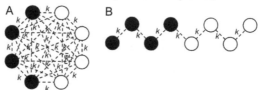

Figure 5. Two haplotype networks depicting simulated population comparisons designed to demonstrate key differences between Φ_{ST} and Φ'_{ST}. In both networks, all haplotypes are unique to a single population. In network A, all haplotypes are equally different, and in network B, the haplotypes are arranged in a linear fashion. Each circle represents a haplotype where the area of the circle is proportional to its abundance in the populations. Each dashed line connection labeled with a k was allowed to vary from one to the maximum number of basepairs possible, given the length of the DNA fragment ($\approx$ 600 bp). The fill color for each haplotype is indicative of population affiliation.

2.2.3 Φ_{ST} *vs.* Φ'_{ST}

We construct two paired population comparisons, where each population has four unique $\approx$ 600 bp haplotypes (fragments can be slightly larger or smaller than 600 bp depending on the network and value of k), as might be obtained from a mtDNA sequence, and we assess how increasing the genetic distance among haplotypes affects Φ_{ST} and Φ'_{ST}. In Figure 5A, all haplotypes are equally different, and in Figure 5B the haplotypes are arranged in a linear array, where haplotypes become increasingly divergent the further they are apart in the network. The maximum possible genetic distance among haplotypes is 599 bp in Figure 5A to accommodate the possibility of more than four maximally equidistant 600 bp haplotypes; the maximum possible distance between adjacent haplotypes in Figure 6B is 86 bp, because the most divergent haplotypes can be no more than 603 bp apart (e.g., the maximum of 7 steps between the most divergent pair of haplotypes with 86 bp differences is $7 \times 86 = 602$).

In order to investigate how Φ_{ST} and $\Phi'ST$ are affected by simple linear haplotype networks with uneven genetic distances among haplotypes, 26 population comparisons were devised (Figure 6). These comparisons explore all possible combinations where some adjacent haplotypes are fixed at being minimally different, while the genetic distance among other adjacent haplotypes is allowed to range from the minimum to the maximum for an $\approx$ 600 bp DNA fragment (left column, Figure 6). An additional 13 paired population comparisons were constructed, for which minimally distant haplotypes were collapsed into a single haplotype in order to demonstrate how the linear networks are effectively collapsed by the incorporation of genetic distance in Φ_{ST} and Φ'_{ST} (right column, Figure 6).

3   RESULTS

3.1   Pairwise comparisons, genetic differentiation = 0%, 25%, 50%, 75%, and 100%

In Figure 3, the mean F_{ST}, F'_{ST}, Φ_{ST}, Φ'_{ST}, and D_{est} are plotted against the actual proportion of genetic differentiation between populations for each comparison. The indices of genetic differentiation, F'_{ST} and D_{est}, scale from approximately 0.0 for identical populations to 1.0 for comparisons between populations that share no haplotypes, indicating that they perform as indicated by Meirmans (2006) and Jost (2008). F_{ST}, Φ_{ST}, and Φ'_{ST}, on the other hand, only scale between approximately zero and one (alternate fixation) in the two allele SNP simulation (Figure 7A). All metrics begin to

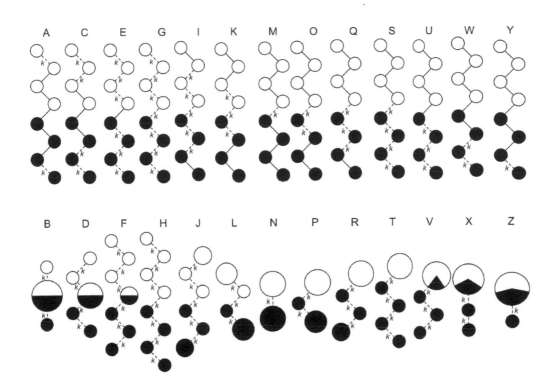

Figure 6. Haplotype networks depicting the simulated population comparisons for a 600 basepair DNA fragment, where genetic distance is manipulated in order to demonstrate differences between Φ_{ST} and Φ'_{ST}, as well as F_{ST}, F'_{ST}, and D_{est}. Each circle represents a haplotype and the areas of the circles are indicative of their proportional representation in the populations. Each solid line connection between haplotypes is a single nucleotide difference, and each dotted line connection labeled with a k was allowed to vary from one to the maximum number of basepairs possible, given the length of the DNA fragment. The color of the fill for each haplotype is indicative of population affiliation. In the upper row of networks, each haplotype is unique to a particular population. The lower row of networks is the result of collapsing haplotypes separated by a single nucleotide in the upper row into a single haplotype—approximating the effect of incorporating genetic distance into AMOVA.

diverge, however, for intermediate levels of population genetic differentiation (Figure 7A) and F_{ST} and Φ_{ST} are the best performing metrics, closely matching the expected level of genetic differentiation. F'_{ST} and Φ'_{ST} overestimate and D_{est} underestimates the expected level of genetic differentiation when the expected level of differentiation is any value other than zero or one in the SNP simulation. No differences are observed between distance-based and equivalent allele/haplotype-based metrics (F_{ST} vs. Φ_{ST} and F'_{ST} vs. Φ'_{ST}), because genetic distances in two allele systems are binary (pairwise genetic distance = 0 or 1).

It is immediately apparent for all comparisons in Figure 7B–E, regardless of haplotype diversity and genetic distance between populations, that D_{est} is the most consistently accurate measure of genetic differentiation. F'_{ST} performs nearly as well as D_{est}, but slightly overestimates genetic differentiation at lower haplotype diversities. F_{ST}, Φ_{ST} and Φ'_{ST} substantially underestimate the level of genetic differentiation when it is not zero. The maximum level of genetic differentiation exhibited by F_{ST} is approximately the average of the inverses of the effective number of haplotypes per popu-

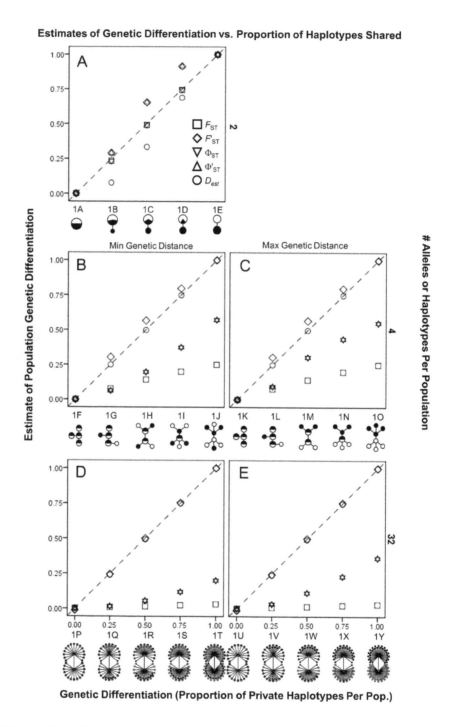

Figure 7. Scatter plots of the mean estimated level of genetic differentiation plotted against the known level of genetic differentiation for F_{ST}, F'_{ST}, Φ_{ST}, Φ'_{ST}, and D_{est} for the population sample comparisons depicted in Figure 3 with two (A), four (B, C), and 32 haplotypes (D, E). For the four and 32 haplotype simulations, haplotypes were arranged within the network to minimize (B, D) and maximize (C, E) genetic distance among haplotypes unique to different populations. The dashed line represents the expected level of genetic differentiation.

lation ($1/4 = 0.25$ and $1/32 = 0.031$). The only metrics taking genetic distance into account, Φ_{ST} and Φ'_{ST}, exhibit the most complex behavior. When genetic diversity is high, both metrics are larger in the cases where genetic distance is maximized (Figure 7D versus Figure 7E), but this effect is muted at lower levels of genetic diversity (Figure 7B versus Figure 7C). Importantly, Φ_{ST} and Φ'_{ST} do not differ markedly, indicating that Φ_{ST} is robust to the phenomenon of increased within-populations genetic diversity that leads to decreases in F_{ST} (Hedrick 1999) when the observed distances among haplotypes are much less than the maximum distance. This is because 600 bp haplotypes that differ by a few nucleotides are not nearly as "diverse" as those that differ by $\approx$ 600 bp (i.e., all haplotypes equally different as in the calculation of F_{ST} and F_{ST}). If the maximum genetic distance is set to anything substantially less than 600 bp, such as the number of variable sites, Φ_{ST} and Φ'_{ST} can often diverge (see Section 3.3). Both, Φ_{ST} and Φ'_{ST} metrics yield lower estimates of genetic differentiation at higher genetic diversities (Figure 7B, C versus Figure 7D, E, respectively), as would be expected for a fixation index, because the condition of 32 haplotypes per population is farther from alternate fixation than two or four haplotypes per population.

The differences in precision among F_{ST}, F'_{ST}, Φ_{ST}, Φ'_{ST}, and D_{est} are fairly small. For example, in Figure 7A, the maximum standard deviations for the bootstrapped estimates of F_{ST}, F'_{ST}, Φ_{ST}, Φ'_{ST}, and D_{est} are 0.072, 0.0914, 0.072, 0.0914, and 0.093, respectively (data not shown). The coefficients of variation (CV, standard deviation divided by the mean) are similar among the metrics, with one exception; in the SNP locus simulation (Figure 7A), the D_{est} CVs at 25–75% genetic differentiation are almost double those of the other metrics. The other interesting pattern in CV is that as the number of haplotypes increases from two to 32 in the simulations in Figure 7, the CVs generally increase. This indicates that the CV generally increases as the number of haplotypes decreases relative to the sample size. Overall, these metrics exhibit similar levels of precision in most cases.

3.2   Pairwise comparisons with unequal shared versus private allelic diversity

Although Figure 4A through 4AB each depict pairs of populations in which 50% of individuals exhibit a haplotype not found in the other population, by manipulating the diversity of haplotypes that are shared or private, we are able to elicit a wide range of genetic differentiation estimates from each of the metrics, ranging from nearly zero to nearly one and rarely the expected 0.5 (Figure 8). Clearly, genetic diversity per se has a substantial effect on all metrics tested here. In general, when the number of shared haplotypes is large (high diversity of shared haplotypes) and the number of private haplotypes is small (low diversity of unique haplotypes), the metrics report more genetic differentiation than when the private haplotypes are diverse and there is a single shared haplotype (Figure 8A vs. Figure 8B). In the asymmetrical case, where one population has a diverse set of private haplotypes (Figure 8C), the metrics report an intermediate level of genetic differentiation, relative to that in Figure 8A and Figure 8B. Overall, F'_{ST} consistently yields the highest estimates of genetic differentiation, followed, in decreasing order, by D_{est}, Φ_{ST} and Φ'_{ST}, and F_{ST}.

In the case where each population has two haplotypes, with one haplotype in common (Figure 4A), D_{est}, Φ_{ST}, and Φ'_{ST} are all approximately 0.5, as expected, but F_{ST} (0.33) and F'_{ST} (0.66) deviate substantially from expectation (Figure 8). Aside from the restricted case of Figure 4A, no single metric can be said to perform better than another. When each population has a single unique haplotype and the diversity of the haplotypes shared by the populations increases (Figure 8A), F'_{ST} and D_{est} asymptote to values greater than 0.9, but F_{ST}, Φ_{ST}, and Φ'_{ST} asymptote to a value less than 0.27. The metrics yield more similar results when the two populations share a common haplotype, and the number of private haplotypes ranges from one to ten (Figure 8B). F_{ST}, Φ_{ST}, and Φ'_{ST} asymptote to approximately 0.04, while D_{est} (0.10) and F'_{ST} (0.14) are slightly higher. In the asymmetrical cases (Figure 8C), the metrics are evenly spread, with F'_{ST} asymptoting to 0.47, D_{est} to 0.36, Φ_{ST} and Φ'_{ST} to 0.27, and F_{ST} to 0.18, as the unshared haplotypic diversity was increased. In this par-

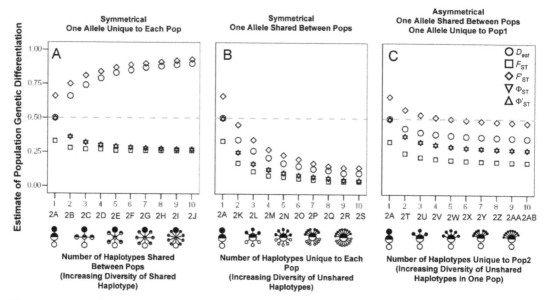

Figure 8. The mean estimated level of genetic differentiation plotted against the number of haplotypes shared between populations (A), the number of haplotypes unique to each population (B), and the number of haplotypes unique to population two (C) for F_{ST}, F'_{ST}, Φ_{ST}, Φ'_{ST}, and D_{est} for the population sample comparisons depicted in Figure 4. In each case, 50% of the individuals in each population have the same haplotypic composition and the dashed line represents the expected level of genetic differentiation.

ticular situation (Figure 8C), F'_{ST} could be said to be the most accurate, but the other simulations demonstrate that this is not a consistent outcome. Moreover, the level of genetic differentiation that is expected, given the population comparisons depicted in the networks in Figure 4 (Figure 8), is highly subjective, depending upon how important common and rare haplotypes are considered to be by the investigator. In other words the "correct" level of genetic differentiation is debatable, since there is no unambiguous standard by which to compare the simulation results in more complex networks.

3.3 Φ_{ST} vs. Φ'_{ST}

For cases where haplotypes are quite similar relative to the maximum possible genetic distance, Φ_{ST} and Φ'_{ST} yield similar estimates of genetic differentiation regardless of the proportion of shared haplotypes (Figure 7) or the level of shared (or unshared) haplotypic diversity (Figure 8). We should anticipate, however, that Φ_{ST} and Φ'_{ST} will yield increasingly divergent estimates of genetic differentiation as the genetic distances among sampled haplotypes increases. We observe that Φ_{ST} remains unchanged and Φ'_{ST} increases as the distance among haplotypes increases and approaches the maximum permissible distance (Figure 9). When all haplotypes are equally different and the distance between haplotypes ranges from 1–599 bp (Figure 5A), Φ_{ST} equals F_{ST} and is approximately the average of the population haplotype diversities (0.25), while Φ'_{ST} ranges from 0.25–1.00. Thus, Φ'_{ST} equals one when no haplotypes are shared and all haplotypes are maximally distant. In the case where the haplotypes are arranged in a strict linear network and the distance between adjacent haplotypes ranges from 1–86 (602 steps across the network, Figure 5B), Φ_{ST} is 0.69 and Φ'_{ST} ranges from 0.69–0.84. As expected, $D_{est} = 1 = F'_{ST}$, and $F_{ST} = 0.25$, and none of them vary with

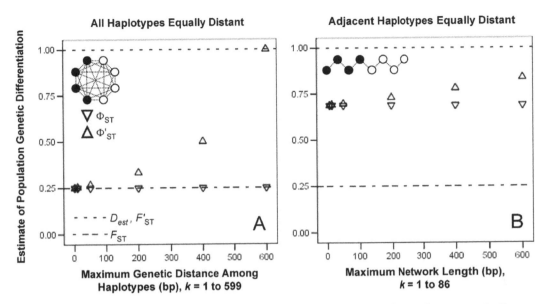

Figure 9. The mean estimated level of genetic differentiation plotted against the maximum genetic distance among haplotypes (or network length) for F_{ST}, F'_{ST}, Φ_{ST}, Φ'_{ST}, and D_{est} for the population sample comparisons depicted in Figure 5. In panel A, all haplotypes are equally different so the number of bp differences (k) ranges from 1–599. In panel B, the haplotypes are arranged in a linear network so the number of bp differences between adjacent haplotypes ranges from 1–86. In both cases, there are no haplotypes shared between the populations.

differing genetic distance or network conformation; Φ_{ST} varies with network conformation, but not with increasing genetic distance; and Φ'_{ST} varies with both the network conformation and genetic distance. Therefore, Φ'_{ST} is the most sensitive of the metrics tested here, if the object is to explore the impact of differing degrees of divergence among haplotypes.

An inherent feature of Φ_{ST} and Φ'_{ST} is that relatively similar haplotypes in a network are effectively collapsed (binned) into a single haplotype and dissimilar haplotypes remain separate (see Figure 6). In Figure 10, estimates of genetic differentiation for F_{ST}, F'_{ST}, Φ_{ST}, Φ'_{ST}, and D_{est} are plotted against the number of steps, in bp across the network. In each pair of panels in Figure 10 (A & B, C & D, etc.), the left panel is a linear network with four unique 603 bp haplotypes per population and the right panel is the corresponding collapsed network where similar haplotypes (those separated by very small genetic distances relative to the largest genetic distance between haplotypes in the network) are grouped into one haplotype. The distance-based metrics, Φ_{ST} and Φ'_{ST}, in the full networks converge upon those in the equivalent collapsed networks at total network lengths of ≈ 50 bp (Figure 10). In networks less than ≈ 50 bp across, both Φ_{ST} and Φ'_{ST} can deviate substantially between the full and collapsed networks, with the largest differences occurring in networks with the most extensive collapsing (Figure 10A and Figure 10B, Figure 10M and Figure 10N, Figure 10Y and Figure 10Z). In strict linear networks, Φ_{ST} and Φ'_{ST} approach 1.0 as alternate fixation is approached (Figure 10N, see progression towards alternate fixation in Figure 10B, D, F, H, J, L, N and Figure 10Z, X, V, T, R, P, N). Φ'_{ST} can also be 1.0 when no haplotypes are shared between populations and the haplotypes are all equally and maximally distant (see Figure 9A).

In contrast to Φ_{ST} and Φ'_{ST}, which utilize variable genetic distances among haplotypes, F_{ST}, F'_{ST}, and D_{est} can be very different between the full and collapsed networks (Figure 10), but remain constant regardless of the length of any particular network (no difference in F_{ST}, F'_{ST}, and D_{est} for

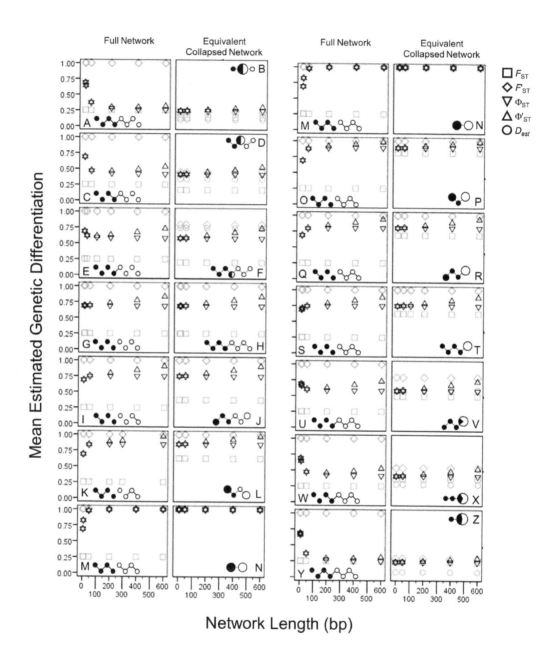

Figure 10. The mean estimated level of genetic differentiation plotted against the network length for F_{ST}, F'_{ST}, Φ_{ST}, Φ'_{ST}, and D_{est} for the population sample comparisons depicted in Figure 6. The data symbols for Φ_{ST} and Φ'_{ST} are darker than those for F_{ST}, F'_{ST}, and D_{est} to highlight the behavior of the distance-based metrics.

Figure 10A, C, E, G, H, I, K, M, O, Q, S, U, W, and Y). In circumstances where Φ_{ST} and Φ'_{ST} do not change between the full and collapsed networks, but the allele/haplotype-based metrics do change, the allele/haplotype-based metrics can be said to have more information than the distance based metrics. In some cases, the differences between the full and collapsed networks are irrelevant (Φ_{ST} and Φ'_{ST} are desirable), but in some cases the differences are important (F_{ST}, F'_{ST}, and D_{est}), see Discussion in Section 4.2.2. Within any particular network (within any panel of Figure 10), however, Φ_{ST} and Φ'_{ST} exhibit more information/resolving power than do the allele/haplotype-based metrics (which do not vary), distinguishing among similar networks with different genetic distances among the haplotypes (Figure 10). In the absence of context and rationale for the choice, there is no a priori reason why distance-based metrics Φ_{ST} and Φ'_{ST} should be considered superior to F_{ST}, F'_{ST}, and D_{est}.

4   DISCUSSION

4.1   What do the index values mean?

The biological meaning of a particular value for an estimate of genetic differentiation can be elusive. We show here that it is possible to have a wide range of estimates of genetic differentiation from the array of standard metrics available (F_{ST}, F'_{ST}, Φ_{ST}, Φ'_{ST}, D_{est}, Figures 7–10). In a sense, the meaning of a particular estimate of genetic differentiation is determined by the conditions that define the maximum value of the index (all indices share the same conditions for a minimum value). F_{ST} and Φ_{ST} attain their maximum (1) when the populations exhibit alternate fixation, thus F_{ST} and Φ_{ST} values are, at the bare minimum, estimates of how close populations are to alternate fixation. F'_{ST} and D_{est} values are estimates of how close populations are to sharing no alleles. Φ'_{ST} is simultaneously an estimate of how close populations are to sharing no alleles that are maximally distant (Figure 9A) and/or alternate fixation (Figure 10N). In simulations involving only small genetic distances among haplotypes, Φ_{ST} and Φ'_{ST} were almost identical, but as the genetic distance among haplotypes approaches the maximum (i.e., all nucleotides are polymorphic), Φ'_{ST} diverges from Φ_{ST} and converges on F'_{ST}. The extreme case of all haplotypes being maximally distant is highly unlikely to be encountered in practice, and under most realistic circumstances (when considering variable and invariable sites in the estimation of maximum genetic distance), Φ_{ST} and Φ'_{ST} will result in similar estimates of population genetic differentiation. Given the similar behavior of Φ_{ST} and Φ'_{ST} and the similarities between F'_{ST} and D_{est}, both of which render within- and among-population diversity measures independent, a trio of F_{ST}, D_{est} (or F'_{ST}) and Φ_{ST} would seem to form a sufficient generic set of criteria for routine reporting, covering both fixation and differentiation.

    We agree with Neigel (2002) that F_{ST} still holds utility. The metric performs as it was originally intended to; the problem derives not from the metric, but from the common interpretation of F_{ST} as an index of genetic differentiation, rather than an index of fixation. Unless populations resemble those in Figure 7 (even allele frequencies and equal numbers of alleles) or unless F_{ST} is applied to a two allele system, differently structured datasets can be expected to yield a variety of results. We can use F'_{ST} to work our way around the scaling limitations. Meirmans and Hedrick (2011) have similarly made the point that G_{ST}, while it has its limitations (Gerlach et al. 2010), can also have its uses, and we can generally use D_{est} to circumvent problems of scaling. In practice, all of the metrics investigated here are sensitive to varying proportions of shared and private diversity. It seems appropriate to consider the biological factors that are driving population structure, and hence the values of F_{ST}, F'_{ST}, Φ_{ST}, Φ'_{ST}, D_{est}.

    An important aspect of any estimate of genetic differentiation is whether it is significantly different from 0.0 (Gerlach et al. 2010), which can be assessed with resampling statistics. Excoffier et al. (1992) use a permutation test to determine whether AMOVA-based F_{ST} or Φ_{ST} values differ from the null expectation, a random distribution of alleles or haplotypes among populations. The same procedure can be employed for D_{est}, where a null distribution is constructed by randomly assigning all individuals sampled to a population, without replacement, and calculating D_{est} 10,000 times (for estimation of P values). Bootstrapping is required to generate estimates of standard error

Table 4. Classification of F_{ST}, F'_{ST}, Φ_{ST}, Φ'_{ST}, and D_{est} as fixation or differentiation indices, and whether they treat genetic distances among alleles as equal or unequal. A fixation index reaches its maximum at alternate fixation, and a differentiation index reaches its maximum when no alleles are shared between populations. Φ'_{ST} behaves as both a fixation and a differentiation index.

Type of distance metrics	Fixation index	Differentiation index
Equal genetic distances	F_{ST}	F'_{ST}, D_{est}
Unequal genetic distances	Φ_{ST}, Φ'_{ST}	Φ'_{ST}

and confidence intervals. To do this, individuals retain their population identity and are randomly sampled with replacement from within their own populations, generating a distribution from which standard errors and confidence intervals can be calculated. It is important to note that all resampling procedures assume one has representative samples from the populations, which requires not only random sampling but also adequate sample sizes (more diverse loci require larger sample sizes to adequately estimate allele frequencies). If these criteria are not met, the assumptions are violated and the P values and standard errors estimated using resampling statistics are likely to be poorly estimated, a potential problem for hypervariable loci.

4.2   Which analysis/index is the most appropriate?

4.2.1   *Fixation or genetic differentiation*

All of the indices investigated here can be classified as either fixation (F_{ST}, Φ_{ST}) or differentiation indices (F'_{ST}, D_{est}, Table 4) except for Φ'_{ST}, which has properties of both a fixation (Figure 10N) and a differentiation index (Figure 9A). Considerable care should be taken in the selection of this metric (setting the maximum distance value) and in its interpretation. Most researchers are currently interested in detecting and quantifying genetic differentiation, where an index will indicate when two or more groups of individuals are completely differentiated (compositionally nonoverlapping), rather than whether single populations are internally fixed for particular alleles or haplotypes. Given that predilection, we would recommend the use of Meirmans' (2006) F'_{ST} (alternatively, G'_{ST}; Hedrick (2005)) or Jost's (2008) D_{est}. There are frequent situations, however, where genetic differentiation indices such as Φ_{ST} (based on genetic distances among alleles) or F_{ST} (two allele systems) are more appropriate than indices of genetic differentiation. Overall, there is no single correct answer, and we recommend that researchers apply both a fixation index and an index of genetic differentiation to their datasets. Where the two methods yield similar results, one can be sure of the conclusion of genetic distinctiveness among the populations. When the two methods yield differing results, the pattern and direction of discord could be diagnostic of a particular phenomenon, potentially driven by confounding effects of genetic diversity or the binning of similar haplotypes.

4.2.2 Φ_{ST} *and* Φ'_{ST} *accentuate either noise or signal*

The paired comparisons in Figure 10 demonstrate the strength of and simultaneously expose a potential issue with distance-based metrics, Φ_{ST} and Φ'_{ST}—the additional information of genetic distance among haplotypes is used to reduce the complexity of genetic population structure and very different population comparisons can yield the same result (for example, Figure 10A vs. Figure 10B). In many of these cases, allele/haplotype-based metrics, F_{ST}, F'_{ST}, and D_{est}, are able to detect these differences; the distance-based metrics are effectively binning similar haplotypes. The decision to incorporate genetic distance among haplotypes into an analysis comes down to a relatively simple question. Does the genetic distance among haplotypes provide additional resolving power for the particular biological/evolutionary question being asked? For example, if the migration rate is much greater than the mutation rate in a species with spatial population partitioning, then the genetic distances among alleles will contribute little meaningful information on the inferred connectivity among locations. This is because the mutational (evolutionary) relationship among alleles has virtually no bearing on where they are found at any particular moment. In this case, each allele is essentially equally different with respect to the system and question being asked, and binning the alleles by their evolutionary relationships introduces noise to the analysis. Under circumstances where the migration rate is much smaller than the mutation rate, however, the evolutionary pattern of interhaplotypic divergence will almost always produce a biogeographic signature, and an analysis that recognizes genetic distance among alleles will provide additional resolution, especially for hypervariable markers. Unfortunately, in real datasets it may not be clear whether the mutation rate or migration rate is greater, resulting in some hints of geographic patterns in haplotypic network. In such cases, which seem to be the rule rather than the exception for most marine species, it will be particularly informative to compare the distance-based and the allele/haplotype-based metrics.

4.2.3   *Compatibility of indices of genetic differentiation with complex sampling designs*

It is generally important to construct a robust sampling or experimental design that will allow relevant hypotheses to be tested. The indices of genetic differentiation reviewed here can only be used with certain sampling designs (Table 5). D_{est} is the most restricted of the indices (due to a lack of software) and can only easily be applied to simple sampling designs which involve two or more simple random samples (SRS), such as pairwise population comparisons. The most common sampling designs implemented in studies investigating genetic differentiation are nested or hierarchical designs, because alleles are nested in individuals, individuals are nested in subpopulations, and subpopulations are nested within populations. It is mathematically possible to extend D_{est} to more complex designs (Tuomisto 2010b, a), but such methods are still being developed, and are not yet widely available as published software.

All AMOVA-based indices (F_{ST}, F'_{ST}, Φ_{ST}, and Φ'_{ST}) can be applied to simple and nested datasets, but standard software such as GENALEX (Peakall & Smouse 2006) and ARLEQUIN (Excoffier & Schneider 2005) can only accommodate a few hierarchical levels. The one exception is PERMANOVA+ for PRIMER 6 (Clarke & Gorley 2006). PERMANOVA+ is a powerful software package aimed at generalized analysis of covariance, of which AMOVA is a special case. Stat et al. (2011) detail how to use PERMANOVA+ for a complex hierarchical AMOVA of genetic variation in dinoflagellate symbionts in corals, using Φ_{ST}. AMOVA-based indices can also be applied to factorial sampling designs (cf., Brown et al. 1996; Irwin et al. 2003) that are common in experimental manipulations. While less common than nested designs, as more researchers seek to interrogate the genetic composition of experimental units, accomplishing such analyses will become more important. In a factorial design, at least two different treatments are independently applied to multiple groups of individuals. Padilla-Gamino et al. (in review) demonstrate how to use PERMANOVA+ on an extremely complex sampling design with factorial, nested, and interaction factors, using Φ_{ST}. It should be noted that it is extremely difficult, to the point of exclusion, to use PERMANOVA+ to calculate F'_{ST} or Φ'_{ST} in complex sampling designs, hence F_{ST} and Φ_{ST} are the only indices that can really be used in complex designs, but this could change with additional software development.

4.2.4   *Matching genetic markers to an index*

The type of genetic marker used in a study can influence which metric (F_{ST}, F'_{ST}, Φ_{ST}, Φ'_{ST}, and D_{est}) realistically can or should be applied. Mitochondrial DNA (mtDNA) sequence data can be analyzed with all of the metrics (Table 5), which is why we simulated analyses of mtDNA. All of

Table 5. Suitability of F_{ST}, F'_{ST}, Φ_{ST}, Φ'_{ST}, and D_{est} for analyses using different sampling designs and genetic markers. A "Y" indicates the metric can be calculated with existing software; "Possible" indicates that it is possible to calculate the metric following primary literature and software intended for another purpose; "SNYA" indicates that software is not yet available for this application; and blank cells indicate that we do not recommend using the metric unless a reliable estimate of genetic distance can be obtained. FLP = fragment length polymorphism.

	F_{ST}	Φ_{ST}	F'_{ST}	Φ'_{ST}	D_{est}
Sampling design					
Simple	Y	Y	Y	Y	Y
Nested	Y	Y	Y	Y	SNYA
Factorial	Possible	Possible	SNYA	SNYA	SNYA
Complex	Possible	Possible	SNYA	SNYA	SNYA
Genetic marker					
mtDNA sequence	Y	Y	Y	Y	Y
nDNA sequence	Y	Y	Y	Y	Y, phased SNYA
Unlinked SNP loci	Y	≥ 3 alleles	Y	≥ 3 alleles	> 1 locus SNYA
SNP haplotypes	Y	Y	Y	Y	Y
Microsatellites	Y		Y		SNYA
FLP	Y		Y		SNYA

the metrics (F_{ST}, F'_{ST}, Φ_{ST}, Φ'_{ST}, and D_{est}) can also be applied to nuclear DNA sequence or SNP haplotypes, but it is difficult to use D_{est} to calculate an inbreeding coefficient (hierarchical nesting) when the gametic phase is unknown and care should be taken when determining the maximum possible genetic distance among SNP haplotypes that consist of only variable sites. For D_{est}, the combination of data across loci for a multilocus analysis can be accomplished by manually taking the Lth root of the product of diversity components from L loci (geometric mean). Φ_{ST} and Φ'_{ST}, being distance-based metrics, can be used on microsatellite or fragment length polymorphism (FLP) data, but we strongly recommend that allele/haplotype-based metrics be used instead. If more than two alleles are observed at neutral unlinked SNP markers, a relatively uncommon occurrence, then it might be useful to distinguish between transitions and transversions, deploying Φ_{ST} and Φ'_{ST}. Otherwise, F_{ST}, F'_{ST}, and D_{est} are the most appropriate for data where genetic distance cannot be reliably estimated.

In the special case of neutral, unlinked SNP data and other categorical data with two equidistant alleles, applying Meirmans' (2006) diversity correction or using D_{est} actually exhibits less accuracy and precision than F_{ST} (Figure 3A). This is the rare case where F_{ST} is a true index of genetic differentiation, because complete genetic differentiation can only be achieved at alternate fixation. With most other markers, it is usually beneficial to use multiple indices to estimate genetic differentiation. While it is not possible to cover all possible scenarios, the simulations here can be used as a guide to gain a deeper understanding of certain aspects of more complex datasets and to guide interpretation of the similarities and differences observed when different metrics are calculated.

4.3   Conclusions

We have tested the performance of F_{ST}, F'_{ST}, Φ_{ST}, Φ'_{ST}, and D_{est} on a range of different simulated population comparisons, designed to demonstrate key differences in the behavior among these metrics, in the presence of varying levels of shared and unshared diversity. Each of the metrics has its own strengths and weaknesses, and there is no single best metric to measure genetic differentiation in all cases. While we agree that D_{est} is a useful metric, in its current configuration, it can be utilized in a limited set of situations, though it will almost certainly be generalized. Based upon the similarities between Φ_{ST} and Φ'_{ST} in simulations, Φ_{ST} is relatively unaffected by high allelic diversity, unlike most fixation indices such as F_{ST}. Additionally, a distance-based D metric, if developed, is not likely to produce estimates of genetic differentiation much different from Φ_{ST}. Moving forward, efforts should be made to modify D_{est} so that it can be applied to more complex hierarchical and factorial sampling designs common to empirical studies. F'_{ST} often yields similar values to D_{est}, though not always, and can be applied much more widely. We reiterate that F_{ST} still holds utility, especially in the analysis of neutral unlinked SNP loci and in their comparison to indices of genetic differentiation. Contrary to popular opinion, the distance based metric, Φ_{ST}, is not automatically superior to non-distance-based metrics, such as F_{ST}. If there is no evolutionary pattern in the biogeographic locations of alleles, distance-based metrics tend to downweight information that could be useful. Finally, diversity of alleles that are shared among populations and that of alleles unique to particular populations have an impact on all estimates of fixation and genetic differentiation. Given our current penchant for second-order metrics, common alleles have a disproportionate influence in all datasets, relative to that of rare alleles, so most allele networks based on complex real datasets can be reduced to much simpler networks of common alleles or (allelic classes) to elucidate what is driving the observed estimates of genetic differentiation.

ACKNOWLEDGEMENTS

We thank Lou Jost, Illiana Baums, Matthew Craig, Jeff Eble, Iria Fernandez, Zac Forsman, Jennifer Schultz, Kimberly Andrews, Brian Bowen, Molly Timmers, Michael Dawson, Michael Hellberg, Ron Eytan, Pat Krug, Louis Bernatchez, and Paul Barber for assistance, guidance and, intellectual discussions that significantly improved this manuscript. This study was supported by NSF grant OCE-0623678, the Seaver Foundation and The Office of National Marine Sanctuaries, Northwestern Hawaiian Islands Coral Reef Ecosystem Reserve Partnership (MOA 2005-008/6882). This is contribution #1447 from the Hawai'i Institute of Marine Biology, #8167 from the School of Ocean and Earth Sciences and Technology (SOEST). PES was supported by NSF-DEB-0514956 and USDA/NJAES-17111.

REFERENCES

Brown, B.L., Epifanio, J.M., Smouse, P.E. & Kobak, C.J. 1996. Temporal stability of mtDNA haplotype frequencies in American shad stocks: to pool or not to pool across years? *Canad. J. Fish. Aquat. Sci.* 53: 2274–2283.

Chao, A., Chazdon, R.L., Colwell, R.K. & Shen, T. 2006. Abundance-based similarity indices and their estimation when there are unseen species in samples. *Biometrics* 62: 361–371.

Chao, A., Jost, L., Chiang, S.C., Jiang, Y.H. & Chazdon, R.. 2008. A two-stage probabilistic approach to multiple-community similarity indices. *Biometrics* 64: 1178–1186.

Charlesworth, B. 1998. Measures of divergence between populations and the effect of forces that reduce variability. *Mol. Biol. Evol.* 15: 538–543.

Clarke, K.R. & Gorley, R.N. 2006. *PRIMER v6: User Manual/Tutorial*. Plymouth: PRIMER-E.

Cockerham, C.C. 1969. Variance of gene frequencies. *Evolution* 23: 72–84.

Cockerham, C.C. 1973. Analyses of gene frequencies. *Genetics* 74: 679–700.

Excoffier, L., Laval, G. & Schneider, S. 2006. *Arlequin ver 3.1 User Manual*. Berne: University of Berne.

Excoffier, L., Smouse, P.E. & Quattro, J.M. 1992. Analysis of molecular variance inferred from metric distances among DNA haplotypes: application to human mitochondrial DNA restriction data. *Genetics* 131: 479–491.

Excoffier, L.G.L. & Schneider, S. 2005. Arlequin (version 3.0): an integrated software package for population genetics data analysis. *Evol. Bioinform.* 1: 47–50.

Gerlach, G., Jueterbock, A., Kraemer, P., Deppermann, J. & Harmand, P. 2010. Calculations of population differentiation based on G_{ST} and D: forget G_{ST} but not all statistics. *Mol. Ecol.* 19: 3845–3852.

Gini, C. 1918. Di una estensione del concetto di scostamento medio e di alcune applicazioni alla misura della variabilità dei caratteri qualitativi. *Atti R. Inst. Veneto Sci., Lett. Arti* 78: 397–461.

Hedrick, P.W. 1999. Highly variable loci and their interpretation in evolution and conservation. *Evolution* 53: 313–318.

Hedrick, P.W. 2005. A standardized genetic differentiation measure. *Evolution* 59: 1633–1638.

Heller, R. & Siegismund, H.R. 2009. Relationship between three measures of genetic differentiation G_{ST}, D_{est}, and G'_{ST}: how wrong have we been? *Mol. Ecol.* 18: 2080–2083.

Irwin, A.J., Hamrick, J.L., Godt, M.J.W. & Smouse, P.E. 2003. A multi-year estimate of the effective pollen donor pool for *Albizia julibrissin*. *Heredity* 90: 187–194.

Jost, L. 2007. Partitioning diversity into independent alpha and beta components. *Ecology* 88: 2427–2439.

Jost, L. 2008. G_{ST} and its relatives do not measure differentiation. *Mol. Ecol.* 17: 4015–4026.

Meirmans, P. G. 2006. Using the AMOVA framework to estimate a standardized genetic differentiation measure. *Evolution* 60: 2399–2402.

Meirmans, P.G. & Hedrick, P.W. 2011. Assessing population structure: F_{ST} and related measures. *Mol. Ecol. Res.* 11: 5–18.

Nei, M. 1973. Analysis of gene diversity in subdivided populations. *Proc. Natl. Acad. Sci. USA* 70: 3321–3323.

Nei, M. 1987. *Molecular Evolutionary Genetics*. New York, NY: Columbia University Press.

Neigel, J.E. 2002. Is F_{ST} obsolete? *Conserv. Genetics* 3: 167–173.

Padilla-Gamino, J.L., Pochon, X., Bird, C., Concepcion, G. & Gates, R.D. (in review). From parent to gamete: vertical transmission of *Symbiodinium* (Dinophyceae) in the scleractinian coral *Montipora capitata*.

Peakall, R. & Smouse, P.E. 2006. GENALEX 6: genetic analysis in Excel. Population genetic software for teaching and research. *Mol. Ecol. Notes* 6: 288–295.

Simpson, E.H. 1949. Measurement of diversity. *Nature* 163: 688.

Stat, M., Bird, C.E., Pochon, X., Chasqui, L., Chauka, L.J., Conception, G.T, Logan, D., Takabayashi, M., Toonen, R.J. & Gates, R.D. 2011. Variation in *Symbiodinium* ITS2 sequence assemblages among coral colonies. *PLoS ONE* 6: e15854.

Tuomisto, H. 2010a. A consistent terminology for quantifying species diversity? Yes, it does exist. *Oecologia* 164: 853–860.

Tuomisto, H. 2010b. A diversity of beta diversities: straightening up a concept gone awry. Part 1. Defining beta diversity as a function of alpha and gamma diversity. *Ecography* 33: 2–22.

Weir, B.S. 1996. *Genetic data analysis II*. Sunderland, MA: Sinauer Associates, Inc.

Weir, B.S. & Cockerham, C.C. 1984. Estimating F–statistics for the analysis of population structure. *Evolution* 38: 1358–1370.

Weir, B.S. & Hill, W.G. 2002. Estimating F–statistics. *Annu. Rev. Genet.* 36: 721–750.

Wright, S. 1943. Isolation by distance. *Genetics* 28: 114–138.

Wright, S. 1951. The genetical structure of populations. *Annu. Eugen.* 15: 323–354.

Wright, S. 1965. The interpretation of population structure by F–statistics with special regard to systems of mating. *Evolution* 19: 395–420.

Wright, S. 1978. *Evolution and the Genetics of Populations, Vol. 4: Variability Within and Among Natural Populations*. Chicago, IL: The University of Chicago Press.

Rethinking the mechanisms that shape marine decapod population structure

BREE K. YEDNOCK & JOSEPH E. NEIGEL

Department of Biology, University of Louisiana at Lafayette, Lafayette, U. S. A.

ABSTRACT

In research on the population genetics of marine decapod crustaceans it is often assumed that planktonic larvae disperse widely and genetic markers are not strongly influenced by natural selection. Although population genetic surveys often seem to corroborate these assumptions, some have revealed genetic differentiation among populations that lack discernable barriers to dispersal. These cases, along with insights from related areas of investigation, suggest it may be time to revise our thinking. We now know that dispersal of meroplanktonic larvae can be strongly limited by oceanographic processes and larval behavior. There is also evidence suggesting that directional selection may be stronger in marine populations than previously thought. These new perspectives have important implications for decapod population genetics and phylogeography. This paper will review the findings that have led to a revised view of larval dispersal and will examine natural selection as a potential modifier of dispersal and influence on genetic population structure in decapod crustaceans.

1 INTRODUCTION

Research on the population genetics of marine decapod crustaceans was long grounded in two widely held assumptions. First was the assumption that a planktonic larval phase ensured widespread dispersal (Thorson 1950; Strathmann 1993). This was based on early observations of the occurrence of larvae in oceanic plankton far from any suitable adult habitats (Scheltema 1971). Such observations also contributed to the idea that the planktonic larval phase is itself an adaptation to promote long-distance dispersal to new habitats. Planktonic larvae were viewed as passive drifters in oceanic currents whose success in ultimately reaching a suitable habitat was largely a matter of chance. Local populations of benthic adults were assumed to be open and dependent on upstream larval sources (Caley et al. 1996). The second assumption, which followed from the first, was that planktonic dispersal results in such high gene flow that it overwhelms natural selection and prevents local adaptation (discussed by Hedgecock 1986). This assumption was bolstered by surveys of genetic markers in marine species in which little population differentiation was found (e.g., Shulman & Bermingham 1995; Brown et al. 2001). Exceptions to this pattern of geographic homogeneity, such as sharp breaks or gradual clines in allele frequencies, were interpreted as imprints of past disruptions of gene flow that were somehow maintained by present day barriers (Hellberg et al. 2002). Heterogeneity on smaller spatial scales as well as temporal shifts in allele frequencies were also explained as consequences of planktonic dispersal. Because chance was thought to play a major role in the success of planktonic larval recruitment, it followed that individual pulses of recruitment might consist of the progeny of just a few lucky individuals that happened to spawn at just the right time and place to deliver their progeny into favorable currents. The sampling error associated with the small number of winners in this reproductive sweepstakes would cause the genetic makeup of the successful recruits to differ from the population as a whole (Hedgecock 1994). Despite scant direct evidence, this model has gained wide acceptance and the term "sweepstakes reproduction" is often used as synonymous with any occurrence of fine scale temporal or spatial heterogeneity in genetic markers.

Over the past fifteen years, a revised view of planktonic larval dispersal has emerged, which places greater emphasis on mechanisms that retain larvae near natal habitats or return them to those habitats (Warner & Cowen 2002). The impressive mobility and complex behaviors of fish and crustacean larvae in particular were found to be incompatible with their characterization as passive drifters. Estimates of larval dispersal distances based on a variety of approaches were consistently much less than would be expected for passive drifters (Shanks et al. 2003; Leis 2006), and tagging studies showed the larvae of reef fish to possess an uncanny ability to return to their natal reefs after weeks spent in the plankton (Jones et al. 1999). This revised view of planktonic larval dispersal has important implications for our understanding of the population genetics of meroplanktonic decapod crustaceans. It implies that we cannot assume that gene flow is always an overwhelming force that homogenizes populations or that strong physical barriers are required to isolate populations. We are now faced with a much richer variety of possibilities, in which gene flow, genetic drift and, natural selection can all play significant roles. In this chapter we will review the evidence that supports a revised view of larval dispersal, and we shall look at natural selection in particular as a potential modifier of dispersal and as a cause of genetic population structure in decapod crustaceans. We will also consider the implications of a revised view of decapod population genetics for phylogeographic studies of decapods. Phylogeography was originally conceived as the application of phylogenetic and biogeographic approaches from systematics to the genealogical structure of mitochondrial DNA (mtDNA) variation within animal species (Avise 2000). As such, it was viewed as "the mitochondrial DNA bridge between population genetics and systematics" (Avise et al. 1987). A central assumption in phylogeography has been that natural selection does not play a significant role in shaping phylogeographic structure, which can thus be interpreted solely in terms of historical patterns of genetic drift and gene flow. However, there is growing evidence to suggest that selection can mimic what are assumed to be historical patterns, especially if mtDNA is the sole genetic marker used. If, as we suggest, selection is an important factor in decapod population genetics, it is likely to be important in decapod phylogeography as well.

2   LARVAL DISPERSAL

2.1   The compexity of oceanographic processes

A central concept in dispersal biology is a species' dispersal kernel, a profile of the probabilities of different dispersal distances. Dispersal kernels have been estimated for many terrestrial species, but none are available for decapods, or for that matter, any meroplanktonic species. Even estimates of mean dispersal distances for planktonic larvae have been difficult to obtain, and are either limited to ecologically atypical circumstances, such as colonization of empty habitats, or based on assumption-laden indirect methods (such as population genetic methods). As a substitute for direct measurements, estimates of larval dispersal kernels and distances are often based on models of planktonic dispersal in oceanographic flows (e.g., Siegel et al. 2003). The accuracy of these models is difficult to test, for the very reason that they are needed in the first place: a lack of empirical data.

The simplest model of planktonic dispersal treats larvae as passive particles advected in a uniform flow; dispersal distance is then the product of the length of the planktonic period and the flow's velocity (e.g., Roberts 1997). However, this model has been found to greatly overestimate actual dispersal distances (Shanks et al. 2003; Shanks 2009). Complex current patterns, eddies, tides, and winds can reduce the transport of planktonic larvae to far below the model's simple predictions

(Sponaugle et al. 2002). Larvae spawned in shallow waters do not immediately enter offshore currents, but may remain in the relatively sluggish coastal boundary layer for days or longer (Largier 2003). In some cases, transport in one direction is followed by transport in the opposite direction. For example, the prevailing current in a well-studied upwelling zone along the coast of California transports larvae tens of km. However, during relaxations of upwelling, the current is reversed and larvae are carried back towards their natal locations (Wing et al. 1995). Similarly, regions of countercurrent flow have been identified as areas where larvae can be retained in natal habitats (Sponaugle et al. 2002). Tidal fluxes can export larvae away from shore on ebb tides and return them on flood tides, and the larvae of some decapods vertically migrate to take advantage of tidal transport shoreward and into estuaries (Hughes 1969; Cronin & Forward 1986; Shanks 1986). High concentrations of brachyuran larvae in tidal fronts at the mouths of estuaries (Epifanio 1987) provide further evidence that decapod larvae can undergo development in coastal waters without being dispersed by offshore currents.

2.2 The compexity of larval behavior

The old supposition that a planktonic larval phase is an adaptation to promote dispersal has not been supported by studies of larval morphology or behavior. Instead, it is now suggested a planktonic phase may allow larvae to exploit particular resources, reduce competition with other life stages, avoid predation (Johannes 1978) or escape from parasites (Strathmann et al. 2002). For example, the projecting spines of crab larvae, although superficially resembling the structures that carry plant seeds on the wind, are unlikely to provide buoyancy or facilitate dispersal. The spines instead deter predators, and are thus an adaptation for survival rather than dispersal (Morgan 1989, 1990). Some behaviors of decapod larvae are directed to reduce, rather than promote dispersal. Despite their small size, decapod larvae are capable of swimming over $10 \, \text{cm s}^{-1}$ (Cobb et al. 1989; Luckenbach & Orth 1992; Fernandez et al. 1994; Chiswell & Booth 1999), which can be enough to make headway against ocean currents. Decapod larvae also possess sensory abilities that enable them to recognize physical and biological cues for orientation and navigation (reviewed in Kingsford et al. 2002). The combination of mobility and the ability to respond to sensory cues allows decapod larvae to direct their movements in ways that increase their chances of reaching suitable habitats for settlement.

2.3 Pelagic duration and dispersal distance

While the pelagic duration (PD) of most fish larvae can be directly estimated by counting otolith growth rings (e.g., Victor 1986; Wellington & Victor 1989), decapod PDs are nearly impossible to measure in the field. As a result, measurements of decapod PDs are generally made in the laboratory (e.g., Costlow & Bookhout 1959; Gore & Scotto 1982; Anger 1991; Nates et al. 1997; Strasser & Felder 1999). These laboratory studies have demonstrated the potential for larval PDs to vary with environmental parameters such as salinity (Anger 1991; Field & Butler 1994) temperature (Anger 1991; Field & Butler 1994; Sulkin et al. 1996) and food availability (Shirley & Zhou 1997). This plasticity could allow decapod larvae to respond to variable field conditions, for example, by delaying settlement until favorable cues indicate a suitable habitat (Christy 1989; Strasser & Felder 1999).

In a compilation of empirical estimates of dispersal distances for marine propagules (spores, eggs as well as larvae), PD was a poor predictor of dispersal distance. Although they were correlated, dispersal distances varied by several orders of magnitude for species with similar PDs (Shanks et al. 2003). However, despite the study's conclusion that dispersal distance is strongly dependent on propagule behavior, Shanks et al. (2003) has been most frequently cited as showing that dispersal distances are determined by PDs. Shanks (2009) has repudiated this interpretation and offered further evidence against it.

2.4   Pelagic duration and genetic population structure

The use of genetic population structure, and in particular F_{ST}, to estimate gene flow has had a long and troubled history. Wright originally defined F_{ST} as the probability of alleles that are identical-by-descent from an ancestral population being combined in zygotes within a subpopulation (Wright 1951). Wright also showed that in a simple island model F_{ST} was determined by the relative strengths of genetic drift (inversely proportional to effective population size, N) and gene flow (the proportion of a subpopulation's gene pool originating from other subpopulations, m) as follows:

$$F_{ST} \approx \frac{1}{4Nm + 1}.$$

An important but generally overlooked aspect of this equation is that F_{ST} is just as dependent on effective population size as it is on the rate of gene flow. Later, with the introduction of allozymes and other genetic markers to population genetics, estimators of F_{ST} based on the frequencies of genetic markers came to replace Wright's parametric definition, for example:

$$\hat{F}_{ST} = \frac{\text{var}(p)}{\overline{p}(1 - \overline{p})}.$$

This estimator, along with all others based on allele frequencies, is only valid when other forces such as selection and mutation are not acting on allele frequencies. Theoretical work by Slatkin and others showed that in small populations (with effective population sizes on the order of 100) F_{ST} would reach equilibrium quickly and estimates of gene flow based upon it would be relatively insensitive to selection or mutation (Slatkin 1987; Slatkin & Barton 1989). However, the critical assumptions behind this conclusion were often overlooked and F_{ST} came to be viewed simply as an estimator of gene flow that could be applied on any scale with any genetic marker. Eventually, the flaws in this approach became painfully obvious and led to a series of papers critiquing the use of F_{ST} as an estimator of gene flow (Neigel 1997; Bossart & Prowell 1998; Whitlock & McCauley 1999).

For marine decapods, it makes little sense to view F_{ST} as an estimator of gene flow because the effect of gene flow is confounded with the effect of genetic drift, and we may know less about genetic drift in marine populations than about gene flow. Although populations of marine benthic species can be very large, high fecundities and other life history characteristics suggest that effective population sizes could be much smaller than census sizes (Orive 1993). Just how much smaller is open to speculation. For oysters, which have extremely high fecundities, estimates of effective population sizes from temporal variation in allozyme frequencies have been as low as 10–100, many orders of magnitude below census sizes (Hedgecock et al. 1992). These remarkably low values led to the hypothesis of "sweepstakes reproduction," an extreme reduction in effective population size that occurs when, by chance, all of the successful progeny in a cohort are produced by a tiny fraction of the entire population (Hedgecock 1994). Sweepstakes reproduction can explain not only temporal changes in allele frequencies, but also microspatial variation in allele frequencies, known as "chaotic patchiness" (Johnson & Black 1982). Genetically distinct patches could represent different cohorts of sweepstakes progeny. However, there is little hard evidence for sweepstakes reproduction apart from the phenomena it is intended to explain. Alternative hypotheses, such as variability in larval sources or selection are not always given full consideration. Decapods generally don't have high enough fecundities for the reproductive output of a few individuals to account for all of the larval settlement in a large population (Corey & Reid 1991; Reid & Corey 1991). Nevertheless, the range of plausible effective population for most decapods is so wide that nothing should be assumed about the relative strengths of genetic drift and gene flow.

There is an important distinction between dispersal and gene flow. Dispersal is an ecological process that increases the distances between organisms or gametes (Baker 1978), whereas gene flow is a population genetic process that changes the genetic composition of populations through the movement of organisms or gametes among populations (Slatkin 1987). Dispersal can only result in gene flow if organisms contribute to the gene pools of the populations they move into. Considering the weakness of the relationship between PD and dispersal distance, the possibility that dispersal does not result in gene flow, and the dependency of F_{ST} on factors other than gene flow, it may be surprising that some studies have found a clear relationship between PD and F_{ST} (reviewed in Bohonak 1999). However, in a large meta-analysis based on a random sample of published studies (Weersing & Toonen 2009) no significant relationship was found between PD and F_{ST} for species with a pelagic phase. Similarly, Shanks (2009) found only a weak relationship between PD and genetic differentiation for species with PDs over 10 h. This dichotomy between the findings of narrow and broad comparative studies suggests that the relationship between PD and F_{ST} may hold for similar species with minimal variation in other factors, but that in a broad comparative setting, PD is not a major factor. F_{ST} continues to be a very useful quantity in population genetics, but we should avoid thinking of it as a measure of gene flow or population connectivity (Neigel 2002).

3   NATURAL SELECTION

3.1   The potential for selection on early life stages

Planktonic larvae experience high rates of mortality, caused primarily by starvation and predation, although variation in environmental conditions such as temperature, salinity, dissolved oxygen, and pollutants can also be important (Thorson 1950; Morgan 1995). Where rates of mortality at early life stages are high, there is the opportunity for strong natural selection (Williams 1975). Presettlement selection during the pelagic larval phase could alter the genetic composition of potential recruits before they reach suitable habitats. Alternatively, postsettlement selection could eliminate individuals that are ill-equipped for the conditions they encounter. If postsettlement selection results from phenotype-environment mismatches rather than intrinsic fitness differences, long-distance dispersal of larvae may not necessarily result in effective gene flow (DeWitt et al. 1998; Marshall et al. 2010). Most studies of decapod populations sample only adults or juveniles, so that what appear to be barriers to dispersal could actually be selective filters on larval survival.

Definitive evidence of post-settlement selection has been documented for some marine invertebrates. For example, Koehn et al. (1976) and Hilbish (1985) observed a cline in allele frequencies of the *Lap* allozyme locus in a population of mussels, *Mytilus edulis*, in Long Island Sound, New York. The cline was found to correspond with a salinity gradient that separated mussels with genotypes specific to oceanic habitats from those with genotypes specific to brackish waters. Although oceanic genotypes were found among recruitment cohorts of mussels in brackish locations of the Sound, the oceanic genotypes failed to persist through subsequent life stages. Therefore, despite larval dispersal between oceanic and brackish water habitats, selection for favorable alleles within recruiting cohorts at brackish locations prevented gene flow from genetically homogenizing these local populations. Biochemical analysis of the allozymes produced by the different *Lap* alleles provided definitive support of selection. The allozymes were found to be functionally different (Koehn & Siebenaller 1981), resulting in lower fitness for mussels with oceanic genotypes in brackish water and consequently higher mortalities.

3.2   The appropriate scale for decapod population genetics

Most studies of genetic population structure in decapods (as well as other marine benthos) have been based on surveys conducted over large distances (several hundred km) to detect pronounced shifts in genetic marker frequencies, typically referred to as "genetic breaks" (Hellberg et al. 2002). This bias probably reflects both the old assumption that planktonic dispersal results in gene flow over large distances and practical considerations about the sample sizes needed to detect small differences in genetic marker frequencies. However, the emphasis on large-scale geographic patterns may be misplaced. Genetic breaks imply deep historical divisions that have been maintained by strong barriers to gene flow. In many cases these phylogeographic divisions might be more appropriately viewed as cryptic taxa than as units of population structure (e.g., Barber et al. 2000). In light of the many factors that can limit the planktonic dispersal of decapod larvae, there is reason to investigate more subtle patterns of population structure over smaller distances. We might expect at these finer scales to identify ecologically relevant subpopulations that are structured by oceanographic and behavioral mechanisms. In these investigations, we should not discount the potential importance of small, but statistically significant differences in genetic marker frequencies. Any significant frequency difference indicates a restriction in gene flow; a small difference could mean either that the rate of migration is only slightly reduced or that large effective populations have made the diversifying force of genetic drift relatively weak.

For decapods, plausible limits for such fundamental population genetic parameters as effective population size and gene flow span at least several orders of magnitudes. These limits encompass radically different possibilities for how decapod populations function. At one extreme, we can envision a scenario in which gene flow is strong as a consequence of planktonic dispersal, and genetic drift is strong because a few highly fecund individuals are responsible for most of the successful reproduction. In this scenario, temporary reductions in gene flow allow genetic drift to produce localized fluctuations in allele frequencies but these are quickly removed when gene flow resumes. Selection is overwhelmed by both gene flow and genetic drift, so local adaptation is prevented. At the other extreme is a scenario in which gene flow is weak because oceanographic and behavioral mechanisms limit dispersal, and genetic drift is weak because local populations consist of large numbers of individuals with limited fecundity. Even without any gene flow, differences in allele frequencies develop slowly because drift is weak, and the differences never become great because there are occasional episodes of moderate gene flow. However, selection is strong enough to overcome both gene flow and genetic drift, so that populations become locally adapted to their environments. These two scenarios, both plausible, obviously have very different implications for the evolution, ecology and management of decapod populations.

4   CASE STUDIES

In this section we review two studies of genetic population structure in marine decapods with benthic adult phases and pelagic larval stages. We choose these examples because they revealed unexpected genetic structure that would have been missed in routine broad range surveys and defy simple interpretation.

4.1   The burrowing ghost shrimp, *Callichirus islagrande*

The ghost shrimp, *Callichirus islagrande*, is restricted to the quartzite beaches and barrier islands of the northern and western Gulf of Mexico. This species' range stretches from Pariso, Tobasco, Mexico, to northwest Florida, with a gap corresponding to the muddy sediments of the Chenier Plain in the northwestern Gulf of Mexico (Staton & Felder 1995). *Callichirus islagrande* excavates burrows that can extend as far as 2 m below the surface and occur in high densities (up to $100 \, \text{m}^{-2}$), which implies large, continuous, and genetically well-mixed local populations. Females can brood several thousand eggs, and the pelagic larval phase is estimated to last from 16 and 20 days, although laboratory experiments indicate some plasticity in PD in response to settlement cues (Strasser & Felder 2000).

A pronounced genetic break has been documented for *C. islagrande* at the Chenier Plain for allozymes (Staton & Felder 1995), mitochondrial DNA (mtDNA) and microsatellites (Bilodeau et al. 2005). Along-shore currents are fast enough to carry the larvae of *C. islagrande* across the Chenier Plain before they complete their pelagic phase, and the location of the break differs from those found for other species in the Gulf of Mexico, which are generally further to the east (Neigel 2009). There is thus nothing to suggest an oceanographic barrier that would prevent gene flow across the Chenier Plain. An alternative possibility is suggested by recent theoretical work that predicts a genetic break will occur, where an environmental gradient in selection crosses a region of reduced dispersal (Pringle & Wares 2007). The division within *C. islagrande* could thus be maintained by selection, but localized by reduced gene flow across the Chenier Plain.

Intensive sampling on the barrier islands off the coast of Louisiana revealed another unexpected level of population structure in *C. islagrande*. Slight but statistically significant differences in microsatellite allele frequencies were found between locales separated by as little as 10 km. Coalescent analysis indicated that this was not due to small effective population size, but rather to limited gene flow. However, this does not necessarily imply that local populations are nearly isolated. A Bayesian analysis indicated that most of the sampled locales contained mixtures of several different source populations. Thus, the differences in allele frequencies could reflect differences in the proportions of larvae they receive from upstream sources (Bilodeau et al. 2005).

4.2 The blue crab, *Callinectes sapidus*

The blue crab, *Callinectes sapidus*, inhabits a wide geographic range spanning temperate, subtropical and tropical regions of the western Atlantic Ocean. The life history of *C. sapidus* suggests the potential for widespread dispersal and mixing of larvae. Females can produce multiple clutches of over three million eggs (Hsueh et al. 1993). Ovigerous females release well-developed zoea in coastal waters that are then transported offshore by currents where they go through seven or eight zoeal stages before molting into a megalopal stage (Costlow & Bookhout 1959) and recruiting to coastal waters. Their entire pelagic development is estimated to take between 37 and 69 days (Costlow & Bookhout 1959). Early assumptions of passive larval dispersal for this species were supported by multiple genetic surveys that found very little or no genetic variation over broad geographic scales (McMillen-Jackson et al. 1994; Berthelemy-Okazaki & Okazaki 1997; McMillen-Jackson & Bert 2004). However, in stark contrast to these findings, Kordos & Burton uncovered significant heterogeneity in allozyme allele frequencies of blue crab adults, juveniles and megalopae collected along a 600-km stretch of coastline in the Gulf of Mexico. Adult blue crabs were found to be genetically differentiated among nearby bays, indicating reduced gene flow over surprisingly short distances. At three different locations, allele frequencies among megalopal recruits also varied spatially and often differed from those of nearby adults. Even more striking, however, were the extreme temporal shifts in allele frequencies among groups of megalopae recruiting to an area, and the significant loss of alleles between megalopal and adult life stages. The temporal genetic variation seen among recruits could be explained by seasonal spawning differences in source populations and changes in current patterns supplying recruits to an area (Kordos & Burton 1993). However, this scenario fails to explain why genotypes common in recruits were underrepresented in later life stages. *Callinectes sapidus* megalopae typically recruit to the mouths of bays and estuaries, molt to an early crab stage within a few days of settling and move further into marshes and bays as they grow (Morgan et al. 1996). If the megalopae recruiting to the beaches in Texas follow the typical pattern for *C. sapidus*, it is reasonable to expect they would become the juveniles and adults in the nearest bay. The fact that Kordos & Burton (1993) did not find similar allele frequencies between megalopal recruits and nearby juveniles and adults indicates a differential loss of alleles within each cohort, which suggests the possibility of postsettlement selection.

5   PHYLOGEOGRAPHY

Phylogeography became possible with the technology available in the early 1980s because of the unique properties of animal mtDNA (Moritz et al. 1987). In contrast to nuclear DNA, mtDNA is circular and relatively small, which facilitates its isolation by ultracentrifugation. This allowed mtDNA sequence variation to be characterized by routine restriction fragment analysis (Brown 1980). And unlike allozymes, which were then the standard marker for population genetics, mtDNA could be used to infer intraspecific phylogenies (Avise et al. 1979). Although phylogeographic analysis is no longer tied to it, mtDNA has remained the marker of choice. The origins of phylogeography in the 1980s also coincided with the growing acceptance of the neutral theory of molecular evolution, which predicted that most of the DNA sequence variation within a population is likely to be selectively neutral (Kimura 1983). Acceptance of the neutral theory allowed phylogeographers to focus on historical biogeographic explanations of the patterns they found without the complications that selection would introduce (Avise et al. 1987). Although the issue of selection acting on mtDNA was raised periodically, it did not become a major concern until recently.

The central project of phylogeography, to assign historical causes to the biogeographical patterns revealed by mtDNA (Avise 2000), continues today although its methods have become more sophisticated. Where once mtDNA phylogenies were constructed by hand from restriction fragment data and phylogeographic patterns were interpreted by eye, large datasets are now generated by direct sequencing of Polymerase Chain Reaction (PCR) products and computers are tied up for weeks with their analysis. Population genetics and molecular evolution have also become more sophisticated. The data provided by PCR and large-scale sequencing of genes and genomes have focused new interest on detecting selection and led to more nuanced forms of the neutral theory of molecular evolution (Austin 2008). Coalescent models have linked the genealogical approaches of phylogeography to a broader theoretical framework that applies to nuclear sequence polymorphisms as well as mtDNA (Avise 2009). A blurring of the boundaries between phylogeography and population genetics developments has opened up new possibilities for phylogeography, but it has also led to sharp debates about the robustness of traditional phylogeographic approaches.

5.1   The debate over methods of phylogeographic analysis

One debate that appears to have lasted far too long concerns the interpretation of phylogeographic patterns as "signatures" of particular historical processes. This approach reached its extreme form in nested clade phylogeographic analysis (NCPA), which is based on the idea that historical processes that generate phylogeographic patterns can be identified much in the same way that plant or animal specimens are identified with a dichotomous key (Templeton et al. 1995). The diagnostic characters used in the NCPA key are statistics that can be easily calculated from phylogeographic data and are intended to capture various predictions of population genetic theory. A major problem with NCPA and similar approaches is that population genetics theory does not predict unique and easily identifiable signatures for distinct historical processes, but rather overlapping ranges of possible outcomes with different probabilities (Nielsen & Wakeley 2001). Support for alternative historical scenarios from phylogeographic data are therefore best evaluated in terms of likelihoods or Bayesian posterior probabilities calculated from probabilistic models (Knowles 2004). The concept of distinct historical signatures can be useful as a heuristic aid, but it is not a sound statistical basis for hypothesis testing. Despite pointed efforts to explain the conceptual flaws of NCPA (Knowles & Maddison 2002; Petit & Grivet 2002; Beaumont & Panchal 2008), which have been backed up by multiple computer simulation studies demonstrating that it is usually wrong (Knowles & Maddison 2002; Panchal et al. 2007; Panchal & Beaumont 2010), NCPA is still used in phylogeographic studies of marine decapods and other taxa. It has been suggested that in spite of its proven inaccuracy, NCPA continues to be used because it is the only method that promises to reconstruct detailed biogeographic histories from modest amounts of data (Knowles 2008).

5.2 The debate over the importance of selection

A second debate that is still to be resolved concerns the role of natural selection in shaping phylogeographic structure. According to the neutral theory of molecular evolution, selection primarily clears deleterious mutations from populations (Kimura 1968), leaving behind selectively neutral polymorphisms (Kimura & Ohta 1971). This is a core assumption of phylogeographic analysis. However, the neutral theory also predicts the appearance of beneficial mutations, although infrequently, that are swept to fixation while displacing the previously neutral polymorphisms (Kimura 1983). Sites in nuclear loci that undergo these selective sweeps will drag nearby tightly linked sites to fixation with them, creating small genomic regions of reduced polymorphism (Smith & Haigh 1974; Gillespie 2000). However, in the case of the animal mitochondrial genome the near absence of recombination means that selective sweeps will carry the entire genome to fixation, eliminating all mtDNA polymorphism. Without unlinked regions for comparisons, the reduction in mtDNA polymorphism from a selective sweep is indistinguishable from one caused by a population bottleneck or founder event. Furthermore, the larger the population, the more often favorable mutations will arise and undergo selective sweeps that eliminate polymorphism. This effect of selective sweeps is the opposite of that expected for purely neutral polymorphisms, which would reach higher levels in larger populations. It could explain why estimates of effective population size based on mtDNA polymorphism are often far below biologically reasonable values (Avise et al. 1988), and why levels of mtDNA polymorphism in different species may be similar despite what are likely to be order-of-magnitude differences in effective population sizes (Bazin et al. 2006). The possibility of selective sweeps has led some to question estimates of historical effective population size based solely on mtDNA polymorphism (Galtier et al. 2009). However, without a reliable method to distinguish the effects of selective sweeps from population size effects this remains a heated subject of debate (Meiklejohn et al. 2007).

Selective sweeps do not necessarily begin with new, random mutations. Two other possibilities are selection of preexisting variants that become favored as a result of environmental change and introgression of selectively favored variants from other species. There is clear evidence of introgression in *Drosophila* and freshwater fish (reviewed by Ballard & Whitlock 2004). For mtDNA, a third possibility exists because mitochondria are maternally inherited in most animal taxa. Just as selection acting on any part of the mitochondrial genome drags along the entire genome because of linkage, so does selection acting on any maternally transmitted factor, such as inherited symbionts. Inherited symbionts are considered to be common in invertebrates and selection on these symbionts has been well documented in several arthropod groups (reviewed by Hurst & Jiggins 2005). One of the best characterized is the alphaproteobacterium *Wolbachia* that has been detected in over 20% of insect species, 50% of spiders, and 35% of isopods. *Wolbachia* alters its host's reproductive system to favor its own transmission, typically by either a distortion of the sex-ratio or by cytoplasmic incompatibility, in which eggs of uninfected females are killed when fertilized by the sperm of infected males. The effects of indirect selection on mtDNA via symbionts are not limited to selective sweeps that reduce polymorphism within populations. They can also increase polymorphism (if there are multiple symbiont strains within a population), and either reduce or increase differentiation between populations (Hurst & Jiggins 2005). Although *Wolbachia* has not been detected in decapod crustaceans (Bouchon et al. 1998), inherited bacteria appear to be common in arthropods (Duron et al. 2008), and the potential for indirect selection on decapod mtDNA should not be ignored.

At present, it is unclear how often or to what extent phylogeographic inferences based on mtDNA alone are compromised by the effects of selection. However, the new view that plank-tonic dispersal does not necessarily imply overwhelming gene flow, along with a new appreciation for the potential of selection on mtDNA suggests the need for a broader view of the causes of phylogeographic structure in decapod crustaceans. Furthermore, we should avoid basing phylogeo-graphic inferences solely on mtDNA data. Nuclear DNA markers are now relatively easy to assay and methods of phylogeographic analysis based on sound statistical principles can be applied to both mitochondrial and nuclear data.

6   DIRECTIONS FOR FUTURE RESEARCH

The hope that allozyme studies would uncover patterns of genetic differentiation caused by selection was based on the fact that allozymes correspond to protein polymorphisms that could conceivably cause differences in fitness (Lewontin 1974). However, the underlying sequence variation respon-sible for observed differences in electrophoretic mobility cannot be determined from traditional allozyme scoring methods. Revolutionary advances in DNA sequencing and other methods since the era of classic allozyme studies now make it relatively inexpensive and straightforward to uncover sequence variation. Single nucleotide polymorphism (SNP) markers offer an exciting opportunity to return to the intellectually fertile ground of allozyme studies with the power of modern molecular methods. SNPs are bi-allelic codominant markers that correspond to single nucleotide substitutions in DNA sequences. They occur in high frequency across animal genomes, and are often considered preferable to other more commonly used frequency-based genetic markers such as microsatellites and Amplified Fragment Length Polymorphisms (AFLPs) because their evolution corresponds to simple mutation models, they are comparatively easy to genotype and are not prone to null alleles (Ranade et al. 2001; Brumfield et al. 2003). These advantages have made SNPs a popular marker for medical and agricultural studies for a number of years, but they have only recently been applied to a broader range of taxa in molecular ecology, evolution and population genetics studies (Morin et al. 2004). Despite increasing acceptance of SNPs for marine conservation and fisheries research, to date, few decapod studies have utilized SNPs (Smith et al. 2005; Zeng et al. 2008).

Within protein-coding genes, SNPs can either be synonymous, which means each allele codes for the same amino acid, or nonsynonymous, which means they correspond to amino acid differ-ences that could potentially affect protein function. In this sense, SNPs can be viewed as highly informative allozyme markers, and comparisons between synonymous and nonsynonymous SNPs can be used to test for selection. For example, population parameter estimates generated from synonymous SNPs can be compared to those generated from nonsynonymous SNPs to test the hy-pothesis that selection on amino acid substitutions has altered those parameters. Such an approach can greatly extend our understanding of how decapods adapt to different environmental conditions and how selection influences population structure. In addition, recognizing and investigating the ex-tent of local adaptation in decapod populations will better inform our conservation and management decisions.

Use of SNPs in investigations of selection in marine species has already produced some re-markable results. For example, a single amino acid residue substitution in the Na^+ channel pore protein sequence in the softshell clam, *Mya arenaria*, is known to confer resistance to saxitonin, a toxin associated with paralytic shellfish poisoning (Bricelj et al. 2005). Selection for alleles associated with saxitonin resistance in regions with a history of toxic algal blooms provides a highly convincing explanation for the observed population level variation of *M. arenaria* along the coast of New England (Connell et al. 2007). However, a prior investigation into the population genetics of *M. arenaria* in the same region found no evidence of genetic population structure based on sequences from the ribosomal internal transcribed spacer (ITS) gene (Caporale et al. 1997). This case highlights the importance of including multiple loci in analyses that aim to describe population connectivity, but more importantly it also emphasizes the need to investigate genetic patterns based on loci from genes that are candidates for natural selection.

The unprecedented amount of genetic data currently being generated for model organisms, as well as an increasing number of nonmodel organisms, is quickly changing the field of population genetics to a genomic-based discipline. These data provide a wealth of opportunities to investigate sequence level variation in protein-coding genes. In particular, alignments of expressed sequence tags (ESTs) generated from cDNA libraries can yield large numbers of SNPs. Creating EST libraries for decapods (and mining those that are already available) will undoubtedly generate countless SNP markers that can be used to better infer population histories and estimate population parameters for species in this group.

Advances in statistical techniques that can be applied to SNP data also further our ability to detect selection. For example, outlier tests are a useful tool for identifying individual loci that may be under selection (Beaumont & Nichols 1996; Antao et al. 2008). These tests utilize simulations to generate an expected neutral distribution based on F_{ST} and heterozygosity values and classify loci that deviate from this distribution as potential candidates for selection. The importance of testing loci for selection in this manner can be illustrated by a study involving two morphotypes of the intertidal snail, *Littorina saxatilis*. Wilding et al. (2001) constructed phylogenies for *L. saxatilis* using AFLP markers and found two highly conflicting patterns. A phylogeny constructed from 290 AFLP loci showed snails grouped by morphotype and habitat despite being collected from locations separated by as much as 300 km of coastline. After removal of 15 outlier loci from the analysis, snails clustered by geographic location. This second scenario is more likely to reflect actual population history given this species' direct development and exceptionally limited vagility as adults. The outlier pattern was observed across all geographic sampling locations and is consistent with a previously reported habitat-based selection gradient for *L. saxatilis* (Johannesson et al. 1995). Clearly, selection can produce complex patterns of population structure in marine species that are difficult to interpret under the assumption of marker neutrality. Therefore, a priori testing of loci for selection should be the first step of any population genetic analysis (Luikart et al. 2003).

7 Conclusions

A growing body of evidence provides overwhelming support for a new view of larval dispersal for meroplanktonic species, including decapods. Oceanographic processes and larval behavior can significantly alter dispersal distances from what would be expected in simplified models assuming passive particles traveling in unidirectional flows. Realized larval dispersal is further complicated by the potential for pre- and postsettlement selection acting on dispersing larvae. In this case, larvae may disperse, but they are not adapted for the conditions they encounter and gene flow does not occur. Several studies of meroplanktonic species suggest selection plays a significant role in structuring populations, and it is likely that investigations of selection in decapod populations will yield similar conclusions. The mounting genetic data currently being generated for decapods and advances in sophisticated statistical techniques offer the promise of numerous, highly informative SNPs that can be used to test hypotheses of neutrality.

ACKNOWLEDGEMENTS

We would like to thank Christoph Schubart, Christoph Held and Stefan Koenemann for their invitation to contribute to this edition of Crustacean Issues, and for their excellent work in editing this volume. We would also like to thank two anonymous reviewers for detailed comments that have improved our chapter.

REFERENCES

Anger, K. 1991. Effects of temperature and salinity on the larval development of the Chinese mitten crab *Eriocheir sinensis* (Decapoda: Grapsidae). *Mar. Ecol. Prog. Ser.* 72: 103–110.

Antao, T., Lopes, A., Lopes, R.J., Beja-Pereira, A. & Luikart, G. 2008. LOSITAN: a workbench to detect molecular adaptation based on a F_{ST}–outlier method. *BMC Bioinform.* 9: 323.

Austin, L.H. 2008. Near neutrality. Leading edge of the neutral theory of molecular evolution. *Ann. N. Y. Acad. Sci.* 1133: 162–179.

Avise, J.C. 2000. *Phylogeography: The History and Formation of Species.* Cambridge, MA: Harvard University Press.

Avise, J.C. 2009. Phylogeography: retrospect and prospect. *J. Biogeogr.* 36: 3–15.

Avise, J.C., Arnold, J., Ball, R.M., Bermingham, E., Lamb, T., Neigel, J.E., Reeb, C.A. & Saunders, N.C. 1987. Intraspecific phylogeography: the mitochondrial DNA bridge between population genetics and systematics. *Annu. Rev. Ecol. Syst.* 18: 489–522.

Avise, J.C., Ball, R.M. & Arnold, J. 1988. Current versus historical population sizes in vertebrate species with high gene flow: a comparison based on mitochondrial DNA lineages and inbreeding theory for neutral mutations. *Mol. Biol. Evol.* 5: 331–344.

Avise, J.C., Giblin-Davidson, C., Laerm, J., Patton, J.C. & Lansman, R.A. 1979. Mitochondrial DNA clones and matriarchal phylogeny within and among geographic populations of the pocket gopher, *Geomys pinetis*. *Proc. Nat. Acad. Sci. USA* 76: 6694–6698.

Baker, R.R. 1978. *The Evolutionary Ecology of Animal Migration.* New York: Holmes and Meier Publishers, Inc.

Ballard, J.W.O. & Whitlock, M.C. 2004. The incomplete natural history of mitochondria. *Mol. Ecol.* 13: 729–744.

Barber, P.H., Palumbi, S.R., Erdmann, M.V. & Moosa, M.K. 2000. Biogeography: a marine Wallace's line? *Nature* 406: 692–693.

Bazin, E., Glémin, S. & Galtier, N. 2006. Population size does not influence mitochondrial genetic diversity in animals. *Science* 312: 570–572.

Beaumont, M.A. & Nichols, R.A. 1996. Evaluating loci for use in the genetic analysis of population structure. *Proc. R. Soc. Lond. B* 263: 1619–1626.

Beaumont, M.A. & Panchal, M. 2008. On the validity of nested clade phylogeographical analysis. *Mol. Ecol.* 17: 2563–2565.

Berthelemy-Okazaki, N.J. & Okazaki, R.K. 1997. Population genetics of the blue crab *Callinectes sapidus* from the northwestern Gulf of Mexico. *Gulf Mex. Sci.* 1: 35–39.

Bilodeau, A.L., Felder, D.L. & Neigel, J.E. 2005. Population structure at two geographic scales in the burrowing crustacean *Callichirus islagrande* (Decapoda, Thalassinidea): historical and contemporary barriers to planktonic dispersal. *Evolution* 59: 2125–2138.

Bohonak, A.J. 1999. Dispersal, gene flow, and population structure. *Q. Rev. Biol.* 74: 21–45.

Bossart, J.L. & Prowell, D.P. 1998. Genetic estimates of population structure and gene flow: limitations, lessons and new directions. *Trends Ecol. Evol.* 13: 202–206.

Bouchon, D., Rigaud, T. & Juchault, P. 1998. Evidence for widespread *Wolbachia* infection in isopod crustaceans: molecular identification and host feminization. *Proc. R. Soc. Lond. B*

265: 1081–1090.

Bricelj, V.M., Connell, L., Konoki, K., MacQuarrie, S.P., Scheuer, T., Catterall, W.A. & Trainer, V.L. 2005. Sodium channel mutation leading to saxitoxin resistance in clams increases risk of PSP. *Nature* 434: 763–767.

Brown, A.F., Kann, L.M. & Rand, D.M. 2001. Gene flow versus local adaptation in the northern acorn barnacle, *Semibalanus balanoides*: insights from mitochondrial DNA variation. *Evolution* 55: 1972–1979.

Brown, W.M. 1980. Polymorphism in mitochondrial DNA of humans as revealed by restriction endonuclease analysis. *Proc. Nat. Acad. Sci. USA* 77: 3605–3369.

Brumfield, R.T., Beerli, P., Nickerson, D.A. & Edwards, S.V. 2003. The utility of single nucleotide polymorphisms in inferences of population history. *Trends Ecol. Evol.* 18: 249–256.

Caley, M.J., Carr, M.H., Hixon, M.A., Hughes, T.P., Jones, G.P. & Menge, B.A. 1996. Recruitment and the local dynamics of open marine populations. *Annu. Rev. Ecol. Syst.* 27: 477–500.

Caporale, D.A., Beal, B.F., Roxby, R. & van Benenden, R.J. 1997. Population structure of *Mya arenaria* along the New England coastline. *Mol. Mar. Biol. Biotechnol.* 6: 33–39.

Chiswell, S.M. & Booth, J.D. 1999. Rock lobster *Jasus edwardsii* larval retention by the Wairarapa Eddy off New Zealand. *Mar. Ecol. Prog. Ser.* 183: 227–240.

Christy, J.H. 1989. Rapid development of megalopae of the fiddler crab *Uca pugilator* reared over sediment: implications for models of larval recruitment. *Mar. Ecol. Prog. Ser.* 57: 259–265.

Cobb, J.S., Wang, D., Campbell, D.B. & Rooney, P. 1989. Speed and direction of swimming by postlarvae of the American lobster. *Trans. Amer. Fish. Soc.* 118: 82–86.

Connell, L.B., MacQuarrie, S.P., Twarog, B.M., Iszard, M. & Bricelj, V.M. 2007. Population differences in nerve resistance to paralytic shellfish toxins in softshell clam, *Mya arenaria*, associated with sodium channel mutations. *Mar. Biol.* 150: 1227–1236.

Corey, S. & Reid, D.M. 1991. Comparative fecundity of decapod crustaceans. I. The fecundity of thirty-three species of nine families of caridean shrimp. *Crustaceana* 60: 270–294.

Costlow, J.D. Jr. & Bookhout, C.G. 1959. The larval development of *Callinectes sapidus* Rathbun reared in the laboratory. *Biol. Bull.* 116: 373–396.

Cronin, T.W. & Forward, R.B. 1986. Vertical migration cycles of crab larvae and their role in larval dispersal. *Bull. Mar. Sci.* 39: 192–201.

DeWitt, T.J., Sih, A. & Wilson, D.S. 1998. Costs and limits of phenotypic plasticity. *Trends Ecol. Evol.* 13: 77–81.

Duron, O., Bouchon, D., Boutin, S., Bellamy, L., Zhou, L., Engelstädter, J. & Hurst, G.D. 2008. The diversity of reproductive parasites among arthropods: *Wolbachia* do not walk alone. *BMC Biol.* 6: 27.

Epifanio, C.E. 1987. The role of tidal fronts in maintaining patches of brachyuran zoeae in estuarine waters. *J. Crust. Biol.* 7: 513–517.

Fernandez, M., Iribarne, O.O. & Armstrong, D.A. 1994. Swimming behavior of Dungeness crab, *Cancer magister* Dana, megalopae in still and moving water. *Estuaries* 17: 271–275.

Field, J.M. & Butler, M.J. 1994. The influence of temperature, salinity, and postlarval transport on the distribution of juvenile spiny lobsters, *Panulirus argus* (Latreille, 1804), in Florida Bay. *Crustaceana* 67: 26–45.

Galtier, N., Nabholz, B., Glémin, S. & Hurst, G.D.D. 2009. Mitochondrial DNA as a marker of molecular diversity: a reappraisal. *Mol. Ecol.* 18: 4541–4550.

Gillespie, J.H. 2000. Genetic drift in an infinite population: the pseudohitchhiking model. *Genetics* 155: 909–919.

Gore, R.H. & Scotto, L.E. 1982. *Cyclograpsus integer* H. Milne Edwards, 1837 (Brachyura, Grapsidae): the complete larval development in the laboratory, with notes on larvae of the genus *Cyclograpsus*. *Fish. Bull.* 80: 501–521.

Hedgecock, D. 1986. Is gene flow from pelagic larval dispersal important in the adaptation and evolution of marine invertebrates? *Bull. Mar. Sci.* 39: 550–564.

Hedgecock, D. 1994. Does variance in reproductive success limit effective population sizes of marine organisms? In: Kawasaki, T., Tanaka, S., Toba, Y., & Taniguchi, A. (eds.) *Long-term Variability of Pelagic Fish Populations and their Environment*: 199–207. Oxford: Pergamon Press.

Hedgecock, D., Chow, V. & Waples, R.S. 1992. Effective population numbers of shellfish broodstocks estimated from temporal variance in allelic frequencies. *Aquaculture* 108: 215–232.

Hellberg, M.E., Burton, R.S., Neigel, J.E. & Palumbi, S.R. 2002. Genetic assessment of connectivity among marine populations. *Bull. Mar. Sci.* 70: 273–290.

Hilbish, T.J. 1985. Demographic and temporal structure of an allele frequency cline in the mussel *Mytilus edulis. Mar. Biol.* 86: 163–171.

Hsueh, P.-W., McClintock, J.B. & Hopkins, T.S. 1993. Population dynamics and life history characteristics of the blue crabs *Callinectes similis* and *C. sapidus* in bay environments of the northern Gulf of Mexico. *Mar. Ecol.* 14: 239–257.

Hughes, D.A. 1969. Responses to salinity change as a tidal transport mechanism of pink shrimp, *Penaeus duorarum. Biol. Bull.* 136: 43–53.

Hurst, G.D.D. & Jiggins, F.M. 2005. Problems with mitochondrial DNA as a marker in population, phylogeographic and phylogenetic studies: the effects of inherited symbionts. *Proc. R. Soc. Lond. B* 272: 1525–1534.

Johannes, R.E. 1978. Reproductive strategies of coastal marine fishes in the tropics. *Environ. Biol. Fishes* 3: 65–84.

Johannesson, K., Johannesson, B. & Lundgren, U. 1995. Strong natural selection causes microscale allozyme variation in a marine snail. *Proc. Nat. Acad. Sci. USA* 92: 2602–2606.

Johnson, M.S. & Black, R. 1982. Chaotic genetic patchiness in an intertidal limpet, *Siphonaria* sp. *Mar. Biol.* 70: 157–164.

Jones, G.P., Milicich, M.J., Emslie, M.J. & Lunow, C. 1999. Self-recruitment in a coral reef fish population. *Nature* 402: 802–804.

Kimura, M. 1968. Evolutionary rate at the molecular level. *Nature* 217: 624–626.

Kimura, M. 1983. *The Neutral Theory of Molecular Evolution*. Cambridge: Cambridge University Press.

Kimura, M. & Ohta, T. 1971. Protein polymorphism as a phase of molecular evolution. *Nature* 229: 467–469.

Kingsford, M.J., Leis, J.M., Shanks, A., Lindeman, K.C., Morgan, S.G. & Pineda, J. 2002. Sensory environments, larval abilities and local self-recruitment. *Bull. Mar. Sci.* 70: 309–340.

Knowles, L.L. 2004. The burgeoning field of statistical phylogeography. *J. Evol. Biol.* 17: 1–10.

Knowles, L.L. 2008. Why does a method that fails continue to be used? *Evolution* 62: 2713–2717.

Knowles, L.L. & Maddison, W.P. 2002. Statistical phylogeography. *Mol. Ecol.* 11: 2623–2635.

Koehn, R.K., Milkman, R. & Mitton, J.B. 1976. Population genetics of marine pelecypods. IV. Selection, migration and genetic differentiation in the blue mussel *Mytilus edulis. Evolution* 30: 2–32.

Koehn, R.K. & Siebenaller, J.F. 1981. Biochemical studies of aminopeptidase polymorphism in *Mytilus edulis*. II. Dependence of reaction rate on physical factors and enzyme concentration. *Biochem. Genet.* 19: 1143–1162.

Kordos, L.M. & Burton, R.S. 1993. Genetic differentiation of Texas Gulf Coast populations of the blue crab *Callinectes sapidus. Mar. Biol.* 117: 227–233.

Largier, J. L. 2003. Considerations in estimating larval dispersal distances from oceanographic data. *Ecol. Appl.* 13: S71–S89.

Leis, J.M. 2006. Are larvae of demersal fishes plankton or nekton? In: Southward, A.J. & Sims,

D.W. (eds.) *Advances in Marine Biology* 51: 57–141. San Diego, CA: Academic Press.

Lewontin, R.C. 1974. *The Genetic Basis of Evolutionary Change.* New York, NY: Columbia University Press.

Luckenbach, M.W. & Orth, R.J. 1992. Swimming velocities and behavior of blue crab (*Callinectes sapidus* Rathbun) megalopae in still and flowing water. *Estuaries* 15: 186–192.

Luikart, G., England, P.R., Tallmon, D., Jordan, S. & Taberlet, P. 2003. The power and promise of population genomics: from genotyping to genome typing. *Nat. Rev. Gen.* 4: 981–994.

Marshall, D.J., Monro, K., Bode, M., Keough, M.J. & Swearer, S. 2010. Phenotype–environment mismatches reduce connectivity in the sea. *Ecol. Lett.* 13: 128–140.

McMillen-Jackson, A.L. & Bert, T.M. 2004. Mitochondrial DNA variation and population genetic structure of the blue crab *Callinectes sapidus* in the eastern United States. *Mar. Biol.* 145: 769–777.

McMillen-Jackson, A.L., Bert, T.M. & Steele, P. 1994. Population genetics of the blue crab *Callinectes sapidus*: modest population structuring in a background of high gene flow. *Mar. Biol.* 118: 53–65.

Meiklejohn, C.D., Montooth, K.L. & Rand, D.M. 2007. Positive and negative selection on the mitochondrial genome. *Trends Genet.* 23: 259–263.

Morgan, S.G. 1989. Adaptive significance of spination in estuarine crab zoeae. *Ecology* 70: 464–482.

Morgan, S.G. 1990. Impact of planktivorous fishes on dispersal, hatching, and morphology of estuarine crab larvae. *Ecology* 71: 1639–1652.

Morgan, S.G. 1995. Life and death in the plankton: larval mortality and adaptation. In: McEdwards, L. (ed.) *Ecology of Marine Invertebrate Larvae*: 279–321. Boca Raton, FL: CRC Press.

Morgan, S.G., Zimmer-Faust, R.K., Heck, K.L.Jr. & Coen, L.D. 1996. Population regulation of blue crabs, *Callinectes sapidus*, in the northern Gulf of Mexico: postlarval supply. *Mar. Ecol. Prog. Ser.* 133: 73–88.

Morin, P.A., Luikart, G. & Wayne, R.K. 2004. SNPs in ecology, evolution and conservation. *Trends Ecol. Evol.* 19: 208–216.

Moritz, C., Dowling, T.E. & Brown, W.M. 1987. Evolution of animal mitochondrial DNA: relevance for population biology and systematics. *Annu. Rev. Ecol. Syst.* 18: 269–292.

Nates, S.F., Felder, D.L. & Lemaitre, R. 1997. Comparative larval development in two species of the burrowing ghost shrimp genus *Lepidopthalmus* (Decapoda: Callianassidae). *J. Crust. Biol.* 17: 497–519.

Neigel, J.E. 1997. A comparison of alternative strategies for estimating gene flow from genetic markers. *Annu. Rev. Ecol. Syst.* 28: 105–128.

Neigel, J.E. 2002. Is F_{ST} obsolete? *Conserv. Gen.* 3: 167–173.

Neigel, J.E. 2009. Population genetics and biogeography of the Gulf of Mexico. In: Felder, D.L. & Camp, C.K. (eds.) *Gulf of Mexico. Origins, Waters, and Biota 1. Biodiversity*: 1353–1370. College Station, Texas: Texas A&M University Press.

Nielsen, R. & Wakeley, J. 2001. Distinguishing migration from isolation: a Markov chain Monte Carlo approach. *Genetics* 158: 885–896.

Orive, M.E. 1993. Effective population size in organisms with complex life-histories. *Theo. Pop. Biol.* 44: 316–340.

Panchal, M. & Beaumont, M.A. 2010. Evaluating nested clade phylogeographic analysis under models of restricted gene flow. *Syst. Biol.* 59: 415–432.

Panchal, M., Beaumont, M.A. & Sunnucks, P. 2007. The automation and evaluation of nested clade phylogeographic analysis. *Evolution* 61: 1466–1480.

Petit, R.J. & Grivet, D. 2002. Optimal randomization strategies when testing the existence of a phylogeographic structure. *Genetics* 161: 469–471.

Pringle, J.M. & Wares, J.P. 2007. Going against the flow: maintenance of alongshore variation in allele frequency in a coastal ocean. *Mar. Ecol. Prog. Ser.* 335: 69–84.

Ranade, K., Chang, M.-S., Ting, C.-T., Pei, D., Hsiao, C.-F., Olivier, M., Pesich, R., Hebert, J., Chen, Y.-D.I., Dzau, V.J., Curb, D., Olshen, R., Risch, N., Cox, D.R. & Botstein, D. 2001. High-throughput genotyping with single nucleotide polymorphisms. *Genome Res.* 11: 1262–1268.

Reid, D.M. & Corey, S. 1991. Comparative fecundity of decapod crustaceans, II. The fecundity of fifteen species of anomuran and brachyuran crabs. *Crustaceana* 61: 175–189.

Roberts, C.M. 1997. Connectivity and management of Caribbean coral reefs. *Science* 278: 1454–1457.

Scheltema, R.S. 1971. Larval dispersal as a means of genetic exchange between geographically separated populations of shallow-water benthic marine gastropods. *Biol. Bull.* 140: 284–322.

Shanks, A.L. 1986. Vertical migration and cross-shelf dispersal of larval *Cancer* spp. and *Randallia ornata* (Crustacea: Brachyura) off the coast of southern California. *Mar. Biol.* 92: 189–199.

Shanks, A.L. 2009. Pelagic larval duration and dispersal distance revisited. *Biol. Bull.* 216: 373–385.

Shanks, A.L., Grantham, B.A. & Carr, M.H. 2003. Propagule dispersal distance and the size and spacing of marine reserves. *Ecol. Appl.* 13: S159–S169.

Shirley, T.C. & Zhou, S. 1997. Lecithotrophic development of the golden king crab *Lithodes aequispinus* (Anomura: Lithodidae). *J. Crust. Biol.* 17: 207–216.

Shulman, M.J. & Bermingham, E. 1995. Early life histories, ocean currents, and the population genetics of Caribbean reef fishes. *Evolution* 49: 897–910.

Siegel, D.A., Kinlan, B.P., Gaylord, B. & Gaines, S.D. 2003. Lagrangian descriptions of marine larval dispersion. *Mar. Ecol. Prog. Ser.* 260: 83–96.

Slatkin, M. 1987. Gene flow and the geographic structure of natural populations. *Science* 236: 787–792.

Slatkin, M. & Barton, N.H. 1989. A comparison of three indirect methods for estimating average levels of gene flow. *Evolution* 43: 1349–1368.

Smith, C.T., Grant, W.S. & Seeb, L.W. 2005. A rapid, high-throughput technique for detecting tanner crabs *Chionoecetes bairdi* illegally taken in Alaska's snow crab fishery. *Trans. Amer. Fish. Soc.* 134: 620–623.

Smith, J.M. & Haigh, J. 1974. The hitch-hiking effect of a favourable gene. *Genet. Res.* 1: 23–35.

Sponaugle, S., Cowen, R.K., Shanks, A., Morgan, S.G., Leis, J.M., Pineda, J., Boehlert, G.W., Kingsford, M.J., Lindeman, K.C., Grimes, C. & Munro, J.L. 2002. Predicting self-recruitment in marine populations: biophysical correlates and mechanisms. *Bull. Mar. Sci.* 70: 341–375.

Staton, J.L. & Felder, D.L. 1995. Genetic variation in populations of the ghost shrimp genus *Callichirus* (Crustacea: Decapoda: Thalassinoidea) in the Western Atlantic and Gulf of Mexico. *Bull. Mar. Sci.* 56: 523–536.

Strasser, K.M. & Felder, D.L. 1999. Larval development in two populations of the ghost shrimp *Callichirus major* (Decapoda: Thalassinidea) under laboratory conditions. *J. Crust. Biol.* 19: 844–878.

Strasser, K.M. & Felder, D.L. 2000. Larval development of the ghost shrimp *Callichirus islagrande* (Decapoda: Thalassinidea: Callianassidae) under laboratory conditions. *J. Crust. Biol.* 20: 100–117.

Strathmann, R.R. 1993. Hypotheses on the origins of marine larvae. *Annu. Rev. Ecol. Syst.* 24: 89–117.

Strathmann, R.R., Hughes, T.P., Kuris, A.M., Lindeman, K.C., Morgan, S.G., Pandolfi, J.M. & Warner, R.R. 2002. Evolution of local recruitment and its consequences for marine populations. *Bull. Mar. Sci.* 70: 377–396.

Sulkin, S.D., Mojica, E. & McKeen, G.L. 1996. Elevated summer temperature effects on megalopal and early juvenile development in the Dungeness crab, *Cancer magister*. *Can. J. Fish. Aquat. Sci.* 53: 2076–2079.

Templeton, A.R., Routman, E. & Phillips, C.A. 1995. Separating population structure from population history: a cladistic analysis of the geographical distribution of mitochondrial DNA haplotypes in the tiger salamander, *Ambystoma tigrinum*. *Genetics* 140: 767–782.

Thorson, G. 1950. Reproduction and larval ecology of marine bottom invertebrates. *Biol. Rev.* 25: 1–45.

Victor, B.C. 1986. Delayed metamorphosis with reduced larval growth in a coral reef fish (*Thalassoma bifasciatum*). *Can. J. Fish. Aquat. Sci.* 43: 1208–1213.

Warner, R.R. & Cowen, R.K. 2002. Local retention of production in marine populations: evidence, mechanisms, and consequences. *Bull. Mar. Sci.* 70: 245–249.

Weersing, K. & Toonen, R. J. 2009. Population genetics, larval dispersal, and connectivity in marine systems. *Mar. Ecol. Prog. Ser.* 393: 1–12.

Wellington, G.M. & Victor, B.C. 1989. Planktonic larval duration of one hundred species of Pacific and Atlantic damselfishes (Pomacentridae). *Mar. Biol.* 101: 557–567.

Whitlock, M.C. & McCauley, D.E. 1999. Indirect measures of gene flow and migration: $F_{ST} \neq 1/(4 N m + 1)$. *Heredity* 82: 117–125.

Wilding, C.S., Butlin, R. K. & Grahame, J. 2001. Differential gene exchange between parapatric morphs of *Littorina saxatilis* detected using AFLP markers. *J. Evol. Biol.* 14: 611–619.

Williams, G.C. 1975. *Sex and Evolution*. Princeton, NJ: Princeton University Press.

Wing, S.R., Botsford, L.W., Largier, J.L. & Morgan, L.E. 1995. Spatial structure of relaxation events and crab settlement in the northern California upwelling system. *Mar. Ecol. Prog. Ser.* 128: 199–211.

Wright, S. 1951. The genetical structure of populations. *Annu. Eugen.* 15: 323–354.

Zeng, D., Chen, X., Li, Y., Peng, M., Ma, N., Jiang, W., Yang, C. & Li, M. 2008. Analysis of *Hsp70* in *Litopenaeus vannamei* and detection of SNPS. *J. Crust. Biol.* 28: 727–730.

Causes of chaos: spatial and temporal genetic heterogeneity in the intertidal anomuran crab *Petrolisthes cinctipes*

ROBERT J. TOONEN[1,2] & RICHARD K. GROSBERG[2]

[1] *Hawai'i Institute of Marine Biology, School of Ocean and Earth Sciences and Technology, University of Hawai'i at Mānoa, Kāne'ohe, U. S. A.*

[2] *College of Biological Sciences, Center for Biology, University of California at Davis, Davis, U. S. A.*

ABSTRACT

Hypotheses to explain chaotic genetic structure (i.e., a surprising degree of nongeographic temporal or spatial population differentiation) include 1) variation in source of larval recruits, 2) self-recruitment and local subdivision, 3) variance in reproductive success (sweepstakes reproduction), and 4) pre- or postsettlement natural selection. We evaluated the relative contribution of each of these four processes to the observed patterns of genetic differentiation among geographic populations of the porcelain crab, *Petrolisthes cinctipes*. We genotyped about 50 individuals of each size class (new recruits, subadults, adults) at each of nine sites across northern California in each of three consecutive years ($N = 3602$). We found significant and consistent population structure ($\theta \approx 0.08$) among sites from each replicate year of sampling (1997–1999). Significant population structure also occurred both among years and among sites, but the pattern of population structuring differed by size class. Among years, the differentiation of new recruits was highest ($\theta = 0.12$) year-to-year, followed by the subadults ($\theta = 0.09$) and finally adults ($\theta = 0.08$). In contrast, the among-sites component was greatest among the adults ($\theta = 0.05$), followed by the subadults ($\theta = 0.03$), and least among new recruits ($\theta < 0.01$, not significant). An overall hierarchical analysis of molecular variance (AMOVA) revealed significant genetic divergence among years ($\theta = 0.05$), and size classes ($\theta = 0.08$), but not among sites ($\theta < 0.01$). Recruits at each site were genetically most similar to a different population of adults each year, and although some recruits were apparently self-recruits, no consistent patterns of assignment or genetic similarity to natal populations emerged across years, indicating that the source of recruitment is unpredictable. Temporal differentiation was stronger than spatial differentiation, and in the full hierarchical analysis, the among-years and among-size class components of variance explain the majority of observed population structure. Most studies of population genetic structuring are snapshots based on single collections that lack explicit temporal components, and may provide an incomplete picture of population structure as a result. Overall, these data suggest that all four hypothesized mechanisms collectively create fine-scale population genetic structure in *P. cinctipes*, and perhaps many other marine species with similar life histories and distributions.

1 INTRODUCTION

Over microevolutionary time scales, rates, and patterns of gene flow determine the potential for, and scale at which, local adaptation can evolve; over macroevolutionary time, rates and patterns of gene flow are related to evolutionary persistence and rates of speciation (Futuyma 1997; Hartl & Clark

1997). Because direct measures of gene flow are nearly impossible for most marine populations, indirect measures of gene flow inferred from the spatial distribution of alleles (or haploypes) present an attractive alternative for marine systems (reviewed by Palumbi 1996; Neigel 1997; Grosberg & Cunningham 2001; Hellberg et al. 2002; Selkoe et al. 2008; Hellberg 2009). Even in cases where tracking of dispersal is possible (e.g., Carlon & Olson 1993; Jones et al. 1999; Planes et al. 2009), direct estimates of gene flow will typically underestimate the importance of rare long-distance dispersal of individuals (e.g., Lewis et al. 1997; Petit & Mayer 1999; reviewed by Hellberg et al. 2002; Hellberg 2009). Although rare long-distance dispersal events are unlikely to be detected by direct methods, they may prove extremely important determinants of the biology and evolutionary history of natural populations (reviewed by Neigel 1997; Ayre 1990; Grosberg & Cunningham 2001; Hellberg et al. 2002; Selkoe et al. 2008; Hellberg 2009).

Another potential problem with direct estimates of gene flow is that data and theory now indicate that marine larval dispersal events often happen in discontinuous, but intense pulses or "spikes" of recruitment. Across a suite of approaches ranging from Lagrangian simulation studies (e.g., Mitarai et al. 2008; Siegel et al. 2008; Pringle et al. 2009) to direct tracking, kinship, and genetic assignment tests (e.g., Jones et al. 2005; Selkoe et al. 2006; Almany et al. 2007; Buston et al. 2009), evidence is accumulating that larval mixing is far less than had been generally assumed, and that dispersal paths within years are highly correlated. Although larvae in a given year may experience similar dispersal probabilities, the annual stochasticity of dispersal also indicates that there is little to no correlation between the non-Gaussian dispersal paths of annual cohorts of larvae (e.g., Mitarai et al. 2008; Siegel et al 2008; Cowen & Sponaugle 2009; Pringle et al. 2009). Patterns of larval dispersal only become a smooth distribution when averaged over many events through time (Siegel et al. 2008; Selkoe et al. 2008; White et al. 2010). Thus, indirect methods of estimating gene flow have the potential advantage of providing estimates of both aspects of dispersal, ranging from spiky annual assignment to the ultimate consequences of dispersal averaged into the distribution of dispersal events themselves (Neigel 1997; Ouborg et al. 1999; Selkoe et al. 2008).

Such indirect approaches generally assume that patterns of genetic structure among geographic populations are stable over ecological time scales (reviewed by Lessios et al. 1994; Neigel 1997; Ouborg et al. 1999; Grosberg & Cunningham 2001). In the absence of natural selection, mutation, and migration, allelic frequencies in an infinitely large and randomly mating population remain constant through time (Hartl & Clark 1997; Futuyma 1997). In nature, however, populations are neither infinite, nor randomly mating, and spatial patterns of genetic structure must change through time as species respond to a suite of stochastic (e.g., reproductive variance and genetic drift) and deterministic (e.g., selection and gene flow) forces (reviewed by Ayre 1990; Hedgecock 1994; Palumbi 1996; Neigel 1997; Ouborg et al. 1999; Grosberg & Cunningham 2001; Hellberg et al. 2002; Selkoe et al. 2008; Hellberg 2009). Despite the operation of both stochastic and deterministic evolutionary forces, does genetic structure remain sufficiently constant that long-term patterns and levels of gene flow can be reliably inferred from a single temporal snapshot of genetic structure?

There are relatively few studies that examine whether the spatial distribution of alleles or allelic frequencies varies significantly through time (e.g., Kordos & Burton 1993; Hedgecock 1994; Viard et al. 1997; Hauser et al. 2002; Turner et al. 2002; Lee & Boulding 2009). For example, a quick search of the BIOSIS database revealed 5126 studies, published between 1985 and 2000, categorized using the keywords "population genetic structure." In contrast, only 360 (7%) of those studies included "temporal" as an additional keyword. The trend has not improved much through time: a similar search in ISI Web of Science returns 2200 studies published between 2000 and 2009, and a search for "temporal" within those returned only 117 (5%) hits. Most studies that assess genetic stability among populations through time focus on terrestrial species, and the handful of marine examples are divided between those documenting that the genetic constitution of populations tends to remain relatively stable (e.g., Berger 1973; Burton & Feldman 1981; Lessios 1985; McClenaghan et al. 1985; Waples 1990; Garant et al. 2000; Palstra et al. 2006; Shen & Tzeng 2007), and those that find significant changes in allele frequencies drawn from the same population over periods of time as short as a few generations (e.g., Powers & Place 1978; Johnson & Black 1984a, b; Piertney & Carvalho 1995; Kusumo & Druehl 2000; Robainas Barcia et al. 2005; Florin & Höglund 2007; Lee & Boulding 2007).

Allele frequencies should show the greatest rate of change in species with a short generation time, small and fluctuating population sizes, and with low levels of gene flow among populations because drift is stronger under such conditions (Lessios et al. 1994; Futuyma 1997; Hartl & Clark 1997). Thus, marine invertebrate taxa with relatively high fecundity, long-lived planktonic larvae, few obvious barriers to dispersal, broad species ranges, and very large and relatively stable population sizes ought to be among the most likely candidates to show temporal stability of allelic frequencies (e.g., Shen & Tzeng 2007). The existing literature generally runs contrary to that prediction, however, with studies of marine taxa demonstrating temporally stable genetic structure more often coming from species from the opposite end of the spectrum, such as the tide pool copepod (e.g., Burton 1997; Burton et al. 1999; Edmands & Harrison 2003), or the intertidal isopod *Excirolana* (e.g., Lessios et al. 1994), with limited dispersal potential and small effective population sizes. Studies of organisms that match the characteristics predicted to reduce temporal variability tend to find that gene frequencies can change significantly through time, even in species that presumably disperse broadly across extensive ranges and maintain large population sizes (e.g., Johnson & Black 1984a, b; Li & Hedgecock 1998; Johnson & Wernham 1999; Robainas Barcia et al. 2005; Lee & Boulding 2007).

The occurrence of temporal variation in allelic frequencies may sometimes be related to another unexpected pattern with respect to spatial structure in the marine environment: many marine species with extensive dispersal potential also exhibit higher genetic differentiation among adjacent sites than between distant sites. A variety of such counterintuitive findings of population genetic structure have come to be known in the marine literature as "chaotic genetic patchiness," following Johnson & Black (1982; 1984a). Chaotic genetic patchiness, defined as population differentiation at fine scales in the absence of larger scale geographic patterns, is also surprisingly common in species with extensive dispersal potential, and likewise begs the question which demographic, oceanographic, and geological processes are responsible for these patterns. Alternative hypotheses to explain chaotic genetic structure generally encompass four alternatives: 1) variation among differentiated source populations from which larvae recruit, 2) self-recruitment and local differentiation of broadly-distributed species, 3) high variance in reproductive success (sweepstakes reproduction), and 4) pre- or postsettlement natural selection (reviewed by Hedgecock et al. 2007; Hellberg 2009).

Here we present a size-structured analysis of spatial and temporal genetic variation in the common and widespread intertidal anomuran crab, *Petrolisthes cinctipes*, in order to evaluate the relative contribution of stochastic (e.g., individual reproductive success, recruitment variation) and deterministic (e.g., selection and gene flow) forces in shaping the distribution of genetic variation in this species. *Petrolisthes cinctipes* is a sedentary marine invertebrate that produces long-lived planktonic larvae with apparently great potential for mixing, has an enormous census population size (likely on the order of 10^8 across the species range), a relatively high fecundity, an essentially continuous geographic range along the linear coastline of North America, and a maximum lifespan on the order of eight years (Jensen 1990). Thus, by all expectations outlined above, *P. cinctipes* seems an obvious choice to expect temporal stability of gene frequencies over ecological time scales. A previous study of geographic structure across the species range (Toonen 2001) revealed significant population structure (overall $\theta = 0.11$), with some pairwise comparisons of genetic differentiation between adjacent sites (e.g., Bodega Bay and Dillon Beach, only 10 km apart, pairwise $F_{ST} = 0.16$) exceeding differences between pairs of populations spanning much of the species range (e.g., Bodega Bay, CA to Neah Bay, WA, nearly 1200 km apart, $F_{ST} = 0.14$). We used polymorphic microsatellite loci with sufficient allelic variation and power (Kalinowski 2002; Ryman et al. 2006) to characterize the year-to-year variation in genetic structure among the recruits, subadults and adults of *P. cinctipes* and to compare the patterns of temporal and spatial genetic structure to that seen in the initial geographic survey of adults.

Our study was designed to characterize the genetic structure among years, sites and among age classes through time in order to evaluate the relative contribution of stochastic (e.g., individual reproductive success, recruitment variation) and deterministic (e.g., selection and gene flow) forces in shaping the distribution of genetic variation in this species. Here we present the data from this multiyear study of age and site structured population sampling to evaluate support for each of the common hypotheses advanced to explain chaotic genetic patchiness in species such as *Petrolisthes cinctipes*.

2   MATERIALS AND METHODS

2.1   Study organism

We focused this study of fine-scale genetic structure on *Petrolisthes cinctipes* for several reasons. First, the species is easily distinguished from co-occurring congeners, even as juveniles (Stillman & Reeb 2001). In addition, *P. cinctipes* has an extensive and basically linear geographic distribution, ranging from the northern limit around the Dixon Entrance (British Columbia, Canada) to around to Point Conception (California, USA). Our field survey located individuals as far north as Prince Rupert, BC (roughly 54.3°N, 130.4°W) and as far south as Morro Bay, CA (roughly 35.3°N, 120.8°W), but we were unable to locate more than a few scattered individuals anywhere south of Morro Bay. Finally, despite being restricted to a relatively narrow vertical range of the rocky intertidal (from roughly 0 to 1.8 m above MLLW, Jensen & Armstrong 1991), *P. cinctipes* is one of the most abundant decapod crustaceans in the eastern North Pacific, typically occurring in densities of 100–600 individuals/m^2 (Donahue 2004), and occasionally as high as 3,933 individuals/m^2 (Jensen 1990). Males and females of all size classes, from newly settled postlarvae (ca. 1.5 mm) to large adults (up to 20 mm carapace width), live together in large aggregations among rocky cobble (Jensen & Armstrong 1991). Dense aggregations result largely from individual substrate preference (lower vertical bound) and physiological tolerance (upper bound) of the adults, reinforced by gregarious settlement of the larvae with conspecific adults (Jensen 1989, 1991; Jensen & Armstrong 1991; Stillman 2002; Akins 2003; Stillman 2004; Donahue 2004, 2006). For example, Akins (2003) found that adult *P. cinctipes* strongly preferred to live under boulders set on cobbles and pebbles, to those under which there was sand; there was a strong correlation ($R^2 = 0.84$) between the overall abundance of *P. cinctipes* at a site and the percent of cobble habitat per boulder. Adults compete for access to high quality feeding spots characterized by high flow rates (Jensen 1990). Individual growth rates decline with increasing conspecific density in both the lab and the field, with juveniles suffering most from competition because adults can monopolize the preferred feeding spots (Donahue 2004, 2006).

Petrolisthes spp. are gonochoric and females brood fertilized eggs until they hatch and begin pelagic larval development. Timing of reproduction in *P. cinctipes* varies with latitude, but in northern California, females typically produce broods from February through mid-April (Morris et al. 1980). However, some individuals may produce a second brood or be delayed, and brooding females can be found at low frequency as late as September (Toonen 2001, 2004). Newly extruded broods may contain up to about 1,300 embryos, each roughly 0.1 mg in dry weight and 800 μm in diameter (Donahue 2004). Although the exact mechanism of copulation in *P. cinctipes* remains unknown, mating occurs in the hard-shelled state, and large males defend territories in an attempt to monopolize access to females (Molenock 1975, 1976). Despite male guarding and attempts to

1) Crescent City
41.75°N, 124.18°W

2) Cape Mendocino
40.45°N, 124.35°W

3) Shelter Cove
40.02°N, 124.07°W

4) Ft. Bragg
39.43°N, 123.80°W

5) Albion
39.22°N, 123.68°W

6) Pt. Arena
38:90°N, 123.68°W

7) Salt Point
38.50°N, 123.23°W

8) Bodega Bay
38.37°N, 123.02°W

9) Dillon Beach
38.25°N, 122.95°W

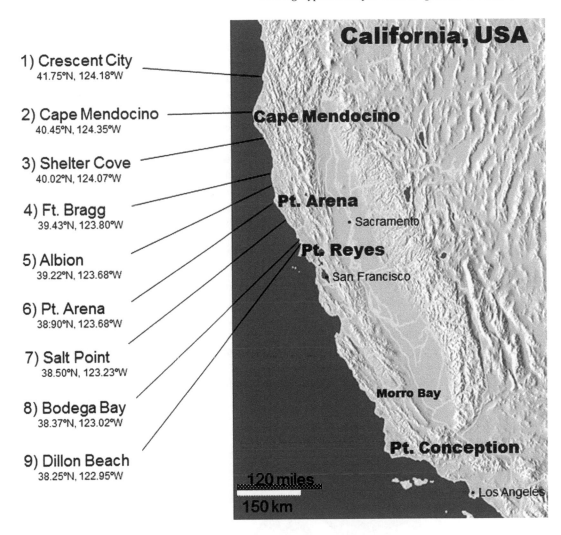

Figure 1. Northern California sampling sites for *Petrolisthes cinctipes*. Each site is numbered from north to south (1–9) and the approximate GPS location for each is listed below the site name.

monopolize fertilization, 71–100% of females produce broods with multiple sires (Toonen 2004).

The exact duration of pelagic larval development is uncertain, but feeding larvae likely spend months in the plankton before becoming competent to settle. Due to the extended pelagic larval duration, larvae have the potential to drift passively farther than 1,000 km, depending on current patterns and velocity. After larvae molt into megalopae, they settle quickly in response to a water-borne cue associated with live conspecific adults (Jensen 1989, 1991). Megalopae of *P. cinctipes* do not molt at the time of settlement, but rather lose the ability to swim due to degeneration of their pleopods; metamorphosis to the pigmented first instar juvenile occurs a week or more after settle-ment (Jensen 1991). This delay of metamorphosis after settlement makes identification of newly settled postlarvae straightforward. New recruits take advantage of the defended space beneath adults for about a year after settlement, until reaching roughly 5 mm carapace width (CW), when adults actively begin to expel them (Jensen 1991). Colonization of new habitats occurs most frequently by crabs of about 5 mm CW, presumably because they are forced to seek new shelter when expelled by adults (Jensen 1991). Additionally, maximal growth and survival of individuals occurs at intermediate densities (ca. 50 per rock), and there is both density- and size-dependent intraspecific competition for preferred feeding sites (Donahue 2004, 2006). Despite the potential motility and evidence of strong intraspecific competition of porcellanid crabs, adults are remarkably sedentary in the field. Mark-release-recapture data show that after two months, more than 50% of crabs were recovered within only a few meters of the point of release (Jensen & Armstrong 1991).

Although the chelae of *P. cinctipes* may account for nearly 50% of the body mass, these crabs are suspension feeders that appear to gain most of their nutrition from capturing diatoms and tiny zooplankton from the water column. The impressive chelae are rarely used even in defense when being harassed, and appear to serve primarily for territorial disputes and mating displays. Chelae are important determinants of mate choice in other crabs (e.g., Sainte-Marie et al. 1999), and the same is likely true of *P. cinctipes* (Jensen, pers. comm.). *Petrolisthes* readily autotomize a cheliped as a defensive strategy when threatened, making nonlethal sampling quite simple.

2.2   Sample collection and DNA extraction

We collected *Petrolisthes cinctipes* from rocky cobble around large boulders at each site shown in Figure 1. We originally sought to visit all sites from Dillon Beach, CA northward to Crescent City, CA on a single low tide series in each sampling year. However, the distance proved too great to reach on a single tidal series, and no collection could be made at Crescent City in the summer of 1997. In addition, winter storms during that year apparently killed the crab population at Dillon Beach, and which did not recover in the following years. Thus, Dillon Beach was only sampled in 1997, then replaced with Crescent City in the remaining (1998 and 1999) sampling seasons. Samples were collected during a low tide series in August of 1997, September of 1998 and July of

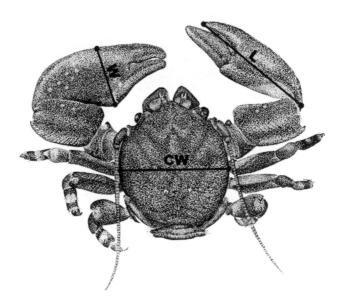

Figure 2. Measurement points for *Petrolisthes cinctipes*. Claw length (L) is measured from the tip of the nonarticulating finger of the chela to the joint with the carpus. Claw width (W) is measured across the widest part of the chela, and carapace width (CW) is measured across the widest dimension of the carapace (drawing by Josh Ferrris).

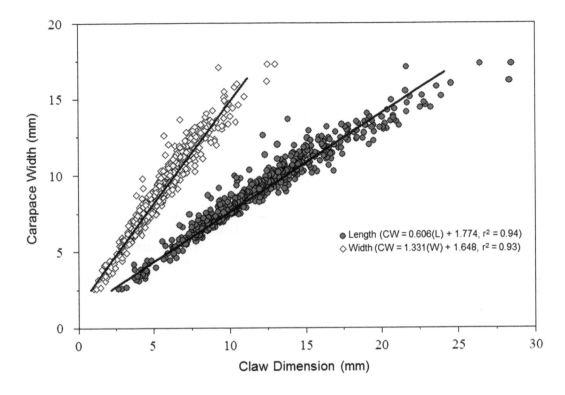

Figure 3. Carapace width as related to claw size in *Petrolisthes cinctipes*. Regression equations for carapace width (CW) based on each claw length (L) and claw width (W) were used to estimate the CW of the released animals, and to bin individuals into size classes for this study, as described in the text.

1999. At each site, boulders were selected based on the presence of dense aggregations of crabs, and the same boulder was used for each annual collection at each site. We collected all individuals encountered at a site, regardless of size, and stored the live crabs in a mesh container filled with fresh kelp (to minimize stress), until at least 50 individuals of each size class were found or the incoming tide ended the collection. After a tidal cycle, the collections were counted, and subadults and adults were individually harassed to obtain a cheliped from each, which was preserved immediately in 95% ethanol (EtOH). Live crabs were then released at the site of collection. New recruits (unpigmented postlarvae) were preserved whole in 95% ethanol. Ethanol-preserved samples were kept on ice in the field for the roughly 4–5 days required to complete the collection, after which the EtOH was decanted and replaced. Vials containing individually preserved chelipeds, from subadult and adult crabs, or whole postlarvae were subsequently stored at $-40\,^\circ$C until DNA could be extracted.

In the laboratory, each cheliped was assigned a unique identification number and when possible 50 individuals per size class from each site were selected randomly for subsequent DNA extraction and genetic analysis. If fewer than 50 individuals of each size class were collected at a site, all samples of that size class were included in the study. Each sample selected for inclusion in the study was removed from the ethanol and the length and width of the chela was measured to the nearest 0.01 mm with digital calipers (Figure 2). After dissection of 3–5 mg of muscle tissue from the claw, the remaining sample was placed back into storage as a backup, and the muscle tissue was labeled only with the six-digit unique ID number. DNA extraction, PCR amplification, electrophoresis and scoring were all done blindly using these unique crab ID numbers. Carapace width (size class) and site of collection for individual samples were reassigned to numbered samples only after all samples had been scored.

Because whole crabs were not available after being released at the site of collection, individuals were assigned to size classes (recruits, subadults, or adults) according to their claw length and claw width, which correlate very well to carapace width (Figure 3). Carapace width (CW) was estimated from the best-fit regression model using each claw width and claw length.

DNA was extracted from the dissected muscle tissue using a modification of the Gentra System PUREGENE marine invertebrate protocol. Extracted DNA pellets were rehydrated in 50 μl of low-TE (10 mM Tris, 0.1 mM EDTA) prior to being stored at $-40\,^\circ$C for subsequent PCR amplification.

2.3   PCR amplification and electrophoresis

The isolation of Simple Sequence Repeat (SSR) loci and development of primers used for amplifying these loci in *Petrolisthes cinctipes* is described in detail in Toonen (2001) using the protocol of Toonen (1997). In brief, we developed primers for 17 putative microsatellite loci, and excluded loci that 1) produced substantial stutter such that the scoring of amplicons was unreliable; 2) had significant null alleles; 3) failed Slatkin's (1994, 1996) exact test for a fit to the Ewens sampling distribution for neutral alleles; 4) failed to amplify a product from the original individual from which the microsatellite library was developed; 5) amplified more than two alleles or did not amplify a product of the expected size; or 6) exhibited a non-Mendelian pattern of inheritance (Toonen 2001). After applying these criteria for quality control (following Selkoe & Toonen 2006), we excluded 15 of the 17 loci from the analysis, and only two loci remained for this study. The two polymorphic microsatellite loci, which each have more than 30 alleles, have sufficient power for the analyses reported here (see Section 4.2 for further details). The primer sequences for amplifying the two highly polymorphic microsatellite loci used in this study, Pc156s and Pc170s, are presented in Table 1.

PCR was performed as outlined in Toonen (2001). In a final volume of 10 μl, PCR reaction mixes contained Perkin Elmer 10× Buffer II at 1× concentration, 2.5 mM MgCl$_2$, 0.1 mM dNTPs, 1× BSA, 0.5 mM of each the forward and reverse primer, 1 unit of Taq polymerase, and 1–50 ng of template DNA. PCR amplifications were performed using a touchdown protocol, beginning with an initial 5 min denaturation at 94 $^\circ$C, followed by 2 cycles with 30 s denaturation at 94 $^\circ$C, 30 s of annealing at 68 $^\circ$C, and 30 s of extension at 72 $^\circ$C, stepping down to 2 cycles of annealing at 65 $^\circ$C,

Table 1. Microsatellite loci developed for *Petrolisthes cinctipes*. Forward and reverse primer sequences (5' to 3') with the fluorescent label used, originally cloned repeat motif, the number of crabs for which the locus was scored (N), number of alleles per locus, and the observed and expected heterozygosities for each locus pooled across residents of all populations studied (see Toonen 2001 for details of development, quality control and marker selection).

Locus	Primer sequence	Repeat motif	N	No. of alleles	Heterozygosity	
					Observed	Expected
Pc156s	**F:** HEX–TTG GCT TTG AAG ACC CTG TGG **R:** CGG GGG ATC ATT GCT TTG TC	(TG)	3602	34	0.60	0.68
Pc170s	**F:** 6FAM–TGG CCG TTG CTG TTG TTG TC **R:** GGC ACC AGT CAT TCC CAG TTG	(TGT)..(TGT)..(TG)	3588	47	0.76	0.83

2 cycles of annealing at 63 °C, and then 24 cycles of annealing at 60 °C, before a final extension at 72 °C for 30 min to ensure all amplicons were +A.

Amplified PCR products were sized by gel electrophoresis on an Applied Biosystems ABI 377 XL automated sequencer and scored using the STRand analysis software (Hughes 1998), as outlined in Toonen & Hughes (2001).

2.4   Analysis of population genetic structure

Standard statistical analyses (ANOVA, posthoc comparisons among means, and linear regression) were all performed using JMP 4.0.2 (SAS Institute 2000). For genetic analyses, STRand stores allelic data generated by the ABI 377XL in the Microsoft Access relational database, which allows direct export of fragment sizes into MS Excel. Allelic data were then translated into ARLEQUIN input format using the microsatellite toolkit macro (Park 2001) for Excel. Summary statistics were all calculated using ARLEQUIN ver. 2 (Schneider et al. 2000). Global tests for deviation from expectations of Hardy-Weinberg equilibrium (HWE) were also performed using ARLEQUIN. These tests for HWE employ a Markov chain method (Guo & Thompson 1992), and chain lengths for these tests were 500,000 steps with a 10,000 step dememorization.

Weir & Cockerham's (1984) unbiased estimator of Wright's $F_{ST}(\theta)$ was calculated locus-by-locus using the Tools for Population Genetic Analyses (TFPGA) software (Miller 1997). TFPGA uses the method and terminology outlined in Weir & Cockerham (1984) and Weir (1996). Because only two loci are included in this dataset, however, bootstrapping and jackknifing are not meaningful, and therefore no mean, standard error or confidence intervals for θ are presented.

We constructed UPGMA (Unweighted Pair Group Method with Arithmetic Mean) dendrograms (Michener & Sokal 1957) using the genetic distance D_A (Nei et al. 1983), based on its superior performance in simulations (Takezaki & Nei 1996). D_A was calculated in ARLEQUIN or DISPAN (Ota 1993), and population dendrograms were drawn using DISPAN or TFPGA. Population dendrograms give an indication of the degree of genetic similarity among populations; the longer the branches separating a pair of populations, the more genetically dissimilar are the populations sampled from those geographic locations. If gene flow depends primarily on the number of dispersal steps between populations, then sites that are geographically proximate should cluster more closely than with distant sites.

Finally, we used ARLEQUIN to perform an analysis of molecular variance (AMOVA), which partitions observed genetic variance into components analogous to traditional analysis of variance (Excoffier et al. 1992; Excoffier 2000; Rousset 2000). Using the AMOVA framework within ARLEQUIN, we partitioned observed genetic variation components and calculated hierarchical F–statistic analogs (Φ) in order to test the relative contributions of year, site, and size class to the observed distribution of genetic variation within and among sampled groups. The significance of each of these values was tested by 100,000 matrix permutations in ARLEQUIN (Excoffier et al. 1992).

2.5   Assignment tests and inference of recruitment sources

As outlined above, we used both Nei's distance D_A and population dendrograms to infer which breeding population was most closely related to the sample of new recruits collected at each site. This approach assumes that genetic distance between recruits and an adult breeding population reflects the likelihood that the recruits originated from a particular source population that produced them, which may not be true in cases of sweepstakes reproduction (Hedgecock 1994). Thus, we also used Bayesian likelihood methods to assign each sampled recruit back to a source population based on its multilocus microsatellite genotype as implemented in GeneClass2 (Piry et al. 2004). Such assignment tests identify migrants, or recent descendents of migrants, through disequilbrium within multilocus genotypes (Wilson & Rannala 2003). Multilocus assignment is obviously a weak test with only two loci on which to assess disequilibrium, but this assignment testing provides an independent approach against which to compare the results based on genetic distances.

3   RESULTS

Determining how genetic variation is distributed among age classes, and among geographic locations through time can provide crucial insights into processes such as gene flow and local adaptation. Thus, in the sections that follow, we report the data analyzed by site, year, and size class to compare the relative contributions of each factor to the overall pattern of geographic population structure. Further details and the data not shown here can be found in Toonen (2001).

3.1   Carapace width measurements and size class assignment

The two preliminary collections made at Ft. Bragg and Salt Point, CA revealed that both claw size dimensions were reliable predictors of crab carapace width (Figure 3), and that male and female crabs showed a similar allometry ($r^2 = 0.99$, $n = 543$, $P < 0.001$), data not shown. Thus, we did not determine the sex of crabs in the field, and we used a single regression model for each claw length (CW = 1.774 + 0.606 L) and claw width (CW = 1.648 + 1.331 W) to estimate crab carapace width for binning into size classes from the preserved chelae (Figure 3). If the CW estimated from the claw length regression model differed from that estimated from the claw-width model by more than 1.0 mm, we excluded the sample from further analysis; if the CW estimates derived from both length and width regressions were within 1.0 mm, the CW was calculated as the average of the two estimates. Recruits were defined as nonpigmented postlarvae that must have settled within at most a few weeks prior to collection (Jensen 1991). Subadults were defined as individuals between 5.0 and 7.0 mm CW, likely corresponding to individuals at least one year, but not more than two years of age. Adults were those individuals with more than 9.0 mm CW, corresponding to individuals likely 3+ years old. We recognize that the relationship between carapace width and age is, at best, an approximation because there is considerable variation in conditions among sites and individual growth rates among locations and even within a site depending on habitat quality and individual density (Donahue 2004).

3.2   Overall patterns of microsatellite variation: all samples pooled

As noted above, Dillon Beach (site 9) was only sampled in 1997, whereas Crescent City (site 1) was only sampled in 1998 and 1999. The samples from Pt. Arena (site 6) and Bodega Bay (site 8) in 1999 were lost, and thus the overall collections included 978 crabs genotyped from the eight sites sampled in 1997, 1,372 crabs genotyped from the eight sites sampled in 1998, and 1,252 crabs genotyped from the six sites sampled in 1999. Expected heterozygosities ranged from 0.68 to 0.83 across all 3,602 crabs genotyped in this study (Table 1).

In most cases, the number of individuals of each size class scored at each site exceeded 50 and was sufficient to minimize bias in estimates of genetic structure and distance (Ruzzante 1998; Kalinowski 2005). Exclusion of those sites with fewer than 50 individuals available did not alter our conclusions (data not shown); therefore, we included all samples in our analyses. The microsatellite loci used in this study are highly polymorphic, with 34 alleles for locus Pc156s, and 47 alleles at locus Pc170s, but most alleles were rare and only three or four common alleles accounting for ≈ 80% of the scored individuals (Figure 4). Expected heterozygosities ranged from 0.68 to 0.83 across all 3,602 crabs genotyped in this study (Table 1). The overall allelic size ranges were 52–124 bp for Pc156s and 86–196 bp for Pc170s and the allele frequency distributions are presented in

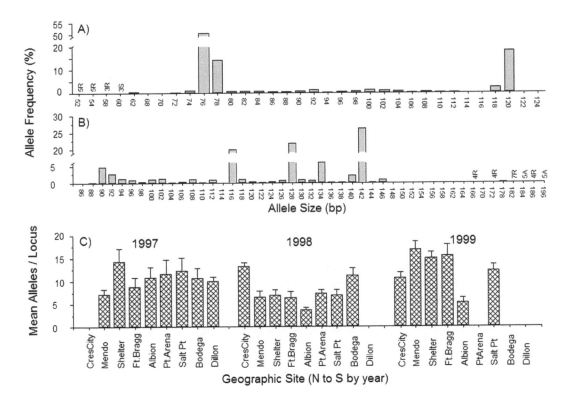

Figure 4. Allele frequency distribution for locus [A] Pc156s and [B] Pc170s from all *Petrolisthes cinctipes* individuals sampled across sites and years. Numbered bars (e.g., 5R) represent private alleles found only in site 5 (Albion) in the R (recruit) size class. The mean number of alleles per locus (error bars represent the maximum value) at each site [C] is also plotted.

Figure 4. Private alleles occurred at very low frequencies at each locus, and seven of the ten private alleles detected in this study were found among recruits, but interestingly all occurred in the central portion of the northern California range, between Shelter Cove (#3) and Salt Point (#7).

Geographic patterns of variation in the number of alleles per locus at each site were similar for both loci (ANOVA, $F = 1.12$, $df = 1$, $P > 0.05$). The mean number of alleles per locus varied from a low of three (1998, residents at Albion) to a high of 22 (1999, residents at Ft. Bragg), but differences among sites (ANOVA, $F = 1.86$, $df = 7$, $P > 0.05$) and size classes (ANOVA, $F = 0.08$, $df = 2$, $P > 0.05$) were not significant (Figure 4). Although there were no significant differences within a sampling year, the mean number of alleles per locus (13.90 ± 1.06 SE in 1997, 9.17 ± 0.95 SE in 1998, and 13.59 ± 1.12 SE in 1999) varied significantly across years (ANOVA, $F = 11.30$, $df = 2$, $P < 0.05$). Detailed locus-by-locus allele frequency distributions by site, size class, and year are available in Toonen (2001).

Because there are only two loci included in these analyses, jackknifing and bootstrapping across loci to calculate confidence intervals and standard errors for estimates of population structure would not be informative. In almost all cases the general trend revealed by both loci was concordant, and differences in estimates of population subdivision and genetic distance generated by each locus were slight (Tables 2 and 3).

3.3   Conformity to HWE among size classes, sites and years: pooled sites

When each size class, site, and year sample was examined individually, there were 25 significant deviations from Hardy-Weinberg expectations, even after correction for multiple tests (data not shown). When sites were pooled within years, however, observed heterozygosities were much closer to expectations, and data generally conformed to expectations of HWE for each size class and each year tested (Table 2). There was one significant deviation from HWE, each for locus Pc156s and Pc170s, but neither was significant after False Discovery Rate (FDR) correction for multiple tests (Benjamini & Yekutieli 2001).

3.4   Allele frequencies among sites and years: size classes pooled

As highlighted in Figure 4, three common alleles account for roughly 80% of the individuals for each locus. Because rare alleles do not contribute much to the overall estimation of F_{ST} and are difficult to visualize, we plot these three most common alleles and a binned category of "other" alleles (Figure 5) to illustrate the significant effects of sampling year across sites in annual allele frequency fluctuations (ANOVA, $F = 12.35$, $df = 7$, $P < 0.01$). In contrast to the sampling-year effect, there are no obvious patterns of variation among sites, and a similar analysis of the effects of sites across years is not significant (ANOVA, $F = 0.96$, $df = 23$, $P > 0.05$).

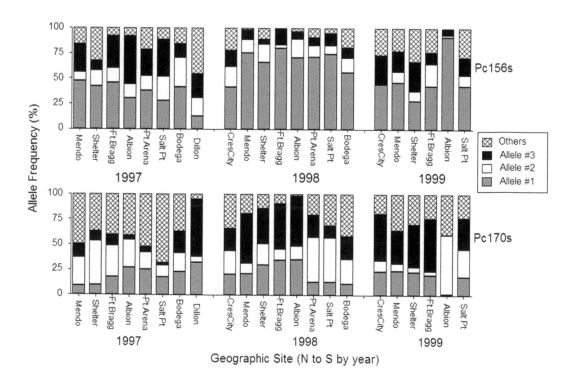

Figure 5. Allele frequency shifts in *Petrolisthes cinctipes* across sites and years. The three most common alleles (76, 78 & 120 for Pc156s; 116, 128 & 142 for Pc170s) are each plotted as Alleles #1, 2 and 3, and all other alleles are binned into the "Others" category.

Table 2. Summary of microsatellite data collected for each size class of *Petrolisthes cinctipes* pooled across sites in northern California in each of 1997–1999. The number of individual crabs genotyped from each size class at each site is presented (N) along with the observed and expected heterozygosities for each locus, and the probability value for conformity to Hardy-Weinberg Equilibrium (see text). All P values less than 0.05 are highlighted in bold, although after False Discovery Rate (FDR) correction for multiple tests none of the deviations from H-W expectation are significant.

Year	Size class	N	Heterozygosity		H-W P value $\pm$ SE
			Observed	Expected	
		Locus Pc156s			
	recruits	186	0.65	0.76	0.25 ± 0.01
1997	subadults	402	0.70	0.74	0.18 ± 0.01
	adults	390	0.54	0.76	$\mathbf{0.03 \pm 0.01}$
	recruits	518	0.51	0.58	0.11 ± 0.01
1998	subadults	428	0.51	0.57	0.35 ± 0.02
	adults	426	0.47	0.50	0.24 ± 0.01
	recruits	320	0.66	0.71	0.27 ± 0.01
1999	subadults	492	0.67	0.77	0.24 ± 0.01
	adults	440	0.65	0.78	0.11 ± 0.01
		Locus Pc170s			
	recruits	184	0.65	0.91	0.10 ± 0.01
1997	subadults	396	0.84	0.87	0.13 ± 0.01
	adults	390	0.81	0.88	0.06 ± 0.01
	recruits	516	0.74	0.78	$\mathbf{0.01 \pm 0.01}$
1998	subadults	428	0.72	0.80	0.06 ± 0.01
	adults	426	0.74	0.80	0.24 ± 0.01
	recruits	316	0.72	0.82	0.20 ± 0.01
1999	subadults	492	0.79	0.78	0.16 ± 0.01
	adults	440	0.80	0.80	0.24 ± 0.02

3.5 Genetic structure among sites and years: adult crabs only

We compared the results from this multi-year study to a previous survey of population structure across the species range which included only adults (Toonen 2001). To do this, we analyzed the adult crab size class collected in each year separately, and compared patterns that emerged from the three years of sampling in this study to that observed in the study reported in Toonen (2001). Estimates of population structure were highly consistent among adults collected in each year, with an estimated θ of 0.07 among the adults collected in 1997, 0.08 for 1998, and 0.08 for 1999. Each of these estimates differs little from that estimated in the original geographic survey across the species range, in which the estimate of θ across these same sites was 0.08, and across the entire species range was 0.11. Based on the relationship $N_e m = (1 - F_{ST})/4F_{ST}$ (Wright 1978), the 95%

confidence intervals for the effective number of migrants per generation was 1.75–3.43 for 1997, 4.39–11.34 for 1998, and 2.78–5.37 for 1999. When the overall dataset is analyzed with samples from all three years pooled, the estimate of $N_e m$ ranges from 11.80 to 24.93, which is significantly higher than any of the single-year estimates.

3.6 Genetic structure by year, site, and size class

Estimates of subdivision among recruits, subadults, and adults are all of roughly the same magnitude in each sampling year, and averaging across size classes in each year returns the consistent estimate for θ of roughly 0.08 seen in previous analyses. There are, however, some interesting trends in the data. First, recruits have significantly more structure from year-to-year at a given site than do the residents (ANOVA, $F = 7.88$, $df = 1$, $P = 0.03$). Likewise, recruits show less geographic structure among sites within any given year than do the residents (ANOVA, $F = 7.23$, $df = 1$, $P = 0.03$).

Subdivision among sites, among size classes, and among years can be estimated simultaneously using θ in the hierarchical nesting scheme of Weir (1996). For clarity, we denote the various hierarchical θ values as θ_Y for among years, θ_S for among sites, and θ_C for among size classes. The hierarchical analysis indicates that for recruits, there is far greater genetic similarity among geographic sites within a year ($\theta_S < 0.01$) than at a single site among years ($\theta_Y = 0.05$). Estimates of genetic differentiation are an order of magnitude higher among years ($\theta_Y = 0.12$) than among sites ($\theta_S = 0.01$) for recruits, whereas for adults the among-year and among-site differentiation are of approximately equal magnitude ($\theta_Y = 0.08$, $\theta_S = 0.05$; Table 3). Overall population subdivision among sites is lowest for recruits ($\theta_S = 0.01$), but the among-sites differentiation increases sequentially for subadults ($\theta_S = 0.03$) and adults ($\theta_S = 0.05$). Estimates of θ_Y among years, on the other hand, are highest for recruits ($\theta_Y = 0.12$), but the among-year differentiation decreases sequentially for subadults ($\theta_Y = 0.09$) and adults ($\theta_Y = 0.08$). It is noteworthy that in each year of sampling, there is more genetic structure among size classes than among sites (ANOVA, $F = 121$, $df = 1$, $P < 0.01$). Comparison of the relative effects of year, site, and size class in a single hierarchical analysis (Table 3) indicates that the observed population subdivision results from significant variation among size classes ($\theta_C = 0.08$) and years ($\theta_Y = 0.05$) rather than among sites ($\theta_S < 0.01$, not significant).

The analysis of molecular variance (AMOVA) is consistent with this pattern. Regardless of the grouping chosen, variation among sites explains less than 1% of the overall genetic variation (Table 4). Although only about 10% of the variation is explained by the best model (which partitions among size classes within years, and does not include the site), the result is consistent across all analyses that differences among size classes and years exceed those among geographic sites (Table 4). The pattern of significantly greater differentiation among recruits than among residents at each site is also consistent whether or not fixation indices are standardized for within population heterozygosity (Hedrick 2005; Meirmans 2006) (Table 5).

3.7 Genetic distance among populations among years

Although the exact clustering of sites and the branch lengths varied among different measures of genetic distance, results were qualitatively similar regardless of whether UPGMA or neighbor-joining (NJ) was used to construct the dendrogram, and whether the analyses used Cavalli-Sforza and Edwards' chord distance (Edwards & Cavalli-Sforza 1965; Cavalli-Sforza & Edwards 1967), Nei's D_A (Nei 1972, 1978), coancestry (Reynolds et al. 1983), or $(\delta\mu)^2$ (Goldstein et al. 1995) (data not shown). Thus, we present only the dendrograms based on Nei's D_A genetic distance here. Overall we found little consistency among sampling years in the patterns of genetic similarity within and

Table 3. Variation in genetic structure among years partitioned by size class for *Petrolisthes cinctipes* populations along the northern California coastline. The overall hierarchical analysis evaluates how the structure is partitioned among years (θ_Y), among sites (θ_S), among size classes (θ_C), and is presented below for comparison.

Locus	N	F (within total)	θ_Y (among years)	θ_S (among sites)	f (within years & sites)
			Recruits		
Pc156s	1024	0.15	0.11	0.03	0.05
Pc170s	1016	0.12	0.12	0.0005	−0.001
Overall	1016	0.13	0.12	0.01	0.02
			Subadults		
Pc156s	1322	0.22	0.11	0.04	0.02
Pc170s	1316	0.05	0.08	0.02	−0.01
Overll	1316	0.13	0.09	0.03	0.002
			Adults		
Pc156s	1256	0.10	0.09	0.06	0.12
Pc170s	1256	0.06	0.07	0.05	−0.03
Overall	1256	0.08	0.08	0.05	0.04

	Overall hierarchical analyis of all 3,588 samples for which both loci were scored				
Locus	F	θ_Y (years)	θ_S (sites)	θ_C (size classes)	f
Pc156s	0.13	0.05	−0.01	0.08	0.06
Pc170s	0.08	0.04	0.01	0.08	−0.02
Overall	0.10	0.05	−0.0001	0.08	0.02

among sites (Figures 6–8). For example, residents (adults and subadults) at Salt Point show little genetic differentiation ($D_A < 0.10$) from residents at most other northern California sites in 1997 and 1999 (Figures 6 and 8). In 1998, however, residents at Salt Point are separated from residents at all other sites sampled by the greatest genetic distance ($D_A > 0.15$) measured in that year (Figure 7). Likewise, residents at Albion in 1998 cluster in the middle of the northern California sampling sites, whereas in 1999 Albion is highly differentiated from residents at any other site sampled (Figures 7 and 8). Neither the magnitude of differentiation among samples (ranging from $D_A \approx 0.17$ in 1998 to > 0.40 in 1999), nor the pattern of clustering of sites or size classes are consistent from year to year (Figures 6–8). In addition, Bayesian assignment tests confirmed annual dispersal variability, with the population of origin inferred from Figures 6–8 being one of the top five scores for assignment of most recruits (data not shown).

4 DISCUSSION

Analysis of temporal changes in the spatial distribution of genetic variation can contribute substantially to our understanding of population dynamics and structure in marine species (e.g., Koehn et al. 1976; Todd et al. 1988; Hedgecock 1994; Johnson & Wernham 1999; Moberg & Burton 2000; Hauser et al. 2002; Flowers et al. 2002; Turner et al. 2002; Lee & Boulding 2009). Nevertheless, studies that compare the genetic composition of populations through time, or of different age classes within a population, remain the exception rather than the rule. For example, Moberg & Burton (2000) documented significant spatial and temporal differentiation among size-stratified samples of the urchin *Strongylocentrotus franciscanus*. Lee & Boulding (2007) found significant shifts in allele frequencies of the northeastern Pacific gastropod *Littorina keenae*, and estimated the effective population size (N_e) at San Pedro, CA to be only 135 individuals from 1996 to 2005, which they interpret as evidence for the sweepstakes recruitment hypothesis. In contrast, Flowers et al. (2002) showed that while some annual recruits of the urchin *Strongylocentrotus purpuratus* were differentiated from one another, there was no evidence for reduced genetic variation among those recruits; although their study had sufficient power to detect large variance in reproductive success, they found no evidence for sweepstakes reproduction and concluded that each generation resulted from a large number of successful parents.

Such debates on the cause of spatial and temporal genetic structure in marine species are longstanding. In their classic 1982 paper, Johnson & Black reported shifting, ephemeral genetic patchiness in a pulmonate limpet (*Siphonaria* sp.) found along the rocky shoreline of Western Australia. Significant genetic differences emerged among adjacent sites approximately 50 m apart, between high and low portions of the shore within those sites, between adults and recruits at each site, and between recruits to each site across replicate years of sampling. The genetic heterogeneity they report does not follow any consistent predictable pattern, such that the ephemeral genetic patchiness is "best described as chaotic" (Johnson & Black 1982). This pattern of chaotic genetic structure (defined as ephemeral spatial or temporal genetic structure that does not follow a predictable pattern)

Table 4. Hierarchical analysis of molecular variance (AMOVA) results for *Petrolisthes cinctipes* examining the relative contribution of size class, sampling year, and site. The relative proportion of genetic variation explained by years, sites, and size class are assessed. Probability of obtaining a more extreme variance component and Φ–statistic by chance alone was determined by permutation test (500,000) in ARLEQUIN, and significant values ($\alpha = 0.01$) are denoted by an asterisk.

Source of variation	df	Variance component	% of variation	Φ–statistic
Sites within years				
Among years	2	19547.953	5.29	$\Phi_{CT} = 0.05^*$
Among sites, within years	18	2660.934	0.52	$\Phi_{SC} = 0.005^*$
Within sites	7170	501213.447	94.19	$\Phi_{ST} = 0.05^*$
Size classes within sites				
Among sites	7	6837.629	0.83	$\Phi_{CT} = 0.008$
Among size classes, within sites	16	6756.074	1.31	$\Phi_{SC} = 0.01^*$
Within size classes	7170	501213.447	94.19	$\Phi_{ST} = 0.05^*$
Size classes within years				
Among years	2	19551.381	4.67	$\Phi_{CT} = 0.05^*$
Among size classes, within years	6	25504.378	5.13	$\Phi_{SC} = 0.05^*$
Within size classes	7194	478379.715	90.20	$\Phi_{ST} = 0.10^*$

turns out to be a relatively common finding among studies of marine populations (e.g., Hedgecock 1986; Watts et al. 1990; David et al. 1997; Moberg & Burton 2000; Flowers et al. 2002; Planes & Lenfant 2002; Selkoe et al. 2006), but there is little consensus on the processes that drive such chaotic patterns that are repeated, rather than accumulated, across broader spatial scales.

A growing number of studies reveal substantial temporal variation in allele frequencies, even in species that do not show significant spatial structure at any given time point (e.g., Toonen 2001; Robainas Barcia et al. 2005; Florin & Höglund 2007; Lee & Boulding 2007, 2009). At least four hypotheses have been proposed to explain such temporal variation even in the absence of strong spatial structure: 1) variation in source of larval recruits, 2) self-recruitment of local populations 3) variance in reproductive success (sweepstakes reproduction), and 4) pre- or postsettlement natural selection. Our size-structured spatial and temporal sampling allows us to evaluate the importance of each of these proposed mechanisms for creating spatially or temporally ephemeral genetic structure in *Petrolisthes cinctipes*. This study reveals roughly equivalent estimates of range wide heterogeneity from each of four replicate surveys of genetic variation in the porcelain crab *Petrolisthes cinctipes*, with an average value of $\theta = 0.08$ (Table 5). Despite this consistent significant genetic heterogeneity, some pairwise comparisons between nearby sites (e.g., Bodega Bay and Dillon Beach, only 10 km apart, pairwise $\theta = 0.16$) exceed those across most of the species range (e.g., Bodega Bay, CA to Neah Bay, WA, nearly 1200 km apart, $\theta = 0.14$). We could find no obvious geographic explanation for this distribution of genetic variation, and below we evaluate the relative role of each of the four primary hypotheses for creating chaotic genetic patchiness in *P. cinctipes*.

4.1 Pelagic larval duration and dispersal

The pelagic larval duration (PLD) of *Petrolisthes cinctipes* exceeds a month, sufficient to promote extensive larval dispersal and mixing. Nevertheless, estimates of $N_e m$ derived from overall F_{ST} values in this study (0.79 to 11.34, mean 4.22) imply that most larvae fall short of their potential. However, when the data are broken down according to size class, inferred levels of gene flow are consistent with more extensive dispersal of larvae within each year of this study. Indeed, θ_S does not significantly differ from zero for the recruit size class (Table 6).

Although several reviews of genetic structure in the sea report a significant relationship between larval dispersal potential and degree of genetic subdivision (reviewed by Waples 1987; Bohonak 1999; Shanks et al. 2003; Siegel et al. 2003; Kinlan et al. 2005), there are numerous cases in which dispersal potential only weakly predicts genetic structure (reviewed by Todd et al. 1998; Grosberg & Cunningham 2001). For example, Todd et al. (1998) reviewed species that showed significant population structure over scales far smaller or far larger than predicted based on their larval dispersal potential, and argued that some pelagic larvae are behaviorally constrained to minimize, rather than facilitate, larval transport. Likewise, Bird et al. (2007) showed that sister species of endemic Hawaiian limpets with similar ecological requirements, life histories, and larval development times show dramatically different patterns of population structure across their range. These exceptions highlight that our understanding of processes generating genetic structure among marine populations, especially the role of contemporary dispersal, is far from complete (e.g., Johnson & Black 1982; Johnson & Black 1984a, b; Watts et al. 1990; Kordos & Burton 1993; Edmands et al. 1996; Moberg & Burton 2000). In fact, several recent meta-analyses find that contrary to earlier reviews reporting a strong correlation, the overall relationship between PLD and population genetic structure is often much weaker than previously assumed (Fauvelot & Planes 2002; Bradbury & Bentzen 2007; Bradbury et al. 2008; Weersing & Toonen 2009; Shanks 2009; Ross et al. 2009; Riginos et al. 2011). Further, differences in population genetic structure among species may be caused by differences in divergence times among populations or changes in effective population sizes rather than by recent dispersal and gene flow resulting from differences in life history (Wares & Cunningham

Table 5. Variation in genetic structure among years partitioned by size class (pooled across loci and sites) for *Petrolisthes cinctipes* populations along the northern California coastline as compared to the initial geographic survey of population structure (Toonen 2001) across the same range. Genetic differentiation standardized for marker heterozygosity (ϕ'_{ST}) is calculated following Meirmans (2006).

Size class	N	F (within total)	θ (among sites)	f (within sites)	ϕ'_{ST} (among sites)
1997					
Recruits	224	0.20	0.08	0.14	0.48
Subadults	402	0.08	0.08	0.00	0.41
Adults	390	0.15	0.07	0.09	0.31
Overall	1016	0.15	0.07	0.09	0.42
1998					
Recruits	462	0.09	0.10	−0.01	0.39
Subadults	428	0.06	0.06	0.02	0.19
Adults	426	0.01	0.08	−0.08	0.25
Overall	1316	0.01	0.08	−0.08	0.25
1999					
Recruits	322	0.09	0.12	−0.03	0.51
Subadults	492	0.04	0.05	−0.01	0.22
Adults	442	0.11	0.08	0.03	0.38
Overall	1256	0.11	0.08	0.03	0.35
Initial geographic survey					
Mixed collection	776	0.08	0.08	0.04	0.33

2001; Marko 2004; Hart & Marko 2010). Together, these studies suggest that other factors, such as the timing and location of larval release, larval behavior, complex patterns of circulation, and a variety of historical effects collectively influence both the dispersal paths of larvae and the genetic structure of populations to potentially obscure any signal generated solely by the length of time larvae spend developing in the plankton.

4.2 Characterizing genetic structure

The use of multiple microsatellite loci to characterize genetic structure provides some insurance of genomic concordance, at least to the extent that each marker can be considered a single sample of the genome, and may reflect a different genealogical history due to recombination, random drift or selective forces (Selkoe & Toonen 2006). However, power to detect population structure is not always increased by the inclusion of more loci; in some cases, increasing numbers of loci can even decrease power (Ryman & Jorde 2001). This counterintuitive result can result from the large number of factors affecting power, including the number of samples, the size and equivalency of those samples, the magnitude of the true divergence, the number and type of loci assayed, the polymorphism of the markers, and allelic or haplotypic frequency distributions (Ryman et al. 2006).

The growing consensus is that, all else being equal, power is a function of the total number of alleles included in the analysis, regardless of how those alleles are distributed (Ryman & Jorde 2001; Kalinowski 2002; Ryman et al. 2006). The equivalent utility of alleles, whether distributed within or across loci, for estimating genetic distances described means that the same power can be attained using few highly polymorphic loci or many loci of low polymorphism; the only requirement is that a sufficiently large number of alleles be examined (Kalinowski 2002).

Only two of the 17 microsatellite loci developed for this project passed our quality control testing (Toonen 2001). For the sample size ($\approx$ 50) and degree of divergence ($F_{ST} = 0.08$) in this study, the power is ≈ 1.0 with greater than 10 alleles at each of two loci (Ryman et al. 2006), and for D_A, the estimate of genetic distance has a minimal variance by 32 total alleles (Kalinowski 2002). Thus, the two polymorphic microsatellite loci, each of which has more than 30 alleles, have substantial power for the analyses reported here. Additionally, initial analyses of mtDNA COI sequences of *Petrolisthes cinctipes* sampled from each of these sites revealed similar levels of genetic structure ($\Phi_{ST} = 0.12$) and patterns of differentiation to the microsatellite results reported here (Toonen, unpubl. data). The patterns and interpretation are similar regardless of whether we apply an F_{ST} correction for within-population heterozygosity (Hedrick 2005; Meirmans 2006) or use D_{est} as a measure of genetic differentiation (Jost 2008); there is far more temporal and spatial genetic heterogeneity among samples than expected. In principle, drift should be a weak force in large populations, and the effects of drift on temporal shifts in allele frequencies should be small (but see Whitlock & McCauley 1999); hence, a single sampling of a population at drift/migration equilibrium is sufficient to characterize allele frequencies in that population. In practice, however, although the census size of many marine populations may be vast, the effective population size, N_e, which determines the strength of drift in those populations, may be orders of magnitude smaller (e.g., Hedgecock 1994; Turner et al. 2002; reviewed by Hedgecock et al. 2007). In the case of *P. cinctipes*, the census size is estimated to be on the order of 10^7 to 10^8 across the species range (Toonen 2001), and even if N_e were three orders of magnitude smaller than the census size, the effective population size would still be sufficient to make drift a weak force. The number of alleles per locus and the prevalence of rare alleles reported herein are both consistent with a large population size (Hartl & Clark 1997). The expected temporal consistency from such population size estimates is not supported, however; there is substantial temporal variation among years within sites (Figure 5). These results beg the question: why is there so much temporal variation?

4.3   Variation in the source of larval recruits

Across the sites we surveyed for this study, the source of recruits varies unpredictably from year-to-year (Figures 6–8). Although assignment tests have little power with so few loci, and assign individuals with confidence into source populations, the results agree with the genetic distance approach (data not shown). A distribution of the most common source populations to which each recruit is assigned is generally concordant with the source populations inferred from the genetic distance figures. Although neither genetic distance, nor assignment tests are entirely convincing here, the concordance of the two different approaches lends some confidence to the conclusion that the source of recruits varies among sites and from year to year.

Interestingly, the major changes to the source populations from which larvae originate each year may reflect changes in currents and coastal upwelling driven by the El Niño Southern Oscillation (ENSO) (Botsford et al. 1994; Wing et al. 1995a, b; Wing et al. 1998; Morgan et al. 2000). For example, the period from 1997 to 1999 coincided with a major ENSO event. The El Niño of 1997 resulted in the weakest upwelling recorded to date in the 54-year database spanning 1946 to 1999 (National Oceanic and Atmospheric Administration, Pacific Fisheries Environmental Group (NOAA, PFEG); www.pfeg.noaa.gov). The El Niño transition in 1998 showed an upwelling index

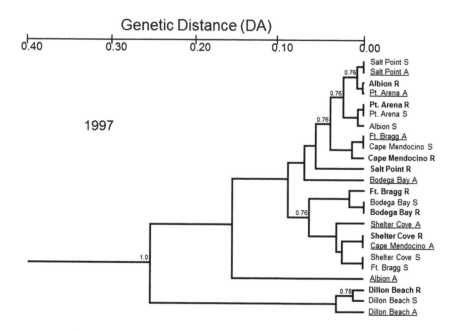

Figure 6. Genetic similarity dendrogram for size classes in *Petrolisthes cinctipes* from each geographic site sampled in 1997, based on the DA genetic distance of Nei. Sites of collection are named, and recruits sampled from each site are denoted by R in bold (**Site R**), adults denoted by A are underlined (<u>Site A</u>) and subadults denoted by S are presented in normal text (Site S). Values from 10,000 bootstraps that are ≥ 0.75 are presented next to the nodes.

that was not significantly different than the rest of the 54-year average (Diehl et al. 2007), although most of the preceding 21 years show anomalously low upwelling due to the warm phase of the Pacific Decadal Oscillation (NOAA PFEG). The 1999 La Niña, in contrast, had stronger and more consistent upwelling than either of the previous years. General predictions of the effect of wind-driven currents on crustacean recruitment throughout California are reviewed in Diehl et al. (2007), but essentially 1997 and the spawning season of 1998 were El Niño years (characterized by weak upwelling and warmer sea water temperatures), whereas in 1999 there was a La Niña event (dominated by unusually low water temperatures and strong upwelling). Thus, there should be a major shift in the likelihood of larval transport alongshore between the El Niño events in 1997 and 1998 (in which frequent relaxation events should bring larvae up the coast from the south) and the consistent upwelling of the La Niña event in 1999 (which should bring larvae down from the north). Despite the unpredictability of the annual source of recruits, our data support the El Niño/La Niña prediction; the D_A trees from 1997 and 1998 do not exhibit cluster of recruits with a single population of adults to the north of the site at which recruits were collected. In striking contrast, none of the recruit samples collected in 1999 cluster with adult populations located to the south, and three of the five cluster with populations to the north. Assignment tests support similar conclusions, with recruits generally assigning as self-recruits or assigning to populations from the south during the El Niño events from 1997 to 1998, and assigning as self-recruits or to populations to the north during the La Niña in 1999. Significant ENSO-induced shifts in population genetic structure have been reported in a variety of other taxa, including tropical butterflies (Fauvelot et al. 2006; Cleary et al. 2006), wild tomato (Sifres et al. 2007), and kelp forest gastropods (Ettinger-Epstein & Kingsford 2008).

There are several different mechanisms by which variation in the source of larval recruits could produce fine-scale chaotic genetic patchiness. First, a localized Wahlund effect, the departure from

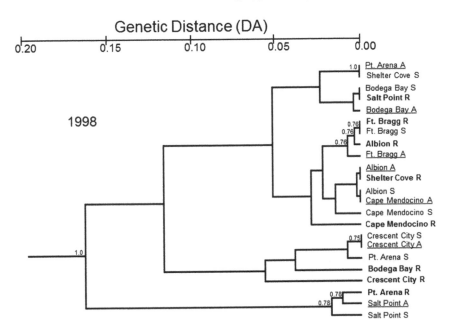

Figure 7. Genetic similarity dendrogram for size classes in *Petrolisthes cinctipes* from each geographic site sampled in 1998, based on the DA genetic distance of Nei. Sites of collection are named, and recruits sampled from each site are denoted by R in bold (**Site R**), adults denoted by A are underlined (Site A) and subadults denoted by S are presented in normal text (Site S). Values from 10,000 bootstraps that are ≥ 0.75 are presented next to the nodes.

Hardy-Weinberg equilibrium due to mixing of individuals from breeding groups with different allelic frequencies, could account for heterozygote deficiencies on small scales in some species (e.g., Johnson & Black 1984b; Kijima et al. 1987; Waples 1990). In the case of the limpet *Siphonaria*, there is little geographic variation in allozyme frequencies at the scale of the species range (Johnson & Black 1982; Johnson & Black 1984b). Despite this large scale uniformity, however, populations less than 50 m apart exhibit fine-scale genetic patchiness, which is repeated, rather than accumulated, on the larger scale (Johnson & Black 1984b). Neither geographic, nor temporal variation in allelic frequencies can explain the pattern. Instead, the fine-scale structure appears to derive from binomial sampling variance among small local breeding groups that result in a localized Wahlund effect (Johnson & Black 1984a, b). This hypothesis predicts that there should be a significant deficiency of heterozygotes across sites that disappears as the scale of sampling is reduced to the point that samples are drawn from only the actual interbreeding group. However, we see no consistent evidence for a Wahlund effect on *Petrolisthes cinctipes* at any scale in our data.

Alternatively, recruits at a given location could originate from genetically distinct source populations, with the identity of the source population varying through time. Differences in seasonal current patterns, spawning season or developmental times among geographic populations may result in temporal variation in the allelic frequencies of recruits arriving at a given site (e.g., Kordos & Burton 1993). Such genetic variation may accumulate within populations and lead to small-scale genetic structure if interpopulation gene flow is not sufficient to overcome population differentiation resulting from genetic drift or natural selection (e.g., Koehn & Williams 1978; Kordos & Burton 1993). The larval source hypothesis predicts that the genetic affinity of recruits to the geographic populations of spawners should vary with time and location, but further requires that adult populations exhibit sufficient differentiation to account for the variation detected among recruits. Temporal

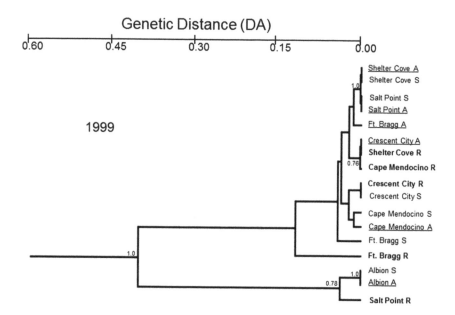

Figure 8. Genetic similarity dendrogram for size classes in *Petrolisthes cinctipes* from each geographic site sampled in 1999, based on the DA genetic distance of Nei. Sites of collection are named, and recruits sampled from each site are denoted by R in bold (**Site R**), adults denoted by A are underlined (Site A) and subadults denoted by S are presented in normal text (Site S). Values from 10,000 bootstraps that are ≥ 0.75 are presented next to the nodes.

variance could also be generated if local recruits originated from genetically differentiated, unsampled source populations (e.g., Kordos & Burton 1993). In the case of *P. cinctipes*, however, the initial geographic survey of population structure across the entire species range revealed spatial differentiation that was only slightly greater ($\theta = 0.11$) than within the smaller geographic area ($\theta = 0.08$) sampled in the present study (Toonen 2001). Additionally, the greatest pairwise values of θ occurred within the geographic area included here, and all private alleles detected across both loci in *P. cinctipes* were found within this central portion of the range (Toonen 2001). Thus, as with the localized Wahlund effect above, the genetic distinctiveness of breeding populations and variation in larval source do not fully explain the observed pattern of differentiation among sites and size classes through time (Table 6).

4.4   Self-recruitment and local subdivision

Data are sometimes consistent with self-recruitment and local retention of larvae (e.g., Dillon Beach, Figure 6). However, the relationship of recruits to potential breeding populations varies among years, such that no site shows a consistent source-sink relationship with any other. When and where it occurs, local retention may be the outcome of retentive gyres formed by the upwelling-relaxation flow mechanism along the West coast of North America (Botsford et al. 1994; Wing et al. 1995 a, b; Wing et al. 1998; Morgan et al. 2000). This upwelling-relaxation hypothesis predicts differential delivery of larvae to sites with increasing distance from a headland because larvae are retained in transient gyres formed by promontories interrupting southward flow during upwelling (Wing et al. 1995a, b). The resulting differential delivery of larvae to sites within embayments between promontories (labeled points on Figure 1) should result in frequent and consistent settlement directly to the north of a headland, while settlement of larvae at sites farther north will be more sporadic and unpredictable, and transport to the coast south of the headland ought to be minimal.

As outlined above, the change in recruit source does generally correspond to the frequency and intensity of ENSO-driven upwelling, but the pattern of similarity among sites does not conform to predictions of the upwelling-relaxation hypothesis. In all years, some sites between headlands that should be well-mixed are more different from one another than sites across headlands that should be isolated from one another (Toonen 2001). For example, in 1997, although recruits appear to be either locally-derived or arrive from the south, they also appear to originate consistently from residents of a different embayment rather than the same embayment. In each year of sampling, the pattern of recruitment differs, but the tendency for recruits to cluster with adults from beyond the region of entrainment predicted by the upwelling-relaxation hypothesis is consistent (Figures 6–8). The data presented herein argue that the effect of these oceanic retention zones is variable by year. Nevertheless, Moberg & Burton (2000) emphasize that data such as these do not constitute a strong test of the upwelling-relaxation flow hypothesis, because the effects of retention zones on genetic structure may only apply at the temporal scale of individual relaxation events, and not over the scale of seasons or years. Regardless, there is little evidence for consistent self-recruitment, and this mechanism alone is insufficient to explain the patterns of spatial and temporal genetic structure reported here.

4.5   Variance in individual reproductive success

In highly fecund species, a small minority of individuals may effectively contribute all recruits for the entire population in each generation by a sweepstakes chance matching of reproductive activity with oceanographic conditions conducive to spawning, fertilization, larval survival and successful recruitment (Hedgecock 1986, 1994). The sweepstakes-reproductive success hypothesis can account for both the widespread, large discrepancies commonly found between effective and census population estimates, as well as for local genetic differentiation in the face of apparently high gene flow (reviewed by Hedgecock 1994; Hedgecock et al. 2007). The sweepstakes hypothesis predicts that, because only a few individuals win the reproductive lottery and provide the majority of recruits for the next generation, allelic frequencies should exhibit significant temporal variation, and that genetic diversity should be substantially lower and temporally variable among those recruits relative to a random sample of the potential spawning population (e.g., Edmands et al. 1996; Li & Hedgecock 1998; Moberg & Burton 2000; Flowers et al. 2002; Planes & Lenfant 2002; Turner et al. 2002; Waples 2002; Selkoe et al. 2006; Lee & Boulding 2009; Buston et al. 2009; Christie et al. 2010). The sweepstakes reproduction hypothesis requires high fecundity (10^6–10^8 eggs/female/reproductive season) and correspondingly high larval mortality, life-history attributes typical of many marine animals (reviewed by Hedgecock et al. 2007). From this perspective, *P. cinctipes* is a low-fecundity species, in which the output of a few individuals clearly cannot contribute to the majority of recruits. Thus, the significant temporal and spatial variation in allele frequencies is difficult to explain by sweepstakes reproduction alone.

Nevertheless, the fact that there are significant temporal fluctuations in allelic frequencies and numbers of alleles in the batches of recruits (Toonen 2001), and that the greatest differentiation occurs among annual batches of recruits (Table 6) argues that successful recruits are not a random draw from the entire potential breeding population each year. In all comparisons, recruits at each site were the most differentiated samples between years, whereas adults at each site showed the lowest differentiation from one year to the next. Furthermore, when we compare each batch of recruits to the most closely-related resident population (Figures 6–8), the heterozygosity of recruits is significantly reduced relative to that of the residents (paired t-tests, $P < 0.05$). In contrast, if recruits are pooled across years and sites, the heterozygosity of recruits is slightly higher than that of

Table 6. Summary of population structure observed for *Petrolisthes cinctipes* by size class among years and sites. Arrows illustrate how the among-sites component of variation increases from nonsignificant among recruits through subadults and is greatest in adults. The converse is true for the among-years component of variation, which is greatest among the recruits and decreases sequentially through subadults, and is least among adults. ns denotes values that do not differ significantly from zero.

N	F (within total)	θ_Y (among years)	θ_S (among sites)	f (within sites)
		Recruits		
1016	0.13	0.12	$< 0.01^{ns}$	0.02
		Subadults		
1316	0.13	0.09	0.03	$< 0.01^{ns}$
		Adults		
1256	0.08	0.08	0.05	0.04

pooled resident populations, although this is not significant (paired t–test, $P < 0.05$). The simulations of Turner et al. (2002) demonstrate that variance in reproductive success among different local populations can be a potent force in reducing effective population size relative to the census population. Although fecundity of *P. cinctipes* is too low for the sweepstakes reproduction hypothesis to explain patterns of allelic variability, our results are consistent with both individual and population variation in adult reproductive success leading to reduced effective population size and contributing to temporal genetic structure in this species.

4.6   Pre- or postsettlement natural selection

Finally, variable small-scale patterns of population genetic structure may result from pre- or postsettlement natural selection within and among populations (e.g., Koehn et al. 1976; Powers & Place 1978; Johnson & Black 1984a; Hilbish & Koehn 1985; Gardner & Palmer 1998). Despite geographic proximity of sites, abiotic (e.g., wave exposure, salinity, nutrient supply, etc.) and biotic (e.g., competitors and predators) differences among sites can exert different selective pressures over very small spatial scales. Although we routinely assume selective neutrality of the genetic markers used in surveys of population genetic structure such as this, markers may be linked or directly subject to natural selection that can either alter or maintain the allelic frequencies among samples obtained from a given site (e.g., Koehn et al. 1976; Powers & Place 1978; Koehn et al. 1976; Hilbish & Koehn 1985; Karl & Avise 1992; McGoldrick et al. 2000; Rand et al. 2002; Schmidt et al. 2008), or even among size classes or tidal heights within a site (e.g., Johannesson & Johannesson 1989; Gardner & Palmer 1998; Schmidt et al. 2000; Schmidt & Rand 2001; Wilding et al. 2001; Dufresne et al. 2002; Grahame et al. 2006). In general, when putatively neutral genetic markers reveal clinal or habitat-specific patterns of allelic frequency change (e.g., Koehn et al. 1976; Powers & Place 1978; Hilbish & Koehn 1985; Schmidt & Rand 2001; reviewed by Sotka & Palumbi 2006), or marked differences among loci or classes of genetic markers (e.g., Reeb & Avise 1990; Karl & Avise 1992; Schmidt & Rand 2001), a reasonable working hypothesis is that natural selection is somehow responsible. Even with the simplest patterns, the ecological and genetic studies required to test this hypothesis are not trivial; with more complex spatial patterns, isolating the specific causes and effects of natural selection can be even more difficult (e.g., Gardner & Palmer 1998; Schmidt et al. 2000; Schmidt & Rand 2001; Rand et al. 2002; Dufresne et al. 2002; Schmidt et al. 2008).

Using Slatkin's exact test (1994, 1996) to compare our observed allele-frequency distributions to neutral expectations, we find no evidence of selection causing a departure of frequency distributions from the Ewens sampling distribution at either of these two loci (Toonen 2001). Despite the lack of direct evidence for selection, it is difficult to envision a neutral mechanism by which recruits show no significant spatial structure in each year of sampling, but spatial genetic differentiation accumulates sequentially in older age classes to a consistent value in each of four replicate studies of the adults (Tables 5 and 6). The probability of randomly selecting two loci under similar selection pressure is small, and populations show significant among-year variation in both number of alleles and the frequencies of alleles within sites (Figures 4 and 5). Furthermore, the ENSO from 1997 to 1999 event was one of the most extreme recorded to date, and the coastal environmental and nearshore oceanography of 1997 and 1999 ought to be maximally different during this study. The genetic patterns we observed do not match that expectation, however, with patterns in 1997 and 1999 being more similar to each other than either is to 1998. Thus, selection alone does not obviously explain the overall results of this study, despite the fact that the pattern of sequentially increasing population structure with age class suggests that postsettlement selection is important.

4.7 Concordant changes and conclusions

The F_{ST} estimates drawn from each of our four independent surveys of *P. cinctipes* are highly consistent (Table 5), but could be misleading relative to the picture drawn from our overall analysis of temporal and spatial sampling of age-structured samples. In sum, these data provide a rare look into the detailed patterns of population genetic structure across space and time in a replicated study structured explicitly by size class and location through time. The spatially and temporally stratified size-specific sampling here provides a much richer interpretation of the relative role of different mechanisms in structuring these populations relative to the standard single-time snapshot survey of population genetic structure; we could not evaluate the relative roles of each of these mechanisms without the dataset we present here. Just as importantly, these findings emphasize that even replicate studies reporting the same overall magnitude of population genetic structure may overlook the underlying variability revealed by a temporally- and spatially-structured analysis.

Our study provides qualified evidence that each of the four primary hypothesized mechanisms may be affecting fine-scale population genetic structure in *P. cinctipes*, but none of the mechanisms alone appears sufficient to explain the overall pattern of population genetic structure among sites, years, and age classes. Thus, while these data are consistent with the action of each of the potential mechanisms, the combined effect of multiple interacting mechanisms, rather than any single process in isolation, is most likely to produce the genetic structure evident in *P. cinctipes*. Until more studies are conducted that simultaneously examine spatial and temporal dynamics of genetic structure, it remains to be seen whether this is a general feature of marine organisms with pelagic dispersal. The question remains whether genetic structure is an idiosyncratic attribute of a particular species, or whether particular combinations of life history, geography, and oceanography elicit predictable genetic patterns that can be generalized among taxa.

ACKNOWLEDGEMENTS

Thanks to C. Fong, K. Batchelor, G. Jensen, P. Jensen, M. Locke, J. Diehl, T. Mai and M. Cahill for assistance with the collections and laboratory work on this project. B. Cameron, M. Hellberg, K. Summers, D. Carlon, J. Diehl, S. Gilman, E. Pearson, L. Borghesi, S. Craig, E. Duffy, D. Levitan,

O. Ellers, M. Frey, M. Hart, P. Marko, N. Tsutsui, J. Wares and A. Wilson all provided technical assistance, moral support, and feedback in the lab and with the data analysis and interpretation at some point during this project. The parade of collaborators, Gros-docs and graduate students that have contributed to this work is substantial, but D. Ayre, L. Botsford, R. Burton, D. Hedgecock, M. Hellberg, J. Stachowicz, J. Gillespie, M. Turelli, C. Langley, M. Donahue, S. Morgan, C. Moritz, J. Neigel, D. Jablonski, E. Pearson, S. Wing, P. Jensen, G. Jensen and J. Stillman all stand out for their comments, suggestions, feedback, and fruitful discussion over the years. Special thanks are due to A. Wilson, J. Wares, P. Sunnucks, M. Stanton, R. Burton, K. Selkoe, B. Bowen and S. Karl for their extensive discussions regarding the data analyses and interpretation. Z. Forsman, C. Bird, K. Selkoe, J. Neigel and two anonymous referees provided much appreciated reviews of the manuscript. In the course of this research, RJT was supported in part by a Natural Sciences & Engineering Research Council of Canada Post-Graduate Scholarship B, a Province of Alberta Sir James Lougheed Award of Distinction, and a National Science Foundation grant OCE99-06741 to RKG. Additional support came from Jastro-Shields, Center for Population Biology, and Graduate Division fellowships to RJT while at the University of California, Davis. The remainder of the funding came from the Andrew Mellon Foundation, a grant from the National Sea Grant College Program, National Oceanic & Atmospheric Administration, U.S. Department of Commerce, under grant number NA06RG0142, project number R/F-177 through the California Sea Grant College System, and NSF award OCE06-23678. This is contribution #1399 from the Hawaii Institute of Marine Biology.

REFERENCES

Akins, L.J. 2003. *The relative effects of top-down and bottom-up processes and abiotic factors on the abundance of shore crabs around a headland.* Masters thesis, University of California, Davis, CA.

Almany, G.R., Berumen, M.L., Thorrold, S.R., Planes, S. & Jones G.P. 2007. Local replenishment of coral reef fish populations in a marine reserve. *Science* 316: 742–744.

Ayre, D.J. 1990. Population subdivision in Australian temperate marine invertebrates: larval connections versus historical factors. *Aust. J. Ecol.* 15: 403–411.

Benjamini, Y. & Yekutieli, D. 2001. The control of the false discovery rate in multiple testing under dependency. *Ann. Stat.* 29: 1165–1188.

Berger, E.M. 1973. Gene-enzyme variation in three sympatric species of *Littorina. Biol. Bull.* 145: 83–90.

Bird, C.E., Holland, B.S., Bowen, B.W. & Toonen, R.J. 2007. Contrasting phylogeography in three endemic Hawaiian limpets (*Cellana* spp.) with similar life histories. *Mol. Ecol.* 16: 3173–3186.

Bohonak, A.J. 1999. Dispersal, gene flow, and population structure. *Q. Rev. Biol.* 74: 21–45.

Botsford, L.W., Moloney, C.L., Hastings, A., Largier, J.L., Powell, T.M., Higgins, K. & Quinn, J.F. 1994. The influence of spatially and temporally varying oceanographic conditions on meroplanktonic metapopulations. *Deep-Sea Res. II* 41: 107–145.

Bradbury I.R. & Bentzen, P. 2007. Non-linear genetic isolation by distance: implications for dispersal estimation in anadromous and marine fish populations. *Mar. Ecol. Prog. Ser.* 340: 245–257.

Bradbury, I.R., Laurel, B., Snelgrove, P.V.R., Bentzen, P. & Campana, S.E. 2008. Global patterns in marine dispersal estimates: the influence of geography, taxonomic category and life history. *Proc. R. Soc. Lond. B* 275: 1803–1809.

Burton, R.S. & Feldman, M.W. 1981. Population genetics of *Tigriopus californicus*. II. Differentiation among neighboring populations. *Evolution* 35: 1192–1205.

Burton, R.S. 1997. Genetic evidence for long term persistence of marine invertebrate populations in an ephemeral environment. *Evolution* 51: 993–998.

Burton, R.S., Rawson, P.D. & Edmands, S. 1999. Genetic architecture of physiological phenotypes: empirical evidence for coadapted gene complexes. *Amer. Zool.* 39: 451–462.

Buston, P.M., Fauvelot, C., Wong, M.Y.L. & Planes, S. 2009. Genetic relatedness in groups of the humbug damselfish *Dascyllus aruanus*: small, similar-sized individuals may be close kin. *Mol. Ecol.* 18: 4707–4715.

Carlon, D.B., & Olson, R.R. 1993. Larval dispersal distance as an explanation for adult spatial pattern in two Caribbean reef corals. *J. Exp. Mar. Biol. Ecol.* 173: 247–263.

Cavalli-Sforza, L.L. & Edwards, A.W.F. 1967. Phylogenetic analysis: models and estimation procedures. *Evolution* 21: 550–570.

Christie, M.R., Johnson, D.W., Stallings, C.D. & Hixon, M.A. 2010. Self-recruitment and sweepstakes reproduction amid extensive gene flow in a coral-reef fish. *Mol. Ecol.* 19: 1042–1057.

Cleary, D.F.R., Fauvelot, C., Genner, M.J., Menken, S.B.J. & Mooers, A.Ø. 2006. Parallel responses of species and genetic diversity to El Niño Southern Oscillation-induced environmental destruction. *Ecol. Lett.* 9: 304–310.

Cowen, R.K. & Sponaugle, S. 2009. Larval dispersal and marine population connectivity. *Ann. Rev. Mar. Sci.* 1: 443–466.

David P., Perdieu, M.-A., Pernot, A.-F. & Jarne, P. 1997. Fine-grained spatial and temporal population genetic structure in the marine bivalve *Spisula ovalis*. *Evolution* 51: 1318–1322.

Diehl, J.M., Toonen, R.J. & Botsford, L.W. 2007. Spatial variability of recruitment in the sand crab *Emerita analoga* throughout California in relation to wind-driven currents. *Mar. Ecol. Prog. Ser.* 350: 1–17.

Donahue, M.J. 2006. Conspecific cueing and growth-mortality tradeoffs jointly lead to conspecific attraction. *Oecologia* 149: 33–43.

Donahue, M.J. 2004. Size-dependent competition in a gregarious porcelain crab *Petrolisthes cinctipes* (Anomura: Porcellanidae). *Mar. Ecol. Prog. Ser.* 267: 219–231.

Dufresne, F., Bourget, E. & Bernatchez, L. 2002. Differential patterns of spatial divergence in microsatellite and allozyme alleles: further evidence for locus-specific selection in the acorn barnacle, *Semibalanus balanoides*? *Mol. Ecol.* 11: 113–123.

Edmands, S., Moberg, P.E. & Burton, R.S. 1996. Allozyme and mitochondrial DNA evidence of population subdivision in the purple sea urchin *Strongylocentrotus purpuratus*. *Mar. Biol.* 126: 443–450.

Edmands, S. & Harrison, J.S. 2003. Molecular and quantitative trait variation within and among populations of the intertidal copepod *Tigriopus californicus*. *Evolution* 57: 2277–2285.

Edwards, A.W.F. & Cavalli-Sforza, L.L. 1965. A method for cluster analysis. *Biometrics* 21: 362–375.

Ettinger-Epstein, P. & Kingsford M.J. 2008. Effects of the El Niño southern oscillation on *Turbo torquatus* (Gastropoda) and their kelp habitat. *Austral Ecol.* 33: 594–606.

Excoffier, L. 2000. Analysis of population subdivision. In: Balding, D., Bishop, M. & Cannings, C. (eds.), *Handbook of Statistical Genetics*: 271–308. New York, NY: John Wiley & Sons, Ltd.

Excoffier, L., Smouse, P.E. & Quattro, J.M. 1992. Analysis of molecular variance inferred from metric distances among DNA haplotypes: application to human mitochondrial DNA restriction data. *Genetics* 131: 479–491.

Fauvelot, C. & Planes, S. 2002. Understanding origins of present-day genetic structure in marine fish: biologically or historically driven patterns? *Mar. Biol.* 141: 773–788.

Fauvelot, C., Cleary, D.F.R. & Menken, S.B.J. 2006. Short term impact of 1997/98 ENSO-induced disturbance on abundance and genetic variation in a tropical butterfly. *J. Hered.* 97: 367–380.

Florin, A.-B. & Höglund, J. 2007. Absence of population structure of turbot (*Psetta maxima*) in the Baltic Sea. *Mol. Ecol.* 16: 115–126.

Flowers, J.M., Schroeter, S.C. & Burton, R.S. 2002. The recruitment sweepstakes has many winners: genetic evidence from the sea urchin *Strongylocentrotus purpuratus*. *Evolution* 56: 1445–1453.

Futuyma, D.J. 1997. *Evolutionary Biology. 3rd ed.* Sunderland, MA: Sinauer Associates, Inc.

Garant, D., Dodson, J.J. & Bernatchez, L. 2000. Ecological determinants and temporal stability of the within-river population structure in Atlantic salmon (*Salmo salar* L.). *Mol. Ecol.* 9: 615–628.

Gardner, J.P.A. & Palmer, N.L. 1998. Size-dependent, spatial and temporal genetic variation at a leucine aminopeptidase (LAP) locus among blue mussel (*Mytilus galloprovincialis*) populations along a salinity gradient. *Mar. Biol.* 132: 275–281.

Goldstein, D.B., Ruiz Linares, A., Cavalli-Sforza, L.L. & Feldman, M.W. 1995. Genetic absolute dating based on microsatellites and the origin of modern humans. *Proc. Nat. Acad. Sci. USA* 92: 6723–6727.

Grahame, J.W., Wilding, C.S. & Butlin, R.K. 2006. Adaptation to a steep environmental gradient and an associated barrier to gene exchange in *Littorina saxatilis*. *Evolution* 60: 268–278.

Grosberg, R.K. & Cunningham, C.W. 2001. Genetic structure in the Sea: from populations to communites. In: Bertness, M.D., Gaines, S.D. & Hay, M.E. (eds.), *Marine Community Ecology*: 61–84. Sunderland, MA: Sinauer Associates, Inc.

Guo, S.W. & Thompson, E.A. 1992. Performing the exact test of Hardy-Weinberg proportions for multiple alleles. *Biometrics* 48: 361–372.

Hart, M.W. & Marko, P.B. 2010. It's about time: divergence, demography, and the evolution of developmental modes in marine invertebrates. *Integr. Comp. Biol.* 50: 643–661.

Hartl, D.L. & Clark, A.G. 1997. *Principles of Population Genetics.* Sunderland, MA: Sinauer Associates, Inc.

Hauser, L., Adcock, G.J., Smith, P.J., Bernal Ramírez, J.H. & Carvalho, G.R. 2002. Loss of microsatellite diversity and low effective population size in an overexploited population of New Zealand snapper (*Pagrus auratus*). *Proc. Nat. Acad. Sci. USA* 99: 11742–11747.

Hedgecock, D. 1986. Is gene flow from pelagic larval dispersal important in the adaptation and evolution of marine invertebrates? *Bull. Mar. Sci.* 39: 550–564.

Hedgecock, D. 1994. Does variance in reproductive success limit effective population sizes of marine organisms? In: Beaumont, A.R. (ed.), *Genetics and Evolution of Aquatic Organisms*: 122–135. London: Chapman & Hall.

Hedgecock, D., Barber, P.H. & Edmands, S. 2007. Genetic approaches to measuring connectivity. *Oceanogr.* 20: 70–79.

Hedrick, P.W. 2005. A standardized genetic differentiation measure. *Evolution* 59: 1633–1638.

Hellberg, M.E., Burton, R.S., Neigel, J.E. & Palumbi, S.R. 2002. Genetic assessment of connectivity among marine populations. *Bull. Mar. Sci.* 70 (Suppl. 1): 273–290.

Hellberg, M.E. 2009. Gene flow and isolation among populations of marine animals. *Annu. Rev. Ecol. Evol. Syst.* 40: 291–310.

Hilbish, T.J. & Koehn, R.K. 1985. Exclusion of the role of secondary contact in an allele frequency cline in the mussel *Mytilus edulis*. *Evolution* 39: 432–443.

Hughes, S. 1998. STRand Nucleic Acid Analysis Software. University of California, Davis, CA. http://www.vgl.ucdavis.edu/STRand

Jensen, G.C. 1989. Gregarious settlement by megalopae of the porcelain crabs *Petrolisthes cinctipes* (Randall) and *P. eriomerus* Stimpson. *J. Exp. Mar. Biol. Ecol.* 131: 223–231.

Jensen, G.C. 1990. *Intertidal zonation of porcelain crabs: resource partitioning and the role of selective settlement.* Ph.D. dissertation, University of Washington, Seattle, WA.

Jensen, G.C. 1991. Competency, settling behavior, and postsettlement aggregation by porcelain crab megalopae (Anomura: Porcellanidae). *J. Exp. Mar. Biol. Ecol.* 153: 49–61.

Jensen, G.C. & Armstrong, D.A. 1991. Intertidal zonation among congeners: factors regulating distribution of porcelain crabs *Petrolisthes* spp. (Anomura: Porcellanidae). *Mar. Ecol. Prog. Ser.* 73: 47–60.

Johannesson, K. & Johannesson, B. 1989. Differences in allele frequencies of AAT between high- and mid-rocky shore populations of *Littorina saxatilis* (Olivi) suggest selection in this enzyme locus. *Genetical Res.* 54: 7–12.

Johnson, M.S. & Black, R. 1982. Chaotic genetic patchiness in an intertidal limpet, *Siphonaria* sp. *Mar. Biol.* 70: 157–164.

Johnson, M.S. & Black, R. 1984a. Pattern beneath the chaos: the effect of recruitment on genetic patchiness in an intertidal limpet. *Evolution* 38: 1371–1383.

Johnson, M.S. & Black, R. 1984b. The Wahlund effect and the geographical scale of variation in the intertidal limpet *Siphonaria* sp. Mar. Biol. 79: 295–302.

Johnson, M.S. & Wernham, J. 1999. Temporal variation of recruits as a basis of ephemeral genetic heterogeneity in the western rock lobster *Panulirus cygnus*. *Mar. Biol.* 135: 133–139.

Jones, G.P., Milicich, M.J., Emslie, M.J. & Lunow, C. 1999. Self-recruitment in a coral reef fish population. *Nature* 402: 802–804.

Jones, G P., Planes, S. & Thorrold, S.R. 2005. Coral reef fish larvae settle close to home. *Curr. Biol.* 15: 1314–1318.

Jost, L. 2008. G_{ST} and its relatives do not measure differentiation. *Mol. Ecol.* 17: 4015–4026.

Kalinowski, S.T. 2002. How many alleles per locus should be used to estimate genetic distances? *Heredity* 88: 62–65.

Kalinowski, S.T. 2005. Do polymorphic loci require larger sample sizes to estimate genetic distances? *Heredity* 94: 33–36.

Karl, S.A. & Avise, J.C. 1992. Balancing selection at allozyme loci in oysters: implications from nuclear RFLPs. *Science* 256: 100–102.

Kijima, A., Taniguchi, N., Mori, N. & Hagiwara, J. 1987. Genetic variability and breeding structure in *Ruditapes phillippinarum*. *Rep. USA Mar. Biol. Inst. Kochi Univ. 1987*: 173–182.

Kinlan, B.P., Gaines, S.D. & Lester, S.E. 2005. Propagule dispersal and the scales of marine community process. *Diver. Distrib.* 11: 139–148.

Koehn, R.K., Milkman, R. & Mitton, J.B. 1976. Population genetics of marine pelecypods. IV. Selection, migration and genetic differentiation in the blue mussel, *Mytilus edulis*. *Evolution* 30: 2–32.

Koehn, R.K. & Williams, G.C. 1978. Genetic differentiation without isolation in the American eel, *Anguilla rostrata*. II. Temporal stability of geographic patterns. *Evolution* 32: 624–637.

Kordos, L.M. & Burton, R.S. 1993. Genetic differentiation of Texas Gulf Coast populations of the blue crab *Callinectes sapidus*. *Mar. Biol.* 117: 227–233.

Kusumo, H.T. & Druehl, L.D. 2000. Variability over space and time in the genetic structure of the winged kelp *Alaria marginata*. *Mar. Biol.* 136: 397–409.

Lee, H.J. & Boulding, E.G. 2007. Mitochondrial DNA variation in space and time in the northeastern Pacific gastropod, *Littorina keenae*. *Mol. Ecol.* 16: 3084–3103.

Lee, H.J. & Boulding, E.G. 2009. Spatial and temporal population genetic structure of four northeastern Pacific littorinid gastropods: the effect of mode of larval development on variation at one mitochondrial and two nuclear DNA markers. *Mol. Ecol.* 18: 2165–2184.

Lessios, H.A. 1985. Genetic consequences of mass mortality in the Caribbean sea urchin *Diadema antillarum*. *Proc. 5th Int. Coral Reef Congr.* 4: 119–126.

Lessios, H.A., Weinberg, J.R. & Starczak, V.R. 1994. Temporal variation in populations of the marine isopod *Excirolana*: how stable are gene frequencies and morphology? *Evolution* 48: 549–563.

Lewis, O.T., Thomas, C.D., Hill, J.K., Brookes, M.I., Crane, T.P.R., Graneau, Y.A., Mallet, J.L.B.

& Rose, O.C. 1997. Three ways of assessing metapopulation structure in the butterfly *Plebejus argus. Ecol. Entomol.* 22: 283–293.

Li, G. & Hedgecock, D. 1998. Genetic heterogeneity, detected by PCR-SSCP, among samples of larval Pacific oysters (*Crassostrea gigas*) supports the hypothesis of large variances in reproductive success. *Can. J. Fish. Aquat. Sci.* 55: 1025–1033.

Marko, P.B. 2004. "What's larvae got to do with it?" Disparate patterns of post-glacial population structure in two benthic marine gastropods with identical dispersal potential. *Mol. Ecol.* 13: 597–611.

McClenaghan, L.R. Jr., Smith, M.H. & Smith, M.W. 1985. Biochemical genetics of mosquitofish. IV. Changes of allele frequencies through time and space. *Evolution* 39: 451–460.

McGoldrick, D.J., Hedgecock, D., English, L.J., Baoprasertkul, P. & Ward, R.D.. 2000. The transmission of microsatellite alleles in Australian and North American stocks of the Pacific oyster (*Crassostrea gigas*): selection and null alleles. *J. Shellfish Res.* 19: 779–788.

Meirmans, P.G. 2006. Using the AMOVA framework to estimate a standardized genetic differentiation measure. *Evolution* 60: 2399–2402.

Michener, C.D. & Sokal, R.R. 1957. A quantitative approach to a problem in classification. *Evolution* 11: 130–162.

Miller, M.P. 1997. Tools for Population Genetic Analyses (TFPGA): a Windows® program for the analysis of allozyme and molecular population genetic data. Northern Arizona University, Flagstaff, AZ. http://www.marksgeneticsoftware.net/tfpga.htm

Mitarai, S., Siegel, D.A. & Winters, K.B. 2008. A numerical study of stochastic larval settlement in the California Current system. *J. Mar. Sys.* 69: 295–309.

Moberg, P.E. & Burton, R.S. 2000. Genetic heterogeneity among adult and recruit red sea urchins, *Strongylocentrotus franciscanus. Mar. Biol.* 136: 773–784.

Morgan, L.E., Wing, S.R., Botsford, L.W., Lundquist, C.J. & Diehl, J.M. 2000. Spatial variability in red sea urchin (*Strongylocentrotus franciscanus*) recruitment in northern California. *Fish. Oceanogr.* 9: 83–98.

Molenock, J. 1975. Evolutionary aspects of communication in the courtship behaviour of four species of anomuran crabs (*Petrolisthes*). *Behaviour* 53: 1–29.

Molenock, J. 1976. Agonistic inetractions of the crab *Petrolisthes* (Crustacea, Anomura). *Zeitschr. Tierpsychol.* 41: 277–294.

Morris, R.H., Abbott, D.P. & Haderlie, E.C. 1980. *Intertidal invertebrates of California.* Stanford, CA: Stanford University Press.

Nei, M. 1972. Genetic distance between populations. *Amer. Naturalist* 106: 283–292.

Nei, M. 1978. Estimation of average heterozygosity and genetic distance from a small number of individuals. *Genetics* 89: 583–590.

Nei, M., Tajima, F. & Tateno, Y. 1983. Accuracy of estimated phylogenetic trees from molecular data. II. Gene frequency data. *J. Mol. Evol.* 19: 153–170.

Neigel, J.E. 1997. A comparison of alternative strategies for estimating gene flow from genetic markers. *Annu. Rev. Ecol. Syst.* 28: 105–128.

Ota, T. 1993. Dispan: genetic distance and phylogenetic analysis. Pennsylvania State University, University Park, PA. ftp://ftp.bio.indiana.edu/molbio/ibmpc/dispan.zip

Ouborg, N.J., Piquot, Y. & Van Groenendael, J.M. 1999. Population genetics, molecular markers and the study of dispersal in plants. *J. Ecol.* 87: 551–568.

Palstra, F.P., O'Connell, M.F. & Ruzzante, D.E. 2006. Temporal stability of population genetic structure of Atlantic salmon (*Salmo salar*) in Newfoundland. *J. Fish Biol.* 69: 241.

Palumbi, S.R. 1996. Macrospatial genetic structure and speciation in marine taxa with high dispersal abilities. In: Ferraris, J.D. & Palumbi, S.R. (eds.), *Molecular Zoology: Advances, Strategies, and Protocols*: 101–117. New York, NY: Wiley-Liss, Inc.

Park, S.D.E. 2001. MSTools: Microsatellite toolkit for MS Excel. Trinity College, Dublin. http://animalgenomics.ucd.ie/sdepark/ms-toolkit/

Petit, E. & Mayer, F. 1999. Male dispersal in the noctule bat (*Nyctalus noctula*): where are the limits? *Proc. R. Soc. B* 266: 1717–1722.

Piertney, S.B. & Carvalho, G.R. 1995. Microgeographic genetic differentiation in the intertidal isopod *Jaera albifrons* Leach. II. Temporal variation in allele frequencies. *J. Exp. Mar. Biol. Ecol.* 188: 277–288.

Piry S., Alapetite, A., Cornuet, J.-M., Paetkau, D., Baudouin, L. & Estoup, A. 2004. GeneClass2: a software for genetic assignment and first-generation migrant detection. *J. Hered.* 95: 536–539.

Planes, S. & Lenfant, P. 2002. Temporal change in the genetic structure between and within cohorts of a marine fish, *Diplodus sargus*, induced by a large variance in individual reproductive success. *Mol. Ecol.* 11: 1515–1524.

Planes, S., Jones, G.P. & Thorrold, S.R. 2009. Larval dispersal connects fish populations in a network of marine protected areas. *Proc. Nat. Acad. Sci. USA* 106: 5693–5697.

Powers, D.A. & Place, A.R. 1978. Biochemical genetics of *Fundulus heteroclitus* (L.). I. Temporal and spatial variation in gene frequencies of *Ldh-B, Mdh-A, Gpi-B*, and *Pgm-A. Biochem. Genet.* 16: 539–607.

Pringle, J.M., Lutscher, F. & Glick, E. 2009. Going against the flow: effects of non-Gaussian dispersal kernels and reproduction over multiple generations. *Mar. Ecol. Prog. Ser.* 377: 13–17.

Rand, D.M., Spaeth, P.S., Sackton, T.B. & Schmidt, P.S. 2002. Ecological genetics of Mpi and Gpi polymorphisms in the acorn barnacle and the spatial scale of neutral and non-neutral variation. *Integr. Comp. Biol.* 42: 825–836.

Reeb, C.A. & Avise, J.C. 1990. A genetic discontinuity in a continuously distributed species: mitochondrial DNA in the American oyster *Crassostrea virginica. Genetics* 124: 397–406.

Reynolds, J., Weir, B.S. & Cockerham, C.C. 1983. Estimation of the coancestry coefficient: basis for a short-term genetic distance. *Genetics* 105: 767–779.

Riginos, C., Douglas, K.E., Jin, Y., Shanahan, D.F. & Treml, E.A. 2011. Effects of geography and life history traits on genetic differentiation in benthic marine fishes. *Ecography* 34: doi: 10.1111/j.1600-0587.2010.06511.x

Robainas Barcia, A., Espinosa López, G., Hernández, D. & García-Machado, E. 2005. Temporal variation of the population structure and genetic diversity of *Farfantepenaeus notialis* assessed by allozyme loci. *Mol. Ecol.* 14: 2933–2942.

Ross, P.M., Hogg, I.D., Pilditch, C.A. & Lundquist, C.J. 2009. Phylogeography of New Zealand's coastal benthos. *NZ J. Mar. Freshw. Res.* 43: 1009–1027.

Rousset, F. 2000. Inferences from spatial population genetics. In: Balding, D., Bishop, M. & Cannings, C. (eds.), *Handbook of Statistical Genetics*: 239–270. New York, NY: John Wiley & Sons, Ltd.

Ruzzante, D.E. 1998. A comparison of several measures of genetic distance and population structure with microsatellite data: bias and sampling variance. *Can. J. Fish. Aquat. Sci.* 55: 1–14.

Ryman, N. & Jorde, P.E. 2001. Statistical power when testing for genetic differentiation. *Mol. Ecol.* 10: 2361–2374.

Ryman, N., Palm, S., André, C., Carvalho, G.R., Dahlgren, T.G., Jorde, P.E., Laikre, L., Larsson, L.C., Palmé, A. & Ruzzante, D.E. 2006. Power for detecting genetic divergence: differences between statistical methods and marker loci. *Mol. Ecol.* 15: 2031–2045.

Sainte-Marie, B., Urbani, N., Sevigny, J.-M., Hazel, F. & Kuhnlein, U. 1999. Multiple choice criteria and the dynamics of assortative mating during the first breeding season of female snow crab *Chionoecetes opilio* (Brachyura, Majidae). *Mar. Ecol. Prog. Ser.* 181: 141–153.

Schmidt, P.S., Bertness, M.D. & Rand, D.M. 2000. Environmental heterogeneity and balancing selection in the acorn barnacle *Semibalanus balanoides*. *Proc. R. Soc. Lond. B* 267: 379–384.

Schmidt, P.S. & Rand, D.M. 2001. Adaptive maintenance of genetic polymorphism in an intertidal barnacle: habitat- and life-stage specific survivorship of MPI genotypes. *Evolution* 55: 1336–1344.

Schmidt, P.S., Serrão, E.A., Pearson, G.A., Riginos, C., Rawson, P.D., Hilbish, T.J., Brawley, S.H., Trussell, G.C., Carrington, E., Wethey, D.S., Grahame, J.W., Bonhomme, F. & Rand, D.M. 2008. Ecological genetics in the North Atlantic: environmental gradients and adaptation at specific loci. *Ecology* 89: S91–S107.

Schneider, S., Roessli, D. & Excoffier, L. 2000. Arlequin: A software for population genetics data analysis. University of Geneva, Geneva. http://anthro.unige.ch/arlequin

Selkoe, K.A. & Toonen, R.J. 2006. Microsatellites for ecologists: a practical guide to using and evaluating microsatellite markers. *Ecol. Lett.* 9: 615–629.

Selkoe, K.A., Gaines, S.D., Caselle, J.E. & Warner, R.R. 2006. Current shifts and kin aggregation explain genetic patchiness in fish recruits. *Ecology* 87: 3082–3094.

Selkoe, K.A., Henzler, C.M. & Gaines, S.D. 2008. Seascape genetics and the spatial ecology of marine populations. *Fish Fisher.* 9: 363–377.

Shanks, A.L., Grantham, B.A. & Carr, M.H. 2003. Propagule dispersal distance and the size and spacing of marine reserves. *Ecol. App.* 13: S159–S169.

Shanks, A.L. 2009. Pelagic larval duration and dispersal distance revisited. *Biol. Bull.* 216: 373–385.

Shen, K.-N. & Tzeng, W.-N. 2007. Population genetic structure of the year-round spawning tropical eel, *Anguilla reinhardtii*, in Australia. *Zool. Stud.* 46: 441–453.

Siegel, D.A., Kinlan, B.P., Gaylord, B. & Gaines, S.D. 2003. Lagrangian descriptions of marine larval dispersion. *Mar. Ecol. Prog. Ser.* 260: 83–96.

Siegel, D.A., Mitarai, S., Costello, C.J., Gaines, S.D., Kendall, B.E., Warner, R.R. & Winters, K.B. 2008. The stochastic nature of larval connectivity among nearshore marine populations. *Proc. Nat. Acad. Sci. USA* 105: 8974–8979.

Sifres, A., Picó, B., Blanca, J.M., De Frutos, R. & Nuez F. 2007. Genetic structure of *Lycopersicon pimpinellifolium* (Solanaceae) populations collected after the ENSO event of 1997–1998. *Genet. Resour. Crop Evol.* 54: 359–377.

Slatkin, M. 1994. An exact test for neutrality based on the Ewens sampling distribution. *Genetical Res.* 64: 71–74.

Slatkin, M. 1996. A correction to the exact test based on the Ewens sampling distribution. *Genetical Res.* 68: 259–260.

Sotka, E.E. & Palumbi, S.R. 2006. The use of genetic clines to estimate dispersal distances of marine larvae. *Ecol.* 87: 1094–1103.

Stillman, J.H. & Reeb, C.A 2001. Molecular phylogeny of eastern Pacific porcelain crabs, genera *Petrolisthes* and *Pachycheles*, based on the mtDNA 16S rDNA sequence: phylogeographic and systematic implications. *Mol. Phylo. Evol.* 19: 236–245.

Stillman, J.H. 2002. Causes and consequences of thermal tolerance limits in rocky intertidal porcelain crabs. *Int. Comp. Biol.* 42: 790–796.

Stillman, J.H. 2004. A comparative analysis of plasticity of thermal limits in porcelain crabs across latitudinal and intertidal zone clines. *Int. Congr. Ser.* 1275: 267–275.

Takezaki, N. & Nei, M. 1996. Genetic distances and reconstruction of phylogenetic trees from microsatellite DNA. *Genetics* 144: 389–399.

Todd, C.D., Havenhand, J.N. & Thorpe, J.P. 1988. Genetic differentiation, pelagic larval transport and gene flow between local populations of the intertidal marine mollusc *Adalaria proxima* (Alder & Hancock). *Funct. Ecol.* 2: 441–451.

Todd, C.D., Lambert, W.J. & Thorpe, J.P. 1998. The genetic structure of intertidal populations of two species of nudibranch molluscs with planktotrophic and pelagic lecithotrophic larval stages: are pelagic larvae "for" dispersal? *J. Exp. Mar. Biol. Ecol.* 228: 1–28.

Toonen, R.J. 1997. Microsatellites for ecologists: non-radioactive isolation and amplification proto-cols for microsatellite markers. University of California, Davis, CA. http://www2.hawaii.edu/~toonen/Msats/index.html

Toonen, R.J. & Hughes, S. 2001. Increased throughput for fragment analysis on an ABI Prism® 377 automated sequencer using a membrane comb and STRand software. *BioTech.* 31: 1320–1324.

Toonen, R.J. 2001. *Molecular Genetic Analysis of Recruitment and Dispersal in the Intertidal Porcelain Crab, Petrolisthes cinctipes.* Ph.D. dissertation, University of California, Davis, CA.

Toonen, R.J. 2004. Genetic evidence of multiple paternity of broods in the intertidal crab *Petrolisthes cinctipes. Mar. Ecol. Prog. Ser.* 270: 259–263.

Turner, T.F., Wares, J.P. & Gold, J.R. 2002. Genetic effective size is three orders of magnitude smaller than adult census size in an abundant, estuarine-dependent marine fish (*Sciaenops ocellatus*). *Genetics* 162: 1329–1339.

Viard, F., Justy, F. & Jarne, P. 1997. Population dynamics inferred from temporal variation at microsatellite loci in the selfing snail *Bulinus truncatus. Genetics* 146: 973–982.

Waples, R.S. 1987. A multispecies approach to the analysis of gene flow in marine shore fishes. *Evolution* 41: 385–400.

Waples, R. S. 1990. Conservation genetics of Pacific salmon: II. Effective population size and the rate of loss of genetic variability. *J. Hered.* 81: 267–276.

Waples, R.S. 2002. Evaluating the effect of stage-specific survivorship on the Ne/N ratio. *Mol. Ecol.* 11: 1029–1037.

Wares, J.P. & Cunningham, C.W. 2001. Phylogeography and historical ecology of the North Atlantic intertidal. *Evolution* 55: 2455–2469.

Watts, R.J., Johnson, M.S. & Black, R. 1990. Effects of recruitment on genetic patchiness in the urchin *Echinometra mathaei* in Western Australia. *Mar. Biol.* 105: 145–151.

Weersing, K.A. & Toonen, R.J. 2009. Population genetics, larval dispersal, and demographic connectivity in marine systems. *Mar. Ecol. Progr. Ser.* 393: 1–12.

Weir, B.S. 1996. *Genetic Data Analysis II: Methods for Discrete Population Genetic Data.* Sunderland, MA: Sinauer Associates, Inc.

Weir, B.S. & Cockerham, C.C. 1984. Estimating F–statistics for the analysis of population structure. *Evolution* 38: 1358–1370.

White, C. Selkoe, K.A., Watson, J., Siegel, D.A., Zacherl, D.C. & Toonen, R.J. 2010. Ocean currents help explain population genetic structure. *Proc. R. Soc. Lond. B* 277: 1685–1694.

Whitlock, M.C. & McCauley, D.E. 1999. Indirect measures of gene flow and migration: $F_{ST} \neq 1/(4Nm + 1)$. *Heredity* 82: 117–125.

Wilding, C.S., Butlin, R.K. & Grahame, J. 2001. Differential gene exchange between parapatric morphs of *Littorina saxatilis* detected using AFLP markers. *J. Evol. Biol.* 14: 611–619.

Wilson, G.A. & Rannala, B. 2003. Bayesian inference of recent migration rates using multilocus genotypes. *Genetics* 163: 1177–1191.

Wing, S.R., Botsford, L.W., Largier, J.L. & Morgan, L.E. 1995a. Spatial structure of relaxation events and crab settlement in the northern California upwelling system. *Mar. Ecol. Prog. Ser.* 128: 199–211.

Wing, S.R., Largier, J.L., Botsford, L.W. & Quinn, J. F. 1995b. Settlement and transport of benthic invertebrates in an intermittent upwelling region. *Limnol. Oceanogr.* 40: 316–329.

Wing, S.R., Botsford, L.W., Ralston, S.V. & Largier, J.L. 1998. Meroplanktonic distribution and circulation in a coastal retention zone of the northern California upwelling system. *Limnol. Oceanogr.* 43: 1710–1721.

Wright, S. 1978. *Evolution and the Genetics of Populations. Vol. 4. Variability Within and Among Natural Populations.* Chicago, IL: University of Chicago Press.

II

Population genetics and
phylogeography of marine crustaceans

Comparative phylogeography of Indo-West Pacific intertidal barnacles

Ling Ming Tsang[1], Tsz Huen Wu[1], Wai Chuen Ng[2], Gray A. Williams[2],
Benny K. K. Chan[3] & Ka Hou Chu[1]

[1] *Simon F. S. Li Marine Science Laboratory, School of Life Sciences, The Chinese University of Hong Kong, Hong Kong*

[2] *The Swire Institute of Marine Science and The School of Biological Sciences, The University of Hong Kong, Hong Kong*

[3] *Biodiversity Research Center, Academia Sinica, Taipei, Taiwan*

ABSTRACT

Previous studies on two common widespread intertidal acorn barnacles, *Tetraclita squamosa* and *Chthamalus malayensis*, in the Indo-West Pacific region have identified genetically distinct lineages, representing several cryptic species. In this paper, we summarize the data on COI divergence of the two species complexes to elucidate the geographic distribution and population genetic structure of each lineage, and compare the phylogeographic patterns and timing of speciation events of the species complexes. We found that the genetic differentiation of both complexes is affected by sea level fluctuations associated with glaciations. *T. squamosa*, however, exhibits a deeper divergence and stronger genetic population structuring which may be attributed to its more stringent habitat requirements (mostly on exposed rocky shores) than *C. malayensis* (found on shores of varying wave exposures). Within each complex, lineages with a continental distribution appear to suffer more in terms of effective population size from sea level changes than those inhabiting islands. Comparative phylogeographic studies on these barnacles suggest that habitat specificity, rather than larval dispersal capability, appears more important in determining population structure and phylogeographic patterns and highlights the importance of islands in serving as refugia for species during glaciation periods.

1 INTRODUCTION

The Indo-West Pacific (IWP) region harbors the highest marine biodiversity in the world (Hughes et al. 2002; Roberts et al. 2002; Mora et al. 2003), which is shaped by the long-term interaction between the processes of speciation and extinction. The past tectonic events and sea level fluctuations during glaciation periods are believed to drive speciation while the contemporary ocean circulation patterns help to maintain and/or promote genetic differentiation (Hewitt 1996, 2000; Grosberg & Cunningham 2001; Waters 2008). Many attempts have been made to elucidate the diversification patterns that lead to the extraordinarily high marine biodiversity observed in the IWP. Phylogeographic studies using molecular data are an effective and common way to achieve this goal (Palumbi 1995; Avise 2000; Hewitt 2004). Early phylogeographic studies have determined that many widespread species show sharp genetic breaks between the Pacific and Indian Oceans (e.g., Lavery et al. 1996; Benzie 1998; Williams & Benzie 1998; Duda & Palumbi 1999; Williams et al. 2002). This IWP phylogeographic break is attributed to reduced gene flow during the Pleistocene period ($\approx$ 20,000 to two million years before present) when repeated glaciation events occurred. These glacial cycles caused dramatic fluctuations in sea level globally, with the lowest sea level (approximately 130 m below present level) being attained during the glacial maxima (Lambeck et al. 2002). The lowering of sea level caused the exposure of shallow regions such as the Sunda Shelf and closure of seaways between islands (Voris 2000). The increased, emergent landmass was likely to be a strong geographic barrier for many marine taxa, reducing habitat availability, and limiting dispersal and thus gene flow between populations in the two oceans for a considerable period of time, resulting in the present-day observed genetic diversification.

However, the IWP phylogeographic break, although evident in many taxonomic groups, does not appear to be a universal phenomenon across all marine species in the region, as many taxa demonstrate high genetic homogeneity in the IWP, e.g., the trumpetfish *Aulostomus chinensis* (Bowen et al. 2001), the sea urchin *Tripneustes* (Lessios et al. 2003), and the gastropod *Echinolittorina reticulata* (Reid et al. 2006). This lack of genetic subdivision is attributed to several reasons including maintenance of a high level of gene flow during the glacial episodes, or rapid postglacial colonization of the Indian Ocean from the Pacific (Grant & Bowen 1998; Lessios et al. 2003). Yet the exact reasons for observed differences in genetic differentiation patterns among taxa remain obscure. Discordance in phylogeographic patterns has been commonly observed between congeneric species, e.g., the gastropods *Echinolittorina* (Reid et al. 2006) and *Nerita* (Crandall et al. 2008), suggesting that relatively subtle differences in biological attributes could result in discrepancies in biological responses to glaciations. Analysis of a single species is therefore unlikely to tell us about the broad effect of environmental changes on the evolution of marine taxa. Comparisons of population genetic patterns across multiple species (i.e., comparative phylogeographic studies), represent a potential solution to investigate the different phylogeographic hypotheses that have been proposed from earlier studies (Bermingham & Moritz 1998; Soltis et al. 2006).

In the marine realm, both biogeography and conservation studies lag far behind their terrestrial counterparts (Beck 2003; Lourie & Vincent 2004), and the majority of these studies have been carried out on marine taxa in North America and Europe (e.g., Wares 2001, 2002; Wares & Cunningham 2001; Dawson et al. 2002; Rocha et al. 2002). Several comparative phylogeographic studies on co-distributed taxa have, however, recently been carried out in Asia, focusing chiefly on the Indo-Australian Archipelago (IAA) within the IWP (Lourie et al. 2005; Barber et al. 2006; Crandall et al. 2008). Many of the proposed hypotheses accounting for phylogeographic patterns await further refinement and at present are limited for a number of reasons. The sampling range of previous studies, for example, has mostly been restricted in geographic scale (concentrating on the IAA and the Coral Triangle) and the alpha taxonomy of most of the marine fauna studied is poorly understood. As a result, despite attempts to evaluate genetic and distribution patterns in the region there are, as yet, few clear generalizations, and major questions concerning the relative roles of historic and contemporary physical and biological factors remain unanswered. Thus, further combined ecological and genetic studies in the region on multiple, co-distributed, taxa for which information on spawning and dispersal characteristics are well known, are needed to identify the extrinsic and intrinsic forces that create the observed biogeographic patterns.

Over the past several years, using intertidal barnacles as a model, our group has attempted to investigate the phylogeography of the IWP region. An important discovery of our studies has been the identification of cryptic species and genetically distinct lineages of taxa that were previously considered to be widely distributed in the IWP (Newman & Ross 1976; Southward & Newman 2003). *Tetraclita squamosa* and *Chthamalus malayensis* are two common intertidal barnacles that were considered to occur throughout the IWP (Newman & Ross 1976; Southward & Newman 2003). Our studies have revealed that *T. squamosa* from Japan to Singapore consists of three genetically and morphologically distinct species (Chan et al. 2007a, b). *Tetraclita squamosa sensu stricto* and the two newly identified species, *T. kuroshioensis* and *T. singaporensis* show restricted

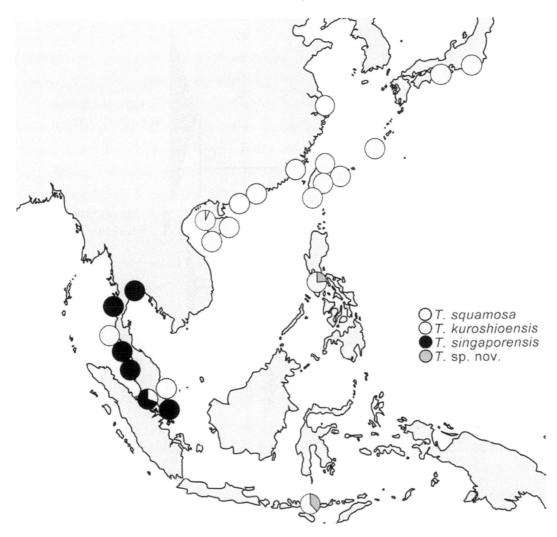

Figure 1. Sampling localities of *Tetraclita squamosa* species complex. Frequency distribution of the individual species in the complex in each locality based on COI analysis is indicated by pie charts.

geographic distributions as compared to the broad range of the entire species complex. In the case of *C. malayensis*, in spite of the low intraspecific morphological differentiation (Southward & Newman 2003), our investigation on populations from East Asia has identified three genetically divergent lineages (the South China Sea clade, Indo-Malay (IM) clade and Taiwan clade; Tsang et al. 2008). Thus, both *T. squamosa* and *C. malayensis* are cryptic species complexes, and the actual geographic range of each species remains uncertain. The identification of cryptic species, in what were once thought to be widely distributed species, once again emphasizes our lack of thorough understanding on IWP biodiversity and highlights critical gaps in both ecological and biological research on the biological communities in the region.

   The arrays of closely related cryptic species in the two complexes provide an excellent model to test biogeographic hypotheses as they exhibit broad, sympatric distributions (Newman & Ross 1976; Southward & Newman 2003). Furthermore, the biology and ecology of most of these barnacles are

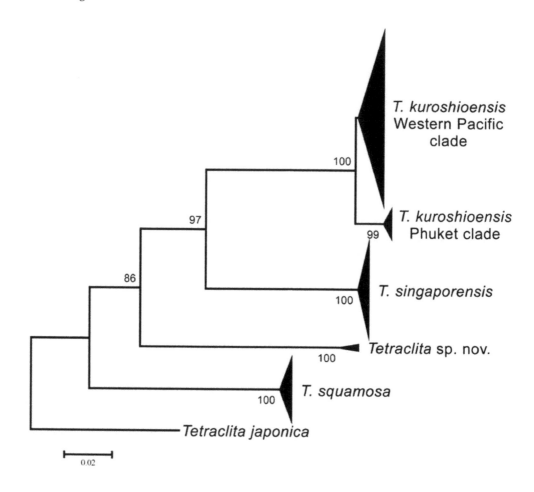

Figure 2. Neighbor-joining tree of COI haplotypes of *Tetraclita squamosa* species complex constructed based on Kimura-2-pairwise distance. The height of triangle corresponds to the number of haplotypes sampled for that clade. The percentage bootstrap support (≥ 75) is shown on each branch.

well documented (e.g., Chan et al. 2001; Yan & Chan 2001; Chan 2003; Koh et al. 2005; Yan et al. 2006). These organisms have a biphasic life cycle. Dispersal of the pelagic larval stage is driven to a large extent by ocean currents while the sessile adult form minimizes the possibility of adult migration affecting gene flow pattern, and facilitates sample collection and field studies. These allow correlation between inferred gene flow and biological and/or physical factors. Concordance in phylogeographic patterns among species from the two complexes would provide strong evidence on the common driving forces exerted on multiple taxa. The phylogeography of closely related species with known distributions of individual lineages can also inform us about the mode and tempo of speciation events, while dissimilarities between these species could tell us how individual biological traits may lead to different responses to a common driving force.

In this paper, we summarize our recent efforts in elucidating the distribution and population genetic structure of the two species complexes, *Tetraclita squamosa* and *Chthamalus malayensis*. Based on comprehensive sampling of the mitochondrial COI sequences over a broad geographic range in the IWP, we aim to estimate the timing of different speciation events and to provide evidence for the driving forces responsible for the observed phylogeographic patterns in the IWP.

2   PHYLOGEOGRAPHY OF THE *TETRACLITA SQUAMOSA* SPECIES COMPLEX

Analysis of samples collected from over 20 locations in the IWP region has refined our understanding of the actual geographic distribution pattern of the barnacles from the two species complexes. For the *Tetraclita squamosa* species complex, *T. kuroshioensis* is the most widely distributed species, covering most of the West Pacific region, in particular areas under the influence of the Kuroshio Current (including Japan, Taiwan, and the Philippines). However, it is interesting that this species also occurs in the Indian Ocean, at Phuket Island, Thailand (Figure 1). Pronounced genetic divergence was detected among populations from the West Pacific and Indian Ocean ($\approx 3\%$ in COI divergence; Figure 2). In contrast, little genetic subdivision is observed among the West Pacific populations, despite the relatively wide distribution of the species. The sharp genetic differentiation between *T. kuroshioensis* individuals from the Indian Ocean and West Pacific region excludes the possibility that the population in Phuket have been introduced through anthropogenic activities.

Tetraclita squamosa sensu stricto inhabits the South China coast and east coast of the Malay Peninsula. Very low abundance of *T. squamosa* is recorded in Hainan Island, a site in between the South China coast and Malay Peninsula. Thus, it appears that *T. squamosa s. str.* is fragmented into two metapopulations, a pattern which is also evident at the molecular level, as two clades in the COI haplotype network of the species were detected (Figure 3A), with one clade found exclusively in the Malay Peninsula and the other predominantly abundant in the South China populations (Figure 3B). The coverage of *T. squamosa s. str.* in the east coast of the Malay Peninsula is low (B.K.K. Chan per. obs.) and this species is not found in the Gulf of Thailand. Therefore, there is a population fragmentation within the range of the species, resulting in a reduced level of gene flow. Unfortunately, since we lack samples from Vietnam and other locations along the east coast of the Malay Peninsula, this hypothesis must be interpreted with caution.

Being restricted to the Malay Peninsula, *T. singaporensis* shows a more confined distribution as compared to the above two species. It occurs in sympatry with *T. squamosa* in southern Malaysia (Figure 1). There are two clades, separated by four mutational steps, in the haplotype network of COI for *T. singaporensis* (Figure 4A). Clade 1 occurs in high abundance along the western coast of Malaysia, while clade 2 can be found on both east and west coasts. All individuals from the east coast belong to clade 2, which only occurs in low abundance in the west (Figure 4B). A gradual decline in the frequency of clade 2 from the Pacific to Indian Ocean is also evident. Hence, the two clades likely originated from allopatric isolation during the last glacial maximum (LGM) when the connection between the Indian and Pacific Oceans was restricted by the exposure of the Sunda Shelf (Voris 2000) as reported in other species (e.g., Lavery et al. 1996; Benzie 1998; Williams & Benzie 1998; Duda & Palumbi 1999; Williams et al. 2002). Secondary admixture occurred following the retreat of glaciers, and the removal of dispersal barriers.

A diverged lineage, presumably representing another cryptic species in the complex, was found in samples from the Philippines and Indonesia, coexisting with *T. kuroshioensis* (Figures 1 and 2). This lineage (hereafter referred to as *Tetraclita* sp. nov.) diverges from the other three species by $> 17\%$ in COI sequences. This value is comparable to the interspecific divergence between *T. squamosa*, *T. kuroshioensis* and *T. singaporensis* (ranging from 17–20%), which exhibit clear morphological differentiation and unique nuclear DNA sequences (Chan et al. 2007a, b). Accordingly, we are confident that further detailed morphological analysis would confirm that the newly identified lineage represents another new species.

The demographic history of the species can be inferred using two coalescence-based methods, the mismatch distribution analysis (Rogers & Harpending 1992) and Bayesian skyline plots (BSP; Drummond et al. 2005). Both *T. kuroshioensis* and *T. squamosa* display a smooth and unimodal mismatch distribution, consistent with predictions under a model of rapid demographic expansion (Figure 5A, B). In contrast, the mismatch distribution of *T. singaporensis* shows a bimodal pattern,

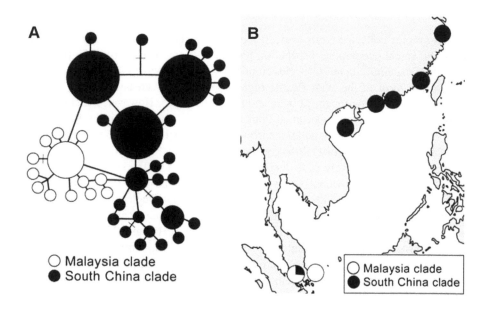

Figure 3. (A) COI haplotype network of *Tetraclita squamosa s. str.* Each branch designates one base difference among haplotypes while slashes indicate additional mutations. The area of the circles is proportional to the number of individuals with the particular haplotype. (B) Map of the West Pacific region with pie charts showing frequency distribution of the two clades in each population.

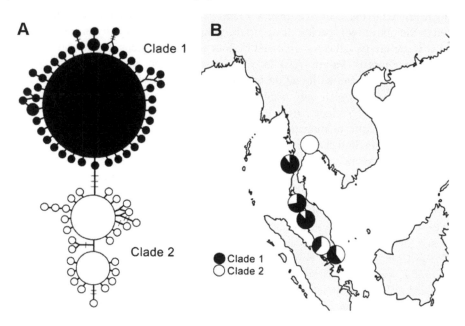

Figure 4. (A) COI haplotype network of *Tetraclita singaporensis*. Each branch designates one base difference among haplotypes while slashes indicate additional mutations. The area of the circles is proportional to the number of individuals with the particular haplotype. (B) Map of the Malay Peninsula with pie charts showing frequency distribution of the two clades in each population.

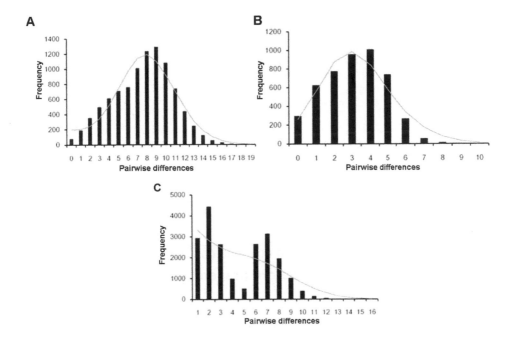

Figure 5. The observed mismatch distribution (bars), and expected mismatch distribution under the model of sudden demographic expansion (solid line) for COI haplotypes. (A) *Tetraclita kuroshioensis*; (B) *T. squamosa s. str.*; (C) *T. singaporensis*.

suggesting an admixture of two diverged lineages (Figure 5C). This corroborates the results of the haplotype network (Figures 3A and 4A). *Tetraclita* sp. nov. is not included in this analysis since it was only collected from two localities with a limited number of individuals.

Results of the Bayesian skyline plots (BSP) are generally concordant with those from the mismatch distribution analyses. Based on a mutation rate of ≈ 1.55% per million years calibrated from the transisthmian *Chthamalus* species (Wares 2001), we have estimated the age of the different demographic events which might have occurred. The three *Tetraclita* species display exponential growth in effective population size within a relative short period of time (e.g., *T. squamosa* shows a > 100-fold increase over the past 100,000 years, Figure 6). The three species, however, differ in the timing of the onset of the population expansion. The demographic expansion of *T. kuroshioensis* began at approximately 250,000 years before the present, prior to the LGM. On the other hand, *T. squamosa* and *T. singaporensis* experienced demographic expansion in the more recent past, at about 100,000 and 50,000 years before the present, respectively.

3   PHYLOGEOGRAPHY OF *CHTHAMALUS MALAYENSIS* SPECIES COMPLEX

Tsang et al. (2008) have defined three clades of *Chthamalus malayensis*. The present study has extended the dataset with new sampling locations. All the *C. malayensis* collected from northeastern India, Sri Lanka, western Malaysia and Singapore cluster together and form the Indo-Malay (IM) clade, while individuals inhabiting eastern Thailand, the South China coast and the Philippines are grouped into another clade, namely the South China Sea (SCS) clade (Figure 7). Some of the individuals from northern Taiwan belong to the SCS clade as well, yet the majority, including the

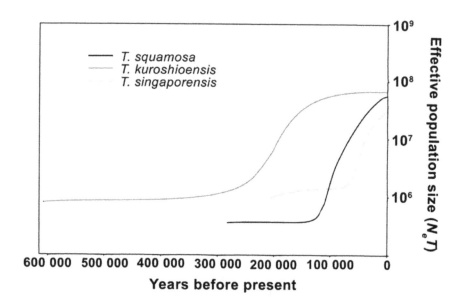

Figure 6. Bayesian skyline plots of species in the *Tetraclita squamosa* species complex based on COI dataset inferred with BEAST (Drummond & Rambaut 2007) (N_eT, where N_e = effective population size and T = generation time).

C. malayensis found in eastern Taiwan, constitute the Taiwan clade, which is endemic to this island. Unlike the *Tetraclita squamosa* species complex, the three lineages of the *C. malayensis* species complex show a largely allopatric distribution, with only the SCS and Taiwan clades co-existing in northern Taiwan. The haplotype network of the IM clade shows a star-like shape with a major dominant haplotype that occurs in all populations (Figure 8). Conversely, the haplotype network of the SCS clade is complicated and diverse. Most of the haplotypes are singletons with no obvious clustering (data not shown). No population genetic structuring could be detected in either of the geographically widespread SCS and IM clades, suggesting a high level of gene flow within each clade over long geographic distances.

The mismatch distribution of the South China Sea and Indo-Malay clades are unimodal, suggesting a rapid demographic expansion (Figure 9). The Bayesian skyline plot, however, implicates a more complicated demographic history for the clades. The SCS clade is inferred to have maintained a relatively stable effective population size throughout the Pliocene and Pleistocene periods (Figure 10). Although population growth is evident, it occurred gradually, as indicated by the gentle slope of the BSP. The IM clade has undergone more rapid and recent population expansion, dated back to approximately 100,000 years before present. The Taiwan clade was not included in the demographic analysis due to the limited number of sequences and relatively restricted range of this clade.

4   IS ADULT HABITAT SPECIFICITY A MORE IMPORTANT FACTOR THAN LARVAL DISPERSAL ABILITY IN DETERMINING SPECIES PHYLOGEOGRAPHIC PATTERNS?

In the present review, we synthesize our recent findings on the phylogeographic patterns of two intertidal acorn barnacle species complexes to shed light on the potential factors governing the ge-

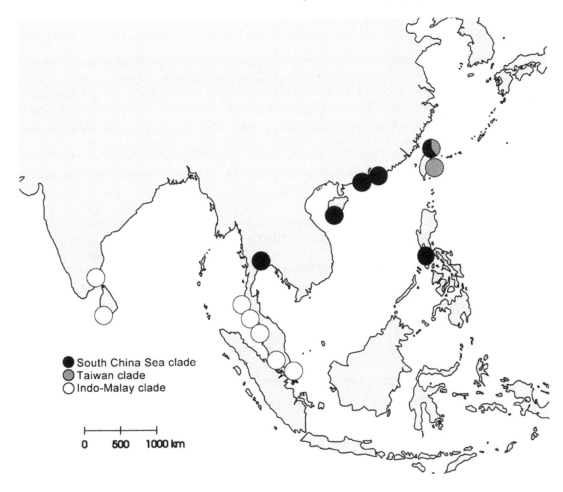

Figure 7. Sampling localities of *Chthamalus malayensis*. Frequency distribution of the different clades of *C. malayensis* in each locality based on COI analysis is indicated by pie charts.

netic structuring and distribution of marine organisms. These species share a number of similarities in their phylogeographic patterns. First, both taxa were previously believed to be widespread and common species in the IWP, yet each actually comprises an array of cryptic species with largely nonoverlapping distributions. The distributions of the individual lineages within the two complexes also display some concordance. Some are more widespread such as *T. kuroshioensis* and the SCS clade of *C. malayensis*, covering a large part of the West Pacific, especially the regions under the influence of the Kuroshio Current. Further south to tropical Asia, these two taxa are replaced by *T. singaporensis* and the IM clade of *C. malayensis* in the Malay Peninsula, respectively (Chan et al. 2007b; Tsang et al. 2008). Both of these southern taxa show low genetic diversity and signs of recent demographic expansion (Tsang et al. 2008), which can be attributed to postglacial recolonization of the Sunda Shelf (Voris 2000). Such similarities in phylogeographic structure of the two species complexes are likely shaped by the same historic and contemporary environmental factors. Discordance in phylogeographic patterns between the two species complexes can, however, also be observed. *Tetraclita* species, in general, display a stronger population structuring than the *Chthamalus* clades, and also appear to have experienced more dramatic demographic expansions.

Possible reasons for the disparities in the population genetic structure between the two taxa will be discussed below.

The pelagic larval phase of marine fauna life histories is believed to facilitate their dispersal, hence homogenizing genetic composition among populations (Palumbi 1994). Yet many population genetic studies have revealed unexpectedly clear genetic subdivision in many marine fauna with seemingly high dispersal potential (e.g., Barber et al. 2000, 2002; Taylor & Hellberg 2003). There is accumulating evidence pointing to the fact that adult habitat specificity, resulting from preferential settlement and postsettlement mortality, plays a more important role than larval dispersal capacity in determining population genetic structuring (Wares & Cunningham 2001; Rocha et al. 2002; Lourie et al. 2005; Ayre et al. 2009). A recent comparative phylogeographic study in southeast Australia on eight intertidal rocky shore species showed that barnacle species with catholic habitat requirements (*Tetraclitella purpurascens*) did not exhibit regional differentiation. In contrast, two rocky shore specialist species (*Tesseropora rosea* and *Catomerus polymerus*), which can only survive on exposed shores, demonstrate clear genetic differentiation across the Ninety Mile Beach, a long coastline of unsuitable soft-bottom habitat that imposes a biogeographic break for these hard substratum-associated species (Ayre et al. 2009). Furthermore, the gene flow of two direct developers (the starfish *Parvulastra exigua* and whelk *Haustrum vinosa*) was not affected by this phylogeographic break (Ayre et al. 2009), possibly due to the ability of these species to utilize sheltered habitats.

Similarly, *C. malayensis* occurs in more diverse habitats than *T. squamosa*. Both taxa are found on exposed rocky shores, but *T. squamosa* are absent from sheltered habitats which *C. malayensis* can also inhabit. *Tetraclita squamosa* could not be found in east India, where most of the natural rocky shore habitats have been destroyed and converted to artificial boulder shores (BKK Chan pers. obs.). Yet *C. malayensis* remains highly abundant in this region (see Fernando 2006). A similar phenomenon is also observed in west Malaysia (e.g., Penang), where *C. malayensis*, but not *T. squamosa*, maintains high population coverage in sites with a high level of anthropogenic disturbance. Taxon-specific habitat requirements may explain the scattered distribution patterns observed in *T. squamosa* as compared to the more continuous and widespread range of each of the *C. malayensis* clades. Among the *Tetraclita* species, only *T. kuroshioensis* shows a wide distribution range, while *T. singaporensis*, *T. squamosa s. str.* and *Tetraclita* sp. nov. are largely confined to

Figure 8. COI haplotype network of the Indo-Malay clade of *Chthamalus malayensis*. Each branch designates one base difference among haplotypes while slashes indicate additional mutations. The area of the circles is proportional to the number of individuals with the particular haplotype.

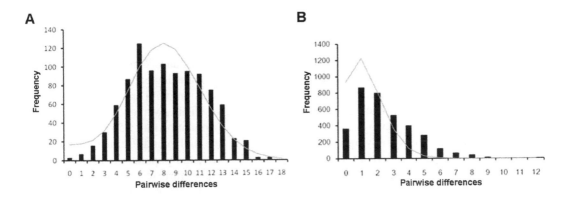

Figure 9. The observed mismatch distribution (bars) and expected mismatch distribution under the model of sudden demographic expansion (solid line) for COI haplotypes. (A) South China Sea clade; (B) Indo-Malay clade of *Chthamalus malayensis*.

relatively restricted geographic regions. Moreover, each of the *C. malayensis* clades is genetically homogeneous within their range. Conversely, apparent population subdivision is detected in three of the four *Tetraclita* species (except *Tetraclita* sp. nov., for which sampling is insufficient). Given that the larval durations under laboratory culture conditions for *T. squamosa s. str.* and *C. malayensis* are similar ($\approx$ 14–20 days depending on temperature; Yan & Chan 2001; Chan 2003), their dispersal capacity is likely to be comparable. The absence of *T. squamosa* on some shores is probably because the larvae do not settle on those shores or they suffer high postsettlement mortality on the unsuitable habitats. Our results, therefore, further support the argument that adult habitat specificity is a more

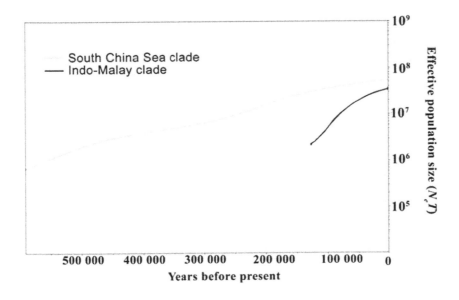

Figure 10. Bayesian skyline plots of the South China Sea and Indo-Malay clades of *Chthamalus malayensis* based on COI dataset inferred with BEAST ($N_e T$, where N_e = effective population size and T = generation time).

important factor in determining gene flow and genetic population structure than larval dispersal capacity (Reid et al. 2006; Ayre et al. 2009).

Among the seven lineages recovered in the two species complexes, three of them (*T. kuroshioensis*, *T. singaporensis*, and the IM clade of *C. malayensis*) exhibit a transoceanic distribution. These taxa, although having overlapping geographic ranges, display contrasting patterns of genetic differentiation. *Tetraclita kuroshioensis* shows a deep genetic subdivision, with approximately 3% COI divergence between the West Pacific and Indian Ocean lineages. This divergence implies a diversification event approximately two million years ago (based on a mutation rate 1.55%/my). *Tetraclita singaporensis* also exhibits a phylogeographic break across the Indian and Pacific Oceans, but the level of divergence is considerably lower (< 0.5%). Moreover, the populations from the Pacific and Indian Oceans do not form reciprocal monophyletic groups. Instead, secondary contact and admixture of the two historically diverged lineages is evident. While phylogeographic breaks are observed in the two *Tetraclita* species, the IM clade of *C. malayensis* shows no sign of any genetic break or population structuring between the Indian and Pacific Ocean. These spatial and temporal disparities in phylogeographic patterns are likely the result of differences of demographic history and biological traits.

Given the habitat specificity of the species in the *T. squamosa* complex and prevalence of population subdivision observed, it is anticipated that these taxa would have been severely affected by sea level changes. The difference in the age of isolation between *T. kuroshioensis* and *T. singaporensis* can probably be attributed to their demographic history. *Tetraclita kuroshioensis*, which shows the deepest divergence, could only be found in a single sampling location (Phuket) in the Indian Ocean. We suspect that this species might once have been widely distributed and common in both the West Pacific and Indian Oceans. Most of the populations inhabiting the Indian Ocean, however, became extinct in the glacial period and only rudimentary populations persist in restricted areas. This long period of isolation results in a high level of divergence between the West Pacific and Indian Ocean populations. On the other hand, following the extinction of the originally dominant *T. kuroshioensis*, *T. singaporensis* would have been free to invade and colonize the vacant habitats during the Pleistocene, and therefore became the common species in the Indian Ocean. During the LGM, however, *T. singaporensis* also suffered from population decline and allopatric isolation due to the reduced current flow between the two oceans. Such scenarios could result in the formation of two lineages, with a low level of divergence.

There are three possible explanations accounting for the lack of a phylogeographic break in the Indo-Malay clade of *C. malayensis*. First, a considerable amount of gene flow could have been maintained during the Pleistocene limiting the speciation process. Another possibility is high contemporary gene flow that would have eroded any signal of historic isolation. Finally, the taxon might have been absent in the Indian Ocean and only attained its present range through postglacial colonization, so that no genetic differentiation would be expected. Given the seemingly high level of contemporary gene flow facilitated by the habitat flexibility of *C. malayensis*, one may speculate the former two hypotheses to be more plausible. However, these two hypotheses do not appear to account for the genetic pattern observed in the IM clade of *C. malayensis*. If this clade has maintained a high level of gene flow during the Pleistocene, it would be reasonable to anticipate that a population would persist in the Indian Ocean, with the preservation of a considerable amount of genetic diversity. Yet very low haplotype and nucleotide diversity is evident in the IM clade. Similarly, if the historic signal of isolation has only recently been eliminated by present-day gene flow, we would expect a higher level of genetic diversity, or admixture of two diverged lineages, as seen in *T. singaporensis*. Since neither of these scenarios is evident, the third hypothesis, i.e., postglacial colonization of the clade to the Indian Ocean, appears to be the most likely explanation. In addition to the low genetic diversity, the star-like haplotype network, unimodal mismatch distribution and demographic expansion dated approximately back to the late Pleistocene (as inferred by BSP) all provide strong evidence to support this claim. Postglacial population expansion and re-colonization have also been postulated as major reasons accounting for the lack of genetic differentiation across phylogeographic breaks in the Mediterranean (Patarnello et al. 2007). Thus, most evidences suggest that the IM clade has only invaded and settled in the Indian Ocean after the LGM.

5   DEMOGRAPHIC HISTORY OF INTERTIDAL FAUNA

The difference in biological traits between the two species complexes can also lead to disparities in their demographic history. Although all the taxa experienced demographic expansion, they differ in the timing and rate of population growth. All three *Tetraclita* species show an exponential growth in effective population size as inferred from BSP. In contrast, the two *C. malayensis* clades only exhibit more gradual and gentle increases in effective population size. This can probably be attributed to the higher flexibility in habitat requirement of *C. malayensis* over *T. squamosa*. A more catholic habitat requirement would allow species to survive better when the availability of habitats is reduced by the emergence of landmasses (Wang 1999; Voris 2000). Accordingly, *C. malayensis* would be more likely to persist during the Pleistocene glaciations as compared to *T. squamosa*.

The differential species responses to glaciations are also evident among taxa with different geographic ranges. Taxa associated with continental landmasses, viz. *T. squamosa s. str.*, *T. singaporensis* and the IM clade of *C. malayensis*, apparently experienced a more recent and rapid population expansion than those with an oceanic/island distribution, i.e., *T. kuroshioensis* and the *C. malayensis* SCS clade. During glaciations, the lowering of sea level caused exposure of landmasses and loss of suitable habitats, yet these changes would have differential effects among different habitats. While habitat availability along the continents was severely reduced by glaciations, outer islands were relatively less affected (Xu & Oda 1999; Voris 2000). Barnacles that inhabited the mainland coastal area would, therefore, be more affected by glaciations, and as a result experience severe population decline, restricted gene flow, or even local extinctions. On the other hand, species that inhabited islands would be affected to a lesser extent. Thus, island regions may serve as refugia for many coastal marine species (Maggs et al. 2008), resulting in relatively stable effective population sizes. This might partly explain the recent finding that many marine species were able to maintain an unexpected large effective population size persisting through the LGM period (Wares & Cunningham 2005; Hickerson & Cunningham 2006; Marko et al. 2010). It appears, therefore, that both micro- and macrohabitat preferences play an important role in determining the demographic and phylogeographic patterns of coastal marine taxa.

On the timing of onset of demographic expansion, *T. singaporensis* and the IM clade of *C. malayensis* started their population expansions at about 100,000 years before present. Both inhabited the Malay Peninsula, which was largely land area due to the exposure of the Sunda shelf during the Pleistocene glaciations (Voris 2000). Hence, it is not surprising that the two taxa suffered severe population crashes, as reported in many other taxa in the region (e.g., Barber et al. 2002; Lourie et al. 2005) and could only attain their present range following the rise of sea level. The concordance in timing and overlapping distribution patterns provide clear evidence for the shared demographic history of these two taxa. The demographic history of the other lineages are more mixed, and relatively little similarity can be seen. This might partly be a result of differences in their geographic ranges. Yet the population expansion of *T. kuroshioensis* and the SCS clade of *C. malayensis* significantly pre-dated the LGM (both occurred > 200,000 years before present). This indicates that taxa with oceanic or island distributions in the West Pacific possibly did not suffer population declines during the LGM. Whether this phenomenon applies to species in other oceans or regions requires further study.

6   CONCLUSIONS

In the present review, we have summarized our recent findings on the phylogeography and genetic diversity of intertidal acorn barnacles in the IWP region. We have revealed cryptic species diversity of "widespread" common species that again calls for more thorough studies on the alpha taxonomy on the IWP biota. Sea level fluctuations with glacial episodes are known to exert considerable effects on the genetic structure of marine animals, resulting in population subdivision and population crashes. However, discordance in population structure is commonly attributed to individual biological traits. We propose that adult habitat specificity, rather than pelagic larval dispersal capacity, plays a more important role in determining the phylogeographic patterns of these intertidal rocky shore fauna. Moreover, taxa with island distributions are less severely affected by the emergence of landmasses during the Pleistocene than those associated with the continents.

ACKNOWLEDGEMENTS

We would like to thank Kee Alfian (National University of Malaysia, Malaysia), Monthon Gamanee (King Mongkut's Institute of Technology Ladkrabang, Thailand), Romy Prabowo (Jendera Soedirman University, Indonesia) and Leena Wong (Universiti Putra Malaysia) in assisting with sampling of barnacles. The work described in this paper was substantially supported by research grants from the Research Grants Council, Hong Kong SAR (Project nos. HKU7597/05M and CUHK4635/09M) and the Research Committee, The Chinese University of Hong Kong (2030394 and 3110040). BKKC was supported by a grant from the Career Development Award, Academia Sinica, Taiwan (AS-98-CDA-L15).

REFERENCES

Avise, J.C. 2000. *Phylogeography: The History and Formation of Species*. Cambridge, Massachusetts: Harvard University Press.

Ayre, D.J., Minchinton, T.E. & Perrin, C. 2009. Does life history predict past and current connectivity for rocky intertidal invertebrates across a marine biogeographic barrier? *Mol. Ecol.* 18: 1887–1903.

Barber, P.H., Palumbi, S.R., Erdmann, M.V. & Moosa, M.K. 2000. A marine Wallace's Line? *Nature* 406: 692–693.

Barber, P.H., Palumbi, S.R., Erdmann, M.V. & Moosa, M.K. 2002. Sharp genetic breaks among populations of *Haptosquilla pulchella* (Stomatopoda) indicate limits to larval transport: patterns, causes, and consequences. *Mol. Ecol.* 11: 659–674.

Barber, P.H., Erdmann, M.V. & Palumbi, S.R. 2006. Comparative phylogeography of three codistributed stomatopods: origins and timing of regional lineage diversification in the Coral Triangle. *Evolution* 60: 1825–1839.

Beck, M.W. 2003. The sea around: conservation planning in marine regions. In: Groves, C.R. (ed.), *Drafting a Conservation Blueprint: A Practitioner's Guide to Planning for Biodiversity*: 319–344. Washington DC: Island Press.

Benzie, J.A.H. 1998. Genetic structure of marine organisms and SE Asian biogeography. In: Hall, R. & Holloway, J.D. (eds.), *Biogeography and Geological Evolution of SE Asia*: 197–209. Leiden: Backbuys Publishers.

Bermingham, E. & Moritz, C. 1998. Comparative phylogeography: concepts and applications. *Mol. Ecol.* 7: 367–369.

Bowen, B., Bass, A.L., Rocha, L.A., Grant, W.S. & Robertson, D.R. 2001. Phylogeography of trumpetfishes (*Aulostomus*): ring species complex on a global scale. *Evolution* 55: 1029–1039.

Chan, B.K.K. 2003. Studies on *Tetraclita squamosa* and *Tetraclita japonica* II: larval morphology and development. *J. Crust. Biol.* 23: 522–547.

Chan, B.K.K., Morritt, D. & Williams, G.A. 2001. The effect of salinity and recruitment on the distribution of *Tetraclita squamosa* and *Tetraclita japonica* (Cirripedia: Balanomorpha) in Hong Kong. *Mar. Biol.* 138: 999–1009.

Chan, B.K.K., Tsang, L.M. & Chu, K.H. 2007a. Morphological and genetic differentiation of the acorn barnacle *Tetraclita squamosa* (Crustacea, Cirripedia) in East Asia and description of a new species of *Tetraclita*. *Zool. Scr.* 36: 79–91.

Chan, B.K.K., Tsang, L.M. & Chu, K.H. 2007b. Cryptic diversity of *Tetraclita squamosa* complex (Crustacea, Cirripedia) in Asia: description of a new species from Singapore. *Zool. Stud.* 46: 46–56.

Crandall, E.D., Frey, M.A., Grosberg, R.K. & Barber, P.H. 2008. Contrasting demographic history and phylogeographical patterns in two Indo-Pacific gastropods. *Mol. Ecol.* 17: 611–626.

Dawson, M.N., Louie, K.D., Barlow, M., Jacobs, D.K. & Swift, C.C. 2002. Comparative phylogeography of sympatric sister species, *Clevelandia ios* and *Eucyclogobius newberryi* (Teleostei, Gobiidae), across the California Transition Zone. *Mol. Ecol.* 11: 1065–1075.

Drummond, A.J. & Rambaut, A. 2007. BEAST: Bayesian evolutionary analysis by sampling trees. *BMC Evol. Biol.* 7: 214.

Drummond, A.J., Rambaut, A., Shapiro, B. & Pybus, O.G. 2005. Bayesian coalescent inference of past population dynamics from molecular sequences. *Mol. Biol. Evol.* 22: 1185–1192.

Duda, T.F. & Palumbi, S.R. 1999. Population structure of the black tiger prawn, *Penaeus monodon*, among western Indian Ocean and western Pacific populations. *Mar. Biol.* 134: 705–710.

Fernando, S.A. 2006. *Monograph on Indian Barnacles.* Kochi: Ocean Science and Technology Cell on Marine Benthos.

Grant, W.S. & Bowen, B.W. 1998. Shallow population histories in deep evolutionary lineages of marine fishes: insights from sardines and anchovies and lessons for conservation. *J. Hered.* 89: 415–426.

Grosberg, R.K. & Cunningham, C.W. 2001. Genetic structure in the sea: from populations to communities. In: Bertness, M., Gaines, S.D. & Hay, M.E. (eds.), *Marine Community Ecology*: 61–84. Sunderland, MA: Sinauer Associates.

Hewitt, G.M. 1996. Some genetic consequences of ice ages, and their role in divergence and speciation. *Biol. J. Linn. Soc.* 58: 247–276.

Hewitt, G.M. 2000. The genetic legacy of the Quaternary ice ages. *Nature* 405: 907–913.

Hewitt, G.M. 2004. The structure of biodiversity—insights from molecular phylogeography. *Front. Zool.* 4: 1–16.

Hickerson, M.J. & Cunningham, C.W. 2006. Nearshore fish (*Pholis gunnellus*) persists across the North Atlantic through multiple glacial episodes. *Mol. Ecol.* 15: 4095–4107.

Hughes, T.P., Bellwood, D.R. & Connolly, S.R. 2002. Biodiversity hotspots, centres of endemicity, and the conservation of coral reefs. *Ecol. Lett.* 5: 775–784.

Koh, L.L., O'Riordan, R.M. & Lee, W.J. 2005. Sex in the tropics: reproduction of *Chthamalus malayensis* Pilsbry (Class Cirripedia) at the equator. *Mar. Biol.* 147: 121–133.

Lambeck, K., Esat, T.M. & Potter, E.K. 2002. Links between climate and sea levels for the past three million years. *Nature* 419: 199–206.

Lavery, S., Moritz, C. & Fielder, D.R. 1996. Indo-Pacific population structure and evolutionary history of the coconut crab *Birgus latro*. *Mol. Ecol.* 5: 557–570.

Lessios, H.A., Kane, J. & Robertson, D.R. 2003. Phylogeography of the pantropical sea urchin *Tripneustes*: contrasting patterns of population structure between oceans. *Evolution* 57: 2026–2036.

Lourie, S.A. & Vincent, A.C.J. 2004. Using biogeography to help set priorities in marine conservation. *Conserv. Biol.* 18: 1004–1020.

Lourie, S.A., Green, D.M. & Vincent, A.C.J. 2005. Dispersal, habitat differences, and comparative phylogeography of Southeast Asian seahorses (Syngnathidae: *Hippocampus*). *Mol. Ecol.* 14: 1073–1094.

Maggs, C.A., Castilho, R., Foltz, D., Henzler, C., Jolly, M.T., Kelly, J., Olsen, J., Perez, K.E., Stam, W., Väinölä, R., Viard, F. & Wares, J. 2008. Evaluating signatures of glacial refugia for North Atlantic benthic marine taxa. *Ecology* 89: S108–S122.

Marko, P.B., Hoffman, J.M., Emme, S.A., McGovern, T.M., Keever, C.C. & Cox, L.N. 2010. The "Expansion-Contraction" model of Pleistocene biogeography: rocky shores suffer a sea change? *Mol. Ecol.* 19: 146–169.

Mora, C., Chittaro, P.M., Sale, P.F., Kritzer, J.P. & Ludsin, S.A. 2003. Patterns and processes in reef fish diversity. *Nature* 421: 933–936.

Newman, W.A. & Ross, A. 1976. Revision of the balanomorph barnacles; including a catalogue of the species. *San Diego Soc. Nat. Hist. Memoir* 9: 1–108.

Palumbi, S.R. 1994. Genetic divergence, reproductive isolation, and marine speciation. *Annu. Rev. Ecol. Syst.* 25: 547–572.

Palumbi, S.R. 1995. Molecular biogeography of the Pacific. *Coral Reefs* 16: S47–S52.

Patarnello, T., Volckaert, F.A.M.J. & Castilho, R. 2007. Pillars of Hercules: is the Atlantic-Mediterranean transition a phylogeographical break? *Mol. Ecol.* 16: 4426–4444.

Reid, D.G., Lal, K., Mackenzie-Dodds, J., Kaligis, F., Littlewood, D.T.J. & Williams, S.T. 2006. Comparative phylogeography and species boundaries in *Echinolittorina* snails in the central Indo-West Pacific. *J. Biogeogr.* 33: 990–1006.

Roberts, C.M., McClean, C.J., Veron, J.E.N., Hawkins, J.P., Allen, G.R., McAllister, D.E., Mittermeier, C.G., Schueler, F.W., Spalding, M., Wells, F., Vynne, C. & Werner, T.B. 2002. Marine biodiversity hotspots and conservation priorities for tropical reefs. *Science* 295: 1280–1284.

Rocha, L.A., Bass, A.L., Robertson, R. & Bowen, B.W. 2002. Adult habitat preferences, larval dispersal, and the comparative phylogeography of three Atlantic surgeonfishes (Teleostei: Acanthuridae). *Mol. Ecol.* 11: 243–252.

Rogers, A.R. & Harpending, H. 1992. Population growth makes waves in the distribution of pairwise genetic differences. *Mol. Biol. Evol.* 9: 552–569.

Soltis, D.E., Morris, A.B., McLachlan, J.S., Manos, P.S. & Soltis, P.S. 2006. Comparative phylogeography of unglaciated eastern Northern America. *Mol. Ecol.* 15: 4261–4293.

Southward, A.J. & Newman, W.A. 2003. A review of some common Indo-Malayan and western Pacific species of *Chthamalus* barnacles (Crustacea: Cirripedia). *J. Mar. Biol. Assoc. UK.* 83: 797–812.

Taylor, M.S. & Hellberg, M.E. 2003. Genetic evidence for local retention of pelagic larvae in a Caribbean reef fish. *Science* 299: 107–109.

Tsang, L.M., Chan, B.K.K., Wu, T.H., Ng, W.C., Chatterjee, T., Williams, G.A. & Chu, K.H. 2008. Population differentiation in the barnacle *Chthamalus malayensis*: postglacial colonization and recent connectivity across the Pacific and Indian Oceans. *Mar. Ecol. Prog. Ser.* 364: 107–118.

Voris, H.K. 2000. Maps of Pleistocene sea levels in Southeast Asia: shorelines, river systems and time durations. *J. Biogeogr.* 27: 1153–1167.

Wang, P. 1999. Response of western Pacific marginal seas to glacial cycles: paleoceanographic and sedimentological features. *Mar. Geol.* 156: 5–39.

Wares, J.P. 2001. Patterns of speciation inferred from mitochondrial DNA in North American *Chthamalus* (Cirripedia: Balanomorpha: Chthamaloidea). *Mol. Phylogenet. Evol.* 18: 104–116.

Wares, J.P. 2002. Community genetics in the northwestern Atlantic intertidal. *Mol. Ecol.* 11:

1131–1144.

Wares, J.P. & Cunningham, C.W. 2001. Phylogeography and historic ecology of the North Atlantic intertidal. *Evolution* 55: 2455–2469.

Wares, J.P. & Cunningham, C.W. 2005. Diversification before the most recent glaciation in *Balanus glandula*. *Biol. Bull.* 208: 60–68.

Waters, J.M. 2008. Marine biogeographical disjunction in temperate Australia: historical land-bridge, contemporary currents, or both? *Diversity Distri.* 14: 692–700.

Williams, S.T. & Benzie, J.A.H. 1998. Evidence of a biogeographic break between populations of a high dispersal starfish: congruent regions within the Indo-West Pacific defined by color morphs, mtDNA, and allozyme data. *Evolution* 52: 87–99.

Williams, S.T., Jara, J., Gomez, E. & Knowlton, N. 2002. The marine Indo-West Pacific break: contrasting the resolving power of mitochondrial and nuclear genes. *Integr. Comp. Biol.* 42: 941–952.

Xu, X. & Oda, M. 1999. Surface-water evolution of the eastern East China Sea during the last 36,000 years. *Mar. Geol.* 156: 285–304.

Yan, Y. & Chan, B.K.K. 2001. Larval development of *Chthamalus malayensis* (Cirripedia: Thoracica) reared in the laboratory. *J. Mar. Biol. Assoc. UK.* 81: 623–632.

Yan, Y., Chan, B.K.K. & Williams, G.A. 2006. Reproductive development of the barnacle *Chthamalus malayensis* in Hong Kong: implications for the life-history patterns of barnacles on seasonal, tropical shores. *Mar. Biol.* 148: 875–887.

Evolution and conservation of marine biodiversity in the Coral Triangle: insights from stomatopod Crustacea

Paul H. Barber[1], Samantha H. Cheng[1,2], Mark V. Erdmann[3], Kimberly Tenggardjaja[4] & Ambariyanto[5]

[1] *Department of Ecology and Evolutionary Biology, University of California at Los Angeles, Los Angeles, U. S. A.*

[2] *Joint Science Department, Scripps College, Claremont, U. S. A.*

[3] *Conservation International, Renon Denpasar, Indonesia*

[4] *Department of Ecology and Evolutionary Biology, University of California at Santa Cruz, Santa Cruz, U. S. A.*

[5] *Faculty of Fisheries and Marine Sciences, Diponegoro University, Semarang, Indonesia*

ABSTRACT

The Coral Triangle is the global epicenter of marine biodiversity as well as one of the world's most imperiled ecosystems. Conserving the biodiversity of the Coral Triangle requires understanding the mechanisms by which this hotspot has evolved. Although there are several theories to explain the origins of this biodiversity hotspot, the center of origin theory has been criticized for lacking credible speciation mechanisms within the center of the Coral Triangle. In this paper, we examine the phylogeography and evolutionary history of coral reef-associated stomatopods. Results indicate significant regional population genetic structure in all 14 species examined. Several species exhibit patterns of Indian-Pacific Ocean differentiation, suggesting vicariance during Pleistocene low sea level stands may promote lineage diversification. However, all species show a pronounced pattern of differentiation between populations east and west of the Maluku Sea, although the precise location of this phylogeographic boundary varies among species. This pattern is most likely explained by limited water flow by the Halmahera Eddy, and thus larval transport and gene flow, across the Maluku Sea, promoting lineage diversification in this region and supporting the notion of a center of origin. In contrast, Bayesian analysis of regional patterns of differentiation suggested that the most basal populations of *Haptosquilla pulchella* and *H. glyptocercus* were found in reefs in the peripheral parts of their Pacific Ocean ranges, potentially suggesting support for the center of accumulation model. However, topology constraint tests indicated that the best trees were not significantly better than alternate topologies, limiting the inferences that can be drawn from this data. Although many processes are likely contributing to the Coral Triangle biodiversity hotspot, data from stomatopods clearly shows evidence for filters to dispersal and gene flow within the center, a result consistent with a center of origin. The identification of these barriers in stomatopods and many other unrelated marine fish and invertebrates is helping define regional limits of connectivity within the Coral Triangle, supporting regional conservation planning in this threatened ecosystem.

1 CORAL TRIANGLE BIODIVERSITY

The Coral Triangle, a region comprised of Malaysia, the Philippines, Indonesia, East Timor, Papua New Guinea, and the Solomon Islands (Hoeksema 2007), is the global epicenter of marine biodi-

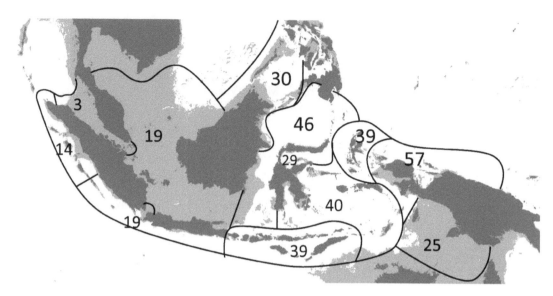

Figure 1. Distribution of reef-associated stomatopod species richness throughout Indonesia and the southern Philippines highlighting elevated biodiversity in eastern Indonesia. Lines correspond to boundaries delineated for Indonesia in the "Marine Ecoregions of the World" (Spalding et al. 2007).

versity. Biodiversity peaks within this region, with species richness decreasing longitudinally and latitudinally away from this center (Bellwood & Wainwright 2002; Mora et al. 2003). While this pattern has been known for decades, the evolutionary origins of this pattern are not well understood. With sharp declines in coral reef ecosystems in the past decade, particularly those in the Coral Triangle (Burke et al. 2002), there is an urgent need to preserve the biodiversity of the Coral Triangle and the processes generating this biodiversity (Briggs 2003, 2005a). As such, there has been a renewed interest in understanding the origins of this biodiversity hotspot (Allen & Werner 2002; Barber 2009; Bellwood & Meyer 2009; Halas & Winterbottom 2009; Malay & Paulay 2010).

There are three major theories proposed to explain elevated marine biodiversity in the Coral Triangle. The center of overlap theory (Woodland 1983) suggests that independently derived Pacific and Indian Ocean faunas overlap in the Coral Triangle. The admixture of these two regional faunas results in elevated biodiversity where these oceans meet in the Coral Triangle. Alternatively, the center of accumulation theory (Ladd 1960) suggests that speciation occurs on the distant islands and atolls of the Pacific and Indian Ocean, promoting allopatric speciation in these peripheral reef environments. Biodiversity generated in the periphery subsequently accumulates in the center as species disperse and expand their ranges, creating a biodiversity hotspot. Last, the center of origin theory (Ekman 1953) suggests that speciation occurs within the Coral Triangle, and that some of this diversity is exported to peripheral habitats in the Pacific and Indian Oceans, resulting in decreasing latitudinal and longitudinal biodiversity gradients.

Recent studies have supported either the center of origin (Briggs 1992, 2000, 2003; Mora et al. 2003; Briggs 2005b) or the center of accumulation (Jokiel & Martinelli 1992; Connolly et al. 2003) and some have found no clear support for any of the three hypotheses (Barber & Bellwood 2005; Bellwood & Meyer 2009; Halas & Winterbottom 2009). Others alternatively suggest that all three processes contribute to elevated marine biodiversity in the Coral Triangle (Palumbi 1997; Wilson & Rosen 1998; Barber & Bellwood 2005; Hoeksema 2007). Furthermore, as the Coral Triangle contains the largest amount of reef area in the world, extinction rates are likely to be lower (MacArthur & Wilson 1967), thus promoting increased survival of biodiversity in this region, creating a "center of survival" (Barber & Bellwood 2005; Bellwood & Meyer 2009).

Table 1. List of reef-associated stomatopod Crustacea known from the Coral Triangle.

Gonodactylidae Giesbrecht 1910

Gonodactylaceus falcatus (Forskål 1775)
Gonodactylaceus glabrous (Brooks 1886)
Gonodactylaceus graphurus (Miers 1875)
Gonodactylaceus ternatensis (de Man 1902)
Gonodactylellus affinis (de Man 1902)
Gonodactylellus annularis Erdmann & Manning 1998
Gonodactylellus barberi Ahyong & Erdmann 2007
Gonodactylellus caldwelli Erdmann & Manning 1998
Gonodactylellus demanii (Henderson 1893)
Gonodactylellus erdmanni Ahyong 2001
Gonodactylellus espinosus (Borradaile 1898b)
Gonodactylellus incipiens (Lanchester 1903)
Gonodactylellus kandi Ahyong & Erdmann 2007
Gonodactylellus lanchesteri (Manning 1967)
Gonodactylellus micronesicus (Manning 1971)
Gonodactylellus rubriguttatus Erdmann & Manning 1998
Gonodactylellus snidsvongi (Naiyanetr 1987)
Gonodactylellus viridis (Scrène 1954)
Gonodactylellus sp. A (*demani* group)
Gonodactylellus sp. B
Gonodactylellus sp. F
Gonodactylellus sp. G (*molyneux* group)
Gonodactyloideus cracens Manning 1984
Gonodactylopsis drepanophora (de Man 1902)
Gonodactylopsis komodoensis Erdmann & Manning 1998
Gonodactylopsis sp. A
Gonodactylopsis sp. B
Gonodactylopsis sp. C
Gonodactylus botti Manning 1975a
Gonodactylus childi Manning 1971
Gonodactylus chiragra (Fabricius 1781)
Gonodactylus platysoma (Wood-Mason 1895)
Gonodactylus smithii Pocock 1893
Gonodactylus sp. A
Hoplosquilla said Erdmann & Manning 1998

Odontodactylidae Manning 1980

Odontodactylus brevirostris (Miers 1884)
Odontodactylus cultrifer (White 1850)
Odontodactylus japonicus (de Haan 1844)*
Odontodactylus latirostris Borradaile 1907
Odontodactylus scyllarus (Linnaeus 1758)
Odontodactylus sp. A

Takuidae Manning 1995

Mesacturus furcicaudatus (Miers 1880b)
Taku spinosocarinatus (Fukuda 1909)

Protosquillidae Manning 1980

Chorisquilla andamanica Manning 1975b
Chorisquilla brooksii (de Man 1888)
Chorisquilla gyrosa (Odhner 1923)
Chorisquilla hystrix (Nobili 1899)
Chorisquilla mehtae Erdmann & Manning 1998
Chorisquilla pococki (Manning 1975b)
Chorisquilla spinosissima (Pfeffer 1888)
Chorisquilla trigibbosa (Hansen 1926)
Chorisquilla sp. A
Chorisquilla sp. B
Echinosquilla guerini (White 1861)
Haptosquilla glabra (Lenz 1905)
Haptosquilla glyptocercus (Wood-Mason 1875)
Haptosquilla moosai Erdmann & Manning 1998
Haptosquilla philippinensis Garcia & Manning 1982
Haptosquilla proxima (Kemp 1915)
Haptosquilla pulchella (Miers 1880a)
Haptosquilla pulchra (Hansen 1926)
Haptosquilla setifera Manning 1969
Haptosquilla stoliura (Müller 1886)
Haptosquilla togianensis Erdmann & Manning 1998
Haptosquilla trispinosa (Dana 1852)
Haptosquilla tuberosa (Pocock 1893)
Haptosquilla sp. A
Haptosquilla sp. B
Haptosquilla sp. C (cf. *stoliura*)
Haptosquilla sp. D
Haptosquilla sp. E (cf. *tuberosa*)
Siamosquilla laevicaudata (Sun & Yang 1998)
Siamosquilla sp. A

Pseudosquillidae Manning 1977

Pseudosquilla ciliata (Fabricius 1787)
Pseudosquilla megalophthalma (Bigelow 1893)
Pseudosquillana richeri (Moosa 1991)
Raoulserenea hieroglyphica (Manning 1972)
Raoulserenea komaii (Moosa 1991)
Raoulserenea ornata (Miers 1880a)
Raoulserenea oxyrhyncha (Borradaile 1898a)

Lysiosquillidae Giesbrecht 1910

Lysiosquilla sulcirostris Kemp 1913
Lysiosquilla tredecimdentata Holthuis 1941
Lysiosquillina lisa Ahyong & Randall 2001
Lysiosquillina maculata (Fabricius 1793)
Lysiosquillina sulcata (Manning 1978)
Lysiosquilloides mapia Erdmann & Boyer 2003
Lysiosquilloides sp. A

* in de Haan (1833–1850).

In an area as large and geologically and oceanographically complex as the Coral Triangle, high biodiversity almost certainly has pluralistic origins. However, the center of origin theory has been repeatedly criticized on the grounds that 1) it lacks a convincing mechanism to create diversity within the Coral Triangle (Jokiel & Martinelli 1992), 2) it must therefore rely on sympatric speciation to create novel biodiversity (Briggs 1999), and 3) studies supporting the center of origin theory have been largely descriptive, rather than analytical in nature (Santini & Winterbottom 2002). In this paper, we use stomatopod crustaceans as a model to test the center of origin hypothesis and determine if lineage diversification is truly occurring within the Coral Triangle. Examining the possibility of speciation occurring within the Coral Triangle not only contributes to our understanding of the evolution of this biodiversity hotspot, but it also addresses key questions regarding the nature of speciation in marine environments. In addition, understanding the evolutionary dynamics driving lineage diversification in the Coral Triangle provides information on regional connectivity patterns that are essential to help conserve existing biodiversity in this region as well as the processes that have created it and maintain it.

1.1   Stomatopod Crustacea

Stomatopods, also known as mantis shrimp, are benthic marine crustaceans that occur predominantly in coral reef and sea grass habitats in the tropical oceans of the world, though some are found in temperate waters and a few species are deep-sea specialists that occur in depths of up to 1,500 meters. Stomatopods are a diverse group, with at least 450 described species representing over 100 genera in 17 families. Though related to the more commonly known decapods such as crabs, shrimp, and lobsters, stomatopods are quite different from these groups in having two enlarged raptorial appendages (much like praying mantis insects) instead of pincers as their primary defensive and predatory appendages.

Stomatopods are generally very common on coral reefs, inhabiting a variety of habitats including live coral, coral rubble, coralline algae cavities on vertical walls, sand burrows, and sponges. Found from the high intertidal to depths exceeding 60 meters, there are 87 species (69 described) of reef-associated stomatopods known from the Coral Triangle (Table 1). Biodiversity of this group peaks in eastern Indonesia (West Papua and Teluk Cenderawasih; Figure 1), matching biodiversity patterns in corals (Veron et al. 2009) although differing slightly from fishes (Carpenter & Springer 2005). The most common stomatopods on coral reefs are in the superfamily Gonodactyloidea. These are largely solitary species that live in cavities within coral and coral rubble, making them quite cryptic. Stomatopods brood their eggs like most crustaceans, and following hatching they undergo a large number of larval instars (Manning & Provenzano 1963; Provenzano & Manning 1978). Stomatopod larvae are presumed to be pelagic as they have been captured in both deepwater (Gurney 1946; Michel 1968) and offshore plankton tows (Michel 1968, 1970). In addition, gonodactylid stomatopods have successfully recolonized Krakatau following the 1883 eruption, a process that required pelagic dispersal (Barber et al. 2002a).

Although stomatopods have fewer taxa than other more commonly studied groups of reef organisms (i.e., reef fishes and scleractinian corals), they are an excellent indicator of biodiversity as their species diversity patterns have been shown to closely parallel those of reef fish and corals (e.g., Wallace et al. 2002). Furthermore, they have the advantage that species accumulation or rarefaction curves tend to approach their asymptotes more quickly due to the lower overall number of species present (Erdmann unpublished data). This result suggests that understanding processes driving diversification of reef-dwelling stomatopods in the Coral Triangle should help inform the processes generating biodiversity in other reef-dwelling taxa.

1.2 Allopatric divergence in the sea

Although speciation may occur in sympatry (Lande 1981; Chesser 1994; Dieckmann & Doebeli 1999; Higashi et al. 1999; Via 2001; Berlocher & Feder 2002) or parapatry via ecological gradients (Schneider et al. 1998, 1999; Moritz et al. 2000), allopatric speciation is the dominant paradigm in evolutionary biology, whether it is through vicariance or the isolation of peripheral populations (Mayr 1942; Bush 1975; Futuyma & Mayer 1980; Lynch 1989; Barraclough et al. 1998; Turelli et al. 2001; Coyne & Orr 2004). In the allopatric model, physical barriers to dispersal limit genetic exchange among populations. In the absence of gene flow, the differential effects of mutation, genetic drift, and selection in these isolated populations will result in regional differentiation, culminating in speciation (Dobzhansky 1937, 1970). The appropriateness of the allopatric model in terrestrial ecosystems is viewed as largely self-evident and is supported by a plethora of biogeographic and phylogeographic studies (see books by Nelson & Platnick 1981; Avise 2000).

Mayr (1954) extended the allopatric model to speciation in marine environments—evidenced clearly in coastal species in response to the closure of the Isthmus of Panama approximately 3.1 Million Years Ago (MYA). As the Isthmus of Panama rose, the Panamanian Seaway was closed, isolating populations in the Caribbean and eastern Pacific, resulting in allopatric speciation (Lessios & Cunningham 1990; Knowlton et al. 1993; Cunningham & Collins 1994; Knowlton & Weigt 1998; Lessios 1998; Marko 2002). Allopatric divergence in the sea has also been suggested to result from founder events (Paulay & Meyer 2002), Pleistocene glacial cycles (McManus 1985; Potts 1985; Fleminger 1986; Woodland 1986; Springer & Williams 1990) and other geographic barriers to dispersal (Terry et al. 2000; Hare et al. 2002). However, the general applicability of the allopatric model in marine environments has been questioned. The majority of marine species have a bipartite life history where adults are either completely or relatively sedentary, and dispersal is achieved through a larval phase that can travel for hours, days, weeks, or even months on ocean currents. Larval duration and dispersal distance are generally correlated (Shanks et al. 2003; Shanks 2009), and some larvae can travel tremendous distances (Scheltema 1971, 1986). Because of this dispersal potential, it has been hypothesized that larval dispersal should facilitate connectivity (demographic and genetic exchange) among distant reefs (Roughgarden et al. 1988; Roberts 1997; Armsworth 2002), and the resulting genetic exchange (gene flow or "genetic connectivity") should mitigate genetic divergence among them (Scheltema 1971, 1986). Thus, in an open marine system, the apparant lack of clear dispersal barriers presents a conundrum to the process of allopatric speciation in the sea (Palumbi 1992).

While this early view of larval mediated connectivity was largely supported by genetic studies (see Rosenblatt & Waples 1986 for review and exceptions; Palumbi 1992, 1994; Lessios et al. 1998), a more complicated picture has gradually immerged. Knowlton and Keller (1986) first showed that larval crustaceans can fail to reach their dispersal potential. Since then, literature has shown that realized dispersal distances are often quite smaller than those estimated from pelagic larval duration (Weersing & Toonen 2009) and inferences from genetic data (Shanks 2009). Thus, while pelagic larval dispersal is the only mechanism to promote connectivity among populations of many benthic marine organisms, the relationship between larval duration and realized connectivity and the processes governing this relationship remains unclear.

1.3 Allopatric divergence in the Coral Triangle

Contemporary oceanography of the Coral Triangle is dominated by the Indonesian Throughflow, which transports water at a rate of up to 20×10^6 m^3/s from the Pacific to the Indian Ocean through the Indonesian Archipelago (Godfrey 1996; Gordon & Fine 1996). Dominant surface currents within Indonesia can achieve speeds of 1 m/s (Wyrtki 1961), and surface drifters and current mea-

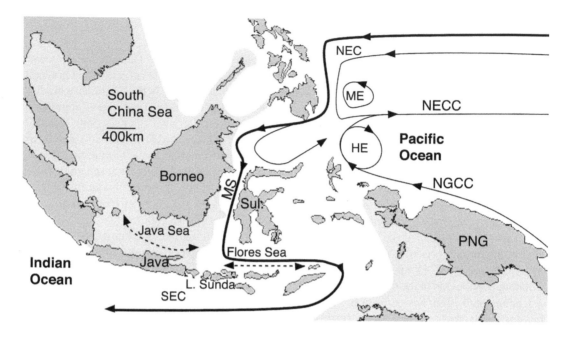

Figure 2. Generalized physical oceanography of the Coral Triangle. The dark line indicates the path of the Indonesian Throughflow, which originates from the North Equatorial Current (NEC), passes through the Makassar Straight (MS) and the Flores Sea before entering the Indian Ocean. The New Guinea Coastal Current is retroflected in the area of Halmahera creating the Halmahera Eddy (HE) before joining the North Equatorial Counter Current (NECC). Dashed lines indicate seasonal current reversals in the Java and Flores Seas. Light shading indicates approximate coastlines during Pleistocene glacial periods when sea levels dropped up to 130 m.

surements in the upper 300 m (Lukas et al. 1991) confirm the potential for the Indonesian Through-flow to disperse plankton larvae a great distance. However, potential barriers to dispersal exist among reef ecosystems in the Coral Triangle limiting dispersal among populations, thus promoting lineage diversification.

First, sea levels dropped 130 m below the present level during Pleistocene glacial periods (Figure 2). As a result, the shallow Sunda and Sahul shelves were exposed, significantly constricting the waterways of the Coral Triangle (Voris 2000). The constriction of the pathway between the Pacific and Indian Ocean was augmented by coldwater upwelling at the head of the Indonesian Throughflow (Fleminger 1986), potentially limiting larval dispersal and gene flow and promoting allopatric divergence of Pacific and Indian Ocean populations. Genetic studies of numerous taxa, including fish (Lacson & Clark 1995), crabs (Lavery et al. 1996), prawns (Duda & Palumbi 1999), sea stars (Williams & Benzie 1998; Benzie 1999), and mangroves (Duke et al. 1998), all indicate strong intraspecific divergence among Indian and Pacific Ocean populations. The recovery of concordant phylogeographic patterns in multiple codistributed taxa strongly suggests a common physical process acting in their formation (Avise 2000), and all of the above studies suggest that the phylogeographic patterns result from vicariance during Pleistocene low sea level stands.

Second, while physical oceanography can facilitate dispersal, it is also possible for currents to constrain dispersal. The New Guinea Coastal Current (NGCC) travels northwest along the northern shores of Papua New Guinea until it is retroflected in the region of Halmahera, creating the Eddy (Lukas et al. 1991, 1996; Nof 1996). These waters are subsequently transported back to the east on the North Equatorial Counter Current, limiting water transport from Papua New Guinea (PNG)

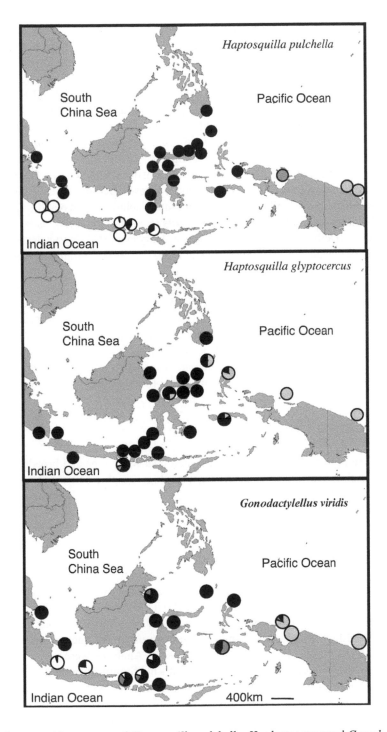

Figure 3. Phylogeographic structure of *Haptosquilla pulchella*, *H. glyptocercus* and *Gonodactylellus viridis* across the Coral Triangle (after Barber et al. 2006) showing both divergence among Pacific and Indian Ocean populations as well as divergence between eastern and central Indonesia (see Figure 3 in Color insert).

to Indonesia across the Maluku Sea (Figure 2). Nearly 90% of Indonesian Throughflow waters originate from the North Equatorial Current (Nof 1995), highlighting the lack of transport of NGCC waters from Papua New Guinea to Indonesia. Limited water transport across the Maluku Sea should result in limited larval transport and gene flow across this region, providing a potential mechanism for allopatric divergence.

Previous studies of stomatopods in the Coral Triangle have revealed strong patterns of phylogeographic structure (Figure 3). Early work by Barber et al. (2000) showed a strong pattern of Pacific-Indian Ocean divergence in *Haptosquilla pulchella*, as well as fine-scale isolation in the South China Sea, Celebes Sea, Flores Sea, and Tomini Bay—a pattern attributed to allopatric divergence during Pleistocene low sea level stands. Subsequent studies on *H. pulchella* also indicated divergence among populations from Papua New Guinea and Indonesia (Barber et al. 2002b), resulting in the hypothesis that this pattern is due to the isolating effects of the Halmahera Eddy, which limits water transport (and thus larval dispersal) across the Maluku Sea. This hypothesis was supported by the recovery of similar patterns of regional isolation in two additional stomatopods, *Haptosquilla glyptocercus* and *Gonodactylellus viridis* (Barber et al. 2006), as well as more recent work on giant clams (DeBoer et al. 2008; Kochzius & Nuryanto 2008; Nuryanto & Kochzius 2009) and echinoderms (Crandall et al. 2008b) that show regional isolation among eastern and central Indonesia.

In this study, we build on previous genetic studies of stomatopod crustaceans from the Coral Triangle. Specifically, we focus on three key questions. First, we examine the degree to which genetic structure is common in species of stomatopod crustaceans distributed throughout the Coral Triangle. Second, we explore the isolating potential of the Halmahera Eddy by examining limits to connectivity associated with this physical oceanographic feature. Third, we examine patterns of directionality in the regional expansion of genetic diversity to explore whether stomatopods can provide additional insights into the mechanisms promoting high marine biodiversity in the Coral Triangle.

2 MATERIALS AND METHODS

To evaluate the extent to which previously published results are repeated broadly across stomatopod crustaceans in the Coral Triangle, we examine phylogeographic patterns in four additional stomatopods representing four different genera: *Haptosquilla pulchra*, *Hoplosquilla said*, *Siamosquilla laevicaudata*, and *Gonodactylopsis* sp. A (Table 2).

We collected data from mtDNA cytochrome c oxidase subunit I (COI) using standard methods for stomatopods described in Barber et al. (2002b). Quality of edited sequences was verified by alignment and successful translation into proteins. Verified sequences were analyzed in an Analysis of Molecular Variance (AMOVA) as implemented in ARLEQUIN (Schneider et al. 2000) enforcing first, no a priori structure, then with enforced structure following the geographic distribution of phylogroups revealed from neighbor-joining in PAUP* 4 (Swofford 2002), Bayesian analyses conducted in MRBAYES 3.0 (Huelsenbeck & Ronquist 2001) as well as from minimum spanning trees created in ARLEQUIN. Bayesian analysis was run using AIC generated model parameters generated from Modeltest 3.7 (Posada & Crandall 1998). Analyses were run for 500,000 generations, sampling every 200 generations and a 10% burn-in. Significance of AMOVA and pairwise F_{ST} calculations was determined using 10,000 random permutations in ARLEQUIN (Schneider et al. 2000).

To further explore the potential isolating effects of the Halmahera Eddy, we collected mtDNA COI sequence data for seven additional stomatopod taxa (Table 3) from select populations from east and west of the Maluku Sea using standard methods for stomatopods described in Barber et

Table 2. List of species and sample sizes used for phylogeographic analyses.

Species	Number of populations	Total sample
Haptosquilla pulchra	16	194
Hoplosquilla said	16	231
Siamosquilla laevicaudata	13	177
Gonodactylopsis sp. A	10	132

al. (2002b). Due to smaller population sizes, analyses were limited to exploring patterns of divergence using simple neighbor-joining analyses to identify the presence and geographic distribution of unique clades within each species.

Lastly, to explore directionality of lineage diversification, for three taxa with extensive geographic coverage and presence of multiple clades (*Haptosquilla pulchella*, *Haptosquilla glyptocercus*, *Hoplosquilla said*), we used Bayesian analaysis to examine the evolutionary origins of the distinct phylogeographic groups. Additional DNA sequence data was collected for *Haptosquilla pulchella* and for the *H. glyptocercus* species group, extending sampling ranges to East Africa and Fiji, respectively. Bayesian phylogenies were generated in MRBAYES 3.0 (Huelsenbeck & Ronquist 2001) using AIC generated model parameters generated from Modeltest 3.7 (Posada & Crandall 1998). Analyses were run for 500,000 generations, sampling every 10 generations and a 25% burnin. To minimize computational complexity, only several exemplars from each clade were included in the analysis. Additional maximum likelihood analyses were conducted in PAUP* (Swofford 2002) using AIC generated model parameters determined by Modeltest 3.7 (Posada & Crandall 1998). Outgroups included multiple species of closely related taxa based on a large phylogenetic analysis that includes nearly 100% of all described Gonodactyloidea (Barber & Erdmann, unpublished). To determine the robustness of these phylogenetic reconstructions, a series of Shimodaira-Hasegawa tests (Shimodaira & Hasegawa 1999) were implemented in PAUP* (Swofford 2002) in which the maximum likelihood tree(s) was/were compared to maximum likelihood trees obtained using two sets of topological constraints: 1) ancestral clades occurring in the Coral Triangle with subsequent step-wise cladogenesis and expansion towards the periphery of the range, and 2) ancestral clades occurring in the periphery, with subsequent step-wise expansion and cladogenesis occurring towards the center of the Coral Triangle.

Table 3. List of species and sample sizes used for phylogeographic analyses.

Species	West of Maluku Sea		East of Maluku Sea	
	N	Population	*N*	Population
Chorisquilla brooksii	12	Sulawesi, Sumbawa, P. Seribu	11	Halmahera, PNG
Chorisquilla spinosissima	8	Sulawesi, Komodo	3	PNG
Gonodactylus affinis	10	Sulawesi, Lombok, P. Seribu	1	PNG
Gonodactylus childi	4	Sulawesi, Komodo	3	Biak
Gonodactylus chiragra	15	Sulawesi, Komodo, Belitung	2	Biak
Gonodactylus smithii	5	Sulawesi	6	Halmahera, Biak, PNG

3 RESULTS—PHYLOGEOGRAPHY

A total of 658 bp of mtDNA COI was sequenced from 194 individuals of *H. pulchra*, 658 bp from 231 individuals of *H. said*, 580 bp, from 177 individuals of *S. laevicaudata*, and 658 bp from 132 individuals of *Gonodactylopsis* sp. A, revealing 105, 167, 164 and 99 haplotypes, respectively. All sequences translated without any stop codons.

Neighbor-joining and Bayesian analysis trees indicated strong evidence of phylogenetic structure in all four species. Haplotypes of *Haptosquilla pulchra* fell into two major clades separated by 21 mutational steps (3.19% sequence divergence). Clade 2 spanned from Bali to West Papua, while Clade 1 occurred in Papua New Guinea (Figure 4A). Although Clade 2 spanned the Maluku Sea, numerous subclades were differentially distributed across this region with seven subclades (2a-2g) found only in populations east of the Maluku Sea and three subclades (2i, 2j and 2k) found only in populations west of the Maluku Sea (Figure 4B). Unstructured AMOVA analyses indicated strong genetic structure among populations ($\Phi_{ST} = 0.64$, $P < 0.00001$). AMOVA results imposing regional groups between Indonesia and Papua New Guinea indicated $\Phi_{ST} = 0.85$, $P < 0.00001$ with 81.96% of the variation between groups, 3.49% among populations within groups, and 14.55% within populations. AMOVA analyses including only Clade 2 populations yielded $\Phi_{ST} = 0.12$, $P < 0.00001$ with 5.4% of the variation between populations east and west of the Maluku Sea, 6.7% among populations within groups, and 87.9% within populations. Average pair-wise F_{ST} among populations east and west of the Maluku Sea was $F_{ST} = 0.16$, while the average among populations in the Philippines and Sulawesi west of the Maluku Sea was $F_{ST} = 0.067$ and averaged $F_{ST} = 0.019$ among population east of the Maluku Sea.

Haplotypes of *Hoplosquilla said* fell into seven clades from Bayesian analysis trees. In general however, individual haplotypes were highly differentiated from each other. Clades 1–3 were found largely west of the Maluku Sea; Clade 1 was the most widely distributed, while Clade 2 was seen in Halmahera and the Togian Islands and Clade 3 was seen only in Bali (Figure 4C). Clade 4 was found largely on the southern shores of West Papua. Clades 5–7 were found east of the Maluku Sea. Unstructured AMOVA analyses indicated strong genetic structure among populations ($\Phi_{ST} = 0.66$, $P < 0.00001$). AMOVA results imposing regional groups between Indonesia and Papua New Guinea indicated $\Phi_{ST} = 0.79$, $P < 0.0001$ with 40.61% of the variation between groups, 38.38% among populations within groups, and 21.01% within populations.

Haplotypes of *Siamosquilla laevicaudata* separated into six distinct clades. All clades, except Clade 5, are found in populations west of the Maluku Sea, although these populations are dominated by Clades 1, 2, and 3. Similarly, all clades are found east of the Maluku Sea, although these populations are dominated by Clades 4, 5, and 6 (Figure 4D). Unstructured AMOVA analyses indicated strong genetic structure among populations ($\Phi_{ST} = 0.24$, $P < 0.00001$). AMOVA results imposing regional groups between Indonesia and Papua New Guinea indicated $\Phi_{ST} = 0.37$, $P < 0.0001$ with 29.78% of the variation between groups, 7.42% among populations within groups, and 62.80% within populations.

Haplotypes of *Gonodactylopsis* sp. A fell out into four distinct clades. Clades 1 and 2 were found throughout Indonesia from Bali to Raja Ampat, spanning the Maluku Sea. Clades 3 and 4 were found from Papua New Guinea to West Papua (Figure 4E). Unstructured AMOVA analyses indicated strong structure ($\Phi_{ST} = 0.33$, $P < 0.00001$). AMOVA results imposing regional groups between Indonesia and Papua New Guinea indicated $\Phi_{ST} = 0.52$, $P < 0.00001$ with 48.57% of the variation between groups, 3.13% among populations within groups, and 48.29% within populations.

A total of 500 bp of mtDNA COI data were collected from 23 *Chorisquilla brooksii*, 11 *Chorisquilla spinosissima*, 11 *Gonodactylus affinis*, 7 *Gonodactylus childi*, 17 *Gonodactylus chiragra*, 11 *Gonodactylus smithii*, and 18 *Haptosquilla stoliura*. In each case, there was evidence for genetic structure between populations spanning the Maluku Sea. For *Gonodactylus chiragra*, *G. childi*, *G.*

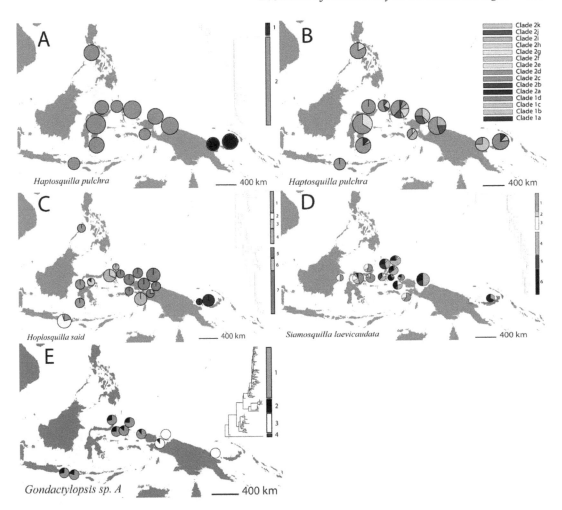

Figure 4. Phylogeographic structure of A) *Haptosquilla pulchra* (broad scale), B) *H. pulchra* (fine scale), C) *Hoplosquilla said*, D) *Siamosquilla laevicaudata* and E) *Gonodactylopsis* sp. A. across the Coral Triangle based on unique mtDNA COI haplotypes (see Figure 4 in Color insert).

smithii, and *Haptosquilla stoliura*, populations were differentiated exactly across the Maluku Sea, although *G. smithii* had two unique clades east of the Maluku Sea. For *Chorisquilla brooksii* the clade containing populations from Sulawesi extended across the Maluku Sea and into the island of Halmahera. For *Chorisquilla spinosissima* and *Gonodactylus affinis*, it was not possible to determine the exact location of the genetic discontinuity, as samples from east of the Maluku Sea were only available from Papua New Guinea. However, these populations were strongly differentiated from populations in Sulawesi (Figure 5).

Additional sampling of *Haptosquilla pulchella* from East Africa yielded a fifth clade, while additional sampling of *H. glyptocercus* from Fiji and the western Pacific uncovered two additional clades, one from Papua New Guinea and the Solomon Islands and the other from Fiji (Figures 6 and 7). Detailed phylogenetic analyses using maximum likelihood and Bayesian analysis were run on three taxa with a sufficient number of well-defined clades and geographic coverage to examine directionality of lineage diversification. Data from *Haptosquilla pulchella* yielded a well-resolved

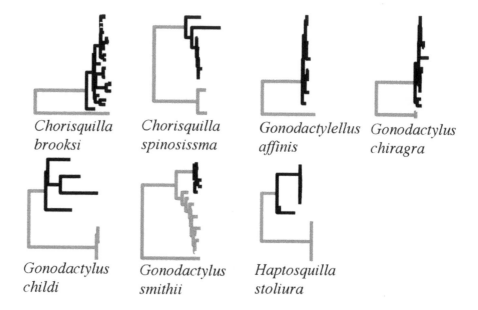

*Chorisquilla
brooksi*

*Chorisquilla
spinosissma*

*Gonodactylellus
affinis*

*Gonodactylus
chiragra*

*Gonodactylus
childi*

*Gonodactylus
smithii*

*Haptosquilla
stoliura*

Figure 5. Phylogenetic structure between populations west (black) and east (gray) of the Maluku Sea. Branch lengths of neighbor-joining trees are not to scale across all taxa.

phylogeny. Papua New Guinea was basal although the posterior probability of this node was only 73%, and the remaining 4 groups fell out into two clades: East African *H. pulchella* were more closely related to Indian Ocean samples from Indonesia (white clade, 99% posterior probability), and samples from Teluk Cenderawasih (red clade) were most closely related to samples from central Indonesia and the Philippines (black clade, 100% posterior probability). For *Haptosquilla glypto-cercus*, the basal taxa are found in Fiji (Figure 7, white clade). Populations from Papua New Guinea and West Papua, Indonesia (gray and blue clades, 93% posterior probability) form a clade, while populations from central Indonesia (black clade) and Bali (red clade) form a clade (88% posterior probably). For *Hoplosquilla said*, three major groupings were recovered, although the relationships among them were ambiguous. Group one consisted of Clade 1 populations (red clade, Figure 4E) from central Indonesia, group 2 consisted of a polytomy of Clades 2, 3, and 4 (orange, yellow, green, Figure 4E), and group 3 included Clades 5, 6, and 7 (blue, purple, and pink, Figure 4E). Within this group, Clade 5 from central Indonesia was basal, while Clades 6 and 7 from West Papua and Papua New Guinea were more derived. Despite the patterns above, Shimodaira-Hasegawa tests indicate that the optimal topologies in all three species are not significantly different than topologies obtained with topological constraints enforcing basal origins in the periphery with subsequent accumulation of diversity in the center, or with central populations being basal and subsequent lineage diversification and expansion towards the periphery.

4   DISCUSSION

4.1   Patterns of diversification of stomatopod Crustacea in the Coral Triangle

Results from 14 different stomatopod species revealed pronounced phylogeographic structure and regional lineage diversification in the Coral Triangle. All species examined had between 2 and 7 divergent clades with unique geographic distributions across this region. Combined with previous studies on stomatopods (Barber et al. 2000, 2002b, 2006) these results indicate that lineage diversification within the Coral Triangle is a ubiquitous features of stomatopod crustaceans and is likely contributing to elevated biodiversity levels of stomatopods in this region.

While previous work on stomatopods (Barber et al. 2000, 2002b, 2006) indicated diversification of Pacific and Indian Ocean populations, likely due to isolation during Pleistocene low sea level stands, results from this study show limited evidence for Pacific-Indian Ocean vicariance. However, this result is likely more a function of the geographic distributions of the focal taxa, rather than biology limiting the isolating effects of lowered sea levels. Despite strong patterns of regional genetic isolation, neither *Haptosquilla pulchra* nor *Gonodactylopsis* sp. A showed evidence of Pacific-Indian Ocean vicariance. This result is likely a function of the western range limit of both species terminating near the island of Bali. Bali is the easternmost limit of the Sunda Shelf, one of the major barriers to dispersal during periods of low sea level. Because these species' ranges do not span this dispersal barrier, it follows that they would not experience Pacific-Indian Ocean vicariance across this region. Similarly, *Siamosquilla laevicaudata* is not known to occur along the margins of the Indian Ocean, so is not expected to be impacted by Pleistocene vicariance. The only species in this study to show divergence in populations along the borders of the Indian Ocean was *Hoplosquilla said*. However, while this pattern could result from Pacific-Indian Ocean vicariance, it might also simply result from strong patterns of fine-scale isolation characteristic of this species (discussed below).

Although the four new species did not support Pacific-Indian Ocean vicariance because they do not span the Sunda Shelf, for stomatopod species that do span the Sunda Shelf, vicariance is observed. For example *Haptosquilla pulchella* populations display a pronounced signature of vicariance in Indian Ocean populations, including those in Bali, Lombok, and Komodo (Barber et al. 2000, 2002b). However, the strongest signals are seen in populations further to the west (Figure 3), whereas populations in Bali, Lombok, and Komodo have a mixed genetic signature, suggesting divergence on either side of the Sunda Shelf with admixture in populations near or on the Sunda Shelf. Similar patterns where the strongest signals of Pacific-Indian differentiation occur further west in Indonesia are seen in *Gonodactylellus viridis* (Figure 3) and have also been reported in giant clams (DeBoer et al. 2008; Kochzius & Nuryanto 2008; Nuryanto & Kochzius 2009), neritid snails (Crandall et al. 2008a) and clownfish (Timm & Kochzius 2008). Together, these results highlight the importance of having geographic ranges that extend into Java and Sumatra for Pacific-Indian Ocean vicariance to occur.

Although not all species exhibited patterns consistent with Pacific-Indian Ocean vicariance, all species did show strong evidence of divergence among populations in eastern Indonesia consistent with previous studies of stomatopods (Barber et al. 2000, 2002b, 2006). The strongest pattern of divergence across the Maluku Sea was seen in *Siamosquilla laevicaudata*. Although there is significant mixing with most clades spanning this region, a clear signal of differentiation is seen with Clades 1, 2, and 3 dominated to the west of the Maluku Sea and Clades 4, 5, and 6 dominated to the east (Figure 4C). A strong pattern of divergence is also seen in *Haptosquilla pulchra* as well as *Gonodactylopsis* sp. A. In the case of the former, the divergence is only seen in comparing Indonesian populations to those in Papua New Guinea (Figure 4A), while in the latter, the genetic break is observed east of the Halmahera Eddy in the region of Teluk Cenderawasih (Figure 4D). However, *H. pulchra* does show evidence of isolation across the Maluku Sea as genetic structure within Clade 2 reveals discontinuous haplotype distributions between eastern and central Indonesia. AMOVA analyses including only Clade 2 haplotypes indicate strong divergence across the Maluku Sea ($\Phi_{ST} = 0.12$, $P < 0.0001$). Average pair-wise F_{ST} values are much higher across this region than among Clade 2 populations on either side of the Maluku Sea, suggesting lower levels of gene flow and larval dispersal across this region. Similar patterns within a single clade are reported in *Haptosquilla pulchella* (Barber et al. 2006), providing further support for isolation across this region. Lower F_{ST} values across the Maluku Sea than among populations on either side indicate that this region must be a regional filter to gene flow and not simply the contact point of previously differentiated lineages.

Of the four species examined, *Hoplosquilla said* was unique in that it exhibited pronounced fine-scale regional divergence across the Coral Triangle where mixing of clades was largely confined to adjacent geographic regions (Figure 4C). As mentioned above, one clade (Clade 3) was found only in Bali. Clade 1 was found in multiple locations in the Makassar Strait as well as in Palawan, Philippines, and Bali. Clade 2 occurred in the Togian Islands as well as western Halmahera. The remainder of western Halmahera was dominated by Clade 6. Eastern Halmahera, in contrast, was dominated by Clade 7, which extended into West Papua and Teluk Cenderawasih. Clade 4 occurred on the southern shores of West Papua with the exception of a few individuals from Teluk Cenderawasih and one individual in Bali. Clade 5 only occurred in Papua New Guinea. Although aspects of these distributions are consistent with Indian-Pacific Ocean vicariance and with isolation across the Maluku Sea, this remarkable pattern of regional differentiation cannot be simply explained as a result of these processes. Instead, these patterns likely result from the fact that *H. said*, as the smallest of all gonodactylid stomatopods (Erdmann 1997), has the most restricted larval dispersal capability. Although the larval forms of *H. said* are unknown, given that adults of this species are smaller than the postlarvae of most other gonodactylid stomatopods, it can be reasonably assumed that the larvae are also quite small. Because stomatopod larval development is isochronal and regular (Hamano et al. 1995), small larval size in this species would strongly suggest a very short larval period, which generally indicates limited realized dispersal (Shanks et al. 2003) and therefore a restricted capacity for dispersal and gene flow, although dispersal could also be limited by behavior as well. Limited dispersal potential would explain many of the patterns seen in this species while the broad range of Clade 1 could be explained by the strong currents of the Indonesian Throughflow promoting greater dispersal. However, the distribution of Clade 4 is particularly anomalous. Not only is this clade distributed on the northern and southern shores of West Papua with none distributed in between, but

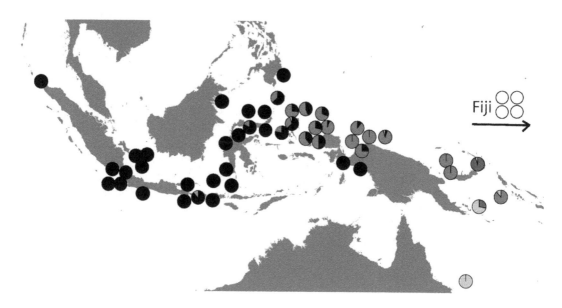

Figure 6. Distribution of five mtDNA clades in *Haptosquilla glyptocercus* across the Coral Triangle and central Pacific. Relationship of clades is shown in Figure 7 (see Figure 5 in Color insert).

it also extends south to Bali, where it is found in one individual. It is unclear how to explain this pattern, but additional sampling in the Banda Sea may provide further insights into this odd distribution.

4.2   Differentiation in eastern Indonesia

Barber et al. (2002b) suggested that one potential mechanism for promoting lineage diversification within the Coral Triangle was limited larval dispersal and gene flow across the Maluku Sea—a process that could contribute to speciation within the Coral Triangle consistent with a center of origin. This hypothesis was supported by Barber et al. (2006), where three species of stomatopods were shown to have strong phylogeographic breaks in this region, although the precise location of these breaks varied among species. In one species, *Haptosquilla pulchella*, where a single clade spanned the Maluku Sea, gene flow was shown to be extremely limited, while gene flow was essentially unlimited on either side of this region, providing further evidence of regional isolation in eastern Indonesia.

The broad-scale phylogeographic analysis of four additional species of stomatopods examined in this study show very similar results. Two species, *Siamosquilla laevicaudata* and *Hoplosquilla said*, show divergence across the Maluku Sea, whereas *Haptosquilla pulchra* and *Gonodactylus* sp. A have a pronounced east-west phylogeographic break further to the east. However, *H. pulchra* also shows limited gene flow among populations of a single clade that span the Maluku Sea, providing additional evidence for the isolating effects of this region. Further support for divergence across this region comes from seven additional species. In four species (*Gonodactylus chiragra*, *G. childi*, *G. smithii*, and *Haptosquilla stoliura*), phylogeographic divergence of populations corresponded exactly to the boundaries of the Maluku Sea, with a pronounced phylogenetic break among populations on either side of this region. However, as with previous studies, the location of the phylogeographic break in *Chorisquilla brooksii* was further to the east. For *Chorisquilla spinosissima* and *Gonodactylus affinis*, it was not possible to determine the exact location of the genetic discontinuity, as samples from east of the Maluku Sea were only available from Papua New Guinea.

The pattern of regional isolation of marine taxa in eastern Indonesia, first reported in stomatopods, is quickly becoming a common pattern in other marine taxa. This pattern has been reported in giant clams (DeBoer et al. 2008; Kochzius & Nuryanto 2008; Nuryanto & Kochzius 2009), in echinoderms (Crandall et al. 2008b) and in clownfish (Timm & Kochzius 2008). There is additional evidence of genetic isolation in this region, coming from a large number of new species being reported from Teluk Cenderawasih (Allen & Erdmann 2006, 2008, 2010). Phylogeographic structure across the Maluku Sea is even seen in human populations (Capelli et al. 2001) attesting to the isolating effects of this region. The recovery of similar phylogeographic isolation in 14 different species of stomatopods as well as many other marine taxa in a very limited region of eastern Indonesia strongly argues for a single physical process acting in their formation (Avise 2000). The most likely explanation is the Halmahera Eddy, which limits westward water movement from the greater island of Papua into central Indonesia. Limited water movement results in limited gene flow, resulting in genetic differentiation, potentially culminating in lineage diversification, contributing to the generation of marine biodiversity in this region.

While the Halmahera Eddy likely limits larval dispersal across the Maluku Sea, it has only done so relatively recently. Geological reconstructions of this region (Hall 1998) show that prior to 6 MYA, the island of Halmahera was further north and east of its current position, creating the Halmahera Seaway. This is significant as the physical location of the island of Halmahera plays a prominent role in creating the Halmahera Eddy (Morey et al. 1999). With a clear pathway for New Guinea Coastal Current (NGCC) waters to travel from the northern coast of the greater island of Papua directly into central Indonesia, there would be a direct pathway for larval transport and gene

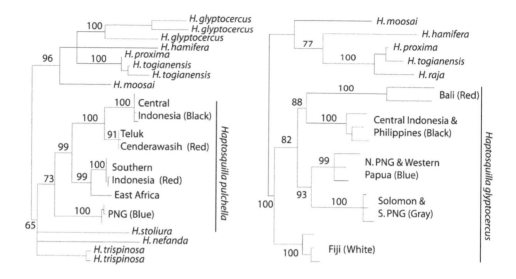

Figure 7. Bayesian phylogenies of major clades *Haptosquilla pulchella* and *H. glyptocercus*. Colors correspond to the clades identified in Figures 3 and 6, respectively. Numbers at nodes indicate posterior probabilities of the node. Maximum likelihood topologies did not differ significantly and analyses are not shown.

flow to unite these regions. Between 5–6 MYA, the movement of Halmahera across the northern shores of Papua constricted the Halmahera Seaway, intensifying the Halmahera Eddy as NGCC waters were shunted to the north. From 5 MYA until 1 MYA, the Halmahera Seaway remained almost completely closed, blocking the pathway of NGCC waters into Indonesia. It is only within the past million years that Halmahera has traveled far enough to the west to allow some NGCC waters to pass between the northern tip of Papua and the Island of Halmahera.

Unfortunately there are no molecular clock estimates for stomatopods. However, if we assume a divergence rate of 1.4% per million years reported from other crustaceans (Knowlton & Weigt 1998) purely as a heuristic, the date of divergence among populations on either side of the Maluku Sea is between 5–6 million years ago for half of the 14 species examined, about the time when the Halmahera Eddy formed. While such estimates are fraught with error due to rate variation and inappropriate calibration (Ho et al. 2005, 2007, 2008), the fact that half of the taxa have similar dates suggests temporal concordance, irrespective of the accuracy of the actual date. Thus, while we cannot confidently say that the dates are concordant with the formation of the Halmahera Eddy, the similarity in genetic divergence among half of the taxa strongly argues for resulting from a common process. Improved rate estimates and calibration methods in the future may allow more robust testing of the above hypothesis.

Another potential explanation for population differentiation in eastern Indonesian may be the physical isolation of populations in Teluk Cenderawasih, a bay located on the northwestern shores of West Papua. There is a very shallow ($\approx$ 30 m) shelf that extends nearly all the way across the mouth of the bay (Allen & Erdmann 2006). Thus, during periods of lowered sea level, this bay becomes almost completely enclosed, potentially driving lineage diversification. Evidence for isolation in this bay comes from phylogeographic studies of multiple stomatopods as well as giant clams (DeBoer et al. 2008; Kochzius & Nuryanto 2008; Nuryanto & Kochzius 2009) and echinoderms (Crandall et al. 2008b). Furthermore, since 2006, eight new species of endemic fish have been described from this Bay (Allen & Erdmann 2006, 2008, 2010), indicating the isolating potential of this region. The combined isolating effects of the Halmahera Eddy and Teluk Cenderawasih may partially explain the variation in the location of genetic breaks within eastern Indonesia.

4.3 Biogeography of lineage diversification

While the analyses above strongly indicate lineage diversification within the Coral Triangle, a key element of the center of origin and center of accumulation theories is the original location of speciation and directionality of export of this novel biodiversity. For three species, *Haptosquilla pulchella*, *H. glyptocercus*, and *Hoplosquilla said*, there are numerous clades distributed broadly across the Coral Triangle allowing for phylogenetic analysis of the directionality of lineage diversification in this region. For *H. pulchella*, the oldest populations were found in Papua New Guinea. These populations are basal to a pair of sister clades: one comprised of Teluk Cenderawasih and central Indonesian populations and the other comprised of East African and southern Indonesian populations that border the Indian Ocean. This result suggests that *H. pulchella* originated on the eastern edge of the Coral Triangle (the eastern extent of the range of *H. pulchella* is the Solomon Islands) with subsequent expansion westward out of the Coral Triangle and into the Indian Ocean. Because of the relationship of the two sister clades, it is not possible to determine how that initial diversification took place. However, it does appear that diversification happened across the Maluku Sea in this species, likely due to the Halmahera Eddy and isolation in Teluk Cenderawasih, as well as across the Indian Ocean, likely as a result of the isolating effects of physical distance across the Indian Ocean.

For *Haptosquilla glyptocercus*, the ancestral lineages are found in Fiji, and these are basal to two pairs of sister clades. One pair includes a clade of haplotypes from populations largely from the southern coast of Papua New Guinea and the Solomon Islands (gray clade, Figure 7) as well as a clade of haplotypes from populations largely from the northern coast of Papua New Guinea, West Papua, and Halmahera. The second pair includes a clade of haplotypes from Bali as well as a clade from populations from central and western Indonesia. An evolutionary scenario consistent with this phylogeny would be an initial diversification and expansion from southern Indonesia, followed by divergence between central and eastern Indonesia. Populations from Papua and Solomon Islands would then diversify on opposite sides of the island of Papua. What is anomalous, however, is the relationship of central and western Indonesian populations to populations in Fiji with no intervening populations.

For *Hoplosquilla said* there was not sufficient resolution to determine which clades were basal with three groups (Clade 1, Clades 2–4, Clades 5–7) forming an unresolved polytomy. Even within Clades 2–4, the relationship of these clades could not be resolved. The only resolution was seen in Clades 5–7, where populations from Papua New Guinea are basal to Clades 6 and 7, which occur in West Papua and Halmahera. Unfortunately, this pattern is insufficient to infer any patterns of lineage diversification, although diversification is clearly occurring within the boundaries of Coral Triangle in this species, supporting the center of origin hypothesis.

As indicated by Kirkendale & Meyer (2004), inferring directionality using phylogenies is very difficult, particularly in marine species where corroborating fossil evidence may be lacking. Not only can multiple evolutionary scenarios lead to the same phylogenetic pattern, but the best phylogenetic reconstructions may also not be significantly different from alternative hypotheses. In the case of this study, well-resolved phylogenies were recovered in *Haptosquilla pulchella* and *H. glyptocercus*. However, these phylogenetic reconstructions are not significantly better than topologies enforced to reflect a center of origin or a center of accumulation. As such, the results are insufficient to distinguish between these two competing hypotheses. Determining the directionality of lineage diversification will likely require much more sophisticated analyses of gene flow based on nuclear markers. That said, the pronounced limits to gene flow across the Maluku Sea in single clades of

Haptosquilla pulchella and *H. pulchra* as well as the fine scale patterns observed in *Hoplosquilla said* unequivocally demonstrate that lineage diversification does occur within the Coral Triangle, consistent with the center of origin theory. This by no means suggests that the center of origin theory solely explains Coral Triangle biodiversity hotspot. Rather, these results simply indicate that, counter to assertions to the contrary (Jokiel & Martinelli 1992), there is convincing evidence of lineage diversification within the Coral Triangle, and there are clear mechanisms to promote allopatric divergence in this region. However, further research will almost certainly continue to show pluralistic origins of the Coral Triangle biodiversity hotspot.

4.4   Limits to gene flow or selection?

Intraspecific phylogeography was introduced as a comparative science, where the discovery of concordance of patterns among multiple unrelated taxa provided evidence of broadly acting physical processes shaping the evolutionary history in codistributed taxa (Avise et al. 1987; Avise 1992). However, as the field evolved, comparative studies were largely displaced by single species mtDNA phylogeographic studies. This plethora of mtDNA studies has resulted in the discovery of examples where mtDNA violates the assumptions of neutrality, recombination, and uniparental inheritance (see Galtier et al. 2009 for review). The discovery of a few instances where genetic structure is determined by selection rather than neutral processes has led many to question the results of any mtDNA only dataset (see Galtier et al. 2009 for review). This has influenced the view that mtDNA results can only be believed if corroborated by nuclear markers, even though the time scale of lineage sorting of nuclear markers generally precludes seeing condordant patterns except in cases of deeply divergent mtDNA lineages (Palumbi et al. 2001). However, this short-sighted view ignores the power of comparative phylogeography, where the recovery of concordant patterns among multiple species, rather than multiple markers, provides corroborating evidence for physical isolation, rather than selection, driving phylogeographic patterns (Avise 2000).

Previous studies on gonodactylid comparative phylogeography (Barber et al. 2006) focused on three codistributed species that are sufficiently similar in their ecology that all three can be found in a single piece of coral rubble. As such, it is difficult to distinguish between concordance resulting from shared barriers to dispersal and shared selective pressures. In contrast, the four species presented here represent a cornucopia of different habitat types. *Haptosquilla pulchra* is found in coral bench in sandy subtidal shallow reef habitats. *Siamosquilla laevicaudata* can often be found codistributed with *H. pulchra*, but can also be found in much deeper coral and coralline algae-dominated habitats. *Hoplosquilla said* is found exclusively in encrusting coralline algal reef habitats, at depths ranging from several to tens of meters. *Gonodactylopsis* sp. A is found in barrel sponge species at depths of 5–40 meters. The additional 7 taxa examined in eastern Indonesia show similar habitat and depth variation. The 14 total species studied herein are almost never found at the exact same localities. As such, while the results of previous studies could potentially result from habitat differences rather than physical limits to dispersal (see Reid et al. 2006) the broad similarity of regional divergence patterns despite the variety of different habitats occupied by the species in this study suggests that physical limits to dispersal is the most likely explanation for the evolution of regional genetic structure in these stomatopods as it is unclear how natural selection could possibly drive the concordant patterns observed.

4.5   Conservation Implications

The Coral Triangle is home to 75% of all coral species and at 2.3 million square miles, contains more than half of the world's coral reef area. It is estimated that the annual value of these ecosystems is $2.3 billion. However, these reefs are some of the most critically threatened in the world (Wilkinson 2002). The ecological, evolutionary, and economic interconnectedness of the reefs of the Coral

Triangle have spawned the Coral Triangle Initiative, a six nation agreement focus on the preservation of reefs of the Coral Triangle and the development of sustainable fisheries to promote economic and food security in this region.

Understanding patterns of connectivity among marine ecosystems has been identified as one of the critical variables for developing effective networks of marine reserves (Sale et al. 2005). The methods of this study are too coarse to provide details on connectivity among reefs, and it is difficult in general to get meaningful information on connectivity from genetic methods because very small amounts of demographic exchange can result in high levels of gene flow and estimates of genetic connectivity (Hedgecock et al. 2007; Hellberg 2009). However, because small amounts of genetic exchange can result in genetic homogenization, the recovery of persistent barriers to gene flow (e.g., phylogeographic structure) indicates the absence of both ecologically (e.g., demographically) as well as evolutionarily significant amounts of connectivity.

Conservation of large areas can be challenging. As such, recent efforts in the marine realm have focused on partitioning large regions of conservation concern into smaller more manageable "sustainable seascapes" (Green & Mous 2004) or "marine ecoregions" (Spalding et al. 2007). The recovery of pronounced phylogeographic structure across multiple stomatopod species in the Coral Triangle indicates that there are clear barriers to genetic exchange within this region, highlighting the evolutionary and demographic independence among several areas. Data strongly suggest that there is limited genetic exchange from eastern Indonesia into central Indonesia, suggesting that these regions are largely evolutionarily and ecologically independent. Even within eastern Indonesia, evidence from stomatopods and other taxa suggest that populations within Teluk Cenderawasih may also be isolated. In addition, a few species highlight the presence of filters to genetic exchange across the Java and Flores Sea. These phylogeographic boundaries are broadly consistent with regional marine ecoregion boundaries proposed by Spalding (Spalding et al. 2007). As marine ecoregions are defined based on species distributions and geology, the correspondence between marine ecoregions and phylogeographic structure suggests that the processes shaping the distribution of genetic diversity in the Coral Triangle are likely contributing to the contemporary distribution of biological diversity as well.

Kirkendale & Meyer (2004) pointed out the importance of avoiding basing management decisions on data from individual species. In this study, numerous species provide insights into the limits of connectivity throughout the Coral Triangle. However, while there was consistency among some species, the exact location of phylogeographic breaks varied slightly. It is not clear whether such variation simply results from the stochasticity of mutation, migration and genetic drift in shaping phylogeographic structure, or whether these differences may be rooted in biological differences among the species examined. However, it is clear that even in comparative studies, generalized conclusions can be elusive. While the three regional filters to gene flow mentioned above affect all species to some degree, species still vary greatly in their response to these filters. In some species, patterns are sharp while in others the barriers are more diffuse. Some species have only two phylogeographic regions, while others like *Hoplosquilla said* have seven. This highlights the challenges facing marine managers in developing management plans that can be generalized across a broad range of species. A hierarchical design to marine reserve networks will almost certainly be required to encompass the range of patterns observed across species. As such, it may be more expeditious and equally effective to simply adopt a hierarchical strategy to the design of marine reserve networks rather than wait for more and better estimates of connectivity among reef ecosystems.

ACKNOWLEDGEMENTS

We thank the governments of Indonesia, Papua New Guinea, the Philippines, Solomon Islands, and Fiji for permitting our collections in these regions. Joshua Drew collected samples from the Solomon Islands and Fiji. Work in Papua New Guinea was supported by Dr. Phillip Crews, with dive support from the Golden Dawn. Dive support in Indonesia was provided by Bali Diving Academy, Murex Divers, and M/V Pelagian. Dr. Kent Carpenter, Annette Menez and Menchie Ablan supported work in the Philippines as part of an NSF PIRE program. Funding for this work was provided NSF awards (Biological Oceanography, OCE-0349177, DEB 0338566) and Conservation International. Samantha Cheng was supported by an A. W. Mellon Environmental Grant from the Joint Science Department of the Claremont Colleges. Eric Crandall assisted with analyses, and Roger Alvillar and Sarah Boyce assisted with data collection. Dr. M. Kasim Moosa supported initial work on stomatopods in Indonesia. Jeannie K. Choi, Brenden Barber-Choi, and Nicholas Barber-Choi assisted with field collections.

REFERENCES

Ahyong, S.T. 2001. Revision of the Australian stomatopod Crustacea. *Rec. Austr. Mus. Suppl.* 26: 1–326.

Ahyong, S.T. & Erdmann, M.V. 2007. Two new species of *Gonodactylellus* from the western Pacific (Gonodactylidae: Stomatopoda). *Raffles Bull. Zool.* 55: 89–95.

Ahyong, S.T. & Randall, E.J. 2001. *Lysiosquillina lisa*, a new species of mantis shrimp from the Indo-West Pacific (Stomatopoda: Lysiosquillidae). *J. South Asian Nat. Hist.* 5: 167–172.

Allen, G.R. & Erdmann, M.V. 2006. *Cirrhilabrus cenderawasih*, a new wrasse (Pisces: Labridae) from Papua, Indonesia. *Aqua.* 11: 125–131.

Allen, G.R. & Erdmann, M.V. 2008. *Pterocaesio monikae*, a new species of fusilier (Caesionidae) from western New Guinea (Papua and Papua Barat provinces, Indonesia). *Aqua.* 13: 163–170.

Allen, G.R. & Erdmann, M.V. 2010. Two new species of *Calumia* (Pisces: Eleotridae) from West Papua, Indonesia. *Aqua.* 16: 71–80.

Allen, G.R. & Werner, T.B. 2002. Coral reef fish assessment in the Coral Triangle' of southeastern Asia. *Env. Biol. Fish.* 65: 209–214.

Armsworth, P.R. 2002. Recruitment limitation, population regulation, and larval connectivity in reef fish metapopulations. *Ecology* 83: 1092–1104.

Avise, J.C. 1992. Molecular population structure and the biogeography of history of a regional fauna: a case history with lessons for conservation biology. *Oikos* 63: 62–67.

Avise, J.C. 2000. *Phylogeography: The History and Formation of Species*. Cambridge, MA: Harvard University Press.

Avise, J.C., Arnold, J., Ball, R.M., Bermingham, E., Lamb, T., Neigel, J.E., Reeb, C.A. & Saunders, N.C. 1987. Intraspecific phylogeography: the mitochondrial DNA bridge between population genetics and systematics. *Ann. Rev. Ecol. Syst.* 18: 489–522.

Barber, P.H. 2009. The challenge of understanding the Coral Triangle biodiversity hotspot. *J. Biogeogr.* 36: 1845–1846.

Barber, P.H. & Bellwood, D.R. 2005. Biodiversity hotspots: evolutionary origins of biodiversity in wrasses (Halichoeres: Labridae) in the Indo-Pacific and new world tropics. *Mol. Phyl. Evol.* 35: 235–253.

Barber, P.H., Erdmann, M. V. & Palumbi, S. R. 2006. Comparative phylogeography of three co-distributed stomatopods: origins and timing of regional diversification in the Coral Triangle. *Evolution* 60: 1825–1839.

Barber, P.H., Moosa, M.K. & Palumbi, S.R. 2002a. Rapid recovery of genetic diversity of stomatopod populations on Krakatau: temporal and spatial scales of marine larval dispersal. *Proc. R. Soc. B* 269: 1591–1597.

Barber, P.H., Palumbi, S.R., Erdmann, M.V. & Moosa, M.K. 2000. A marine Wallace's line? *Nature* 406: 692–693.

Barber, P.H., Palumbi, S.R., Erdmann, M.V. & Moosa, M.K. 2002b. Sharp genetic breaks among populations of a benthic marine crustacean indicate limited oceanic larval transport: patterns, causes, and consequences. *Mol. Ecol.* 11: 659–674.

Barraclough, T.G., Vogler, A.P. & Harvey, P.H. 1998. Revealing the factors that promote speciation. *Proc. R. Soc. B* 353: 241–249.

Bellwood, D.R. & Meyer, C.P. 2009. Searching for heat in a marine biodiversity hotspot. *J. Biogeogr.* 36: 569–576.

Bellwood, D.R. & Wainwright, P.C. 2002. The history and biogeography of fishes on coral reefs. In: Sale, P.F. (ed.), *Coral Reef Fishes. Dynamics and Diversity in a Complex Ecosystem*: 5–32. San Diego, CA: Academic Press.

Benzie, J.A.H. 1999. Major genetic differences between crown-of-thorns starfish (*Acanthaster planci*) populations in the Indian and Pacific Oceans. *Evolution* 53: 1782–1795.

Berlocher, S.H. & Feder, J.L. 2002. Sympatric speciation in phytophagous insects: moving beyond controversy? *Ann. Rev. Ent.* 47: 773–815.

Bigelow, R.P. 1893. Preliminary notes on the Stomatopoda of the Albatross collections and other specimens in the National Museum. *J. Hopkins Univ. Circ.* 12: 100–102.

Borradaile, L.A. 1898a. On some crustaceans from the South Pacific. Part I. Stomatopoda. *Proc. Zool. Soc. Lond.* 1898: 32–38, pls. V–VI.

Borradaile, L.A. 1898b. On some crustaceans from the South Pacific. Part III. *Macrura. Proc. Zool. Soc. Lond.* 1898: 1000–1015, pls. LXIII–LXV.

Borradaile, L.A. 1907. Stomatopoda from the western Indian Ocean. The Percy Sladen Trust Expedition to the Indian Ocean in 1905, under the leadership of Mr. Stanley Gardiner. *Trans. Linn. Soc. Lond.* (2)12: 209–216, pl. 22.

Briggs, J.C. 1992. The marine East-Indies: center of origin? *Glob. Ecol. Biogeogr. Lett.* 2: 149–156.

Briggs, J.C. 1999. Modes of speciation: marine Indo-West Pacific. *Bull. Mar. Sci.* 65: 645–656.

Briggs, J.C. 2000. Centrifugal speciation and centres of origin. *J. Biogeogr.* 27: 1183–1188.

Briggs, J.C. 2003. Marine centres of origin as evolutionary engines. *J. Biogeogr.* 30: 1–18.

Briggs, J.C. 2005a. Coral reefs: conserving the evolutionary sources. *Biol. Cons.* 126: 297–305.

Briggs, J.C. 2005b. The marine East Indies: Diversity and speciation. *J. Biogeogr.* 32: 1517–1522.

Brooks, W.K. 1886. Report on Stomatopoda collected by H.M.S. Challenger during the years 1873–76. *Rep. Sci. Res. Expl. Voyage Challenger, Zool.* 16: 1–116, pls. I–XVI.

Burke, L., Selig, E. & Spalding, M. 2002. *Reefs at Risk in Southeast Asia*. Washington, DC: World Resources Institute.

Bush, G.L. 1975. Modes of animal speciation. *Ann. Rev. Ecol. Syst.* 6: 339–365.

Capelli, C., Wilson, J.F., Richards, M., Stumpf, M.P., Gratrix, F., Oppenheimer, S., Underhill, P., Pascali, V.L., Ko, T.M. & Goldstein, D.B. 2001. A predominantly indigenous paternal heritage for the Austronesian-speaking peoples of insular Southeast Asia and Oceania. *Am. J. Hum. Gen.* 68: 432–443.

Carpenter, K.E. & Springer, V.G. 2005. The center of the center of marine shorefish biodiversity: the Philippine Islands. *Env. Bio. Fish.* 72: 467–480.

Chesser, R. 1994. Modes of speciation in birds: a test of Lynch's method. *Evolution* 48: 490–497.

Connolly, S.R., Bellwood, D.R. & Hughes, T.P. 2003. Indo-Pacific biodiversity of coral reefs: deviations from a mid-domain model. *Ecology* 84: 2178–2190.

Coyne, J.A. & Orr, H.A. 2004. *Speciation*. Sunderland, MA: Sinauer Associates.

Crandall, E.D., Frey, M.A., Grosberg, R.K. & Barber, P.H. 2008a. Contrasting demographic history and phylogeographical patterns in two Indo-Pacific gastropods. *Mol. Ecol.* 17: 611–626.

Crandall, E.D., Jones, M.E., Muñoz, M.M., Akinronbi, B., Erdmann, M.V. & Barber, P.H. 2008b. Comparative phylogeography of two seastars and their ectosymbionts within the Coral Triangle. *Mol. Ecol.* 17: 5276–5290.

Cunningham, C.W. & Collins, T.M. 1994. Developing model systems for molecular biogeography: vicariance and interchange in marine invertebrates. In: Schierwater, B., Streit, B., Wagner,

150   *Barber et al.*

G.P., DeSalle, R. (eds.), *Molecular Ecology and Evolution: Approaches and Applications*: 405–434. Basel: Birkauser Verlag.

Dana, J.D. 1852. Crustacea. Part I. *U. S. Explor. Exped. 1838–1842* 13: 1–685.

de Haan, W. 1833–1850. Crustacea. In: de Siebold, P.F. (ed.), *Fauna Japonica, sive descriptio animalium, quae in itinere per Japoniam, jussu et auspiciis superiorum, qui summum in India Batava imperium tenent, suscepto, annis 1823–1830 collegit, notis, obervationibus et adumbrationibus illustravit Ph. Fr. de Siebold. Conjunctis studiis C. J. Temminck et H. Schlegel pro vertebratis atque W. de Haan pro invertebratis elaborata*: 1–243. Lugundi Batavorum: Auctore.

de Man, J.G. 1888. Bericht über die von Herrn Dr. G. Brock im indischen Archipel gesammelten Decapoden und Stomatopoden. *Arch. Naturgesch.* 53: 215–600, pls. VII–XXIIa.

de Man, J.G. 1902. Die von Herrn Professor Kükenthal im Indischen Archipel gesammelten Decapoden und Stomatopoden. *Abh. Senckenb. Naturf. Ges.* 25: 467–929, pls. XIX–XXVII.

DeBoer, T.S., Subia, M.D., Ambariyanto, Erdmann, M.V., Kovitvongsa, K. & Barber, P.H. 2008. Phylogeography and limited genetic connectivity in the endangered boring giant lam across the Coral Triangle. *Cons. Biol.* 22: 1255–1266.

Dieckmann, U. & Doebeli, M. 1999. On the origin of species by sympatric speciation. *Nature* 400: 354–357.

Dobzhansky, T. 1937. *Genetics and the Origin of Species.* New York, NY: Columbia University Press.

Dobzhansky, T. 1970. *Genetics of the Evolutionary Process.* New York, NY: Columbia University Press.

Duda, T.F. & Palumbi, S.R. 1999. Population structure of the black tiger prawn, *Penaeus monodon*, among western Indian Ocean and western Pacific populations. *Mar. Biol.* 134: 705–710.

Duke, N.C., Benzie, J.A.H., Goodall, J.A. & Ballment, E.R. 1998. Genetic structure and evolution of species in the mangrove genus *Avicennia* (Avicenniaceae) in the Indo-West Pacific. *Evolution* 52: 1612–1626.

Ekman, S. 1953. *Zoogeography of the Sea.* London: Sidgwick and Jackson, Ltd.

Erdmann, M.V.N. 1997. *The ecology, distribution, and bioindicator potential of Indonesian coral reef stomatopod communities.* Ph.D. dissertation, University of California, Berkeley, CA.

Erdmann, M.V. & Boyer, M. 2003. *Lysiosquilloides mapia*, a new species of stomatopod crustacean from northern Sulawesi (Stomatopoda: Lysiosquillidae). *Raffles Bull. Zool.* 51: 49–59.

Erdmann, M.V. & Manning, R.B. 1998. Preliminary descriptions of nine new stomatopod crustaceans from coral reef habitats in Indonesia and Australia. *Raffles Bull. Zool.* 56: 615–626.

Fabricius, J.C. 1781. *Species Insectorum exhibentes eorum differentias specificas, synonyma auctorum, loca natalia, metamorphosin adiectis observationibus, descriptionibus. Tom. I.* Hamburgi et Kilonii: Carol. Ernest. Bohnii.

Fabricius, J.C. 1787. *Mantissa Insectorum sistens eorum species nuper detectas adiectis characteribus genericus, differentiis specificus, emendationibus, observationibus. Tom. I.* Hafniae: Christ. Gottl. Proft.

Fabricius, J.C. 1793. *Entomologia systematica emendata et aucta. Secundum classes, ordines, genera, species adjectis synonymis, locis, observatiosnibus, descriptionibus. Tom. II.* Hafniae: Christ. Gottl. Proft.

Fleminger, A. 1986. The Pleistocene equatorial barrier between the Indian and Pacific Oceans and a likely cause for Wallace's Line. In: Pierrot-Bults, A.C., van der Spoel, S., Zahuranec, B.J. & Johnson, R.K. (eds.), *Pelagic biogeography: proceedings of an international conference, the Netherlands, 29 May-5 June 1985*: 84–97. Paris: UNESCO.

Forskål, P. 1775. *Descriptiones Animalium, Avium, Amphibiorum, Piscium, Insectorum, Vermium; quae in itinere orientali observavit Petrus Forskål. Prof. Haun. Post mortem auctoris ed.*

Carsten Neibuhr. Adjuncta est materia medica Kahirina atque tabula Maris Rubri geographica. Hauniae: Officina Mölleri.

Fukuda, T. 1909. The Stomatopoda of Japan. *Dobutsugaku Zasshi* 21: 54–62, 4 pls.

Futuyma, D.J. & Mayer, G.C. 1980. Non-allopatric speciation in animals. *Syst. Zool.* 29: 254–271.

Galtier, N., Nabholz, B., Glémin, S. & Hurst, G.D.D. 2009. Mitochondrial DNA as a marker of molecular diversity: a reappraisal. *Mol. Ecol.* 18: 4541–4550.

Garcia, R.G. & Manning, R.B. 1982. Four new species of stomatopod crustaceans from the Philippines. *Proc. Biol. Soc. Wash.* 95: 537–544.

Giesbrecht, W. 1910. Stomatopoden. Erster Theil. *Faun. Flor. Golf Neapel* 33: I–VII, 1–239.

Godfrey, J.S. 1996. The effect of the Indonesian throughflow on ocean circulation and heat exchange with the atmosphere: a review. *J. Geophys. Res.* 101: 12217–12237.

Gordon, A.L. & Fine, R.A. 1996. Pathways of water between the Pacific and Indian oceans in the Indonesian seas. *Nature* 379: 146–149.

Green, A. & Mous, P.J. 2004. *Delineating the Coral Triangle, its ecoregions and functional seascapes.* Sanur, Bali, Indonesia: The Nature Conservancy, Southeast Asia Center for Marine Protected Areas.

Gurney, R. 1946. Notes on stomatopod larvae. *Proc. R. Soc. B* 116: 133–175.

Halas, D. & Winterbottom, R. 2009. A phylogenetic test of multiple proposals for the origins of the East Indies coral reef biota. *J. Biogeogr.* 36: 1847–1860.

Hall, R. 1998. The plate techtonics of Cenozoic SE Asia and the distribution of land and sea. In: Hall, R. & Holloway, J.D. (eds.), *Biogeography and Geological Evolution of SE Asia*: 99–132. Leiden: Backhuys Publishers.

Hamano, T., Kikkawa, T., Ueno, S. & Hayashi, K.I. 1995. Use of larval size, instead of larval stage, to study the ecology of a stomatopod crustacean *Oratosquilla oratoria*. *Fish. Sci.* 61: 165–166.

Hansen, H.J. 1926. The Stomatopoda of the Siboga Expedition. *Siboga Exped. Monogr.* 35: 1–48, pls. 1–2.

Hare, M.P., Cipriano, F. & Palumbi, S.R. 2002. Genetic evidence on the demography of speciation in allopatric dolphin species. *Evolution* 56: 804–816.

Hedgecock, D., Barber, P.H. & Edmands, S. 2007. Genetic approaches to measuring connectivity. *Oceanography* 20: 70–79.

Hellberg, M.E. 2009. Gene flow and isolation among populations of marine animals. *Ann. Rev. Ecol. Evol. Syst.* 40: 291–310.

Henderson, J.R. 1893. A contribution to Indian Carcinology. *Trans. Linn. Soc. Lond. (Zool.)* (2)5: 325–458, pls. XXXVI–XL.

Higashi, M., Takimoto, G. & Yamamura, N. 1999. Sympatric speciation by sexual selection. *Nature* 402: 523–526.

Ho, S.Y.W., Phillips, M.J., Cooper, A. & Drummond, A.J. 2005. Time dependency of molecular rate estimates and systematic overestimation of recent divergence times. *Mol. Biol. Evol.* 22: 1561–1568.

Ho, S.Y.W., Saarma, U., Barnett, R., Haile, J. & Shapiro, B. 2008. The effect of inappropriate calibration: three case studies in molecular ecology. *Plos One* 3: e1615.

Ho, S.Y.W., Shapiro, B., Phillips, M.J., Cooper, A. & Drummond, A.J. 2007. Evidence for time dependency of molecular rate estimates. *Syst. Bio.* 56: 515–522.

Hoeksema, B.W. 2007. Delineation of the Indo-Malayan centre of maximum marine biodiversity: the Coral Triangle. In: Renema, W. (ed.), *Biogeography, Time, and Place: Distributions, Barriers, and Islands*: 117–178. New York, NY: Springer.

Holthuis, L.B. 1941. Biological results of the Snellius Expedition XII. The Stomatopoda of the Snellius Expedition. *Temminckia* 6: 241–294.

Huelsenbeck, J.P. & Ronquist, F. 2001. MRBAYES: Bayesian inference of phylogenetic trees. *Bioinformatics* 17: 754–755.

Jokiel, P. & Martinelli, F.J. 1992. The vortex model of coral reef biogeography. *J. Biogeogr.* 19: 449–458.

Kemp, S. 1913. An account of the Crustacea Stomatopoda of the Indo-Pacific region based on the collection in the Indian Museum. *Mem. Indian Mus.* 4: 1–217.

Kemp, S. 1915. On a collection of stomatopod Crustacea from the Philippine Islands. *Philippine J. Sci. D* 10: 169–187, pl. I.

Kirkendale, L.A. & Meyer, C.P. 2004. Phylogeography of the *Patelloida profunda* group (Gastrapoda: Lottidae): diversification in a dispersal driven marine system. *Mol. Ecol.* 13: 2749–2762.

Knowlton, N. & Keller, B.D. 1986. Larvae which fall far short of their potential: highly localized recruitment in an alpleid shrimp with extended larval development. *Bull. Mar. Sci.* 39: 213–223.

Knowlton, N. & Weigt, L.A. 1998. New dates and new rates for divergence across the Isthmus of Panama. *Proc. R. Soc. B* 265: 2257–2263.

Knowlton, N., Weigt, L.A., Solorzano, L.A., Mills, D.K. & Bermingham, E. 1993. Divergence in proteins, mitochondrial DNA, and reproductive compatibility across the Isthmus of Panama. *Science* 260: 1629–1632.

Kochzius, M. & Nuryanto, A. 2008. Strong genetic population structure in the boring giant clam, *Tridacna crocea*, across the Indo-Malay Archipelago: implications related to evolutionary processes and connectivity. *Mol. Ecol.* 17: 3775–3787.

Lacson, J.M. & Clark, S. 1995. Genetic divergence of Maldivian and Micronesian demes of the damselfishes *Stegastes nigricans*, *Chrysiptera biocellata*, *C. glauca* and *C. leucopoma* (Pomacentridae). *Mar. Biol.* 121: 585–590.

Ladd, H.S. 1960. Origin of the Pacific island molluscan fauna. *Am. J. Sci.* 258(A): 137–150.

Lanchester, W.F. 1903. Marine crustaceans. VIII. Stomatopoda, with an account of the varieties of *Gonodactylus chiragra*. In: Gardiner, J.S. (ed.), *The fauna and geography of the Maldive and Laccadive Archipelagoes. Being an account of the work carried on and of the collections made by an expedition during the years 1899 and 1900. Vol. I. Part IV*: 444–459, pl. XXIII. Cambridge: Cambridge University Press.

Lande, R. 1981. Models of speciation by sexual selection on polygenic traits. PNAS, USA. 78; 3721–3725.

Lavery, S., Moritz, C. & Fielder, D.R. 1996. Indo-Pacific population structure and evolutionary history of the coconut crab *Birgus latro*. *Mol. Ecol.* 5: 557–570.

Lenz, H. 1905. Ostafrikanische Dekapoden und Stomatopoden. *Abh. Senckenberg. Naturforsch. Ges.* 27: 339–392, pls. XLVII–XLVIII.

Lessios, H.A. & Cunningham, C.W. 1990. Gametic incompatibility between species of the sea urchin *Echinometra* on the two sides of the Isthmus of Panama. *Evolution* 44: 933–941.

Lessios, H.A. 1998. The first stage of speciation as seen in organisms separated by the Isthmus of Panama. In: Howard, D. & Berlocher, S. (eds.), *Endless Forms: Species and Speciation*: 186–201. Oxford: Oxford University Press.

Lessios, H.A., Kessing, B.D. & Robertson, D.R. 1998. Massive gene flow across the world's most potent marine biogeographic barrier. *Proc. R. Soc. B* 265: 583–588.

Linnaeus, C. 1758. *Systema naturae per regna tria naturae, secundum classes, ordines, genera, species cum characteribus, differentiis, synonymis, locis. Tomus I. Editio decima, reformata.* Holmiae: Laurentius Salvius.

Lukas, R., Firing, E., Hacker, P., Richardson, P.L., Collins, C.A., Fine, R. & Gammon, R. 1991. Observations of the Mindanao Current during the western Equatorial Pacific Ocean circulation study. *J. Geophys. Res.* 96: 7089–7104.

Lukas, R., Yamagata, T. & McCreary, J.P. 1996. Pacific low-latitude western boundary currents and the Indonesian throughflow. *J. Geophys. Res.* 101: 12209–12216.

Lynch, J.D. 1989. The gauge of speciation: on the frequencies of modes of speciation. In: Otte, D. & Endler, J.A. (eds.), *Speciation and Its Consequences*: 527–556. Sunderland, MA: Sinauer Associates.

MacArthur, R.H. & Wilson, E.O. 1967. *The Theory of Island Biogeography.* Princeton, NJ: Princeton University Press.

Malay, M.C.D. & Paulay, G. 2010. Peripatric speciation drives diversification and distributional pattern of reef hermit crabs (Decapoda: Diogenidae: *Calcinus*). *Evolution* 64: 634–662.

Manning, R.B. 1967. Notes on the *demanii* section of the genus *Gonodactylus* Berthold with descriptions of three new species (Crustacea: Stomatopoda). *Proc. U. S. Nat. Mus.* 123): 1–27.

Manning, R.B. 1969: Notes on the *Gonodactylus* section of the family Gonodactylidae (Crustacea, Stomatopoda), with descriptions of four new genera and a new species. *Proc. Biol. Soc. Wash.* 82: 143–166.

Manning, R.B. 1971. Two new species of *Gonodactylus* (Crustacea, Stomatopoda), from Eniwetok Atoll, Pacific Ocean. *Proc. Biol. Soc. Wash.* 84: 73–80.

Manning, R.B. 1972. Two new species of *Pseudosquilla* (Crustacea, Stomatopoda) from the Pacific Ocean. *Am. Mus. Novits.* 2484: 1–11.

Manning, R.B. 1975a. *Gonodactylus botti*, a new stomatopod crustacean from Indonesia. *Senckenbergiana biol.* 56: 289–291.

Manning, R.B. 1975b. Two new species of the Indo-West Pacific genus Chorisquilla (Crustacea, Stomatopoda), with notes on *C. excavata* (Miers). *Proc. biol. Soc. Wash.* 88: 253–262.

Manning, R.B. 1977. A monograph of the West African Stomatopod Crustacea. *Atlantide Rep.* 12: 1–181.

Manning, R.B. 1978. Synopses of the Indo-West-Pacific species of *Lysiosquilla* Dana, 1852 (Crustacea: Stomatopoda: Lysiosquillidae). *Smiths. Contr. Zool.* 259: 1–16.

Manning, R.B. 1980. The superfamilies, families, and genera of recent stomatopod crustacea, with diagnoses of six new families. *Proc. Biol. Soc. Wash.* 93: 362–372.

Manning, R.B. 1984. *Gonodactyloideus cracens* n. gen. n. sp., a new stomatopod crustacean from Western Australia. *Beagle* 1: 83–86.

Manning, R. B. 1995. Stomatopod Crustacea of Vietnam: the legacy of Raoul Serne. *Crust. Res., Special No.* 4: 1–339.

Manning, R.B. & Provenzano, A.J. 1963. Studies on development of stomatopod Crustacea. I. Early larval stages of *Gonodactylus oerstedii* Hansen. *Bull. Mar. Sci. Gulf Caribbean* 13: 467–487.

Marko, P.B. 2002. Fossil calibration of molecular clocks and the divergence times of geminate species pairs separated by the Isthmus of Panama. *Mol. Bio. Evol.* 19: 2005–2021.

Mayr, E. 1942. *Systematics and the Origin of Species.* New York, NY: Columbia University Press.

Mayr, E. 1954. Geographic speciation in tropical echinoids. *Evolution* 8: 1–18.

McManus, J.W. 1985. Marine speciation, tectonics and sea-level changes in southeast Asia. *Proc. 5th Int. Coral Reef Congr. Tahiti* 4: 133–138.

Michel, A. 1968. Drift of stomatopod larvae in eastern Indian Ocean. *Cahiers Orstom Oceanogr.* 6: 13–41.

Michel, A. 1970. Pelagic larvae and post-larvae of genus *Odontodactylus* (Crustacea: Stomatopoda) in south and equatorial tropical Pacific. *Cahiers Orstom Oceanogr.* 8: 111–126.

Miers, E.J. 1875. On some new or undescribed species of Crustacea from the Samoa Islands. *Ann. Mag. Nat. Hist.* (4)16: 341–344.

Miers, E.J. 1880a. On a collection of Crustacea from the Malaysian region. Part III. Crustacea

Anomura and Macrura (except Penaeidea). *Ann. Mag. Nat. Hist.* (5)5: 370–384.

Miers, E. J. 1880b. On a collection of Crustacea from the Malaysian Region. Part IV. Penaeidea, Stomatopoda, Isopoda, Suctoria and Xiphosura. *Ann. Mag. Nat. Hist.* (5)5: 457–467.

Miers, E.J. 1884. Crustacea. In: Günther, A. (ed.), *Report on the zoological collections made in the Indo-Pacific Ocean during the voyage of H.M.S. "Alert" 1881–2*: 178–322, 513–575. pls. XVIII–XXIV, XLVI–LII. London: British Museum.

Moosa, M.K. 1991. The Stomatopoda of New Caledonia and Chesterfield Islands. In: Richer de Forges, B. (ed.), *Le benthos des fonds meubles des lagons de Nouvelle-Calédonie. Volume 1*: 147–219. Montpellier: IRD Editions.

Mora, C., Chittaro, P.M., Sale, P.F., Kritzer, J.P. & Ludsin, S.A. 2003. Patterns and processes in reef fish diversity. *Nature* 421: 933–936.

Morey, S.L., Shriver, J.F. & O'Brien, J.J. 1999. The effects of Halmahera on the Indonesian throughflow. *J. Geophys. Res.* 104: 23281–23296.

Moritz, C., Patton, J.L., Schneider, C.J. & Smith, T.B. 2000. Diversification of rainforest faunas: an integrated molecular approach. *Annu. Rev. Ecol. Syst.* 31: 533–563.

Müller, F. 1886. Zur Crustaceenfauna von Trincomali. *Verh. naturforsch. Ges. Basel* 8: 470–479, pl. IV.

Naiyaneter, P. 1987. Two new stomatopod crustaceans from Thailand with a key to the genus Manningia Serène, 1962. *Crustaceana* 53: 237–242.

Nelson, G.J. & Platnick, N. 1981. *Systematics and Biogeography: Cladistics and Vicariance*. New York, NY: Columbia University Press.

Nobili, G., 1899. Contribuzioni alla conoscienza della fauna carcinologica della Papuasia, delle Molucche e dell'Australia. *Ann. Mus. Civ. Stor. Nat. Genova* (2)20: 230–282.

Nof, D. 1995. Choked flows from the Pacific to the Indian-Ocean. *J. Phys. Oceanogr.* 25: 1369–1383.

Nof, D. 1996. What controls the origin of the Indonesian throughflow? *J. Geophys. Res.* 101: 12301–12314.

Nuryanto, A. & Kochzius, M. 2009. Highly restricted gene flow and deep evolutionary lineages in the giant clam *Tridacna maxima*. *Coral Reefs* 28: 607–619.

Odhner, T. 1923. Indopazifische Stomatopoden. *Medd. Gotb. Mus. Zool. Avd.* 30: 3–16, 1 pl.

Palumbi, S.R. 1992. Marine speciation on a small planet. *Trend Ecol. Evol.* 7: 114–118.

Palumbi, S.R. 1994. Genetic-divergence, reproductive isolation, and marine speciation. *Ann. Rev. Ecol. Syst.* 25: 547–572.

Palumbi, S.R. 1997. Molecular biogeography of the Pacific. *Coral Reefs* 16: S47–S52.

Palumbi, S.R., Cipriano, F. & Hare, M.P. 2001. Predicting nuclear gene coalescence from mitochondrial data: the three-times rule. *Evolution* 55: 859–868.

Paulay, G. & Meyer, C. 2002. Diversification in the tropical pacific: comparisons between marine and terrestrial systems and the importance of founder speciation. *Int. Comp. Biol.* 42: 922–934.

Pfeffer, G. 1888. Übersicht der von Herrn Dr. F. Stuhlmann in Ägypten, auf Sansibar und dem gegenüberliegenden Festlande gesammelten Reptilien, Amphibien, Fische, Mollusken und Krebse. *Jahrb. Hamburg. wiss. Anst.* 6: 1–36.

Pocock, R.I. 1893. Report upon the stomatopod crustaceans obtained by P.W. Basset-Smith, Esq., Surgeon R.N., during the cruise, in the Australian and China seas, of H.M.S. "Penguin," Commander W.U. Moore. *Ann. Mag. Nat. Hist.* (6)11: 473–479, pl. XXb.

Posada, D. & Crandall, K.A. 1998. Modeltest: testing the model of DNA substitution. *Bioinformatics* 14: 817–818.

Potts, D.C. 1985. Sea-level flucuations and speciation in Scleractinia. *Proc. 5th Int. Coral Reef Congr. Tahiti* 4: 127–132.

Provenzano, A.J. & Manning, R.B. 1978. Studies on development of stomatopod Crustacea. II. The later larval stages of *Gonodactylus oerstedii* Hansen reared in the laboratory. *Bull. Mar. Sci.* 28: 297–315.

Reid, D.G., Lal, K., Mackenzie-Dodds, J., Kaligis, F., Littlewood, D.T.J. & Williams, S.T. 2006. Comparative phylogeography and species boundaries in *Echinolittorina* snails in the central Indo-West Pacific. *J. Biogeogr.* 33: 990–1006.

Roberts, C.M. 1997. Connectivity and management of Caribbean coral reefs. *Science* 278: 1454–1457.

Rosenblatt, R.H. & Waples, R.S. 1986. A genetic comparison of allopatric populations of shore fish species from the eastern and central Pacific Ocean: dispersal of vicarience? *Copeia* 2; 275–284.

Roughgarden, J., Gaines, S. & Possingham, H. 1988. Recruitment dynamics in complex life cycles. *Science* 241: 1460–1466.

Sale, P.F., Cowen, R.K., Danilowicz, B.S., Jones, G.P., Kritzer, J.P., Lindeman, K.C., Planes, S., Polunin, N.V.C., Russ, G.R., Sadovy, Y.J. & Steneck, R.S. 2005. Critical science gaps impede use of no-take fishery reserves. *Trends Ecol. Evol.* 20: 74–80.

Santini, F. & Winterbottom, R. 2002. Historical biogeography of Indo-western Pacific coral reef biota: is the Indonesian region a centre of origin? *J. Biogeogr.* 29: 189–205.

Scheltema, R.S. 1971. Larval dispersal as a means of genetic exchange between geographically separated populations of shallow-water benthic marine gastropods. *Biol. Bull.* 140: 284–322.

Scheltema, R.S. 1986. On dispersal and planktonic larvae of benthic invertebrates—an eclectic overview and summary of problems. *Bull. Mar. Sci.* 39: 290–322.

Schneider, C.J., Cunningham, M. & Moritz, C. 1998. Comparative phylogeography and the history of endemic vertebrates in the wet tropics rainforests of Australia. *Mol. Ecol.* 7: 487–498.

Schneider, C.J., Smith, T.B., Larison, B. & Moritz, C. 1999. A test of alternative models of diversification in tropical rainforests: ecological gradients vs. rainforest refugia. *Proc. Natl. Acad. Sci.* 96: 13869–13873.

Schneider, S., Roessli, D. & Excoffier, L. 2000. ARLEQUIN ver. 2.000: A software for population genetics data analysis. Geneva: Genetics and Biometry Laboratory, University of Geneva, Switzerland.

Serène, R. 1954. Observations biologiques sur les stomatopodes. *Mém. Inst. Océanogr. Nhatrang* 8: 1–93, pls. 1–10.

Shanks, A. 2009. Pelagic larval duration and dispersal distance revisited. *Biol. Bull.* 216: 373–385.

Shanks, A.L., Grantham, B.A. & Carr, M.H. 2003. Propagule dispersal distance and the size and spacing of marine reserves. *Ecol. App.* 13: S159–S169.

Shimodaira, H. & Hasegawa, M. 1999. Multiple comparisons of log-likelihoods with applications to phylogenetic inference. *Mol. Bio. Evol.* 16: 1114–1116.

Spalding, M.D., Fox, H.E., Halpern, B.S., McManus, M.A., Molnar, J., Allen, G.R., Davidson, N., Jorge, Z.A., Lombana, A.L., Lourie, S.A., Martin, K.D., McManus, E., Recchia, C.A. & Robertson, J. 2007. Marine ecoregions of the world: a bioregionalization of coastal and shelf areas. *Bioscience* 57: 573–583.

Springer, V.G. & Williams, J.T. 1990. Widely distributed pacific plate endemics and lowered sea-level. *Bull. Mar. Sci.* 47: 631–640.

Sun, X. & Yang, S. 1998. Studies on the stomatopod Crustacea from Nansha Islands, China. Part 1, Protosquillidae and Pseudosquillidae, with descriptions of a new genus and two new species. *Stud. Mar. Faun. Flor. Biogeogr. Nansha Is. Neighb. Waters* 3: 142–155.

Swofford, D.L. 2002. PAUP*. Phylogenetic Analysis Using Parsimony (*and Other Methods). Sunderland, MA: Sinauer Associates.

Terry, A., Bucciarelli, G. & Bernardi, G. 2000. Restricted gene flow and incipient speciation in disjunct Pacific Ocean and Sea of Cortez populations of a reef fish species, *Girella nigricans*. *Evolution* 54: 652–659.

Timm, J. & Kochzius, M. 2008. Geological history and oceanography of the Indo-Malay Archipelago shape the genetic population structure in the false clown anemonefish (*Amphiprion ocellaris*). *Mol. Ecol.* 17: 3999–4014.

Turelli, M., Barton, N.H. & Coyne, J.A. 2001. Theory and speciation. *Trends Ecol. Evol.* 16: 330–343.

Veron, J.E.N., DeVantier, L., Turak, E., Green, A.L., Kininmonth, S., Stafford-Smith, M. & Peterson, N. 2009. Delineating the Coral Triangle. *Galaxea* 11: 91–100.

Via, S. 2001. Sympatric speciation in animals: the ugly duckling grows up. *Trends Ecol. Evol.* 16: 381–390.

Voris, H.K. 2000. Maps of Pleistocene sea levels in Southeast Asia: shorelines, river systems and time durations. *J. Biogeogr.* 27: 1153–1167.

Wallace, C.C., Paulay, G., Hoeksema, B.W., Bellwood, D.R., Hutchings, P.A., Barber, P.H., Erdmann, M. & Wolstenholme, J. 2002. Nature and origins of unique high diversity reef faunas in the Bay of Tomini, Central Sulawesi: the ultimate "centre of diversity"? In: Moosa, M.K., Soemodihardjo, S., Soegiarto, A., Romimohtarto, K., Nontji, A., Soekarno & Suharsono (Eds.), *Proceedings of the Ninth International Coral Reef Symposium, Bali. 23–27 Oct. 2000. Vol. 1*: 185–192. Bali: Ministry of Environment, Indonesian Institute of Sciences & International Society for Reef Studies.

Weersing, K. & Toonen, R.J. 2009. Population genetics, larval dispersal, and connectivity in marine systems. *Mar. Ecol. Progr. Ser.* 393: 1–12.

White, A. 1850. Descriptions of two species of Crustacea in the British Museum. *Proc. Zool. Soc. Lond.* 18: 95–97, pls. XV–XVI.

White, A. 1861. Descriptions of two species of Crustacea belonging to the families Callianassidae and Squillidae. *Proc. Zool. Soc. Lond.* 1861: 42–44, pls. 6–7.

Wilkinson, C. 2002. *Status of Coral Reefs of the World: 2002*. Townsville, Australia: Australian Institute of Marine Science.

Williams, S.T. & Benzie, J.A.H. 1998. Evidence of a phylogeographic break between populations of a high-dispersal starfish: congruent regions within the Indo-West Pacific defined by colour morphs, mtDNA and allozyme data. *Evolution* 52: 87–99.

Wilson, M.E.J. & Rosen, B.R. 1998. Implications of paucity of corals in the Paleogene of SE Asia: plate tectonics or centre of origin? In: Hall, R. & Holloway, J.D. (eds.), *Biology and Geological Evolution of SE Asia*: 165–195. Leiden: Backhuys Publishers.

Wood-Mason, J. 1875. On new or little-known crustaceans. *Proc. Asiat. Soc. Bengal* 1875: 230–232.

Wood-Mason, J. 1895. *Figures and Descriptions of Nine Species of Squillidae from the Collection in the Indian Museum*. Calcutta: Indian Museum.

Woodland, D.J. 1983. Zoogeography of the Siganidae (Pisces): an interpretation of distribution and richness patterns. *Bull. Mar. Sci.* 33: 713–717.

Woodland, D.J. 1986. Wallace's Line and the distribution of marine inshore fishes. In: Uyeno, T., Arai, R., Taniuchi, T. & Matsuura, K. (eds.), *Indo-Pacific Fish Biology: Proceedings of the Second International Conference on Indo-Pacific Fishes*: 453–460. Tokyo: Ichthyological Society of Japan.

Wyrtki, K. 1961. *Physical Oceanography of the Southeast Asian Waters*. La Jolla, CA: University of California, Scripps Institution of Oceanography.

Comparative phylogeography of three achelate lobster species from Macaronesia (northeast Atlantic)

Elsa Froufe[1], Patricia Cabezas[2], Paulo Alexandrino[3] & Marcos Pérez-Losada[3]

[1] CIIMAR, Centro Interdisciplinar de Investigação Marinha e Ambiental, Rua dos Bragas 289, 4050-123 Porto, Portugal

[2] MNCN, Museo Nacional de Ciencias Naturales (CSIC), Biodiversidad y Biología Evolutiva, José Gutiérrez Abascal 2, 28006 Madrid, Spain

[3] CIBIO, Centro de Investigação em Biodiversidade e Recursos Genéticos, Universidade do Porto, Campus Agrário de Vairão, 4485-661 Vairão, Portugal

ABSTRACT

The northeast Atlantic region has been the focus of intense phylogeographic research over the last 20 years. However, most of those studies have been centered on the Atlantic-Mediterranean junction and have consistently neglected the Macaronesian archipelagos (Cape Verde, Madeira, Selvagens, Azores, and Canary Islands). Here we used mtDNA cytochrome c oxidase subunit I (COI) sequences to infer common patterns of differentiation in three achelate lobster species (Decapoda: Palinura: Achelata): *Scyllarides latus* ($N = 26$) from the Cape Verde Islands, Azores, and Atlantic coast; *Palinurus elephas* ($N = 247$) from the Azores, Atlantic, and Mediterranean; and *Panulirus regius* ($N = 19$) from the Cape Verde and southwestern Africa. Two competing phylogeographic hypotheses were tested: panmixia due to long-distance dispersion and population structuring due to environmental factors (e.g., oceanographic barriers and currents). Our analyses suggest: 1) panmixia in the Azores populations and between the Azores and the continental populations of *P. elephas* and *S. latus*; and 2) weak genetic differences, in relation to oceanographic fronts, between the Atlantic and Mediterranean populations of *P. elephas* and a trend for differentiation, in relation to marine currents, between Cape Verde and southwestern African populations of *P. regius*.

1 INTRODUCTION

Decapods are the most species-rich group of Crustacea consisting of $\approx 15,000$ species (De Grave et al. 2009). Within them, lobsters of the infraorder Achelata are among the most commercially and socially important (Bowman & Abele 1982). The average catch of these lobsters in the northwestern Atlantic for 2002 exceeded 82,000 tons (FAO 2005). In fact, the overexploitation of this marine resource may have led to a severe crisis in commercial fisheries around the world, since catches per year have dramatically declined over the last decades (e.g., Pauly & Maclean 2003).

Achelate lobsters are classified in three families, namely Scyllaridae (slipper lobsters), Palinuridae (spiny lobsters), and Synaxidae (furry or coral lobsters). They are all characterized by the presence of a phyllosoma larval stage (Dixon et al. 2003), which shows a planktonic zoea phase of up to 24 months before settlement, one of the longest reported in crustaceans (Díaz et al. 2001). These larvae are adapted to a long offshore drifting life and are not capable of swimming large distances, so it is thought that oceanographic currents are a determining factor in their dispersal (Goñi & Latrouite 2005). Within the family Scyllaridae, the species *Scyllarides latus* (Latreille 1802) occurs in depths of 4 to 100 m, mainly associated with rocky and sandy bottoms. It is distributed across the Mediterranean and eastern Atlantic from the coast of Portugal to Senegal, including the Macaronesian archipelagos Azores, Cape Verde, Madeira, Selvagens, and Canary Islands (Holthuis 1991). Although the species is captured and consumed across its entire distribution range, its occurrence is rare and so there is not a significant commercial fishery. There is a considerable lack of information on the reproductive biology of this species, but it is known that berried females appear from May to July, the spawning season extends from August to September and there is normally only one generation per year (Spanier & Lavalli 1998). Phyllosoma larvae are assumed to spend many months in the open sea (up to 11 months according to Martins (1985); however, as for other lobster species, the great plasticity of its larval development raises questions about the span of its larval periods (Goñi & Latrouite 2005).

The spiny lobster *Palinurus elephas* (Fabricius 1787) shows preference for shallow waters, being typically recorded at depths of 10 to 70 m in association with rocky bottoms. It occurs in the eastern Atlantic (including the Azores), from Norwegian waters to Morocco, and the Mediterranean (Zariquiey-Alvárez 1968). Differences in mating and spawning season have been observed between populations from the Atlantic and West Mediterranean, which occur from June to October in the Atlantic and from July to September in the Mediterranean (Goñi & Latrouite 2005). The duration of the pelagic larval stage has been estimated to last 5–6 months for Mediterranean populations (Marin 1985) and 10–12 months for the Atlantic ones (Mercer 1973), but, as already mentioned, these estimates are questioned. *Palinurus elephas* has been overfished because of its great commercial value, which has led to an alarming decrease of its average annual catches and triggered management conservation plans (Díaz et al. 2005; Goñi et al. 2006). This species has been the subject of multiple studies focused on its ecology, population dynamics, phylogeography, and evolutionary relationships (e.g., Díaz et al. 2001; Patek & Oakley 2003; Goñi & Latrouite 2005; Palero et al. 2008, 2009a, 2009b; Babbucci et al. 2010), which have revealed population structuring within and between stocks from the Atlantic and Mediterranean waters and reduced species diversity compared to other lobsters.

Finally, the royal spiny lobster *Panulirus regius* (De Brito Capello 1864) is distributed along the eastern Atlantic, including the west coast of Africa between Morocco and Angola, the Macaronesian archipelago of Cape Verde and the western part of the Mediterranean (Holthuis 1991). It inhabits rocky substrata from shallow waters to depths of 55 m. This lobster is the most economically important crustacean species found in West Africa, being heavily exploited in Cape Verde (Dias 1993; Reis 1997), where it accounts for the majority of the shallow-water lobster catches in the area (Freitas et al. 2007). Little is known about its basic biology and, up to date, only one study carried out in Cape Verde has attempted to clarify some aspects of its reproductive biology (Freitas et al. 2007). Ovigerous females are observed between April and December and although a single spawning season is assumed from June to September, multiple spawning may take place in large females (Freitas et al. 2007). No previous studies exist on the larval development of *P. regius*, but long phyllosoma larval stages have been reported in other species of *Panulirus* (e.g., up to 10 months in *P. longipes*) (Matsuda 2000); which suggest *P. regius* may also show a large dispersal.

Over the last 20 years, the marine regions where these three lobster species occur have been the focus of intense phylogeographic research (see review in Paternello et al. 2007 and references therein). Previous studies have shown a) lack of or weak population structuring (i.e., panmixia) (e.g., Domingues et al. 2006; Reuschel & Schubart 2007; Bargelloni et al. 2008; Comesaña et al. 2008; Sotelo et al. 2009), b) population structuring unrelated to the Mediterranean-Atlantic separation (e.g., Kotoulas et al. 1995; Pérez-Losada et al. 2007), or c) genetic differences between the Mediterranean and Atlantic with or without further substructuring within each basin (e.g., Sá-Pinto et al. 2005; Triantafyllidis et al. 2005; Magoulas et al. 2006; Reuschel et al. 2010). Thereby, three nonexclusive hypotheses of population structuring have been proposed to explain the phylogeo-

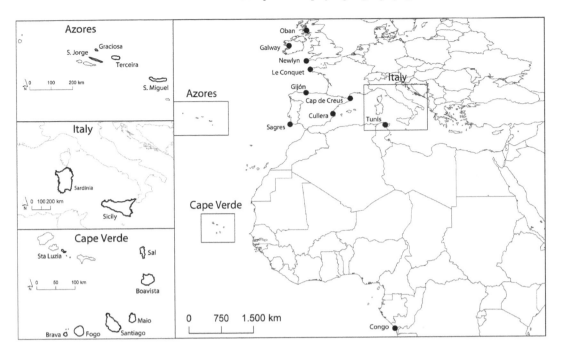

Figure 1. Map showing the locations of *Scyllarides latus*, *Palinurus elephas*, and *Panulirus regius* included in this study.

graphic patterns observed: 1) fragmentation due to historical (e.g., Gibraltar Strait) and present-day (e.g., Almería-Oran oceanographic front) barriers, 2) isolation by distance, and 3) differentiation due to life-history traits (e.g., limited dispersal ability).

Such phylogeographic studies have mainly focused on the northeast (NE) Atlantic and Mediterranean regions and have consistently neglected the marine fauna from Macaronesia (Figure 1). To our knowledge, only a recent paper by Babbucci et al. (2010) has looked at the population structure and evolution history of *P. elephas* in the Azores. So far, no other information is available regarding the genetic diversity of the lobster populations from Macaronesia and their relationships to their continental relatives, despite their commercial importance and threatened status.

Several currents and oceanographic barriers have been described in the NE Atlantic and western Mediterranean. The Gulf Stream (a western boundary current) moves from the equator northward bringing warm equatorial waters up past the east coast of the United States. It then connects with the transverse North Atlantic Current, which moves to the east until it reaches Europe and turns south forming the Canary Current (an eastern boundary current), which then moves south along the coast of Africa bringing cold water with it. These three currents merge to make up the major subtropical gyre of the North Atlantic Ocean and work to move warm water from the tropics toward the poles, and cold water from the poles toward the tropics (e.g., Clarke et al. 1980; Krauss et al. 1987; Johnson & Stevens 2000). Therefore, although no oceanographic barrier has been described separating the Azores and Cape Verde from the continental landmasses, powerful currents separate Cape Verde from southwestern Africa. Two oceanographic barriers have been described in the Atlantic-Mediterranean junction, one historical (the repeated aperture and closure of the Gibraltar Strait) and another present-day barrier (the Almería-Oran front). The Almería-Oran oceanographic front is associated with steep temperature (1.4 °C) and salinity (2 psu) gradients over a distance of 2 km, and with strong water currents (40 cm/s average speed) that flow in a south-easterly direction away from the Spanish coast towards North Africa (Tintoré et al. 1988), whilst the Strait of Gibraltar experiences one-way surface current flows (Parrilla & Kinder 1992).

In this study we analyze several populations of *Scyllarides latus* from the Azores and Cape Verde archipelagos, *Palinurus elephas* from the Azores and *Palinurus regius* from Cape Verde in combination with available data from the eastern Atlantic and Mediterranean, in order to determine their population diversity and structure and infer patterns of genetic differentiation within the NE Atlantic. Based on the biology, dispersal ability, and distribution of these lobster species, overall low levels of genetic subdivision should be expected; however, considering the oceanographic currents and barriers described in the NE Atlantic and western Mediterranean, there is potential for divergence across those marine subdivisions. Here, we will test these two competing hypotheses.

2   MATERIALS AND METHODS

2.1   Lobster samples

We collected (i) 18 specimens of the slipper lobster *Scyllarides latus* from the archipelago of the Azores (S. Miguel, Terceira, and Graciosa islands), five from Cape Verde (Sal, Boavista, and Brava islands) and two specimens from Portugal (Sagres); (ii) 19 specimens of the spiny lobster *Palinurus elephas* from the Azores (S. Miguel, S. Jorge, Terceira, and Graciosa islands) and one specimen from Portugal (Sagres); and (iii) 17 specimens of the royal lobster *Panulirus regius* from Cape Verde (Sal, Maio, Santiago, Boavista, S. Luzia, and Fogo islands) (Table 1 and Figure 1). All specimens were acquired in commercial catches and stored at the Centro Interdisciplinar de Investigação Marinha e Ambiental (CIIMAR).

2.2   Sequence data

Genomic DNA was extracted from gill tissue using the DNeasy Blood & Tissue Kit (Qiagen). Partial fragments of the cytochrome c oxidase subunit I (COI) gene were amplified and sequenced using primers from Folmer et al. (1994). PCR conditions were as follows: initial denaturation at 94 °C (5 min), then denaturation at 94 °C (30 sec), annealing at 53 °C (45 sec) and extension at 72 °C (45 sec) repeated for 35 cycles, and final extension at 72 °C (7 min). We used PCR reaction volumes of 20 μl and Platinum Taq (Invitrogen, Carlsbad, CA, USA) for amplification. PCR products (658 basepairs (bp)) were purified and sequenced in both directions. Our COI sequences were then combined with available COI data from GenBank corresponding to one *S. latus* from Morocco (FJ174947), 227 *P. elephas* from England, France, Ireland, Italy, Portugal, Spain, and Tunisia (Palero et al. 2008), and two *P. regius* from Congo and southwestern Africa (AF339470.1 and FJ174962, respectively) (Table 1 and Figure 1).

2.3   Genetic analysis

Sequences were aligned in MAFFT (Katoh et al. 2005) and each species' dataset was analyzed separately. Population structure in *S. latus* and *P. regius* was inferred using methods of maximum likelihood (ML) and Bayesian phylogenetic inference (BI). Best-fit models of nucleotide substitution were estimated using JModelTest 0.1.1 (Posada 2008) and in both species the GTR $+\Gamma+$ I model was chosen. ML trees were built in RAxML 7.2.6 (Stamatakis 2006) running 1,000 bootstrap replicates and searching for the best-scoring ML tree. Phylogenetic BI was performed in MrBayes 3.1.2 (Huelsenbeck & Ronquist 2001). Two independent runs 5×10^6 generations long were run and sampled every 1,000 generations. The program Tracer 1.5 (Rambaut & Drummond 2009) was used to assess convergence and determine the burn-in. Trees resulting from separate analyses were com-

Table 1. *Scyllarides latus*, *Palinurus elephas*, and *Panulirus regius* samples and specimens (*N*) included in this study. Samples from previous studies ([1] = Palero et al. 2009a, [2] = Palero et al. 2008, [3] = Ptacek et al. 2001) are indicated. Voucher numbers for the specimens sequenced in this study are also included.

Species	Archipelago/Country	Island/Location	*N*	CIIMAR voucher
Scyllarides latus	Azores	S. Miguel	2	149, 158
Scyllarides latus	Azores	Terceira	14	47, 50, 51, 65–67, 76, 77, 112–117
Scyllarides latus	Azores	Graciosa	2	108, 109
Scyllarides latus	Cape Verde	Sal	1	39
Scyllarides latus	Cape Verde	Boavista	3	82, 84, 87
Scyllarides latus	Cape Verde	Brava	1	130
Scyllarides latus[1]	Morocco		1	–
Scyllarides latus	Portugal	Sagres	2	185, 186
Palinurus elephas	Azores	Terceira	8	2–4, 15, 21, 23, 26, 48
Palinurus elephas	Azores	S. Miguel	6	31, 32, 80, 150, 151, 153
Palinurus elephas	Azores	S. Jorge	4	40–43
Palinurus elephas	Azores	Graciosa	1	110
Palinurus elephas[2]	England	Newlyn	18	–
Palinurus elephas[2]	France (Atlantic)	Le Conquet	21	–
Palinurus elephas[2]	Ireland	Galway	19	–
Palinurus elephas[2]	Italy	Sicily	19	–
Palinurus elephas[2]	Italy	Sardinia	19	–
Palinurus elephas	Portugal	Sagres	1	187
Palinurus elephas[2]	Portugal	Sagres	24	–
Palinurus elephas[2]	Scotland	Oban	22	–
Palinurus elephas[2]	Spain (Atlantic)	Gijón	15	–
Palinurus elephas[2]	Spain (Mediterranean)	Cullera	26	–
Palinurus elephas[2]	Spain (Mediterranean)	Cap de Creus	24	–
Palinurus elephas[2]	Tunisia	Tunis	20	–
Panulirus regius	Cape Verde	Sal	5	16, 25–27, 36
Panulirus regius	Cape Verde	Maio	3	17–19
Panulirus regius	Cape Verde	Santiago	1	20
Panulirus regius	Cape Verde	Boavista	6	27, 28, 30, 88–90
Panulirus regius	Cape Verde	Sta. Luzia	1	83
Panulirus regius	Cape Verde	Fogo	1	128
Panulirus regius[1]	Southwestern Africa		1	–
Panulirus regius[1,3]	Congo	Pointe-Noire	1	–

bined and summarized as a 50% majority-rule consensus tree. *Scyllarides herklotsii* (FJ174946) was used to root the *S. latus* phylogenetic tree, while *Panulirus homarus* (FJ174963) and *Panulirus inflatus* (FJ174964) were used to root the *P. regius* tree.

Following Palero et al. (2008), *P. elephas* evolutionary relationships were inferred using the unrooted network approach of Templeton et al. (1992) as implemented in TCS (Clement et al. 2000). Population structure in this species was further explored using AMOVA (Excoffier et al. 1992) and Φ_{ST} statistics in ARLEQUIN 3.5 (Excoffier & El Lischer 2010). Such tests were not carried out in *S. latus* and *P. regius* due to their small sample sizes. Significance was assessed by 1,000 permutations. Both tests were performed incorporating the Tamura-Nei genetic distance with gamma

Table 2. *Scyllarides latus*, *Palinurus elephas*, and *Panulirus regius* COI diversity. Sample size (N), number of haplotypes (h), number of segregating sites (S), current (θ_π) and historical (θ_W) genetic diversity, and Tajima's D and Fu's F_S estimates (* = $P < 0.05$; ** = $P < 0.01$; *** = $P < 0.001$) are indicated.

Species	Region	N	h	S	θ_π	θ_W	D	D_S
Scyllarides latus	Azores	18	14	34	0.010	0.017	−1.748*	−5.448**
	All	26	22	44	0.010	0.020	−1.943*	−14.913***
Palinurus elephas	Azores	19	7	10	0.003	0.006	−1.590	−2.057
	Atlantic	120	23	24	0.002	0.009	−2.330**	−26.949**
	Mediterranean	108	17	14	0.002	0.005	−1.917	−15.585**
	All	247	37	37	0.002	0.013	−2.440**	−54.630***
Panulirus regius	Cape Verde	17	10	20	0.009	0.010	−0.298	−1.219
	All	19	12	26	0.011	0.013	−0.730	−2.026

correction. Pairwise genetic differences among *P. elephas* populations were also estimated using Φ_{ST} statistics. For the latter analysis, the Terceira, S. Jorge, and Graciosa islands belonging to the Central Island group were pooled together and analyzed separately from the S. Miguel island, which belongs to the Eastern Island group (Figure 1).

Population demographics were examined in all lobster species using Tajima's D (Tajima 1989) and Fu's F_S (Fu 1997) neutrality tests and mismatch distribution analysis as implemented in DnaSP 5 (Librado & Rozas 2009). The validity of the expansion model was tested using parametric boot-strap (1,000 replicates) in ARLEQUIN. The fit to the expected mismatch distribution was quantified by the sum of squared deviations (SSD) and the raggedness statistic (r) between the observed and simulated distributions on one hand and the expected distribution on the other. Current (θ_π) and historical (θ_W) genetic diversities were also estimated in the DNA Sequence Polymorphism (DnaSP) program.

The time to most recent common ancestor (TMRCA) was estimated in all lobster species using the Bayesian coalescent approach implemented in BEAST 1.5.4 (Drummond & Rambaut 2007). Preliminary analyses using a lognormal relaxed clock yielded an estimation close to zero for the coefficient of variation parameter, hence justifying the use of a clock-like model. A mean substitution rate of 0.009–0.011 substitutions/My has been proposed for COI in isopod species (Ketmaier et al. 2003). However, it has been indicated that intraspecific rates could be three to ten times faster (Emerson 2007). Therefore, as in Palero et al. (2008), TMRCAs were estimated using three different substitution rates (0.01/0.03/0.1 substitution/My). All analyses were performed using the HKY + Γ + I model of nucleotide substitution, a strict clock and constant population size. For each dataset we ran two independent analyses 10^7 generations long and with parameter values sampled every 1,000 generations. Appropriated effective sample sizes for each parameter and burn-ins were determined in TRACER 1.5.

3   RESULTS

3.1   *Scyllarides latus*

Twenty-five new sequences with a length of 583 bp long were generated (JF928170–JF928194). A total of 44 segregating sites were detected defining 22 haplotypes, with most specimens representing a unique haplotype (Table 2). The overall estimate of current genetic diversity ($\theta_\pi = 0.01$) was lower than the historical one ($\theta_W = 0.02$) (Table 2), which indicates recent population decline.

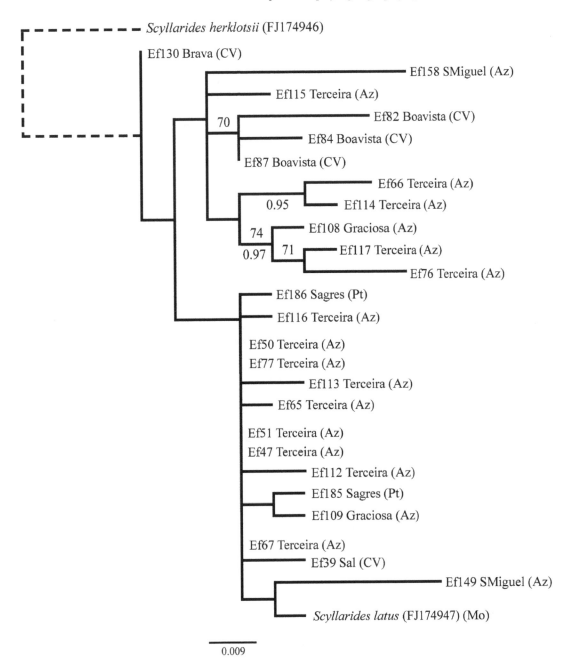

Figure 2. *Scyllarides latus* maximum likelihood tree (583 bp, COI). Bootstrap proportions (after 1,000 reiterations, if ≥ 70%) and posterior probabilities (if ≥ 0.95) are shown below and above nodes, respectively. Az = Azores, CV = Cape Verde, Pt = Portugal, Mo = Morocco. Dashed tree branches are not represented to scale. See Table 1 and Figure 1 for details.

Both ML and Bayesian phylogenetic inference (BI) trees (Figure 2) yielded similar topologies with low nodal support in most cases. Two clades without statistical support were recovered, one

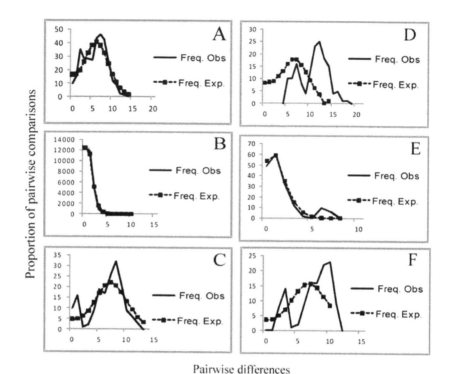

Figure 3. Mismatch distribution analyses for *Scyllarides latus* (A), *Palinurus elephas* (B), *Panulirus regius* (C), *Scyllarides latus* from the Azores (D), *Palinurus elephas* from the Azores (E), and *Panulirus regius* from Cape Verde (F).

composed of an admixture of continental (Sagres) and the Azores and Cape Verde insular haplotypes, and another of only insular haplotypes. This result suggests panmixia in *S. latus* from the NE Atlantic.

Both Tajima's D and Fu's F_S estimates were negative and overall significant and for the Azores samples (Table 2). The mismatch distribution analysis showed overall unimodal distributions (Figure 3A) and a bimodal distribution for the Azores samples (Figure 3D). The SDD and r estimates were nonsignificant in both cases. This evidence can be interpreted as indicative of sudden demographic expansion.

The estimated TMRCA was 1.02 (0.63–1.48) My when a mean substitution rate of 0.01 substitutions/My was used and accordingly were three and ten times lower when 0.03 and 0.1 substitution/My rates, respectively, were used. TMRCAs estimated for *P. elephas* and *P. regius* were similar (see Discussion) to those reported here for *S. latus*.

3.2   *Palinurus elephas*

Twenty new sequences of a length of 500 bp long were generated (JF928195–JF928214). The analysis of the entire dataset (Table 1; $N = 247$) revealed a total of 37 segregating sites defining 37 haplotypes (Table 2). The overall estimate of current genetic diversity ($\theta_\pi = 0.002$) is much lower than the historical one ($\theta_W = 0.013$), which indicates a severe recent population decline (Table 2).

The TCS haplotype network (Figure 4) presents a star-like shape, which suggests that *P. elephas* experienced a bottleneck and subsequently a population expansion. Two main haplotypes representing 153 and 30 individuals (from the Azores 10 and 3, from the Atlantic 77 and 11, and from the Mediterranean 66 and 16, respectively) could be distinguished with only one basepair difference between them. Seven haplotypes were recovered from the Azores of which four were unique. The samples from Sagres (Portugal; $N = 25$) showed six haplotypes, five common and one unique, which was seven mutations apart from the central most abundant haplotype. This and another unique haplotype from the Azores were the most divergent haplotypes in the network; all others were one or two mutations apart from the most common haplotype.

Pairwise Φ_{ST} population comparisons did not reveal a clear differentiation pattern between the Azores and the continental samples. Five comparisons, of which four involved western Ireland samples, showed $P < 0.05$; however, none of them remained significant after applying sequential Bonferroni correction (data not shown). The overall Φ_{ST} for these samples ($\Phi_{ST} = 0.010$) did not result significant either.

The hierarchical Analysis of Molecular Variance (AMOVA) indicated that most of the variation ($> 98\%$) was within populations (Table 3) and detected significant differences among the Azores (2 populations), Atlantic (6 populations) and Mediterranean (5 populations) regions ($\Phi_{ST} = 0.011$; $P = 0.05$), but no significant differences were observed between the Azores and the continental (Atlantic and Mediterranean combined) regions.

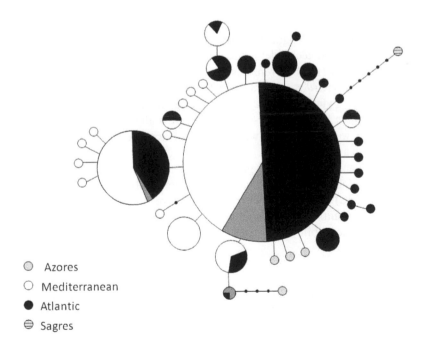

Azores
Mediterranean
Atlantic
Sagres

Figure 4. *Palinurus elephas* parsimony network (500 bp, COI). Lines represent a mutational step, circles represent the number of individuals per haplotype, and dots represent missing haplotypes. See Table 1 and text for details.

Table 3. *Palinurus elephas* hierarchical analyses of molecular variance (AMOVA).

Structure tested	Source of Variation	% Variance	Φ–statistics	P
Continental + Azores	Among regions	1.31	−0.002	0.386
	Among localities/Within regions	1.03	0.010	0.094
	Within localities	99.14	0.009	0.092
Atlantic + Mediterranean + Azores	Among regions	1.05	0.011	0.050
	Among localities/Within regions	0.35	0.004	0.318
	Within localities	98.60	0.014	0.089

Both Tajima's D and Fu's F_S estimates were negative and significant overall and for the Atlantic and Mediterranean (F_S) samples (Table 2). They were negative and nonsignificant for the Azores and Mediterranean (D) samples (Table 2). The mismatch distribution analysis showed positively skewed unimodal distributions overall (Figure 3B) and for the Azores samples (Figure 3E). The SDD and r estimates were nonsignificant. This evidence can be interpreted as indicative of very sudden demographic expansion.

3.3 *Palinurus regius*

Nineteen new sequences with a length of 589 bp long were generated (JF928153–JF928169). A total of 26 segregating sites were detected defining 12 haplotypes, hence almost each individual presented a unique haplotype (Table 2). The overall estimate of current genetic diversity ($\theta_\pi = 0.011$) was lower than the historical one ($\theta_W = 0.013$), which indicates recent population decline (Table 2).

Both ML and BI trees yielded similar topologies, but the latter was less resolved (Figure 5). All samples from Cape Verde formed a well-supported clade with some internal structure, although no reciprocal monophyly was observed between islands. The two continental samples were the most basal and did not form a clade.

Both Tajima's D and Fu's F_S estimates were negative but nonsignificant overall and for the Cape Verde samples (Table 2). The mismatch distribution analysis showed unimodal distributions for the former (Figure 3C) and bimodal for the latter (Figure 3F) and the SDD and r estimates were nonsignificant. These results cannot unambiguously distinguish between sudden demographic expansion and demographic stability.

4   DISCUSSION

The Macaronesian Islands comprise the northeastern Atlantic archipelagos of the Azores, Madeira, Selvagens, Cape Verde, and Canary Islands. These oceanic islands are the result of intense submarine volcanic activity and were never connected to continental landmasses. Their biota are the product of dispersal or migration from other geographic areas and in situ diversification (Cowie & Holland 2006). This study represents the first attempt to explore the genetic diversity and structure of the lobsters *Scyllarides latus*, *Palinurus elephas*, and *Panulirus regius* in the Macaronesian archipelagos of the Azores and Cape Verde, and expands previous results for the continental *P. elephas* by Palero et al. (2008). We generated 64 new partial sequences for the mtDNA COI gene. Our results are based on a single marker, which can be problematic since mtDNA genes are particularly sensitive to selective sweeps (Bazin et al. 2006). Nonetheless, many studies have demonstrated the

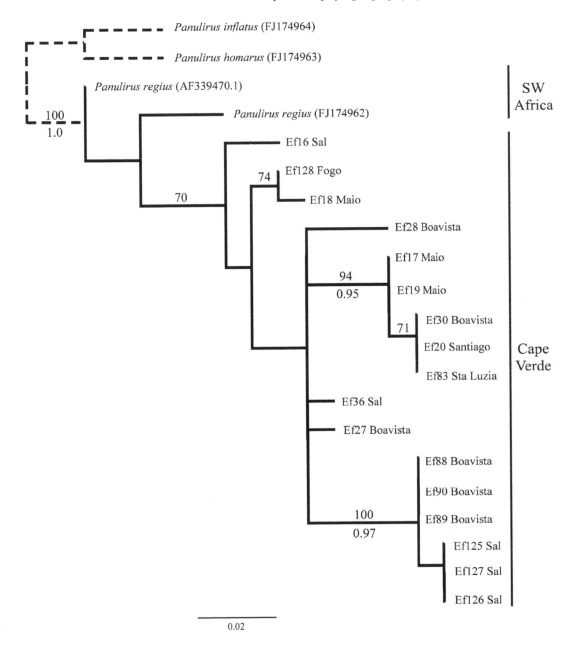

Figure 5. *Panulirus regius* maximum likelihood tree (589 bp, COI). Bootstrap proportions (after 1,000 reiterations, if ≥ 70%) and posterior probabilities (if ≥ 0.95) are shown below and above nodes, respectively. Dashed tree branches are not represented to scale. See Table 1 and Figure 1 for details.

efficiency of COI to detect population structure and estimate genetic diversity and divergence times in marine species from this and other areas (e.g., Roman & Palumbi 2004; Papadopoulos et al. 2005; Remerie et al. 2006; Pérez-Losada et al. 2007), including lobsters (e.g., Palero et al. 2008).

In general, marine species living in open waters are expected to exhibit little intraspecific genetic structuring even over large geographic distances (Palumbi 1994; Avise 2000). The influence of ocean currents in combination with the apparent lack of physical barriers in the marine realm seems to greatly facilitate extensive gene flow among populations (Palumbi 1994). Moreover, the potential for long-distance dispersal and large population sizes of larvae and adults may generate low levels of genetic subdivision and eventually lead to panmictic populations (Palumbi 1992, 1994). Several studies, at both large and small geographic scales, support this view (Pujolar et al. 2002; Gilbert-Horvath et al. 2006). Other studies, however, have shown genetic differentiation in marine species between, for example, insular and continental populations, due to isolation caused by oceanographic currents and fronts (e.g., Emerson 2002; Schönhuth et al. 2005; Domingues et al. 2007; Sá-Pinto et al. 2008; González-Wangüemert et al. 2011). Here, we tested these two competing hypotheses in three lobster species.

The slipper lobster *S. latus* presented a high level of genetic diversity within both the Azores and Cape Verde archipelagos with almost all individuals sampled presenting unique haplotypes (Table 2). However, both phylogenetic trees failed to reveal differentiation between insular and continental populations (Figure 2). On the other hand, the spiny lobster *P. elephas* showed lower genetic diversity than *S. latus* (Table 2) and a homogeneous structure, as indicated by a phylogenetic network estimation using the program TCS (Figure 4 and see Palero et al. 2008), but weakly significant ($P = 0.05$) genetic differences among populations from the NE Atlantic, Mediterranean and the Azores were observed (Table 3). An Atlantic-Mediterranean separation of *P. elephas* populations was also detected by Palero et al. (2008) and Babbucci et al. (2010) using mtDNA and microsatellite markers. As in this study, Babbucci et al. (2010) did not find significant differences between insular and continental *P. elephas* populations. Finally, the royal lobster *P. regius* showed intermediate levels of genetic diversity but relatively greater structure in Cape Verde, as indicated by our phylogenetic trees (Figure 5).

Three main factors have been generally invoked to explain population structuring in marine taxa: life-history traits (e.g., low dispersal ability), environmental factors (e.g., oceanographic fronts and currents), and increasing geographical distance. Both Scyllaridae and Palinuridae families have a long larval phase (lasting up to 24 months in some species) (Phillips & Sastry 1980; Booth et al. 2005). Such a long larval stage is unusual among marine invertebrates, as only 5% of the benthic invertebrates have larval developments that last longer than 12 weeks (Thorson 1950). Therefore, one would expect long-distance dispersion in these lobster species and, concomitantly, lack of genetic differentiation across the NE Atlantic, including the Macaronesian archipelagos. Our results seem to confirm that hypothesis for the three lobster species within the Macaronesian archipelagos and for *S. latus* and *P. elephas* in the NE Atlantic.

No oceanographic barriers have been described separating the Azores and Cape Verde from the continental landmasses, but powerful currents separate Cape Verde from southwestern Africa (Clarke et al. 1980; Johnson & Stevens 2000; Alves et al. 2002). These same currents also connect these archipelagos to the nearby continent. Such currents could have facilitated larval migration between the archipelagos and the Iberian Peninsula in *S. latus* and *P. elephas*, but caused isolation between Cape Verde and southwestern Africa in *P. regius*. Two oceanographic barriers, however, have been described in the Atlantic-Mediterranean junction: the Gibraltar Strait and the Almería-Oran front (Tintoré et al. 1988; Parrilla & Kinder 1992). In agreement with previous studies (Palero et al. 2008), our results indicate that these barriers may have influenced population structuring in *P. elephas*, causing weak isolation of the Mediterranean populations (Table 3).

Alternatively, the differences observed among the three lobster species may be also due to historical factors such as a more recent establishment of their populations in the Macaronesian archipelagos, to effects of genetic drift, or to selection. In this regard, it is interesting to note that in *P. elephas* the neutrality tests rejected equilibrium expectations (Table 2). Departures of the insular populations from the expected levels of genetic variation may occur if these populations experienced rapid expansion in the past, as our mismatch results and previous studies (Palero et al. 2008; Babbucci et al. 2010) seem to point out. Such expansion seems to be faster for *P. elephas* (Figure 3), which may explain its lower genetic diversity in the Azores.

Our results also have important implications for the management of these lobster species in the NE Atlantic and Mediterranean. Genetic diversity has decreased in all species and populations within, being particularly dramatic in *P. elephas* (Table 2). This latter result could be a consequence of the intensive exploitation this species is being subjected to across its entire distribution range (Díaz et al. 2005; Goñi et al. 2006). Interestingly, *P. regius* genetic diversity has remained almost constant over time (Table 2), despite it is also heavily fished in Cape Verde (Freitas et al. 2007). This may indicate recruiting from nearby African populations.

Coalescence times for all lobsters species were ≈ 1.0 (0.53–1.60) My when a substitution rate of 1% substitutions/My was assumed. However, there has been recently some debate about intraspecific mutation rates being higher than interspecific substitution rates (Ho et al. 2005). Taking these observations into account and using substitution rates three and ten times faster (Emerson 2007), the estimated coalescence times for these lobsters decreased to 0.33 (0.17–0.52) and 0.10 (0.05–0.16) My, respectively. Previous estimates by Palero et al. (2008) using a nonphylogenetic coalescent approach suggested a more recent divergence for *P. elephas* (0.27–0.36 My under a 1% substitutions/My rate), but other studies (Groeneveld et al. 2007; Palero et al. 2009b) prompted older divergences for *Palinurus*. Such discrepancies have considerable impact on phylogeographic hypothesis, so further investigations including more taxa, loci, external calibrations (e.g., fossils), and accurate substitution rates need to be implemented.

5 CONCLUSIONS

Genetic differentiation in *S. latus*, *P. elephas*, and *P. regius* lobsters from the NE Atlantic (Azores, Cape Verde, and continent) and Mediterranean was assessed using COI sequences. Two competing hypotheses of population structuring were tested in this area: one assuming panmixia due to long-distance dispersion and another suggesting fragmentation of populations due to life-history traits or environmental factors. Our results suggest panmixia in the archipelagos for all the species and across the NE Atlantic for *S. latus* and *P. elephas*. They, however, reveal shallow structuring between the Atlantic and Mediterranean due to historical or present-day barriers for *P. elephas* and a trend of differentiation between the Cape Verde and southwestern Africa, possibly due to ocean currents for *P. regius*.

ACKNOWLEDGEMENTS

We would like to thank Albano Beja, Luis Gouveia, Sonia Ferreira, Raquel Vasconcelos, and Ana Perera for providing tissue samples and Pedro Sousa for the map. We also thank the two referees and editor for their constructive comments. This study was partially supported by a grant from the Fundação para a Ciência e a Tecnologia (PTDC/BIA-BEC/098553/2008) to MP-L.

REFERENCES

Alves, M., Gaillard, F., Sparrow, M., Knoll, M. & Giraud, S. 2002. Circulation patterns and transport of the Azores Front-Current system. *Deep-Sea Res. Pt. II* 49: 3983–4002.
Avise, J.C. 2000. *Phylogeography: The History and Formation of Species*. Cambridge, MA: Harvard University Press.
Babbucci, M., Buccoli, S., Cau, A., Cannas, R., Goñi, R., Díaz, D., Marcato, S., Zane, L. & Patarnello, T. 2010. Population structure, demographic history, and selective processes: contrasting evidences from mitochondrial and nuclear markers in the European spiny lobster *Palinurus elephas* (Fabricius, 1787). *Mol. Phylogenet. Evol.* 56: 1040–1050.
Bargelloni, L., Alarcon, J.A., Alvarez, M.C., Penzo, E., Magoulas, A., Reis, C. & Patarnello, T. 2008. Discord in the family Sparidae (Teleostei): divergent phylogeographical patterns across

the Atlantic-Mediterranean divide. *J. Evolution. Biol.* 16: 1149–1158.

Bazin, E., Glémin, S. & Galtier, N. 2006. Population size does not influence mitochondrial genetic diversity in animals. *Science* 312: 570–572.

Booth, J.D., Webber, W.R., Sekiguchi, H. & Coutures, E. 2005. Diverse larval recruitment strategies within the Scyllaridae. *N. Z. J. Mar. Freshw. Res.* 39: 581–592.

Bowman, T.E. & Abele, L.G. 1982. Classification of the recent Crustacea. In: Abele, L.G. (ed.), *Biology of Crustacea. Volume 1. Systematics, the fossil record and biogeography*: 1–27. New York, NY: Academic Press.

Clarke, R.A., Hill, H.W., Reiniger, R.F. & Warren, B.A. 1980. Current system south and east of the Grand Banks of Newfoundland. *J. Phys. Oceanogr.* 10: 25–65.

Clement, M., Posada, D. & Crandall, K.A. 2000. TCS: a computer program to estimate gene genealogies. *Mol. Ecol.* 9: 1657–1659.

Comesaña, A.S., Martínez-Areal, M. & Sanjuan, A. 2008. Genetic variation in the mitochondrial DNA control region among horse mackerel (*Trachurus trachurus*) from the Atlantic and Mediterranean areas. *Fish. Res.* 89: 122–131.

Cowie, R.H. & Holland, B.S. 2006. Dispersal is fundamental to biogeography and the evolution of biodiversity on oceanic islands. *J. Biogeogr.* 33: 193–198.

De Brito Capello, F. 1864. Descripção de tres especies novas de Crustaceos da Africa occidental e observações ácerca do *Penoeus Bocagei*. Johnson. Especie nova dos Mares de Portugal. *Mem. Acad. Real Sci. Lisboa (Sci. Mat. Phys. Nat.)* 2: 1–11.

De Grave, S., Pentcheff, N.D., Ahyong, S.T., Chan, T.Y., Crandall, K.A., Dworschak, P.C., Felder, D.L., Feldmann, R.M., Fransen, C.H.J.M., Goulding, L.Y.D., Lemaitre, R., Low, M.E.Y., Martin, J.W., Ng, P.K.L., Schweitzer, C.E., Tan, S.H., Tshudy, D. & Wetzer, R. 2009. A classification of living and fossil genera of decapod crustaceans. *Raffles Bull. Zool.* 1: 1–109.

Dias, J.M.A. 1993. *A pesca das lagostas costeiras em Cabo Verde. Boletim Técnico-Científico. Nº*. Mindelo: Instituto Nacional de Desenvolvimento das Pescas.

Díaz, D., Mari, M., Abelló, P. & Demestre, M. 2001. Settlement and juvenile habitat of the European spiny lobster *Palinurus elephas* (Crustacea: Decapoda: Palinuridae) in the western Mediterranean Sea. *Sci. Mar.* 65: 347–356.

Díaz, D., Zabala, M., Linares, C., Hereu, B. & Abelló, P. 2005. Increased predation of juvenile European spiny lobster (*Palinurus elephas*) in a marine protected area. *N. Z. J. Mar. Freshw. Res.* 39: 447–453.

Dixon, C.J., Ahyong, S.T. & Schram, F.R. 2003. A new hypothesis of decapod phylogeny. *Crustaceana* 76: 935–975.

Domingues, V.S., Santos, R.S., Brito, A., Alexandrou, M. & Almada, V.C. 2007. Mitochondrial and nuclear markers reveal isolation by distance and effects of Pleistocene glaciations in the northeastern Atlantic and Mediterranean populations of the white seabream (*Diplodus sargus*, L.). *J. Exp. Mar. Biol. Ecol.* 346: 102–113.

Domingues, V.S., Santos, R.S., Brito, A. & Almada, V.C. 2006. Historical population dynamics and demography of the eastern Atlantic pomacentrid *Chromis limbata* (Valenciennes, 1833). *Mol. Phylogenet. Evol.* 40: 139–147.

Drummond, A.J. & Rambaut, A. 2007. BEAST: Bayesian evolutionary analysis by sampling trees. *BMC Evol. Biol.* 7: 214.

Emerson, B.C. 2002. Evolution on oceanic islands: molecular phylogenetic approaches to understanding pattern and process. *Mol. Ecol.* 11: 951–966.

Emerson, B.C. 2007. Alarm bells for the molecular clock? No support for Ho et al.'s model of time-dependent molecular rate estimates. *Syst. Biol.* 56: 337–345.

Excoffier, L. & El Lischer, H. 2010. Arlequin suite ver 3.5: a new series of programs to perform population genetics analyses under Linux and Windows. *Mol. Ecol. Resour.* 10: 564–567.

Excoffier, L., Smouse, P.E. & Quattro, J.M. 1992. Analysis of molecular variance inferred from metric distances among DNA haplotypes: application to human mitochondrial DNA restriction data. *Genetics* 131: 479–491.

Fabricius, J.C. 1787. *Mantissa Insectorum sistens eorum species nuper detectas adiectis characteribus genericis, differentiis specificis, emendationibus, observationibus. Tom. I.* Hafniae: Christ. Gottl. Proft.

FAO. 2005. *Review of the state of world marine fishery resources. FAO fisheries technical paper 457.* Rome: Food and Agriculture Organization of the United Nations.

Folmer, O., Black, M., Hoeh, W., Lutz, R. & Vrijenhoek, R. 1994. DNA primers for amplification of mitochondrial cytochrome c oxidase subunit I from diverse metazoan invertebrates. *Mol. Mar. Biol. Biotech.* 3: 294–299.

Freitas, R., Medina, A., Correia, S. & Castro, M. 2007. Reproductive biology of spiny lobster *Panulirus regius* from the north-western Cape Verde Islands. *Afr. J. Mar. Sci.* 29: 201–208.

Fu, Y.X. 1997. Statistical tests of neutrality of mutations against population growth, hitchhiking and background selection. *Genetics* 147: 915–925.

Gilbert-Horvath, E.A., Larson, R.J. & Garza, J.C. 2006. Temporal recruitment patterns and gene flow in kelp rockfish (*Sebastes atrovirens*). *Mol. Ecol.* 15: 3801–3815.

González-Wangüemert, M., Froufe, E., Pérez-Ruzafa, A. & Alexandrino, P. 2011. Phylogeographical history of the white seabream *Diplodus sargus* (Sparidae): Implications for insularity. *Mar. Biol. Res.* 7: 250–260.

Goñi, R. & Latrouite, D. 2005. Review of the biology, ecology and fisheries of *Palinurus* spp. species of European waters: *Palinurus elephas* (Fabricius, 1787) and *Palinurus mauritanicus* (Gruvel, 1911). *Cah. Biol. Mar.* 46: 127–142.

Goñi, R., Quetglas, A. & Reñones, O. 2006. Spillover of spiny lobsters *Palinurus elephas* from a marine reserve to an adjoining fishery. *Mar. Ecol. Prog. Ser.* 308: 207–219.

Groeneveld, J.C., Gopal, K., George, R.W. & Matthee, C.A. 2007. Molecular phylogeny of the spiny lobster genus *Palinurus* (Decapoda: Palinuridae) with hypotheses on speciation in the NE Atlantic/Mediterranean and SW Indian Ocean. *Mol. Phylogenet. Evol.* 45: 102–110.

Ho, S.Y.W., Phillips, M.J., Cooper, A. & Drummond, A.J. 2005. Time dependency of molecular rate estimates and systematic overestimation of recent divergence times. *Mol. Bio. Evol.* 22: 1561–1568.

Holthuis, L.B. 1991. *Marine lobsters of the world. An annotated and illustrated catalogue of species of interest to fisheries known to date. FAO species Catalogue. Vol. 13.* Rome: FAO.

Huelsenbeck, J.P. & Ronquist, F. 2001. MrBayes: a program for the Bayesian inference of phylogeny. *Bioinformatics* 17: 754–755.

Johnson, J. & Stevens, I. 2000. A fine resolution model of the eastern North Atlantic between the Azores, the Canary Islands and the Gibraltar Strait. *Deep-Sea Res. Pt. I* 47: 875–899.

Katoh, K., Kuma, K., Toh, H. & Miyata, T. 2005. MAFFT version 5: improvement in accuracy of multiple sequence alignment. *Nucleic Acids Res.* 33: 511.

Ketmaier, V., Argano, R. & Caccone, A. 2003. Phylogeography and molecular rates of subterranean aquatic stenasellid isopods with a peri-Tyrrhenian distribution. *Mol. Ecol.* 12: 547–555.

Kotoulas, G., Bonhomme, F. & Borsa, P. 1995. Genetic structure of the common sole *Solea vulgaris* at different geographic scales. *Mar. Biol.* 122: 361–375.

Krauss, W., Fahrbach, E., Aitsam, A., Elken, J. & Koske, P. 1987. The North Atlantic Current and its associated eddy field southeast of Flemish Cap. *Deep-Sea Res. Pt. A* 34: 1163–1185.

Latreille, P.A. 1802. *Histoire naturelle générale et particulière des Crustacés et des Insectes. Ouvrage faisant suite à l'Histoire naturelle générale et particulière, composée par Leclerc de Buffon, et rédigée par C. S. Sonnini, membre de plusieurs Sociétés savantes. Familles naturelles des Genres. Tome troisième.* Paris: F. Dufart.

Librado, P. & Rozas, J. 2009. DnaSP v5: a software for comprehensive analysis of DNA polymorphism data. *Bioinformatics* 25: 1451–1452.

Magoulas, A., Castilho, R., Caetano, S., Marcato, S. & Patarnello, T. 2006. Mitochondrial DNA reveals a mosaic pattern of phylogeographical structure in Atlantic and Mediterranean populations of anchovy (*Engraulis encrasicolus*). *Mol. Phylogenet. Evol.* 39: 734–746.

Marin, J. 1985. La langouste rouge: biologie et exploitation. *La Pêche Maritime* 64: 105–113.

Martins, H.R. 1985. Biological studies of the exploited stock of the Mediterranean locust lobster *Scyllarides latus* (Latreille, 1803) (Decapoda: Scyllaridae) in the Azores. *J. Crustacean Biol.* 5: 294–305.

Matsuda, H.I. 2000. The complete development and morphological changes of larval *Panulirus longipes* (Decapoda, Palinuridae) under laboratory conditions. *Fisheries Sci.* 66: 278–293.

Mercer, J.P. 1973. *Studies on the spiny lobsters (Crustacea: Decapoda: Palinuridae) of the West Coast of Ireland with particular reference to Palinurus elephas, Fabricius 1787*. Ph.D. dissertation, University College, Galway, Ireland.

Palero, F., Abelló, P., Macpherson, E., Gristina, M. & Pascual, M. 2008. Phylogeography of the European spiny lobster (*Palinurus elephas*): influence of current oceanographical features and historical processes. *Mol. Phylogenet. Evol.* 48: 708–717.

Palero, F., Crandall, K.A., Abelló, P., Macpherson, E. & Pascual, M. 2009a. Phylogenetic relationships between spiny, slipper and coral lobsters (Crustacea, Decapoda, Achelata). *Mol. Phylogenet. Evol.* 50: 152–162.

Palero, F., Lopes, J., Abelló, P., Macpherson, E., Pascual, M. & Beaumont, M.A. 2009b. Rapid radiation in spiny lobsters (*Palinurus* spp) as revealed by classic and ABC methods using mtDNA and microsatellite data. *BMC Evol. Biol.* 9: 263.

Palumbi, S.R. 1992. Marine speciation on a small planet. *Trends Ecol. Evol.* 7: 114–118.

Palumbi, S.R. 1994. Genetic divergence, reproductive isolation, and marine speciation. *Annu. Rev. Ecol. Syst.* 25: 547–572.

Papadopoulos, L.N., Peijnenburg, K. & Luttikhuizen, P.C. 2005. Phylogeography of the calanoid copepods *Calanus helgolandicus* and *C. euxinus* suggests Pleistocene divergences between Atlantic, Mediterranean, and Black Sea populations. *Mar. Biol.* 147: 1353–1365.

Parrilla, G. & Kinder, T. 1992. The physical oceanography of the Alborán Sea. *Rep. Meteorol. Oceanogr.* 40: 143–184.

Patarnello, T., Volckaert, F.A.M.J. & Castilho, R. 2007. Pillars of Hercules: is the Atlantic-Mediterranean transition a phylogeographical break? *Mol. Ecol.* 16: 4426–4444.

Patek, S.N. & Oakley, T.H. 2003. Comparative tests of evolutionary trade-offs in a palinurid lobster acoustic system. *Evolution* 57: 2082–2100.

Pauly, D. & Maclean, J. (eds.) 2003. *In a Perfect Ocean: The State of Fisheries and Ecosystems in the North Atlantic Ocean.* Washington, DC: Island Press.

Pérez-Losada, M., Nolte, M.J., Crandall, K.A. & Shaw, P.W. 2007. Testing hypotheses of population structuring in the Northeast Atlantic Ocean and Mediterranean Sea using the common cuttlefish *Sepia officinalis*. *Mol. Ecol.* 16: 2667–2679.

Phillips, B.F. & Sastry, A.N. 1980. Larval ecology. In: Cobb, J.S. & Phillips, B.F. (eds.), *The Biology and Management of Lobsters. Vol. II*: 11–57. New York, NY: Academic Press.

Posada, D. 2008. jModelTest: phylogenetic model averaging. *Mol. Bio. Evol.* 25: 1253–1256.

Ptacek, M.B., Sarver, S.K., Childress, M.J. & Herrnkind, W.F. 2001. Molecular phylogeny of the spiny lobster genus *Panulirus* (Decapoda: Palinuridae). *Mar. Freshwater Res.* 52: 1037–1048.

Pujolar, J.M., Roldan, M.I. & Pla, C. 2002. A genetic assessment of the population structure of swordfish (*Xiphias gladius*) in the Mediterranean Sea. *J. Exp. Mar. Biol. Ecol.* 276: 19–29.

Rambaut, A. & Drummond, A.J. 2009. Tracer: MCMC trace analysis tool 1.5. Institute of Evolutionary Biology, Edinburgh. http://trio.bio.ed.ac.uk/software/tracer.

Reis, D.C.C. 1997. *Estudo da Pescaria de Lagosta verde (Panulirus regius De Brito Capello, 1864) do Arquipélago de Cabo Verde.* Relatrio de Estgio de Licenciatura em Biologia Marinha e Pescas, Universidade do Algarve, Portugal.

Remerie, T., Bourgois, T., Peelaers, D., Vierstraete, A., Vanfleteren, J. & Vanreusel, A. 2006. Phylogeographic patterns of the mysid *Mesopodopsis slabberi* (Crustacea, Mysida) in Western Europe: evidence for high molecular diversity and cryptic speciation. *Mar. Biol.* 149: 465–481.

Reuschel, S. & Schubart, C.D. 2007. Contrasting genetic diversity with phenotypic diversity in coloration and size in *Xantho poressa* (Brachyura: Xanthidae), with new results on its ecology. *Mar. Ecol.* 28: 1–10.

Reuschel, S., Cuesta, J.A. & Schubart, C.D. 2010. Marine biogeographic boundaries and human introduction along the European coast revealed by phylogeography of the prawn *Palaemon elegans. Mol. Phylogenet. Evol.* 55: 765–775.

Roman, J. & Palumbi, S.R. 2004. A global invader at home: population structure of the green crab, *Carcinus maenas*, in Europe. *Mol. Ecol.* 13: 2891–2898.

Sá-Pinto, A., Branco, M., Harris, D.J. & Alexandrino, P. 2005. Phylogeny and phylogeography of the genus *Patella* based on mitochondrial DNA sequence data. *J. Exp. Mar. Biol. Ecol.* 325: 95–110.

Sá-Pinto, A., Branco, M., Sayanda, D. & Alexandrino, P. 2008. Patterns of colonization, evolution and gene flow in species of the genus *Patella* in the Macaronesian Islands. *Mol. Ecol.* 17: 519–532.

Schönhuth, S., Álvarez, Y., Rico, V., González, J.A., Santana, J.I., Gouveia, E., Lorenzo, J.M. & Bautista, J.M. 2005. Molecular identification and biometric analysis of Macaronesian archipelago stocks of *Beryx splendens. Fish. Res.* 73: 299–309.

Sotelo, G, Posada, D. & Morán, P. 2009. Low-mitochondrial diversity and lack of structure in the velvet swimming crab *Necora puber* along the Galician coast. *Mar. Biol.* 156: 1039–1048.

Spanier, E. & Lavalli, K.L. 1998. Natural history of *Scyllarides latus* (Crustacea: Decapoda): a review of the contemporary biological knowledge of the Mediterranean slipper lobster. *J. Nat. Hist.* 32: 1769–1786.

Stamatakis, A. 2006. RAxML-VI-HPC: maximum likelihood-based phylogenetic analyses with thousands of taxa and mixed models. *Bioinformatics* 22: 2688–2690.

Tajima, F. 1989. Statistical method for testing the neutral mutation hypothesis by DNA polymorphism. *Genetics* 123: 585–595.

Templeton, A.R., Crandall, K.A. & Sing, C.F. 1992. A cladistic analysis of phenotypic associations with haplotypes inferred from restriction endonuclease mapping and DNA sequence data. III. Cladogram estimation. *Genetics* 132: 619.

Thorson, G. 1950. Reproductive and larval ecology of marine bottom invertebrates. *Biol. Rev.* 25: 1–45.

Tintoré, J., La Violette, P.E., Blade, I. & Cruzado, G. 1988. A study of an intense density front in the eastern Alboran Sea: the Almería-Oran front. *J. Phys. Oceanogr.* 18: 1384–1397.

Triantafyllidis, A., Apostolidis, A.P., Katsares, V., Kelly, E., Mercer, J., Hughes, M., Jørstad, K.E., Tsolou, A., Hynes, R. & Triantaphyllidis, C. 2005. Mitochondrial DNA variation in the European lobster (*Homarus gammarus*) throughout the range. *Mar. Biol.* 146: 223–235.

Zariquiey-Alvárez, R. 1968. Crustáceos decápodos Ibéricos. *Invest. Pesq.* 32: 1–510.

Genetic variation and differentiation of *Fenneropenaeus merguiensis* in the Thai Peninsula

WARAPOND WANNA & AMORNRAT PHONGDARA

Center for Genomics and Bioinformatics Research, Faculty of Science, Prince of Songkla University, Hat-Yai, Songkhla, Thailand

ABSTRACT

Knowledge about DNA polymorphisms found in *Fenneropenaeus merguiensis* in Thailand may represent an important contribution to the study of the phylogeographic history of this widespread genus and provide important genetic information, which can be useful in the management of this resource. In this study, the genetic diversity and geographic differentiation of the banana prawn, *F. merguiensis*, in the Thai Peninsula region were examined using an intron polymorphism of the gene locus PvAmy. Our survey of 163 samples of *F. merguiensis* from five populations collected from the Gulf of Thailand (Trad, Surat Thani, and Songkhla) and the Andaman Sea (Satun and Trang) revealed a great and highly significant differentiation between the Gulf of Thailand and the Andaman Sea ($F_{ST} = 0.324$, $P < 0.001$). Significant population differentiation was also found within the Gulf of Thailand. These results were compared to those of a previous mitochondrial DNA survey spanning the same geographical range, and in which two divergent mitochondrial clades were reported. This study provides support for the hypothesis that the existence of these two clades is not due to a mixture of cryptic species but, rather, reflects their phylogeographic origin. The strong genetic structure of *F. merguiensis* on each side of the Thai Peninsula that was observed based on both mitochondrial and nuclear genes could, thus, be linked to the phylogeographic divide between the Indian Ocean and Pacific forms on the west and east sides of the Peninsula, respectively.

1 *FENNEROPENAEUS MERGUIENSIS* (DE MAN 1888)

Fenneropenaeus merguiensis, also known as banana prawns or white shrimp, are native to the Indo-Pacific region (Figure 1) and represent one of the most extensive types of prawn population found in Southeast Asia and other places in this region, such as Australia. The life cycles of all penaeid prawn species found on the coast zone follow the same general pattern with a few variations (Dall et al. 1990). Prawns reach sexual maturity at an age of about six months in tropical waters and live to between 18 and 24 months of age. After 24 months, they migrate offshore into the fishing grounds at a time when the salinity of estuarine water has substantially decreased because of monsoon rains. *Fenneropenaeus merguiensis* is most abundant in shallow areas less than 20 m deep. According to Somers (1994), the major spawning periods for *F. merguiensis* occur in the autumn and spring, from March to April and September to October. Each female releases 100,000–200,000 eggs per spawn. Within 24 hours, the eggs hatch into larvae. After passing through three stages (nauplius, zoea, and mysis stages) in a period of about 12 days, the larvae develop into young shrimp known as postlarvae. The postlarvae of *F. merguiensis* concentrate in mangrove-surrounded, muddy estuaries and may swim upriver for up to 85 km, to a region where salinity is very low (Staples 1980), to grow from postlarvae to larger juveniles/subadults. The offshore habitats of the subadult and adult

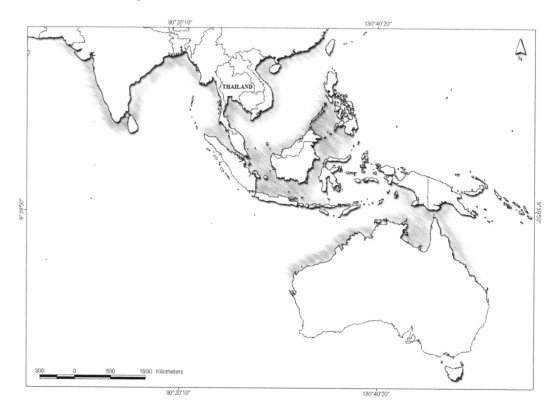

Figure 1. Geographic distribution (shaded area) of *Fenneropenaeus merguiensis* in the Indo-West Pacific.

prawns also differ between species, mostly in relation to substrate type and depth (Somers 1994). Banana prawns generally inhabit mangrove-lined creeks and rivers (Staples & Vance 1985; Vance et al. 1990). However in arid environments, juvenile *F. merguiensis* may be found in mangrove-lined inlets, where salinity may reach 60% (Staples & Vance 1987). Thus, for this species, habitat preference, rather than salinity, appears to be the main factor that determines where the post-larvae will settle.

This species is often morphologically confused with *Fenneropenaeus indicus*, *Fenneropenaeus silasi*, or *Fenneropenaeus penicillatus* of the same genus. The adult males of these species can be differentiated based on the third maxilliped (Carpenter & Niem 1998). The morphological characters that have been used to differentiate between adults of *F. merguiensis*, *F. indicus*, and *F. penicillatus* also include the rostral crest, adrostral groove, and gastro-orbital carina (Chong & Sasekumar 1982; Grey et al. 1983).

2 STUDIES ON THE GENETIC VARIATION OF *FENNEROPENAEUS MERGUIENSIS*

Variations that occur in genes produce individuals that are different either at the molecular or organismal level. These individuals may form separate groups within a species, and such groups are the fundamental genetic units of evolution. Studying polymorphic genetic loci, which provide useful data that can provide insight into the processes shaping the genetic structure of natural populations, is necessary to understand the history and evolution of populations. Intraspecific groups act as stocks that fishery biologists have started using to manage commercially important marine organisms. Tracking individuals under field conditions has led many population biologists to consider DNA-based markers (Avise 1994). Benzie (2000) concluded that the DNA-based markers in penaeid prawns reveal far greater levels of variation compared with allozyme data.

2.1 Allozymes

Most of the early studies on genetic variation in *F. merguiensis* were done using allozyme analysis. For example, Mulley & Latter (1980) found a low level of genetic variation in *F. merguiensis* from Australia (mean observed heterozygosity $H_o = 0.008 \pm 0.004$). The authors investigated genetic variation in specimens from two sampling sites, and their allozyme analysis was based on 6 polymorphic loci using starch gel electrophoresis. Tam & Chu (1993) used four polymorphic loci to investigate the genetic variation of *F. merguiensis* in Hong Kong and found a mean observed heterozygosity of 0.045 ± 0.025. Allozymes have also been used in Thailand to examine the population structure of *F. merguiensis*. Five polymorphic loci out of 26 loci were screened from samples collected from three geographic regions (i.e., Chonburi, Surat Thani, and Satun), and the results revealed an average observed heterozygosity (H_o) of 0.067 ± 0.029, suggesting a low level of genetic diversity in this taxon (Sodsuk & Sodsuk 1999).

2.2 Cytochrome c oxidase subunit I (COI)

Assessing the genetic diversity within populations of *F. merguiensis* using markers with higher levels of variation than allozymes has been performed through mitochondrial DNA (mtDNA) analyses. In animals, mtDNA genes, such as COI, are particularly suitable for the determination of intraspecific genetic diversity due to a high evolutionary rate found in this maternally transmitted genome (Avise et al. 1987). Therefore, COI has been widely used as a genetic marker to assess the phylogeography and phylogeny of numerous species and genera. Examination of the polymorphism of COI has been carried out in studies on both intraspecific differentiation within populations and phylogenetic relationships of penaeid shrimp (Benzie 2000; Lavery et al. 2004; Khamnamtong et al. 2009).

Hualkasin et al. (2003) used mitochondrial COI sequences to examine the phylogeny of *F. merguiensis* from eight locations in the Gulf of Thailand and the Andaman Sea. Their COI sequence analysis revealed the presence of two distinct clades. In clade A, most samples were from the east

Table 1. Collection sites and mtDNA clades of *Fenneropenaeus* samples used in this study. Abbreviations: *F.* = *Fenneropenaeus*, *L.* = *Litopenaeus*, *P.* = *Penaeus*.

Species	Collection site (abbreviation)	Sample Size	mtDNA clade
F. merguiensis	Songkhla (SKE)	12	A
F. merguiensis	Surat Thani (SRE)	12	A (10), B (2)
F. merguiensis	Trad (TDE)	6	A
F. merguiensis	Satun (STW)	10	A (3), B (7)
F. merguiensis	Trang (TRW)	5	B
F. silasi	Nakon Si Thammarat (NKE)	6	C
F. indicus	Tanzania, Africa	4	–
P. monodon	Tungkan, Taiwan (from GenBank)	1	–
L. vannamei	Centre Océanologique du Pacifique, Tahiti (from GenBank)	1	–

coast (Gulf of Thailand), while clade B contained mostly samples from the west coast (Andaman Sea), with the exception of five individuals representing mismatching genotypes. These included samples SRE1 and SRE2 in clade B, and samples STWA5, STWA6 and STWA7 in clade A (see Figure 4). These two clades exhibited 5% sequence divergence, suggesting evidence of substantial intraspecific genetic differentiation (Avise 2000). This was interpreted either as evidence of the existence of two cryptic species within *F. merguiensis* or as a single species with phylogeographic subdivision (Hualkasin et al. 2003). However, mtDNA does not yield a complete picture of these phenomena and can even result in a misleading interpretation of the isolation between populations if there is also some form of male-mediated gene flow occurring (FitzSimmons et al. 1997a, b).

3   WHAT WAS DONE IN THIS STUDY?

Relatively little was previously known about the population structure of *F. merguiensis* in Thailand (see above). The objectives of this study were to test for the existence of two divergent mitochondrial clades and to examine the genetic structure among five populations of *F. merguiensis* along the Thai Peninsula using nuclear DNA markers.

Since mtDNA is maternally inherited and does not reflect mating processes, we wanted to test which of the interpretations proposed by Hualkasin (2003) is correct, the existence of two cryptic species within *F. merguiensis*, or the presence of a single species with phylogeographic subdivision. Therefore, we examined genetic variability at the level of nuclear DNA (nDNA) loci. Moreover, a comparison of nuclear and mitochondrial genotypes may provide a more complete understanding of the population genetic structure of this species.

3.1   Materials and methods

3.1.1   *Sample collection*

163 living individuals of *F.merguiensis* were collected from Satun (STW, $N = 27$) and Trang (TRW, $N = 33$), located in the Andaman Sea, and from Trad (TDE, $N = 39$), Songkhla (SKE, $N = 40$) and Surat Thani (SRE, $N = 24$), located in the Gulf of Thailand (Figure 2). In addition, eleven samples of *F. silasi* and *F. indicus*, and published sequences of *Penaeus monodon* and *Litopenaeus vannamei* were used in the analysis as outgroup (see also Appendix, Table A1). Samples were stored frozen at $-70\,°C$ until processed. For DNA extraction, 0.1–0.2 g of frozen muscle tissue were dissected from each individual of *F. merguiensis* and digested at $55\,°C$ for 2 h in a microcentrifuge tube containing extraction buffer (1 mM Tris-HCl, pH 7.5, 100 mM EDTA, 1% SDS, 1 μg/ml proteinase K and 0.05 μg/μl RNase A). The homogenate was extracted with phenol/chloroform/isoamyl alcohol and then precipitated in 100% cold ethanol.

3.1.2   *Exon-primed intron-crossing PCR (EPIC-PCR)*

In this study, the EPIC-PCR (exon-primed intron-crossing) nuclear DNA marker was used. The advantage of this strategy is that exon sequences are relatively more conservative, and therefore the primer designed in exon may have more extensive applications than those designed in noncoding sequences. A total of 13 exonic primers that were described by Bierne et al. (2000) were tested for their ability to amplify polymorphisms in introns in 10 individuals of *F. merguiensis*. Only PvAmy (PvAmyF: 5'-TGG AAG TGG TCG GAC ATC GC-3'; PvAmyR: 5'-CGG AGC GAG TGA CGA GTT TAT AGG-3'), which is based on amylase genes, was informative and resulted in an interpretable length polymorphism (Figure 3). PCR was carried out in a 10–μl reaction mixture consisting of 100 ng of DNA, 0.25 mM each of dNTPs, 1.5 mM MgCl$_2$, 0.8 μM reverse primer,

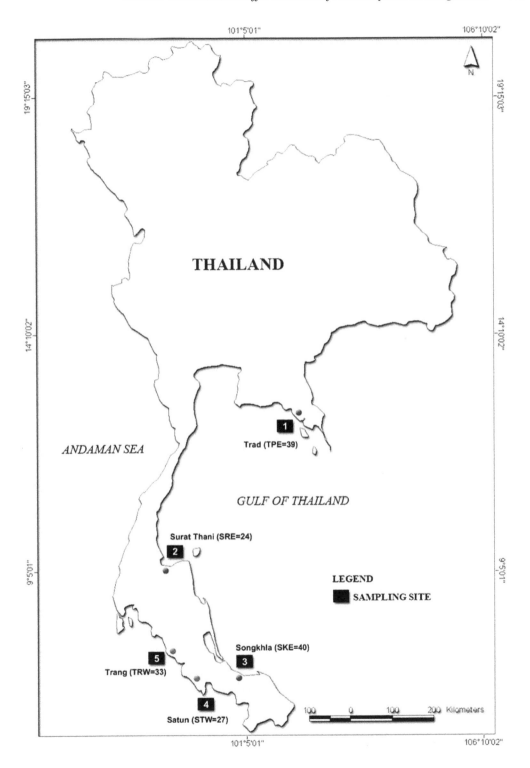

Figure 2. Map of Thailand showing sampling locations for the five geographic populations of *Fenneropenaeus merguiensis*. Numbers in parentheses indicate the number of individuals analyzed at each sampling location.

0.8 μM fluorescently labeled forward primer and 0.05 units of Taq DNA polymerase. The PCR stages were: an initial denaturation step for 10 min at 94 °C followed by 35 cycles of amplification (1 min at 94 °C, 1 min at the optimal annealing temperature of the primer, 1 min at 72 °C), and a final extension step for 10 min at 72 °C. The amplified products were analyzed on 6% denaturing polyacrylamide sequencing gel and visualized using FMBIO II Fluorescence Imaging System (HITACHI).

3.1.3   *mtDNA analyses*

A 558–bp fragment of the COI gene was amplified from 45 individuals of *F. merguiensis* (24 individuals from this study and 21 individuals from Hualkasin et al. (2003) (Table 1; Appendix, Table A1). The sequence divergence between pairs of sequences was estimated using Kimura's (1980) two-parameter model. Data were bootstrapped 1,000 times. A phylogenetic tree was constructed using a neighbor-joining method (Saitou & Nei 1987) in the PHYLIP package (Felsenstein 1995).

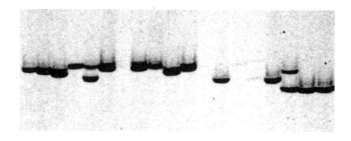

Figure 3. Length polymorphisms of the PCR products of an intron of the PvAmy gene in samples of *Fenneropenaeus merguiensis*.

3.1.4   *Statistical analyses*

All five populations were tested for departure from the Hardy-Weinberg equilibrium. Most statistical analyses were performed using the GENETIX program (Belkhir et al. 2001). Standard parameters were estimated including allele frequencies, the observed (H_o) and unbiased expected heterozygosity (H_e) of Nei (1978) and Weir & Cockerham (1984), and estimation of F_{IS} and F_{ST}. Departure of F_{IS} and F_{ST} from zero was tested using a permutation approach provided by this software. Moreover, the probabilities of deviation from the Hardy-Weinberg equilibrium were computed using the Markov chain method (Guo & Thompson 1992) as implemented in the GENEPOP package (Raymond & Rousset 1995). To assess significance at the 5% level, P values were adjusted for the number of simultaneous tests using the sequential Bonferroni procedure (Rice 1989). Reynolds et al.'s (1983) coancestry ($D_{Reynold} = -\ln(1 - F_{ST})$) coefficient was also calculated between population pairs. The genetic proximity between the five populations was illustrated in the form of a tree using the NEIGHBOR program of the PHYLIP package.

3.2   Results and discussion

3.2.1   *Association of mtDNA and nuclear DNA*

To address the questions raised by previous COI sequence analyses (i.e., the presence of individuals collected at the same site distributed between two otherwise geographically well-differentiated

Table 2. Summary of statistics for the five populations; see Table 1 for abbreviations of collection sites. Allele frequencies at each locus. Abbreviations: N = number of prawns per population, n = number of alleles, H_o = observed heterozygosity, H_e = expected heterozygosity, F_{IS} = Wright's fixation index, $P(HW)$ = probability of deviation from Hardy-Weinberg equilibrium, – = allele absent in the sample, bp = nucleotide basepairs; * = F_{IS} value found highly significant by permutation test (Belkhir et al. (2001), 500 permutations).

Locus	Allele (bp)	Samples of *Fenneropenaeus merguiensis*				
		TDE	SKE	SRE	STW	TRW
PvAmy	360	–	–	–	–	–
	365	–	0.15	–	–	–
	370	–	–	–	–	0.16
	371	0.02	0.02	0.04	0.63	0.69
	373	0.11	0.06	0.08	0.17	0.06
	375	–	0.03	0.02	–	–
	377	0.03	0.15	–	0.02	0.02
	379	–	0.59	0.42	0.02	0.16
	381	0.69	0.09	0.42	0.15	0.05
	389	0.16	0.05	0.02	–	–
N		39	40	24	27	33
n		5	8	6	5	6
H_e		0.498	0.622	0.657	0.560	0.494
H_o		0.375	0.485	0.485	0.478	0.452
F_{IS}		0.249*	0.223*	0.307*	0.149	0.087
$P(HW)$		0.023	0.016	0.157	0.229	0.330

clades), 45 individuals of *F. merguiensis* (24 individuals from this study and 21 individuals from Hualkasin et al. 2003) were analyzed using both nuclear and mtDNA markers. In principle, this should have allowed us to investigate an association between these genomic compartments that could have resulted from the existence of two cryptic species. Our phylogenetic tree based on COI sequence analysis showed that the vast majority of individuals from the east were retrieved in clade A, and most individuals from the west in clade B (Figure 4). The only exceptions to this pattern were two individuals (SRE1 and SRE2) from Surat Thani (appearing in clade B) and three individuals (STWA5, STWA6, and STWA7) from Satun (appearing in clade A). Given the small sample size, no statistical test could be implemented, though our results do not seem to suggest a particular association between mtDNA and nuclear DNA. When we examined in detail the nuclear genotypes of the individuals of clade B (west) that are represented in the populations of clade A (east), or the reverse situation, no particular association between the mtDNA and nuclear genes could be found. Thus, the hypothesis of two cryptic species that overlap sympatrically seems very unlikely. Therefore, the origin of the two divergent mitochondrial clades is likely the result of phylogeographic differentiation in *F. merguiensis*, with an Indian Ocean form in one region and a Pacific form in the other. The presence of an apparently low level of admixture between the two clades may be caused by a limited amount of genetic exchange or a human-mediated admixture due to transportation.

3.2.2   *Hardy-Weinberg equilibrium*

The PvAmy intron polymorphism was also used to study the population genetics of *F. merguiensis*. A total of 163 wild *F. merguiensis* individuals from five locations in Thailand were compared genetically (Figure 2). There were 10 alleles observed at the PvAmy locus. This PvAmy intron

Figure 4. A bootstrapped neighbor-joining tree showing relationships of *Fenneropenaeus merguiensis* samples (PMER) and other penaeid prawns in Thai waters based on nucleotide sequence divergence of COI. GB indicates sequences obtained from GenBank. Asterisk (*) denotes five samples (SRE1, SRE2, STWA5, STWA6, and STWA7) exhibiting mismatching genotypes (see 1.2.1 above). Numbers above branches indicate bootstrap values; see Table 1 for abbreviations of collection sites, and clade labels. Bootstrap values from 1000 pseudoreplicates are shown above 50%.

polymorphism of *F. merguiensis* exhibited observed heterozygosity (H_o) for each population ranging from 0.375–0.485 (Table 2). The intrapopulation F_{IS} value for the PvAmy locus examined was significant in the samples collected from the three populations in the Gulf of Thailand, but not in those from the Andaman Sea. Significant deviations from the Hardy-Weinberg expectations indicate that a heterozygote deficit is found in the Gulf of Thailand. This finding is consistent with a previous report on a heterozygote deficit in *P. monodon* collected in the Gulf of Thailand (Supungul et al. 2000). Several explanations for the heterozygous deficit have been proposed, including mating of close relatives (inbreeding), null alleles, technical artifacts, population mixing (Wahlund effect), and others (Gaffney et al. 1990; Castric et al. 2002).

Additionally, the possibility of the deficit from the null allele must be considered first. The possibility of null alleles to create a heterozygote deficit can be checked by calculating the frequency of null alleles explaining $F_{IS} = 0.3$ (leading to $P \approx 0.20$) in the Surat Thani (SRE) population. For 24 individuals, the expected number of null homozygotes is only $24 \times p^2 \approx 1$. Thus, it is very possible that simply by chance none were observed, so the existence of a nonamplifying allele in moderate frequency on the east side of the Thai Peninsula remains a reasonable explanation for the observed heterozygote deficiency.

3.2.3 *Population differentiation between the Thai Peninsula populations*

Data from the PvAmy intron polymorphisms in *F. merguiensis* showed a marked gene frequency difference between populations from the Gulf of Thailand and the Andaman Sea. Estimation of the F_{ST} between the populations using the PvAmy locus showed a highly significant value between the two sides of the Peninsula ($F_{ST} = 0.324$, $P < 0.001$). The pairwise genetic distances computed at this intron polymorphism showed clear-cut genetic differentiation in *F. merguiensis* between the Andaman Sea and the Gulf of Thailand and within the Gulf of Thailand (Figure 5). The strong genetic structure of *F. merguiensis* on each side of the Thai Peninsula for all markers can, therefore, be linked to the phylogeographic divide between the Indian and the Pacific Ocean. The patterns of genetic differentiation between the two sides that we observed are consistent with those previously described for *P. monodon* (Klinbunga et al. 2001; Khamnamtong et al. 2009). Benzie (2000) noted that the significant genetic differences found between the populations of *P. monodon* in the Andaman Sea and the Gulf of Thailand suggest that genetic disjunction between the Indian Ocean populations and those from the Pacific occurs along the margin of the Malay Peninsula and the Indonesian Ocean.

Strong genetic differences in the Indo-Pacific have been reported between the Indian and the Pacific Ocean populations of a variety of marine organisms, such as in two species of starfish, *Linckia laevigata* (Williams & Benzie 1998), *Acanthaster planci* (Benzie 1999), coconut crabs (Lavery et al. 1996), butterfly fish (McMillan & Palumbi 1995), and snapping shrimp (Williams et al. 2002). The major hypotheses that have been advanced to explain the high biodiversity found in the Southeast Asia (SA) region distinguish two major categories (Rosen 1978): (i) the region may be acting as a zone of accumulation of species that originate peripherally in the Indian and Pacific Oceans, and then disperse into the SA region. In this model, diversity arises because the region is a zone of overlap (dispersal model). (ii) The region is a center of generation of biological diversity, the many islands, bays, and seas, in combination with the sea level changes, are considered to provide the high level of subdivision of populations required for allopatric speciation and the diversity of habitats that is required to maintain high species numbers (vicariance model). Regarding these two models, mtDNA-surveys in *P. monodon* provided the first clear evidence that the Indo-West Pacific region is, in fact, a site of accumulation of genetic diversity, rather than a site of the origin of genetic diversity (Benzie 2000).

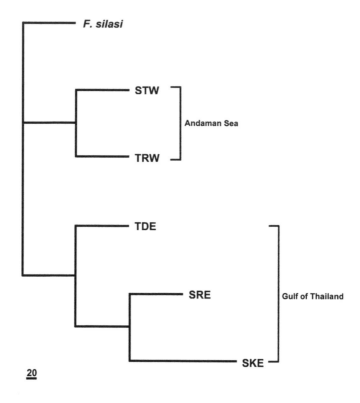

Figure 5. Phylogenetic tree generated by a neighbor-joining algorithm using Reynolds' genetic distance for the PvAmy locus to illustrate the relationships between populations of *Fenneropenaes merguiensis* from the Thai Peninsula (modified from Wanna et al. 2004). *Fenneropenaeus silasi* was used as an outgroup.

3.2.4   *Population differentiation within the Gulf of Thailand*

Significant genetic structure was also found within populations from the Gulf of Thailand in addition to the strong genetic differences that were found in populations between the Gulf of Thailand and the Andaman Sea. Our study, using PvAmy intron polymorphisms, revealed genetic differentiation among three locations in the Gulf of Thailand (mean $F_{ST} = 0.213$, $P < 0.01$), whereas no genetic subdivision was detected between two Andaman Sea samples. These results are also consistent with those previously obtained for *P. monodon* (Tassanakajon et al. 1997, 1998a, b; Klinbunga et al. 2001; Khamnamtong et al. 2009), in which significant substructuring was observed within the Gulf of Thailand. This pattern may suggest that larvae encounter environmental effects within the Gulf of Thailand that limit the gene flow among populations. The role of the environment in larval and post-arval mortality is likely to be complex. Intensive studies on the *F. merguiensis* in the southeastern Gulf of Carpentaria, Australia, have shown that the recruitment dynamics at each life history stage are controlled by complex interactions between seasonal hydrological and meteorological events. These events affect postlarval and juvenile recruitment patterns, which in turn affect the structure and fecundity of the adult population. *Fenneropenaeus merguiensis* has a relatively shallow spawning distribution. Its postlarvae use brackish water, such as mangrove-lined estuaries, as nursery areas (Staples 1979; Staples & Vance 1987). Staples & Vance (1986) found that timing and the amount of rainfall directly affect the number and size of juvenile prawns that emigrate offshore to the adult population through an apparent behavioral stimulus response. Differences in the time of maturation in relation to environmental optima may become established during development and cause shrimp populations to exhibit significant degrees of differentiation. The rainy seasons in the northern and southern parts of the Gulf occur at different times (August–October in the North and December–February in the South), affecting the ocean-surface temperature. This creates additional obstacles to gene flow among these populations, and will help to preserve their distinct genetic compositions if different populations of these shrimp have adapted their reproductive maturation to different time optima. Furthermore, a current separates the northern and southern parts of the Gulf during the monsoon period and acts as an oceanographic barrier to the migration of populations.

4 CONCLUSIONS

Population genetic studies of Thai populations of *F. merguiensis* between the Gulf of Thailand and the Andaman Sea and within the Gulf of Thailand based on available data had contradictory results regarding patterns of genetic differentiation. The strong genetic structure in populations of *F. merguiensis* on each side of the Thai Peninsula can be linked to the phylogeographic divide between the Indian and the Pacific Ocean. The occurrence of strong patterns of geographic variation in wild stocks suggests that more detailed planning will be required to maintain this diversity and to determine how best to capture its benefits for aquaculture purposes. The observation of the existence of population differentiation between wild stocks of *F. merguiensis* from Thailand indicates that each genetic population of Thai *F. merguiensis* should be treated as a separate management unit because it may display unique demographic and dynamic properties.

REFERENCES

Avise, J.C., Arnold, J., Ball, R.M., Bermingham, E., Lamb, T., Neigel, J.E., Reeb, C.A. & Saunders, N.C. 1987. Intraspecific phylogeography: the mitochondrial DNA bridge between population genetics and systematics. *Ann. Rev. Ecol. Syst.* 18: 489–522.

Avise, J.C. 1994. *Molecular Markers: Natural History and Evolution.* London: Chapman and Hall.

Avise, J.C. 2000. *Phylogeography: The History and Formation of Species.* Cambridge, MA: Harvard University Press.

Belkhir, N., Borsa, P., Goudet, J., Chikhi, L. & Bonhomme, F. 2001. *GENETIX Version 4.02: logiciel sous Windows*^{TM} *pour la génétique des populations*, Laboratoire Génome et Populations, Université Montpellier, France.

Benzie, J.A.H. 1999. Genetic structure of coral reef organisms: ghosts of dispersal past. *Amer. Zool.* 39: 131–145.

Benzie, J.A.H. 2000. Population genetic structure in penaeid prawns. *Aquacult. Res.* 31: 95–119.

Bierne, N., Lehnert, S.A., Bédier, E., Bonhomme, F. & Moore, S.S. 2000. Screening for intron-length polymorphisms in penaeid shrimps using exon-primed intron-crossing (EPIC)-PCR. *Mol. Ecol.* 9: 233–235.

Carpenter, K.E. & Niem, V.H. (eds.) 1998. *The Living Marine Resources of the Western Central Pacific. 2nd ed.* Rome: Food and Agriculture Organization of the United Nations.

Castric, V., Bernatchez, L., Belkhir, K. & Bonhomme, F. 2002. Heterozygote deciencies in small lacustrine populations of brook charr *Salvelinus fontinalis* Mitchill (Pisces, Salmonidae): a test of alternative hypotheses. *Heredity* 89: 27–35.

Chong, V.C. & Sasekumar, A. 1982. On the identification of three morphospecies of prawns - *Penaeus merguiensis* de Man, *Penaeus indicus* H. Milne Edwards and *Penaeus penicillatus* Alcock (Decapoda, Penaeoidea). *Crustaceana* 42: 127–141.

Dall, W., Hill, B.J., Rothlisberg, P.C. & Staples, D.J. (eds.) 1990. *Advances in Marine Biology, Vol. 27. The Biology of the Penaeidae.* London: Academic Press.

de Man, J.G. 1888. Report on the podophthalmous Crustacea of the Mergui Archipelago, collected for the trustees of the Indian Museum, Calcutta, by Dr. John Anderson. Part V. *J. Linn. Soc. Lond. (Zool.)* 22: 241–305.

Felsenstein, J. 1995. *PHYLIP (Phylogeny Inference Package) Version 3.57c.* Department of Genetics, University of Washington, Seattle, USA.

FitzSimmons, N.N., Moritz, C., Linmpus, C.J., Pope, L. & Prince, R. 1997a. Geographic structure of mitochondrial and nuclear gene polymorphisms in Australian green turtle populations and male-biased gene flow. *Genetics* 147: 1843–1854.

FitzSimmons, N.N., Limpus, C.J., Norman, J.A., Goldizen, A.R., Miller, J.D. & Moritz, C. 1997b. Philopatry of male marine turtles inferred from mitochondrial DNA markers. *Proc. Natl. Acad. Sci. USA* 94: 8912–8917.

Gaffney, P.M., Scott, T.M., Koehn, R.K. & Diehl, WJ. 1990. Interrelationships of heterozygosity, growth rate heterozygote deciencies in the coot clam, *Mulinia lateralis. Genetics* 124: 687–699.

Grey, D.L., Dall, W. & Baker, A. (eds.) 1983. *A Guide to the Australian Penaeid Prawns.* Darwin: Department of Primary Production of the Northern Territory.

Guo, S.W. & Thompson, E.A. 1992. Performing the exact test of Hardy-Weinberg proportion of multiple alleles. *Biometrics* 48: 361–372.

Hualkasin, W., Sirimontaporn, P., Chotigeat, W., Querci, J. & Phongdara, A. 2003. Molecular phylogenetic analysis of white prawns species and the existence of two clades in *Penaeus merguiensis. J. Exp. Mar. Biol. Ecol.* 296: 1–11.

Kimura, M. 1980. A simple method for estimating evolutionary rate of base substitutions through comparative studies of nucleotide sequences. *J. Mol. Evol.* 16: 111–120.

Klinbunga, S., Siludjai, D., Wuthijinda, W., Tassanakajon, A., Jarayabhand, P. & Menasveta, P. 2001. Genetic heterogeneity of the giant tiger shrimp (*Penaeus monodon*) in Thailand revealed by RAPD and mtDNA-RFLP analyses. *Mar. Biotechnol.* 3: 428–438.

Khamnamtong, B., Klinbunga, S. & Menasveta, P. 2009. Genetic diversity and geographic differentiation of the giant tiger shrimp (*Penaeus monodon*) in Thailand analyzed by mitochondrial COI sequences. *Biochem. Genet.* 47: 42–55.

Lavery, S., Moritz, C. & Fielder, D.R. 1996. Indo-Pacific population structure and evolutionary history of the coconut crab *Birgus latro. Mol. Ecol.* 5: 557–570.

Lavery, S., Chan, T.Y., Tam, Y.K. & Chu, K.H. 2004. Phylogenetic relationships and evolutionary history of the shrimp genus *Penaeus* s. l. derived from mitochondrial DNA. *Mol. Phylogenet. Evol.* 31: 39–49.

McMillan, W.O. & Palumbi, S.R. 1995. Concordant evolutionary patterns among Indo-West Pacific butterflyfishes. *Proc. R. Soc. Lond. B* 260: 229–236.

Mulley, J.C. & Latter, B.D.H. 1980. Genetic variation and evolutionary relationships within a group of thirteen species of penaeid prawns. *Evolution* 34: 904–916.

Nei, M. 1978. Estimation of average heterozygosity and genetic distance from a small number of individuals. *Genetics* 89: 583–590.

Raymond, M. & Rousset, F. 1995. GENEPOP (Version 1.2): population genetics software for exact tests and ecumenicism. *J. Hered.* 86: 248–249.

Reynolds, J., Weir, B.S. & Cockerham, C.C. 1983. Estimation of the coancestry coefficient: a basis for a short-term genetic distance. *Genetics* 105: 767–779.

Rice, W.R. 1989. Analyzing tables of statistical tests. *Evolution* 43: 223–225.

Rosen, D.E. 1978. Vicariant patterns and historical explanation in biogeography. *Syst. Zool.* 27: 159–188.

Saitou, N. & Nei, M. 1987. The neighbor-joining method: a new method for reconstructing phylogenetic trees. *Mol. Biol. Evol.* 4: 406–425.

Sodsuk, S. & Sodsuk, P. 1999. Genetic diversity and structure of banana prawn from three locations of Thailand. *Proceeding of the 1st National Symposium on Marine Shrimps, Songkhla, Thailand, 15–17 December*: 194–203.

Somers, I.F. 1994. Species composition and distribution of commercial penaeid prawn catches in the Gulf of Carpentaria, Australia, in relation to depth and sediment type. *Aust. J. Mar. Freshw. Res.* 45: 317–335.

Staples, D.J. 1979. Seasonal migration patterns of postlarval and juvenile banana prawns, *Penaeus merguiensis* de Man, in the major rivers of the Gulf of Carpentaria, Australia. *Aust. J. Mar. Freshw. Res.* 30: 143–157.

Staples, D.J. 1980. Ecology of juvenile and adolescent banana prawns, *Penaeus mergulensis*, in a mangrove estuary and adjacent off-shore area of the Gulf of Carpentaria; II. Emigration, population structure and growth of juveniles. *Aust. J. Mar. Freshw. Res.* 31: 653–665.

Staples, D.J. & Vance, D.J. 1985. Short-term and long-term influences on the immigration of postlarval banana prawns, *Penaeus merguiensis*, into a mangrove estuary of the Gulf of Carpentaria, Australia. *Mar. Ecol. Prog. Ser.* 23: 15–29.

Staples, D.J. & Vance, D.J. 1986. Emigration of juvenile banana prawns *Penaeus merguiensis* from a mangrove estuary and recruitment to offshore areas in the wet-dry tropics of the Gulf of Carpentaria, Australia. *Mar. Ecol. Prog. Ser.* 27: 239–252.

Staples, D.J. & Vance, D.J. 1987. Comparative recruitment of the banana prawn, *Penaeus merguiensis*, in five estuaries of the south-eastern Gulf of Carpentaria. *Aust. J. Mar. Freshw. Res.* 38: 29–45.

Supungul, P., Sootanan, P., Klinbunga, S., Kamonrat, W., Jarayabhand, P. & Tassanakajon, A. 2000. Microsatellite polymorphism and the population structure of the black tiger shrimp (*Penaeus monodon*) in Thailand. *Mar. Biotechnol.* 2: 339–347.

Tam, Y.K. & Chu, K.H. 1993. Electrophoretic study of the phylogenetic relationships of some species of *Penaeus* and *Metapenaeus* (Decapoda: Penaeidae) from the South China Sea. *J. Crust. Biol.* 13: 697–705.

Tassanakajon, A., Pongsomboon, S., Rimphanitchayakit, V., Jarayabhand, P. & Boonsaeng, V. 1997. Random amplified polymorphic DNA (RAPD) markers for determination of genetic variation in wild populations of the black tiger prawn (*Penaeus monodon*) in Thailand. *Mol. Mar. Biol. Biotechnol.* 6: 110–115.

Tassanakajon, A., Pongsomboon, S., Jarayabhand, P., Klinbunga, S. & Boonsaeng, V. 1998a. Genetic structure in wild populations of black tiger shrimp (*Penaeus monodon*) using randomly amplified polymorphic DNA analysis. *J. Mar. Biotechnol.* 6: 249–254.

Tassanakajon, A., Tiptawonnukul, A., Supungul, P., Rimphanitchayakit, V., Cook, D., Jarayabhand, P., Klinbunga, S. & Boonsaeng, V. 1998b. Isolation and characterization of microsatellite markers in the black tiger prawn *Penaeus monodon*. *Mol. Mar. Biol. Biotechnol.* 7: 55–61.

Vance, D.J., Haywood, M.D.E. & Staples, D.J. 1990. Use of a mangrove estuary as a nursery area for postlarval and juvenile banana prawns, *Penaeus merguiensis* de Man, in Northern Australia. *Estuar. Coast. Shelf Sci.* 31: 689–701.

Wanna, W., Rolland, J.-L., Bonhomme, F. & Phongdara, A. 2004. Population genetic structure of *Penaeus merguiensis* in Thailand based on nuclear DNA variation. *J. Exp. Mar. Biol. Ecol.* 311: 63–78.

Weir, B.S. & Cockerham, C.C. 1984. Estimating F-statistics for the analysis of population structure. *Evolution* 38: 1358–1370.

Williams, S.T. & Benzie, J.A.H. 1998. Evidence of a biogeographic break between populations of a high dispersal starfish: congruent regions within the Indo-West Pacific defined by color morphs, mtDNA and allozyme data. *Evolution* 52: 87–99.

Williams, S.T., Jara, J., Gomez, E. & Knowlton N. 2002. The marine Indo-West Pacific break:

contrasting the resolving power of mitochondrial and nuclear genes. *Integr. Comp. Biol.* 42: 941–952.

APPENDIX

Table A1. List of samples used in molecular analyses, including GenBank Accession numbers for COI sequences, sampling sites in Thailand, and other regions. Samples in bold print indicate newly generated data.

Sample Name	Sampling Sites	GenBank Accession Number
	Thailand	
NKE1	Nakhon Si Thammarat	HQ206386
NKE9	Nakhon Si Thammarat	HQ206387
SKE26	Songkhla	HQ206388
SKE43	Songkhla	HQ206389
SKE92	Songkhla	HQ206390
SKE96	Songkhla	HQ206391
SKEB1	Songkhla	HQ206392
SKEB5	Songkhla	HQ206393
SRE1	Surat Thani	HQ206394
SRE2	Surat Thani	HQ206395
SRE30	Surat Thani	HQ206396
TDE1	Trad	HQ206397
TDE2	Trad	HQ206398
TDE12	Trad	HQ206399
TDE23	Trad	HQ206400
RNW55	Ranong	HQ206401
RNW66	Ranong	HQ206402
PKW7	Phuket	HQ206403
STWA5	Satun	HQ206404
STWA6	Satun	HQ206405
STWA7	Satun	HQ206406
TRW11	Trang	HQ206407
TRW15	Trang	HQ206408
TRW7	Trang	HQ206409
STW4	Satun	HQ206410
STW31	Satun	HQ206411
SRE6	Surat Thani	HQ206412
SRE13	Surat Thani	HQ206413
SRE17	Surat Thani	HQ206414
SRE28	Surat Thani	HQ206415
SRE14	Surat Thani	HQ206416
SRE3	Surat Thani	HQ206417
SRE25	Surat Thani	HQ206418
SRE16	Surat Thani	HQ206419
SRE19	Surat Thani	HQ206420
SKE12	Songkhla	HQ206421

Table A1. Continuation.

Sample Name	Sampling Sites	GenBank Accession Number
	Thailand (continuation)	
SKE36	Songkhla	HQ206422
SKE28	Songkhla	HQ206423
SKE18	Songkhla	HQ206424
SKE22	Songkhla	HQ206425
SKE13	Songkhla	HQ206426
STW8	Satun	HQ206427
STW12	Satun	HQ206428
STW18	Satun	HQ206429
STW22	Satun	HQ206430
STW26	Satun	HQ206431
TRW30	Trang	HQ206432
TRW18	Trang	HQ206433
TDE4	Trad	HQ206434
TDE40	Trad	HQ206435
SRE24	Surat Thani	HQ206436
NKE2	Nakhon Si Thammarat	HQ206437
NKE5	Nakhon Si Thammarat	HQ206438
NKE8	Nakhon Si Thammarat	HQ206439
NKE17	Nakhon Si Thammarat	HQ206440
NKE18	Nakhon Si Thammarat	HQ206441
NKE19	Nakhon Si Thammarat	HQ206442
	Other Regions	
Africa1	Tanzania, Africa	HQ206443
Africa2	Tanzania, Africa	HQ206444
Africa3	Tanzania, Africa	HQ206445
Africa4	Tanzania, Africa	HQ206446
P. monodon (GB)	Tungkang, Taiwan	AF014377
L. vannamei (GB)	Centre Océanologique du Pacifique, Tahiti	AF014383
F. indicus (GB)	Tungkang, Taiwan	AF014378

Abbreviations: *F.* = *Fenneropenaeus*, *L.* = *Litopenaeus*, *P.* = *Penaeus*, GB = GenBank.

Population genetics in the rocky shore crab *Pachygrapsus marmoratus* from the western Mediterranean and eastern Atlantic: complementary results from mtDNA and microsatellites at different geographic scales

SARA FRATINI[1], CHRISTOPH D. SCHUBART[2] & LAPO RAGIONIERI[1]

[1] *Department of Evolutionary Biology, University of Florence, Florence, Italy*
[2] *Biology 1, University of Regensburg, 93040 Regensburg, Germany*

ABSTRACT

Until recently, it was generally believed that marine species with planktonic larval dispersal should be genetically homogeneous across their geographic range. Nowadays, however, there is increasing evidence for genetically structured marine populations among larval-dispersed species, and thus a higher degree of intraspecific local variability than expected. The sometimes very complex patterns of intraspecific genetic differentiation may be due to historical environmental factors related to habitat, currents and sea level fluctuations, and/or to present-day species-specific traits and ecological factors.

Studies on population genetic structure are commonly based on the geographical distribution of mitochondrial haplotypes or, alternatively, on polymorphism at microsatellite loci. These genetic markers have different molecular and evolutionary properties, and, as a consequence, they may reveal different distribution patterns of the recorded genetic variation. Mitochondrial DNA is a haploid genetic marker, exclusively maternally inherited, and with a medium to low level of genetic variability. It thus enables the reconstruction of ancient processes. In contrast, microsatellites represent a co-dominant genetic marker with high variability, suitable for the investigation of recent colonization and migration events.

The aim of the current study is to investigate the population genetic structure on both sides of the Atlanto-Mediterranean biogeographic boundary of the intertidal crab *Pachygrapsus marmoratus* (Brachyura: Grapsidae) by means of a 598 basepair region of the mitochondrial DNA gene cytochrome c oxidase subunit I of 238 individuals from 15 populations throughout the western Mediterranean Sea and eastern Atlantic Ocean. Moreover, for a subset of western Mediterranean populations, we also genotyped 295 individual crabs at 6 microsatellite loci. While the mitochondrial data show a recent and weak subdivision of genetic variation at a medium to large scale (based on rare haplotypes), the microsatellite loci reveal genetic differentiation among populations at a local scale. The relative advantages of these two genetic marker systems for studies of population dynamics of highly dispersive marine organisms and the importance of using these methods complementarily is discussed.

1 INTRODUCTION

The lack of obvious barriers to dispersal of marine planktonic larvae has given rise to a central dogma in marine ecology, namely that marine species with planktonic larval phases are able to

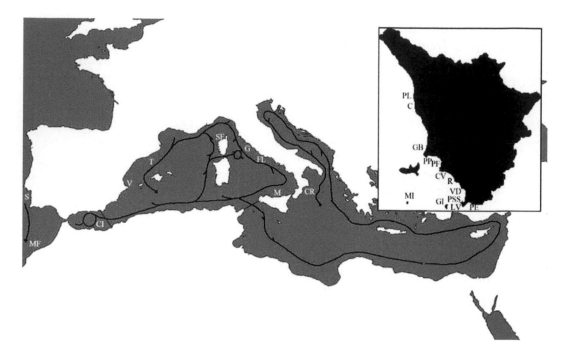

Figure 1. Analyzed western Mediterranean and eastern Atlantic populations of *Pachygrapsus marmoratus* with an enlarged map of the coast of Tuscany. Abbreviations for populations correspond to those given in Table 1. The arrows roughly indicate the main directions of currents, and the dotted lines mark the two main geographic breaks present in the study area.

disperse over wide distances (Hedgecock 1986; Avise 2004). The majority of marine species, especially invertebrates, have a reproductive r–strategy with thousands of larvae per spawn, a large effective population size, external fertilization, and extensive larval phases. This has led to an expectation of populations in equilibrium and with low genetic diversity among even very distant (in the order of thousands of kilometres) populations (Hedgecock 1986; Avise 2004). Undeniably, many studies confirm this hypothesis. However, extensive genetic population structure has also been recorded for many larval-dispersed species at different geographic scales (for a review see Avise 2004). Nowadays, dispersal of marine organisms is thus recognized as a complex process, significantly affecting the demographic history, the dynamics, and the evolution of marine populations (Patarnello et al. 2007; Hauser & Carvalho 2008). Moreover, its entity is most often unpredictable when considering the duration of larval phases and the direction/intensity of main currents and winds (see Weersing & Toonen 2009). In the marine realm, a series of physical, ecological, physiological, and behavioral mechanisms, both historical, such as sea level fluctuations during glaciations, as well as recent, such as species-specific traits and present-day ecological factors contribute to generate and maintain intraspecific genetic diversity. Moreover, each of these factors can leave a footprint on population genetic structure and distribution of genetic variability. To add to the complication, different ecological and environmental circumstances can result in similar genetic patterns, leading to alternative interpretation of the evolutionary processes really involved (Rousett 2004; Patarnello et al. 2007).

The extensive development of molecular genetics in the last decades, thanks to the establishment of new experimental techniques, is providing researchers with numerous genetic markers able to describe with high accuracy the genetic structure of natural populations. Among these, the two predominant marker types for population studies are mitochondrial DNA (mtDNA) sequences and microsatellite loci.

Mitochondrial DNA represents a haploid maternally inherited genetic marker. The near-absence of genetic recombination in mtDNA makes it a useful source of information for scientists involved in population genetics and evolutionary biology, since it enables reconstruction of a matrilineal phylogeographic model of the analyzed individuals, represented as a gene tree (Avise 2000, 2004). Mitochondrial haplotypes exhibit a medium to low level of genetic variability, thus they are able to provide a deep historical perspective in the investigation of the population dynamics and incipient speciation.

Microsatellite loci are usually characterized by a high degree of polymorphism due to variation in repeat copy number and represent a neutral and co-dominant genetic marker (Queller et al. 1993; Tóth et al. 2000). Notwithstanding microsatellites have proven to be versatile molecular markers employed in many fields soon after their description (Tautz 1989), they are not without limitations. The major drawback is that they need to be isolated de novo for most species, since the percentage of successfully amplified loci described for another allied species may decrease with increasing genetic distance. A number of methods for their isolation have been developed in the last years (Zane et al. 2002; Leese et al. 2008).

According to the above description, it becomes clear that mtDNA and microsatellite DNA differ in their molecular properties, levels of polymorphism, ease and cost of use, statistical approach and spatial-temporal scale of application. A combination of the two techniques is thus highly recommended, since it is the most powerful approach to investigate the population structure in marine organisms (see Ballard & Whitlock 2004). An important aspect to be considered is that complementary results can be obtained in comparative studies of these two markers, as a consequence of their different molecular behavior and level of variation. Some studies reported a separation among populations using microsatellites, which were not revealed by mtDNA, probably due to their greater sensitivity in the detection of population structure. For example, Duran et al. (2004a, b), when investigating the genetic population structure of the sponge *Crambe crambe* across its Atlantic-Mediterranean distribution range, recorded a very low level of intra- and interpopulation genetic variation of mtDNA (only two haplotypes of the cytochrome c oxidase subunit I sequences, COI) (Duran et al. 2004b), but a strong genetic separation among populations at macro- and micro geographic scales with microsatellite loci (Duran et al. 2004a). Another example comes from *Carcinus maenas*, for which Roman & Palumbi (2004) recorded restricted gene flow only across longer distances along the Atlantic coast of Europe with the mtDNA COI; while Pascoal et al. (2009) found a genetic separation between northern and central Portuguese populations when genotyping 3 microsatellite loci. In contrast, other studies found greater genetic heterogeneity at the mtDNA level than with microsatellites. This is, for example, the case of the European sardine *Sardina pilchardus* (see Gonzalez & Zardoya 2007) and of the landlocked Atlantic salmon *Salmo salar* (see Tessier et al. 1995), studied over an extensive geographic range. This phenomenon may be due to the fact that microsatellites are generally less useful at larger geographic and temporal scales due to homoplasy than at finer scales, as suggested by Shaw et al. (1999) in his study on the Atlantic herring from Norwegian waters and the Barents Sea.

We here present data on the population genetic structure of the intertidal crab *Pachygrapsus marmoratus* at macro- to mesogeographic scales by mtDNA as well as at microgeographic scales by microsatellites. At a broader geographic scale we included European Atlantic and Mediterranean populations. We tested two alternative hypotheses: on the one hand a separation among populations under the influence of phylogeographic breaks known for this area, i.e., the Gibraltar Strait and the Sicily Strait, potentially separating Atlantic from Mediterranean as well as eastern from western Mediterranean populations, respectively. On the other hand, larvae of *P. marmoratus* may be able to form a widespread metapopulation within the Mediterranean Basin and along the European Atlantic coasts. At a small geographic scale, we focused on populations from the central Italian coast, aiming to test the existence of fine population structure due to subtle barriers to gene flow, such as local currents and winds. The comparisons of the results obtained using the two markers permit us to discuss the importance of combining the two approaches in the study of connectivity among marine populations.

2   MATERIAL AND METHODS

2.1   Study species

Pachygrapsus marmoratus (Fabricius 1787) (Decapoda: Brachyura: Thoracotremata: Grapsidae) is widespread within the entire Mediterranean Basin, the Black Sea and along the European and African eastern Atlantic coasts, including the Canary and Azores Islands (Zariquiey Álvarez 1968; Ingle 1980). Recently, as a consequence of climate change, its distribution has been expanding northward (Dauvin 2009). It is one of the most common macro-invertebrates of the upper and middle levels of rocky shores, where it thrives in natural crevices and cleavages. It is also abundant in ports, marinas, and breakwaters. The species is omnivorous, feeding mainly on algae and small animals (Cannicci et al. 2002). Adults are relatively sedentary, occupying a specific area, actively defended from con-specific intrusion, the extent of which depends on the sex and size of the owner (Cannicci et al. 1999). Otherwise, connectivity among populations is potentially guaranteed by the larval stages, lasting at least four weeks in the marine plankton (Cuesta & Rodríguez 2000).

    P. marmoratus breeds from late spring to late summer (Zariquiey Álvarez 1968; Ingle 1980), thus megalopal settlement has a peak at the end of the summer and seems to follow a semilunar cycle in correspondence with the spring tides, at least along the open Atlantic coast where tidal excursions are substantial in comparison to the Mediterranean Sea (Flores et al. 2002). Moreover, hydrological features operating in coastal areas (i.e., the magnitude of the across-shore wind component and of the onshore currents) have been shown to be the main factors affecting settlement on a microgeographic scale, while on a meso- and macrogeographic scale, oceanographic processes are involved (Flores et al. 2002). Thus, nearby areas under the influence of divergent hydrological conditions may be more or less separated in their recruitment and demographic processes.

2.2   Sampling area

Specimens of *P. marmoratus* were collected from a total of 23 sites of the western Mediterranean Sea and eastern Atlantic Ocean (Figure 1 and Table 1), from 2004 to 2009. In particular, 15 of the populations distributed over a large geographic scale, were used for analyzing the genetic variation of mitochondrial haplotypes (Table 1), while 10, concentrated along the Tuscan coast (Italy), were used for assessing population differentiation at a microgeographic scale by means of microsatellite loci (Table 1). Microsatellite data for 8 of these 10 Tuscan populations were previously reported in Fratini et al. (2008), while the entire mtDNA dataset is new. From each population 15 to 30 crabs were collected, a chela or a pereiopod was removed from each individual and preserved in absolute ethanol for successive genetic analyses. The Mediterranean Sea is a semienclosed marginal sea which communicates with the World oceans through the narrow Strait of Gibraltar and more recently the man-made Suez Channel. Within the Mediterranean, the circulation is complex and is made up of more or less steady currents, combined with a high energetic mesoscale field (eddies, gyres, and fronts), which in turn shows an important interseasonal and interannual variability (Korres et al. 2000; Lascaratos et al. 1999; Lascaratos & Nittis 1998; Skliris et al. 2007). Its main subbasins (western Mediterranean, Ionian Sea, Aegean-Levantine Sea) have independent circulation regimes,

Table 1. Samples of *Pachygrapsus marmoratus* from western Mediterranean and eastern Atlantic populations, including collecting sites that represent populations, countries/regions, Mediterranean basins, GPS coordinates, and the number of individuals used for mtDNA and/or microsatellite analyses.

Collecting sites	Abbr.	Country/Region	Basin	GPS coordinates		number of individuals	
						mtDNA	microsatellite
Porto di Livorno	PL	Italy	Mediterranean/Ligurian	43°32.70′N	10°17.70′E	–	30
Calafuria	C	Italy	Mediterranean/Ligurian	43°28.16′N	10°20.02′E	15	30
Golfo di Baratt	GB	Italy	Mediterranean/Ligurian-Tyrrhenian	42°99.90′N	10°51.52′E	–	26
Porto di Piombino	PP	Italy	Mediterranean/Ligurian-Tyrrhenian	42°92.42′N	10°53.03′E	–	30
Porto di Follonica	PF	Italy	Mediterranean/Tyrrhenian	42°91.40′N	10°75.83′E	–	30
Cala Violina	CV	Italy	Mediterranean/Tyrrhenian	42°81.75′N	10°76.71′E	–	29
Rocchette	R	Italy	Mediterranean/Tyrrhenian	42°77.42′N	10°79.09′E	14	–
Villa Domizia	VD	Italy	Mediterranean/Tyrrhenian	42°43.63′N	11°13.15′E	–	30
Porto Santo Stefano	PSS	Italy	Mediterranean/Tyrrhenian	42°43.61′N	11°12.26′E	–	30
Le Viste	LV	Italy	Mediterranean/Tyrrhenian	42°38.56′N	11°20.77′E	–	30
Porto Ercole	PE	Italy	Mediterranean/Tyrrhenian	42°39.45′N	11°20.88′E	15	30
Giglio Island	GI	Italy	Mediterranean/Tyrrhenian	42°35.54′N	10°92.65′E	11	–
Montecristo Island	MI	Italy	Mediterranean/Tyrrhenian	42°34.13′N	10°32.23′E	15	–
Gaeta	G	Italy	Mediterranean/Tyrrhenian	41°19.51′N	13°58.45′E	15	–
Fusaro Lagoon	FL	Italy	Mediterranean/Tyrrhenian	40°80.95′N	14°04.63′E	15	–
Crotone	CR	Italy	Mediterranean/Ionian	39°07.03′N	17°13.31′E	16	–
Messina	M	Italy	Mediterranean/Ionian	38°20.36′N	15°57.31′E	16	–
St. Florent	SF	Corsica	Mediterranean/Ligurian	42°69.05′N	9°26.14′E	15	–
Tarragona	T	Spain	Mediterranean/Balearic	41°11.24′N	1°26.41′E	20	–
Valencia	V	Spain	Mediterranean/Balearic	39°43.99′N	−0°31.79′E	20	–
Cala Iris	CI	Morocco	Mediterranean/Alboran	35°15.08′N	4°35.47′E	15	–
Sesimbra	S	Portugal	Atlantic	38°40.19′N	−9°06.37′E	18	–
Matorral	MF	Fuerteventura	Atlantic	28°04.28′N	−14°32.89′E	18	–

interlinked through the Atlantic surface waters. On small scales, the circulation is mainly dependent on local forcing of winds and eddies. Considering the Tuscan coast, its northern region is washed by the Ligurian Sea and its southern one by the Tyrrhenian Sea, but they are connected to each other via the Corsican Channel (Millot 1987; Artale et al. 1994; Astraldi et al. 1994). This channel is affected by two major surface currents, the West Corsican Current bringing mixed Atlantic water from the Algerian Basin, and the Tyrrhenian Current, which, after having lapped on the coasts of Latium and Tuscany, brings water northward through the Corsican Channel from the Tyrrhenian Sea (Millot 1987; Artale et al. 1994; Astraldi et al. 1994).

The palaeo-oceanography of the Mediterranean Sea is rather complex. One of the most dramatic events was the Messinian Salinity Crisis occurring during the late Miocene, about 5.5 Million years ago (Mya), during which the Mediterranean Sea was completely separated from the Atlantic Ocean and experienced an almost total desiccation. After that, during the Cenozoic era, Atlantic waters reinvaded the Mediterranean Basin and many Atlantic species colonized this area. Successive short periods of separations between the Atlantic and the Mediterranean Seas also occurred afterwards, during the Quaternary, an era characterized by the alternation of glacial and interglacial periods. All these climatic events have been shown to influence the evolutionary history of Atlantic-Mediterranean marine species (see Patarnello et al. 2007; Reuschel et al. 2010).

2.3   DNA extraction and amplification

Total genomic DNA was extracted from muscle tissue using the Puregene Kit (Gentra Systems). DNA was then resuspended in sterile distilled water and stored at $4\,°C$ for routine use, or at $-20\,°C$ for long-term storage.

Mitochondrial DNA genetic analysis was based on a total of 238 individuals. Selective amplification of a 658 basepair (bp) fragment of the mtDNA cytochrome c oxidase subunit I gene (COI) was performed using the primers COL6b (5'-ACA AAT CAT AAA GAT ATY GG-3'; Schubart & Huber 2006) and HCO2198 (5'-TAA ACT TCA GGG TGA CCA AAA AAT CA-3'; Folmer et al. 1994). The PCR amplification was performed with the following PCR conditions: 40 cycles with 45 s at $94\,°C$ for denaturation, 1 min at $48\,°C$ for annealing, 1 min at $72\,°C$ for extension, followed by 10 min at $94\,°C$ for initial denaturation and 10 min at $72\,°C$ for final extension. Subsequently, PCR products were visualized on an agarose gel, purified by precipitation with Sure Clean (Bioline) and then resuspended in water. The sequence reactions were performed with the Big Dye terminator mix (Big Dye Terminator® V 1.1 Cycle Sequencing kit; Applied Biosystems) followed by electrophoresis in an ABI Prism automated sequencer (ABI Prism™ 310 Genetic Analyzer; Applied Biosystems). The sequences were corrected manually with the program CHROMAS version 1.55 (Technelysium Pty Ltd., Queensville, Australia) and aligned by eye (no indels) with Bioedit (Hall 1999).

Regarding the nuclear marker analyses, a total of 295 individuals were screened for polymorphism at six microsatellite loci specifically isolated for *P. marmoratus* (see Fratini et al. 2006). Information on microsatellite features and amplification are reported in Fratini et al. (2006, 2008). For detection of polymorphisms, the forward primer for each locus was 5'-labelled and then labelled amplicons from the six loci were divided into two sets (pm-101 TET + pm-108 HEX; and pm-79 HEX + pm-183 FAM + pm-187 NED + pm-99 FAM). For each set, 1 μll to 5 μll of each PCR product was combined with water in a final volume of 10 μl for successive dimensional analysis. Sizing was done in an ABI Prism 310 Genetic Analyzer (Applied Biosystems) with reference to an internal size standard (TAMRA500 for the set TET + HEX, and ROX400 for the set HEX + FAM + NED) using GENOTYPER ver. 3.7 and GENESCAN ver. 3.7 (Applied Biosystems).

Sequences were deposited in GenBank (accession numbers JF930650–JF930682). A single voucher specimen (male) from Tuscany was deposited in the collection of the Museo Zoologico Università di Firenze (MZUF 2660).

2.4 Statistical analysis of mt DNA data

For each population and for the overall dataset, we calculated the number of haplotypes, the haplotype diversity (h: the probability that two randomly chosen haplotypes are different in a single population) and the nucleotide diversity ($\theta\pi$: the percentage mean number of differences between all pairs of haplotypes in a population; Nei 1987), using ARLEQUIN ver. 3.1 (Excoffier et al. 2005).

An analysis of molecular variance (AMOVA) performed in ARLEQUIN ver. 3.1 (Excoffier et al. 2005), was used to examine population genetic structure for the COI sequences. Φ_{ST} values were computed using haplotypic frequencies only as well as genetic distances (Tajima-Nei model, suggested for unequal nucleotide frequencies, as observed in our dataset; Tajima & Nei 1984). Significance levels of pairwise Φ_{ST}, under the null hypothesis of no differentiation between populations, were computed by permutation tests from 20,000 random permutations of haplotypes between populations. In addition, the geographical distribution of haplotypes among populations was investigated by contingency χ^2 tables without pooling rare haplotypes, using a Monte Carlo simulation (Rolf & Bentzen 1989), as implemented in the program CHIRXC (Zaykin & Pudovkin 1993). Probability of heterogeneity in haplotypic distribution was determined by comparison between observed and simulated χ^2 values, obtained from 20,000 random permutations of the original data.

A minimum spanning network was built with TCS version 1.13 (Clement et al. 2000), with a connection limit fixed at 95%, to assess the intraspecific evolutionary relationships among the COI haplotypes of *P. marmoratus* found within both the Mediterranean and Atlantic populations.

Departures from neutral expectations were assessed for each population and for the overall population, using the Tajima's D test (Tajima 1989) and Fu's F_S test (Fu & Li 1993; Fu 1997), as implemented in ARLEQUIN vers. 3.1. These two neutrality tests estimate the deviations from neutral molecular evolution assessing departure of their parameters from zero, i.e., the value expected under the hypothesis of selective neutrality and population equilibrium, by generating random samples. Significant negative D and F_S values can be interpreted as signatures of population expansion. For better depicting the demographic history of our populations we also applied the R_2 test (Ramos-Onsins & Rozas 2002) proven to be particularly effective in revealing population expansions (Ramírez-Soriano et al. 2008). The R_2 test significance level was estimated by coalescent simulations using DnaSP v5 (Librado & Rozas 2009), based on 1000 simulated re-sampling replicates.

Additionally, the demographic history of each population and of the overall sample was reconstructed using mismatch distribution analysis (Rogers & Harpending 1992). ARLEQUIN 3.1 compares the observed distribution of nucleotide site differences between pairs of haplotypes with that expected under a model of population expansion (Rogers 1995), based on the raggedness index. This index is larger for multimodal than unimodal distributions and its significance is tested by a parametric bootstrap approach (10,000 replicates) under the null hypothesis of population expansion. A unimodal and approximately Poisson-like distribution is expected for populations that have experienced demographic expansion in the recent past, while multimodal frequency distribution may be interpreted as evidence for populations at equilibrium (Rogers 1995; Patarnello et al. 2007). The graphs of each observed mismatch distributions was obtained using DnaSP v5.

The mismatch distribution analysis also provides an estimation for the expansion parameters Tau (τ), Theta 0 (θ_0) and Theta 1 (θ_1), under a demographic or spatial expansion hypothesis. To avoid underestimation of τ, the parameters are estimated by a generalized nonlinear least-square approach as suggested by Schneider & Excoffier (1999). The value of τ can be used to calculate the time (t) at which the demographic or spatial expansion began, by the formula $t = \tau/2\mu$ (with μ as the mutation rate per site per year; Li 1977). Approximate confidence intervals for the demographic parameters were obtained by 1000 parametric bootstrap replicates.

Table 2. Variable sites among 33 mitochondrial haplotypes of *Pachygrapsus marmoratus* from western Mediterranean and eastern Atlantic populations (*n* = 238). All haplotypes are compared with haplotype 1. Symbols and abbreviations: * = identical nucleotides; P = parsimonious sites, aa1 and aa2 = amino acids in haplotype 1 (corresponding to nonsilent mutations); V = valine; I = isoleucine; A = alanine.

Haplotype	9	69	102	114	129	141	144	147	150	156	168	183	198	234	246	249	267	279	297	303	315	336	376	388	462	467	561
1	T	A	T	T	G	T	G	T	A	G	G	A	G	C	C	A	A	G	C	A	C	C	G	C	G	T	T
2	*	*	*	*	*	*	A	*	*	*	*	*	*	*	*	*	*	*	*	*	*	*	*	*	*	*	*
3	*	*	*	*	*	*	A	*	*	*	*	*	*	*	*	*	G	*	*	*	*	*	*	*	*	*	*
4	*	*	*	*	*	*	A	*	*	*	*	*	*	*	*	*	*	*	*	*	*	*	*	*	*	*	*
5	*	G	*	*	*	*	A	*	*	*	*	*	*	*	*	G	*	*	*	*	*	*	*	*	*	*	*
6	*	*	*	*	*	*	A	*	*	*	*	*	*	*	*	*	*	*	*	*	*	*	*	*	*	*	*
7	*	*	*	*	*	*	A	*	*	*	*	*	*	*	*	G	*	*	*	*	*	*	*	*	*	*	*
8	*	*	*	*	*	*	A	*	*	*	*	*	*	*	*	*	*	*	*	*	*	*	*	*	*	*	*
9	*	*	*	*	*	*	A	*	*	A	*	*	*	*	*	*	G	*	*	*	*	*	*	*	*	*	*
10	*	*	*	*	*	*	A	*	*	*	*	*	*	*	*	*	G	*	*	*	T	*	*	*	*	*	*
11	*	G	*	*	*	*	A	*	*	*	*	*	A	*	*	*	*	*	*	*	*	*	*	*	*	*	*
12	*	*	*	*	*	*	A	*	*	*	*	*	*	*	*	*	*	*	*	*	*	T	*	*	*	*	*
13	*	*	C	C	*	C	A	*	*	*	*	*	*	*	*	*	*	*	*	*	*	*	*	*	*	*	*
14	*	*	*	*	*	*	A	*	*	*	*	*	*	*	*	*	G	*	*	*	*	*	*	*	*	*	*
15	*	G	C	*	*	*	A	*	*	*	*	*	*	*	*	*	G	*	*	*	*	*	*	*	*	*	*
16	*	*	*	*	*	*	A	*	*	*	*	*	*	*	*	*	*	*	A	*	*	*	*	*	*	*	*
17	*	*	*	*	*	*	A	*	*	*	*	*	*	T	*	*	G	*	*	*	*	*	*	*	*	*	*
18	*	*	*	*	*	*	A	*	*	*	*	*	*	*	T	*	G	*	*	*	*	*	*	*	*	*	*
19	*	*	*	*	*	*	A	*	G	*	*	*	*	*	*	*	*	*	*	*	*	*	*	*	*	*	*
20	*	*	*	*	*	*	A	C	*	*	*	*	*	*	*	*	G	*	*	*	*	*	*	*	*	*	*
21	*	*	*	*	*	*	A	*	*	*	*	*	*	*	*	*	G	*	*	*	*	*	*	*	*	*	*
22	*	*	*	*	*	*	A	*	*	*	*	*	*	*	*	*	G	*	*	*	*	*	A	*	A	*	C
23	*	*	*	*	*	*	A	*	*	*	*	*	*	*	*	*	G	*	*	*	*	*	A	*	*	*	*
24	*	*	*	*	*	*	A	*	*	*	A	*	*	*	*	*	*	*	*	*	*	*	*	*	*	*	*
25	*	*	*	*	*	*	A	*	*	*	*	*	*	*	*	*	*	*	*	G	*	*	*	*	*	*	*
26	*	G	*	*	*	*	A	*	*	*	*	*	*	*	*	*	*	C	*	*	*	*	*	*	*	*	*
27	*	*	*	*	*	*	A	*	*	*	*	*	*	*	*	*	*	A	*	*	*	*	*	*	*	*	*
28	*	*	*	*	*	*	A	*	*	A	*	T	*	*	*	*	*	*	*	*	*	*	*	*	*	*	*
29	*	*	*	*	*	*	A	*	*	A	G	*	*	*	*	*	G	*	*	*	*	*	*	*	*	*	*
30	C	*	*	*	*	*	A	*	*	*	*	*	*	*	*	*	G	*	*	*	*	*	*	*	*	*	*
31	*	*	*	*	*	*	A	*	*	A	*	*	*	*	*	*	G	*	*	*	*	*	*	*	*	*	*
32	*	*	*	*	*	*	A	*	*	*	*	*	*	*	*	*	*	*	*	*	*	*	*	T	*	*	*
33	*	*	*	*	*	*	A	*	*	A	*	*	*	*	*	*	G	*	*	*	*	*	*	*	*	C	*
parsimony site		P	P				P			P						P	P										
codon position	3	3	3	3	3	3	3	3	3	3	3	3	3	3	3	3	3	3	3	3	3	3	1	1	2	3	3
aa1																							V		V		
aa2																							I		A		

Table 3. Number of haplotypes (N_a), haplotype (h), and nucleotide (π) diversities, Tajima's D parameter, Fu's F_S, R_2 parameter, and mismatch distribution raggedness index (rg) for each population of *Pachygrapsus marmoratus* for the entire sample. Significant values are in bold. See Table 1 for abbreviations of populations.

Abbr.	N_a	h	π	D	F_S	R_2	rg
C	5	0.64 ± 0.13	0.20 ± 0.20	-0.28	-0.67	0.13	0.12
R	7	0.81 ± 0.09	0.31 ± 0.21	-0.14	$\mathbf{-2.38}$	0.13	0.07
PE	3	0.45 ± 0.13	0.13 ± 0.11	0.63	0.36	0.19	0.31
GI	5	0.78 ± 0.09	0.24 ± 0.18	0.14	-1.20	0.16	0.19
MI	9	0.85 ± 0.09	0.32 ± 0.21	-1.17	$\mathbf{-4.71}$	0.08	0.07
G	7	0.77 ± 0.10	0.28 ± 0.20	-1.17	$\mathbf{-2.44}$	**0.09**	**0.37**
FL	5	0.63 ± 0.13	0.23 ± 0.17	-0.36	-0.74	0.13	**0.52**
CR	5	0.53 ± 0.14	0.19 ± 0.14	-0.89	-1.18	**0.11**	0.20
M	7	0.82 ± 0.73	0.29 ± 0.20	-0.66	-2.19	0.12	**0.35**
SF	7	0.72 ± 0.12	0.25 ± 0.18	-0.72	$\mathbf{-2.90}$	0.11	0.03
T	9	0.82 ± 0.73	0.28 ± 0.19	-0.90	$\mathbf{-4.14}$	**0.09**	0.11
V	5	0.44 ± 0.13	0.16 ± 0.13	-1.42	-1.28	0.10	0.19
CI	6	0.57 ± 0.15	0.15 ± 0.12	-1.45	$\mathbf{-3.23}$	**0.09**	0.06
S	3	0.22 ± 0.12	0.06 ± 0.06	$\mathbf{-1.71}$	-1.02	0.17	0.47
MF	6	0.67 ± 0.11	0.21 ± 0.15	-0.96	-1.81	**0.10**	0.13
Overall	33	0.65 ± 0.03	0.22 ± 0.15	$\mathbf{-2.01}$	$\mathbf{-28.30}$	**0.02**	0.13

2.5   Statistical analysis of microsatellite data

The number of alleles and the allelic richness for each locus and population were calculated using FSTAT ver. 2.9.3 (Goudet 1995). Linkage equilibrium among loci and Hardy-Weinberg equilibrium (HWE) were assessed for each population using GENEPOP ver. 3.4 (Raymond & Rousset 1995; wbiomed.curtin.edu.au/genepop). Since all populations but two (Cala Violina (CV) and Port of Piombino (PP)) showed heterozygote deficiencies, we assessed the existence of null alleles by means of equation 2 from Brookfield (1996) as implemented in MICROCHECKER 2.2.3 (van Oosterhout et al. 2004). We then used the software FREENA (www.montpellier.inra.fr/URLB) to compute a genotype dataset corrected for null alleles following the Including Null Alleles (INA) method described in Chapuis & Estoup (2007). This new dataset was used in all further analyses, and for recalculating the measures of population variability.

Genetic differentiation was estimated by the exact test of population differentiation (Raymond & Rousset 1995) implemented in GENEPOP. This test verifies the existence of differences in allele frequencies at each locus and for each population. Single locus P values were calculated using a Markov chain with 1000 batches and 1000 iterations per batch, combined over loci using the Fisher method.

The existence of population structure was also assessed by one level AMOVA (Excoffier et al. 1992) using ARLEQUIN ver. 3.1 (Excoffier et al. 2005). Significance Φ_{ST} values were computed by permutation tests from 10,000 random permutations. Additional AMOVAs with grouped populations were performed on the basis of the results of pairwise Φ_{ST} comparisons, thus testing specific biogeographic hypotheses.

Finally, we also applied the Pearson's χ^2 method as implemented in CHIFISH (Ryman 2006). Correlations between genetic (Φ_{ST} values) and geographical (calculated as effective distance by sea) distances were evaluated by the Mantel test, as implemented in ARLEQUIN, with 20,000 random permutations.

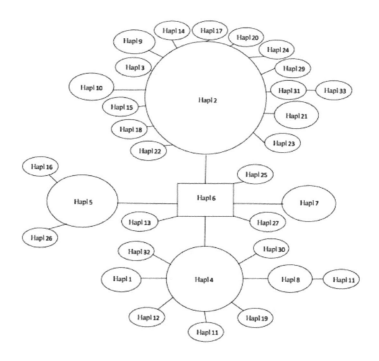

Figure 2. Minimum spanning network showing the relationships among haplotypes in *Pachygrapsus marmoratus* recorded in the western Mediterranean Sea and eastern Atlantic Ocean. Each line between two points represents one mutational step. Circles denote haplotypes that are scaled to their frequencies. The square represents the postulated ancestral haplotype.

3   RESULTS

3.1   Mitochondrial DNA data

In our mtDNA dataset, with a length of 596 basepairs, we found 29 variable and 9 parsimony-informative sites, resulting in a total of 33 haplotypes. All but two mutations are silent (Table 2). The fragment nucleotide composition is unbalanced, as expected for invertebrate mtDNA (Simon et al. 1994), being $C = 21\%$, $T = 35.9\%$, $A = 25.1\%$, and $G = 18\%$. The haplotype and nucleotide diversities for each population and for the whole sample are reported in Table 3: all populations except three (Porto Ercole (PE), Valencia (V), and Sesimbra (S)) are characterized by haplotype diversity values greater than 0.5; while all populations without any exception have nucleotide diversity values lower than 0.35.

Most of the haplotypes differ by few mutations, corresponding to an average nucleotide diversity of about 0.22. Overall, the haplotype network has a star-like shape, centered around two main haplotypes, haplotypes 2 and 4 (Figure 2). Only haplotype 2, found in 136 individuals, is present in all the populations (Figure 2 and Table 4). Haplotype 4, found in 36 individuals is lacking in the Portuguese population Sesimbra (S in Table 4). The proposed ancestral haplotype is haplotype 6, that connects haplotypes 2 and 4 (Figure 2). Twenty-two haplotypes are rare, being present in 1–2 individuals of a single population (Table 4).

The AMOVA showed lack of population genetic structure among populations ($\Phi_{ST} = 0.023$, $P = 0.054$, from sequence divergence data, calculated as the Tajima-Nei distance, as suggested for unbalanced nucleotide composition; $\Phi_{ST} = 0.019$, $P = 0.062$, from haplotypic frequencies) as well as the exact test of differentiation based on haplotype frequencies ($P = 0.18 \pm 0.04$). Notwithstanding, some pairwise Φ_{ST} comparisons are significant, mainly those involving the Atlantic population S from Portugal (Table 5) and the Mediterranean population Messina (M in Table 5). Otherwise, the contingency χ^2 tables of haplotype frequencies reveal highly significant differentiation among all populations ($\chi^2 = 548.28$, $df = 448$, $P < 0.001$), which seems to indicate that genetic separation between our localities is occurring, although it is recent and based on rare haplotypes. Based on this test, almost all pairwise comparisons were significant (data not shown); which is due to the high number of rare alleles present exclusively in single populations.

The population demographic history was reconstructed using the mismatch distribution analysis and three neutrality tests, for each population separately and for the entire sample. All separate populations as well as the whole sample (Figures 3 and 4) have a mismatch distribution associated with a typical sudden population expansion (Rogers & Harpending 1992; Patarnello et al. 2007). The value of the raggedness index for each population and for the whole sample is reported in Table 3; for all populations, except three (Gaeta (G), Fusano Lake (FL), and Messina (M)). This index does not discount the null hypothesis of a population out of equilibrium, in accordance with the shape of the curve. All the applied neutrality tests also revealed significant deviations from mutation-drift equilibrium for the whole sample (Table 3). Moreover, a total of nine populations (S by the Tajima D test; Rochette (R), Montecristo Island (MI), G, St. Florent (SF), Tarragona (T), Cala Iris (CI) by the Fu's F_S test; and G, Crotone (CR), T, CI, and Mattoral, Fuerteventura (MF) by R_2 test) seem to have experienced a recent event of population expansion (Table 3).

Under the assumption of a sudden demographic expansion of a population, the value of τ for the whole sample is 2.27 (0.08–4.81 for a 95% confidence interval), and the value of Theta 1 is equal to 2.65. Thus, assuming a COI mutation rate of 1.66% per million years (Schubart et al. 1998), the last population expansion began at around 0.7 Mya. Under the spatial expansion hypothesis, the value of τ decreases to 1.97, corresponding to an expansion time of approximately 0.6 Mya.

3.2   Microsatellite data

All loci were highly polymorphic, except the dimorphic locus pm-79. Each population showed relatively high levels of molecular variation, with allelic richness varying from 8.36 to 9.57 (Table 6).

Significant linkage disequilibrium was recorded across all populations only for the pairwise comparison between locus pm-183 and pm-99, due to a significant deviation recorded in the population PSS. Based on the original dataset, all populations except two (CV and PP) deviated from HWE, with an excess of homozygotes, mainly due to locus pm-101 (not in equilibrium in all populations except CV) and locus pm-99 (not in equilibrium in populations except in Golfo di Baratti (GB), Porto Santo Stefano (PSS), and Villa Domizia (VD)). For these two loci, the software MICROCHECKER suggested the presence of null alleles. After the creation of a new dataset corrected for null alleles, all populations across all loci as well as at each specific locus were in HWE (Table 6).

The hypothesis of genetic homogeneity among populations was consistently rejected by all statistical tests applied (Fisher's exact test: $\chi^2 = 29.96$, $df = 12$, $P = 0.003$; AMOVA test: $F_{ST} = 0.005$, $df = 591$, $P = 0.005$; Pearson's χ^2 test: $\chi^2 = 1152.5$, $df = 990$, $P = 0.0002$). Considering the significant pairwise population comparisons (Table 7), we performed additional AMOVAs with grouped populations on the basis of two biogeographic hypotheses: hypothesis-1 created with 3 groups, corresponding to northern (Port of Livorno (PL) + Calafuria (C)), central (PP + GB + Port of Follonica (PF) + CV) and southern (VD + PSS + LV + PE) Tuscan populations;

Table 4. Geographic distribution of the 33 haplotypes of *Pachygrapsus marmoratus* recorded at the 15 sampling sites within the western Mediterranean Sea and eastern Atlantic Ocean. See Table 1 for abbreviations of populations.

Abbr.	1	2	3	4	5	6	7	8	9	10	11	12	13	14	15	16	17	18	19	20	21	22	23	24	25	26	27	28	29	30	31	32	33
C	2	9	1	2	1	0	0	0	0	0	0	0	0	0	0	0	0	0	0	0	0	0	0	0	0	0	0	0	0	0	0	0	0
R	0	6	0	2	1	0	0	2	0	0	0	0	0	0	0	0	0	0	0	0	1	0	0	0	0	0	0	0	0	1	1	0	0
PE	0	11	0	3	0	1	0	0	0	0	0	0	0	0	0	0	0	0	0	0	0	0	0	0	0	0	0	0	0	0	0	0	0
GI	0	4	0	4	0	1	1	0	0	0	0	0	0	0	0	0	0	0	0	0	0	0	0	0	0	0	0	0	0	0	0	0	0
MI	0	6	0	1	1	1	1	1	0	2	1	0	1	0	0	0	0	0	0	0	0	0	0	0	0	0	0	0	0	0	0	0	0
G	0	7	0	3	0	0	0	0	1	0	0	1	0	0	0	1	0	0	0	0	0	0	0	1	1	1	1	0	0	0	0	0	0
FL	0	9	0	3	1	0	1	0	0	0	0	0	0	0	0	0	0	0	0	0	0	0	0	0	0	0	0	0	0	0	0	0	0
CR	0	11	0	2	0	0	1	1	0	0	0	0	0	0	0	0	0	0	0	0	0	0	0	0	0	0	0	0	0	0	0	0	1
M	0	6	0	4	1	1	2	0	0	0	0	0	0	1	1	0	0	0	1	1	0	0	0	0	0	0	0	0	0	0	0	1	0
SF	0	8	0	1	1	1	1	2	0	0	0	0	0	0	0	0	0	1	1	0	0	0	0	0	0	0	0	0	0	0	0	0	0
T	1	8	0	4	1	1	0	1	0	0	0	0	0	0	0	0	1	0	0	0	1	0	1	0	0	0	0	0	0	0	0	0	0
V	0	15	0	2	0	0	0	0	1	0	0	0	0	0	0	0	0	0	0	1	0	0	1	0	0	0	0	0	0	0	0	0	0
CI	0	10	0	1	1	1	0	0	0	0	0	0	0	0	0	0	0	0	0	0	1	0	0	0	0	0	0	0	0	0	0	0	0
S	0	16	0	0	1	0	0	0	0	0	0	0	0	0	0	0	0	0	0	0	0	1	0	0	0	0	0	0	0	0	0	0	0
MF	0	10	0	1	4	0	0	0	0	0	0	0	0	0	0	0	0	0	0	0	1	1	1	0	0	0	0	0	0	0	0	0	0
Overall	3	136	1	33	14	6	8	6	2	3	1	1	1	1	1	1	1	1	1	1	4	2	2	1	1	1	1	0	0	1	1	1	1

while hypothesis-2 split the central populations into two groups, northern-central (PP + GB) and southern-central (PF + CV) populations (Table 8). The two hypotheses do not consistently differ from each other in explaining the genetic separation among groups, even if hypothesis-2 records a lightly higher F_{CT} value than hypothesis-1.

Using the Mantel test, we found no relationship between pairwise genetic differentiation and geographic distance ($r = 0.04$, $P = 0.37$), meaning that the hypothesis of isolation-by-distance is not appropriate for explaining our data and other factors have to be considered in order to explain the recorded population differentiation.

4   DISCUSSION

This study investigates the population genetic structure of the rocky intertidal crab *P. marmoratus* at macro- to mesogeographic scales using a mtDNA marker, and at microgeographic scale using nuclear markers. The two different classes of molecular markers were applied to reconstruct different temporal and geographic scenarios. Heterogeneity in microsatellite allele frequency among populations separated by a few hundred kilometers is coupled with relatively homogeneous mtDNA haplotype composition among populations, which are separated by several thousands of kilometres and belong to distinct oceanic basins. In our opinion, these results reflect the different properties of the applied molecular markers, with mtDNA able to represent a deeper historical perspective in the description of population genetic structure compared to microsatellites, which are more suitable for the investigation of actual gene flow. We also contend that in this and other cases in the marine environment, mtDNA-derived dispersal estimates do not exclusively represent female dispersal, but most likely are representative of overall dispersal, since sex-biased dispersal is highly unlikely for planktonic larvae (see for example Barber et al. 2006). In our opinion, this study is therefore another example underlining the importance of mitochondrial data for investigating population history and the usefulness of highly variable microsatellite loci for recording present-day or very recent gene flow.

Based on the mtDNA dataset, the AMOVA test, the exact test of differentiation as well as the star-like shape of the network reveal high genetic similarities among western Mediterranean and eastern Atlantic populations of *P. marmoratus*. The larvae of this intertidal crab thus seem to disperse over long distances, forming a widespread metapopulation, at least in this area of its geographic distribution range. The Gibraltar Strait and the Almería-Oran Front therefore do not seem

Table 5. Pairwise Φ_{ST} values from sequence divergence data (below diagonal) and from haplotypic frequencies (above diagonal). Significant values are in bold (P threshold < 0.05). See Table 1 for abbreviations of populations.

	C	R	PE	GI	MI	G	FL	CR	M	SF	T	V	CI	S	MF
C	–	0.009	0.018	0.035	0.005	−0.016	−0.040	−0.026	0.013	−0.021	−0.012	−0.003	−0.031	0.087	−0.013
R	−0.009	–	−0.042	−0.024	−0.020	−0.027	−0.015	0.010	−0.020	−0.041	−0.028	0.059	0.012	**0.176**	−0.006
PE	−0.035	0.019	–	0.070	0.0663	0.013	−0.035	−0.042	0.056	0.009	0.041	−0.034	−0.033	0.052	0.039
GI	0.032	−0.026	0.117	–	0.010	−0.033	−0.000	0.058	−0.058	0.015	−0.034	**0.133**	0.075	**0.307**	0.073
MI	−0.030	−0.004	−0.021	0.020	–	−0.017	−0.000	0.021	−0.014	−0.0104	−0.0184	**0.078**	0.0144	**0.184**	0.004
G	−0.030	−0.009	−0.012	0.001	−0.036	–	−0.036	−0.003	−0.036	−0.013	−0.031	0.043	0.006	**0.169**	0.015
FL	−0.041	−0.009	−0.026	0.029	−0.041	−0.048	–	−0.040	−0.015	−0.021	−0.019	−0.007	−0.030	0.097	−0.011
CR	−0.031	0.008	−0.054	0.097	−0.029	−0.009	−0.022	–	0.040	−0.019	0.029	−0.033	−0.036	0.041	0.017
M	−0.009	−0.024	0.038	−0.055	−0.015	−0.033	−0.024	0.031	–	0.006	−0.041	**0.099**	0.042	**0.231**	0.030
SF	−0.030	−0.034	−0.036	0.044	−0.025	−0.025	−0.037	−0.040	−0.001	–	−0.009	0.023	−0.026	**0.111**	−0.014
T	−0.030	−0.016	0.0097	−0.025	−0.021	−0.0355	−0.036	0.016	−0.038	−0.009	–	0.075	0.0208	**0.186**	0.004
V	0.015	0.061	−0.029	**0.196**	0.029	0.048	0.031	−0.031	**0.110**	−0.003	0.078	–	−0.025	0.003	0.038
CI	0.011	0.077	−0.024	**0.205**	0.004	0.027	0.005	−0.022	**0.099**	−0.006	0.063	−0.023	–	0.031	−0.014
S	**0.131**	**0.208**	0.100	**0.411**	**0.117**	0.158	**0.137**	0.067	**0.250**	**0.101**	0.195	0.012	0.003	–	0.101
MF	0.017	0.063	0.016	**0.171**	0.010	0.0168	−0.013	0.016	**0.078**	0.002	0.045	0.028	−0.016	0.051	–

Table 6. Populations of *Pachygrapsus marmoratus* genotyped for the 6 microsatellites, each with mean allelic richness (N_a), the observed heterozygosity (H_o), the expected heterozygosity (H_e), and the P value of departure from the Hardy-Weinberg equilibrium (P) (after correction for null alleles). See Table 1 for abbreviations of populations.

Abbr.	N_a	H_o	H_e	P
PL	9.14	0.71	0.68	0.94
C	9.11	0.71	0.70	0.32
GB	9.30	0.71	0.73	0.47
PP	8.36	0.68	0.67	0.45
PF	9.27	0.77	0.73	0.05
CV	8.63	0.73	0.72	0.63
VD	9.57	0.69	0.69	0.83
PSS	8.99	0.71	0.70	0.55
LV	8.98	0.71	0.69	0.77
PE	9.34	0.70	0.70	0.70

to act as a pronounced barrier for gene flow in *P. marmoratus*, as recorded for other marine species with planktonic larval phases (see Patarnello et al. 2007). Nevertheless, a certain degree of separation of the Portuguese population from many Mediterranean populations is discernible (see also Silva et al. 2009 for genetic patchiness along the Portuguese coast). Specific life-history traits as well as historical factors may promote gene flow of these species across the Atlantic-Mediterranean transition (Patarnello et al. 2007). Furthermore, larger sample sizes of 25–30 individuals per population and more populations from the Atlantic coast may change our current view of gene flow patterns across the Strait of Gibraltar.

Other possible barriers to mtDNA gene flow, such as the Sicily Strait separating the Ionian Sea from the Tyrrhenian Sea, do not seem to interrupt the connectivity among populations within the Mediterranean Basin, even if the population from Messina, located on the Sicily Strait itself, is differentiated from some other Mediterranean populations. An increased sample size from the eastern Mediterranean and further investigations including nuclear markers should clarify the level of gene flow in this area.

The recorded pattern of mtDNA genetic variation of *P. marmoratus*, commonly found in marine species with good dispersal ability and large effective population sizes, is a clear indication of a recent population expansion (for a review see Avise 2000). Also the mismatch distribution analysis, the small value of t (2.27), and the neutrality tests confirm the expectations of a model of recent and sudden expansion, for the whole sample as well as for most of the studied populations. Moreover, the high number of private alleles coupled with a significant population differentiation by the contingency χ^2 tables of haplotype frequencies, as well as rather high haplotype diversities and low nucleotide diversities among haplotypes, is another residual effect of a recent evolutionary history.

Based on the demographic parameters of the mismatch distribution analysis, this presumed expansion event seems to have occurred less than 1 Mya, i.e., during the Mid-Pleistocene. This is a period characterized by strong climate cycling and sea-level fluctuations for the Mediterranean Sea (including sporadic isolation from the Atlantic Ocean), and it is known to have affected the genetic population structure of many marine species (see Hewitt 2000; Patarnello et al. 2007). The last demographic historical events affecting the distribution of genetic variation of *P. marmoratus* within the western Mediterranean Basin and across the Mediterranean-Atlantic transition thus seems to have occurred relatively recently.

Table 7. Pairwise comparisons of nuclear genetic differentiation, estimated from the pairwise F_{ST} values (below diagonal), and from the χ^2 exact test of differentiation (above diagonal). Significant values are in bold (P threshold < 0.05). See Table 1 for abbreviations of populations.

	PL	C	GB	PP	PF	CV	VD	PSS	LV	PE
PL	–	18.18	14.17	**21.76**	**25.76**	18.14	20.88	19.10	**21.97**	17.57
C	−0.000	–	9.50	5.44	19.45	15.70	14.53	10.55	10.37	10.90
GB	0.011	−0.003	–	13.16	15.50	8.56	15.14	12.37	11.74	17.27
PP	0.009	−0.007	0.004	–	**21.37**	15.59	**32.28**	**21.99**	14.88	11.26
PF	0.022	0.008	0.009	0.016	–	12.70	**38.75**	**31.12**	**20.95**	**29.18**
CV	0.008	0.002	0.003	0.001	0.007	–	17.11	**25.18**	18.10	15.55
VD	0.004	−0.003	−0.002	0.002	**0.017**	0.007	–	11.12	11.60	14.75
PSS	**0.008**	−0.001	0.002	0.003	**0.010**	0.007	−0.0032	–	12.17	17.90
LV	0.002	−0.004	0.003	−0.004	**0.009**	0.002	−0.0022	−0.0062	–	12.44
PE	0.007	−0.002	0.005	0.008	**0.021**	0.0072	−0.0002	−0.0002	0.0032	–

Overall, results obtained by the mitochondrial marker indicate that an equilibrium between gene flow, genetic drift, and mutation is rather unexpected, and that the genetic structure should be affected by population history (Palumbi 1994). Moreover, as a consequence of its low mutation rate, mtDNA may retain ancestral population polymorphisms. This means that the population genetic structure we recorded today by mtDNA may not reflect neither contemporary genetic exchange nor the larval dispersal abilities of the species. In contrast, due to the greater effective population size of nuclear alleles compared to mtDNA alleles, the effect of genetic drift is less evident at microsatellite loci, allowing a more realistic estimate of present-day gene flow. This may explain, at least partially, why at local geographic scales we recorded a lower level of gene flow than at meso- and macro-geographic scales.

In fact, along the Tuscan coast we found a moderate but clear partitioning of genetic variation in *P. marmoratus*, with the evidence of distinct geographic groups of populations, corresponding to northern, central (possibly also divided in north-central and central-southern) and southern Tuscany. As discussed in Fratini et al. (2008) from a biogeographic point of view, this genetic structure is coherent with the sea circulation occurring in this Mediterranean region. Notwithstanding, the reproductive biology of *P. marmoratus* (i.e., an r–strategy species with a planktonic larval phase lasting about one month), and the results obtained at a meso- and macrogeographic scale by mtDNA, could suggest a mixed metapopulation, widespread along this restricted tract of the Italian coast. Genetic separation among populations is not correlated with geographic distance, but it depends, at least to a degree, on the part of the Mediterranean Basin to which they belong to. In fact, northern populations belong to the Ligurian Basin, central-southern and southern populations to the Tyrrhenian Basin, while northern-central ones are placed at the boundaries between the two basins. The Ligurian and Tyrrhenian Sea are connected to each other by a northward current lapping the Italian coast. However, on the other side, they are known to be two ecologically and chemophysically distinct basins of the Mediterranean Sea, where bottom and surface currents have diverse directions and intensities (Millot 1987; Artale et al. 1994; Astraldi et al. 1994). Another feature affecting the interchange among populations seems to be the degree of communication with the open sea; all four central populations are, in fact, located within creeks (GB and CV) and small protected marinas (PP and PF). This should hamper the dispersal of local populations and, consequently, enhance their partial isolation.

An alternative possible explanation of the recorded genetic separation among populations may be the so-called "sweepstakes effect" (Hedgecock 1994), in which only few adults reproduce per population, i.e., a small proportion of the available gene pool contributes to the next generations.

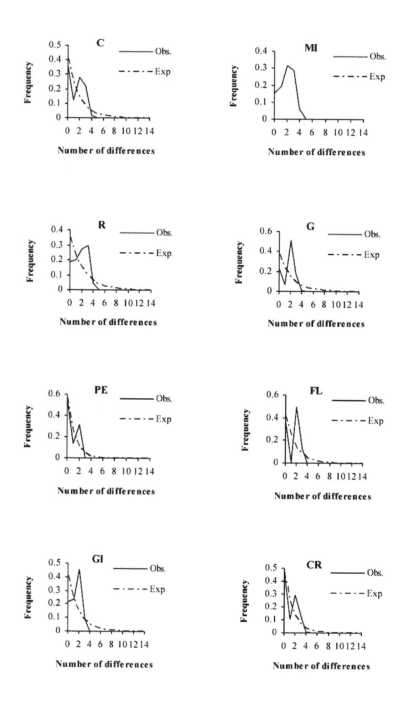

Figure 3. Observed mismatch distribution (dotted line) and expected curve (continous line) under a past population expansion model for each population and for the whole dataset. See Table 1 for abbreviations of populations.

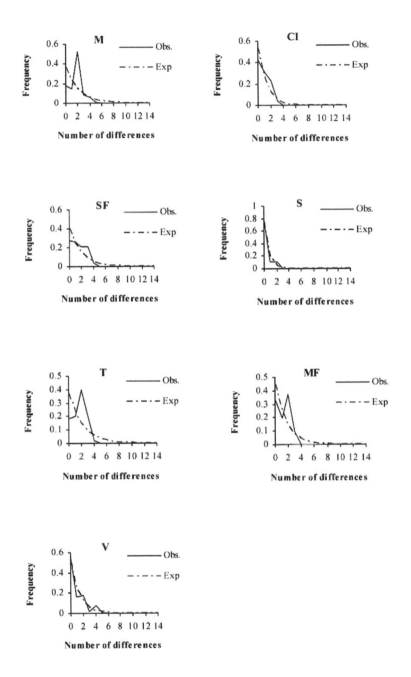

Figure 4. Observed mismatch distribution (dotted line) and expected curve (continous line) under a past population expansion for each population and for the whole dataset. See Table 1 for abbreviations of populations.

The sweepstakes effect predicts that, as a consequence of a small effective size, chaotic genetic patchiness in the composition of recruits may occur, creating spatial and temporal genetic variation (Hedgecock et al. 2007; Christie et al. 2010). Genetic comparisons of pre- and postsettlement larvae with adults should confirm this explanation.

A certain degree of genetic heterogeneity at a local scale was also recorded for Portuguese populations of *P. marmoratus* by Silva et al. (2009), using our same microsatellite set. The authors noted that gene flow was strong enough to interconnect populations as far as several hundred kilometers apart. On the other hand, they found a genetic separation of specific populations that was not linked to any geographic gradient, but followed a "chaotic genetic patchiness" (*sensu* Avise 2004). Independently from the differences in the patterns of population genetic structure found by us and by Silva et al. (2009), these two studies seem to generally indicate that fine dispersal and recruitment of *P. marmoratus* is presumably under the influence of local coastal hydrological systems.

In general, Tuscan populations show a good level of genetic variation (expressed as allelic richness, with a mean value equal to 9) and are in HWE. This seems to indicate that these populations are not imminently suffering from population reductions and loss of genetic diversity due to coastal erosion and flooding under a climate change scenario or due to the anthropogenic alteration and pollution of coastal waters, which unfortunately are affecting the Mediterranean Sea (Airoldi & Beck 2007). More in-depth analyses for assessing the effects of heavy metals on genetic diversity of *P. marmoratus* populations supported the "genetic erosion" hypothesis for contaminant exposure in natural environments, evidencing less genetic variability, measured as mean standardized $d2$, and a significantly lower percentage of unrelated individuals in populations from polluted sites than in those from unpolluted ones (Fratini et al. 2008).

5   Conclusions

The investigation of population genetic variation of marine species has nowadays been recognized to be a complex task, because of some peculiarities of species-specific biology, such as dispersal abilities and effective population size, as well as the presence of barriers in the marine realm, even if not evidently visible, created by chemophysical factors, currents and winds. As a consequence, the molecular markers for inferring the evolutionary processes in marine species have to be chosen rationally, keeping in mind the different molecular properties they exhibit and the evolutionary questions they are appropriate to detect.

In accordance with other authors (for example, Tessier et al. 1995; Shaw et al. 1999; Duran et al. 2004a, b; Roman & Palumbi 2004; Pascoal et al. 2009), our study clearly underlines that the two most common genetic markers, mtDNA and microsatellite loci, are not interchangeable in the study of connectivity among marine populations, and that they may give us a complementary perspective of the population dynamics and genetic structures. In highly dispersive species, mtDNA has proven to be particularly useful for inferring phylogeographic patterns and revealing historical events (bottlenecks, expansions, and speciation; Avise 2000), while microsatellites are the best approach for detecting fine-scale population structure (Shaw et al. 1999) and for facing the influence of ecological factors in shaping the contemporary partitioning of genetic variation (Gonzalez & Zardoya 2007). This is related to the different level of genetic variation of these two markers. Furthermore, microsatellite loci and mtDNA are inherited in different ways, and this permits to resolve questions of population structure for species with differential migration or dispersal rates between sexes. Finally, as a consequence of different effective population sizes, these two markers experience different effects of genetic drift and mutations.

Table 8. Analysis of molecular variance testing for distribution of genetic variation under two biogeographic hypotheses, using 6 microsatellite loci. Degrees of freedom (*df*), sum of squares (*SS*), variance, percentage of total variation, *F*–statistics and *P* values are reported. Significant *P* values are shown in bold. See Table 1 for abbreviations of populations.

Hypothesis	Source of variation	*df*	SS	Variance	Variance (%)	*F*–statistics	*P*
1) northern (PL, C) vs. central (GB, PP, PF, CV) vs. southern (PSS, VD, LV, PE)	Among groups	2	6.69	0.0054	0.26	$F_{CT} = 0.003$	**0.030**
	Among population/groups	7	16.20	0.0050	0.22	$F_{SC} = 0.002$	0.078
	Within populations	582	1188.10	2.04	99.51	$F_{ST} = 0.005$	**0.006**
2) northern (PL, C) vs. northern-central (GB, PP) vs. southern-central (PF, CV) vs. Southern (PSS, VD, LV, PE)	Among groups	3	9.89	0.0080	0.39	$F_{CT} = 0.005$	**0.010**
	Among population/groups	6	13.00	0.0020	0.10	$F_{SC} = -0.001$	0.213
	Within populations	582	1188.10	2.04	99.51	$F_{ST} = 0.005$	**0.005**

ACKNOWLEDGEMENTS

Our warm thanks are due to Carolina Biagi and the other students, who assisted us with crab collections and lab work. Mitochondrial DNA sequencing analysis and microsatellite sizing were performed at the C. I. B. I. A. C. I. (University of Florence); we thank Manule Goti of this center for his technical support. Crabs from Sesimbra were collected during an invitation of CDS by Jose Paula; southern Italian populations were collected during the German-Italian academic exchange program DAAD-Vigoni D/04/47157 or by our esteemed colleague Carlo LoPresti. Spanish populations were collected during the German-Spanish exchange program DAAD-Acciones Integradas D/03/40344. Research was funded by the "Cassa di Risparmio di Firenze" and "Monte dei Paschi di Siena" foundations, and by Fondi dAteneo to S. Cannicci and M. Vannini (University of Florence), whom we also kindly thank for their invaluable scientific comments. The manuscript profited from comments by two anonymous referees and the subject editors of this volume of *Crustacean Issues*.

REFERENCES

Airoldi, L. & Beck, M.W. 2007. Loss, status and trends for coastal marine habitats of Europe. *Oceanogr. Mar. Biol. Annu. Rev.* 45: 345–405.

Artale, V., Astraldi, M., Buffoni, G. & Gasparini, G.P. 1994. Seasonal variability of the gyre-scale circulation in the northern Tyrrhenian Sea. *J. Geophys. Res.* 99 (C7): 14127–14137.

Astraldi, M., Gasparini, G.P. & Sparnocchia, S. 1994. The seasonal and interannual variability in the Ligurian-Provençal Basin. *Coast. Estuar. Stud.* 46: 93–113.

Avise, J.C. 2000. *Phylogeography: The History and Formation of Species.* Cambridge, MA: Harvard University Press.

Avise, J.C. 2004. *Molecular Markers, Natural History, and Evolution. 2nd ed.* Sunderland, MA: Sinauer Associates.

Ballard, J.W.O. & Whitlock, M.C. 2004. The incomplete natural history of mitochondria. *Mol. Ecol.* 13: 729–744.

Barber, P.H., Erdmann M.V & Palumbi, S.R. 2006. Comparative phylogeography of three codistributed stomatopods: origins and timing of regional lineage diversification in the coral triangle. *Evolution* 60: 1825–1839.

Brookfield, J.F.Y. 1996. A simple new method for estimating null allele frequency from heterozygote deficiency. *Mol. Ecol.* 5: 453–455.

Cannicci, S., Gomei, M., Boddi, B. & Vannini, M. 2002. Feeding habits and natural diet of the intertidal crab *Pachygrapsus marmoratus*: opportunistic browser or selective feeder? *Estuar. Coast. Shelf Sci.* 54: 983–1001.

Cannicci, S., Paula, J. & Vannini, M. 1999. Activity pattern and spatial strategy in *Pachygrapsus marmoratus* (Decapoda: Grapsidae) from Mediterranean and Atlantic shores. *Mar. Biol.* 133: 429–435.

Chapuis, M.-P. & Estoup, A. 2007. Microsatellite null alleles and estimation of population differentiation. *Mol. Biol. Evol.* 24: 621–631.

Christie, M.R., Johnson, D.W., Stalling, C.D. & Hixon, M.A. 2010. Self-recruitment and sweepstakes reproduction amid extensive gene flow in a coral-reef fish. *Mol. Ecol.* 19: 1042–1057.

Clement, M., Posada, D. & Crandall, K.A. 2000. TCS: a computer program to estimate gene genealogies. *Mol. Ecol.* 9: 1657–1659.

Cuesta, J.A. & Rodríguez, A. 2000. Zoeal stages of the intertidal crab *Pachygrapsus marmoratus* (Fabricius, 1787) (Brachyura, Grapsidae) reared in the laboratory. *Hydrobiologia* 436: 119–130.

Dauvin, J.C. 2009. New record of the marbled crab *Pachygrapsus marmoratus* (Crustacea: Brachyura:

Grapsoidea) on the coast of northern Cotentin, Normandy, western English Channel. *Mar. Biodiv. Rec.* 2: e92.

Duran, S., Pascual, M., Estoup A. & Turon, X. 2004a. Strong population structure in the marine sponge *Crambe crambe* (Poecilosclerida) as revealed by microsatellite markers. *Mol. Ecol.* 13: 511–522.

Duran, S., Pascual, M. & Turon, X. 2004b. Low levels of genetic variation in mtDNA sequences over the western Mediterranean and Atlantic range of the sponge *Crambe crambe* (Poecilosclerida). *Mar. Biol.* 144: 31–35.

Excoffier, L., Laval, G. & Schneider, S. 2005. Arlequin (version 3.0): an integrated software package for population genetics data analysis. *Evol. Bioinform. Online* 1: 47–50.

Excoffier, L., Smouse, P.E. & Quattro, J.M. 1992. Analysis of molecular variance inferred from metric distances among DNA haplotypes: application to human mitochondrial DNA restriction data. *Genetics* 131: 479–491.

Fabricius, J.C. 1787. *Mantissa Insectorum sistens eorum Species nuper detectas adiectis Characteribus genericus, Differentiis specificus, Emendationibus, Observationibus. Tom. I.* Hafniae: Christ. Gottl. Proft.

Flores, A.A.V., Cruz, J. & Paula, J. 2002. Temporal and spatial patterns of settlement of brachyuran crab megalopae at a rocky coast in central Portugal. *Mar. Ecol. Progr. Ser.* 229: 207–220.

Folmer, O., Black, M., Hoeh, W., Lutz, R. & Vrijenhoek, R. 1994. DNA primers for amplification of mitochondrial cytochrome c oxidase subunit I from diverse metazoan invertebrates. *Mol. Mar. Biol. Biotechnol.* 3: 294–299.

Fratini, S., Ragionieri, L., Papetti, C., Pitruzzella, G., Rorandelli, R., Barbaresi, S. & Zane, L. 2006. Isolation and characterization of microsatellites in *Pachygrapsus marmoratus* (Grapsidae; Decapoda; Brachyura). *Mol. Ecol. Notes*: 6: 179–181.

Fratini, S., Ragionieri, L., Zane, L., Vannini, M. & Cannicci, S. 2008. Relationship between heavy metal accumulation and genetic variability decrease in the intertidal crab *Pachygrapsus marmoratus* (Decapoda; Grapsidae). *Estuar. Coast. Shelf Sci.* 79: 679–686.

Fu, Y.-X. 1997. Statistical tests of neutrality of mutations against population growth, hitchhiking and background selection. *Genetics* 147: 915–925.

Fu, Y.-X. & Li, W.-H. 1993. Statistical tests of neutrality of mutations. *Genetics* 133: 693–709.

Gonzalez, E.G. & Zardoya, R. 2007. Relative role of life-history traits and historical factors in shaping genetic population structure of sardines (*Sardina pilchardus*). *BMC Evol. Biol.* 7: 197.

Goudet, J. 1995. FSTAT (version 1.2): a computer program to calculate F-statistics. *J. Hered.* 86: 485–486.

Hall, T.A. 1999. BioEdit: a user-friendly biological sequence alignment editor and analysis program for Windows 95/98/NT. *Nucleic Acids Symp. Ser.* 41: 95–98.

Hauser, L., Carvalho, G.R. & Pitcher, T.J. 1998. Genetic population structure in the Lake Tanganyika sardine, *Limnothrissa miodon. J. Fish Biol.* 53: 413–429.

Hedgecock, D. 1986. Is gene flow from pelagic larval dispersal important in the adaptation and evolution of marine invertebrates? *Bull. Mar. Sci.* 39: 550–564.

Hedgecock, D. 1994. Does variance in reproductive success limit effective population size of marine organisms? In: Beaumont, A.R. (ed.), *Genetics and Evolution of Aquatic Organism*: 122–135. London: Chapman and Hall.

Hedgecock, D., Launey, S., Pudovkin, A.I., Naciri, Y., Lapègue, S. & Bonhomme, F. 2007. Small effective number of parents (Nb) inferred for a naturally spawned cohort of juvenile European flat oysters *Ostrea edulis. Mar. Biol.* 150: 1173–1182.

Hewitt, G.M. 2000. The genetic legacy of the Quaternary ice ages. *Nature* 405: 907–913.

Ingle, R.W. 1980. *British Crabs*. London: British Museum (Natural History), Oxford University

Press.

Korres, G., Pinardi, N. & Lascaratos, A. 2000. The ocean response to low frequency interannual atmospheric variability in the Mediterranean Sea. Part I: sensitivity experiments and energy analysis. *J. Climate* 13: 705–731.

Lascaratos, A. & Nittis, K. 1998. A high-resolution three-dimensional numerical study of intermediate water formation in the Levantine Sea. *J. Geophys. Res.* 103: 18497–18511.

Lascaratos, A., Roether, W., Nittis, K. & Klein, B. 1999. Recent changes in deep water formation and spreading in the eastern Mediterranean Sea. *Prog. Oceanogr.* 44: 5–36.

Leese, F., Mayer, C. & Held, C. 2008. Isolation of microsatellites from unknown genomes using known genomes as enrichment templates. *Limnol. Oceanogr. Meth.* 6: 412–426.

Li, W.-H. 1977. Distribution of nucleotide differences between two randomly chosen cistrons in a finite population. *Genetics* 85: 331–337.

Librado, P. & Rozas, J. 2009. DnaSP v5: a software for comprehensive analysis of DNA polymorphism. *Bioinformatics* 25: 1451–1452.

Millot, C. 1987. Circulation in the western Mediterranean Sea. *Oceanol. Acta* 10: 143–149.

Nei, M. 1987. *Molecular Evolutionary Genetics*. New York, NY: Columbia University Press.

Palumbi, S.R. 1994. Genetic divergence, reproductive isolation, and marine speciation. *Annu. Rev. Ecol. Evol. Syst.* 25: 547–572.

Pascoal, S., Creer, S., Taylor, M.I., Queiroga, H., Carvalho, G, & Mendo, S. 2009. Development and application of microsatellites in *Carcinus maenas*: genetic differentiation between northern and central Portuguese populations. *PLoS ONE* 4: e7268.

Patarnello, T., Volckaert, F.A.M. & Castilho, R. 2007. Pillars of Hercules: is the Atlantic-Mediterranean transition a phylogeographical break? *Mol. Ecol.* 16: 4426–4444.

Queller, D.C., Strassman, J.E. & Hughes, C.R. 1993. Microsatellites and kinship. *Trends Ecol. Evol.* 8: 285–288.

Ramírez-Soriano, A., Ramos-Onsins, S.E., Rozas, J., Calafell, F. & Navarro, A. 2008. Statistical power analysis of neutrality tests under demographic expansions, contractions and bottlenecks with recombination. *Genetics* 179: 555–567.

Ramos-Onsins, S.E. & Rozas, J. 2002. Statistical properties of new neutrality tests against population growth. *Mol. Biol. Evol.* 19: 2092–2100.

Raymond, M. & Rousset, F. 1995. GENEPOP (version 1.2): population genetics software for exact tests and ecumenicism. *J. Hered.* 86: 248–249.

Reuschel, S., Cuesta, J.A. & Schubart, C.D. 2010. Marine biogeographic boundaries and human introduction along the European coast revealed by phylogeography of the prawn *Palaemon elegans*. *Mol. Phylogenet. Evol.* 55: 765–775.

Rolf, D.A. & Bentzen, P. 1989. The statistical analysis of mitochondrial DNA polymorphisms: χ^2 and the problem of small samples. *Mol. Biol. Evol.* 6: 539–545.

Rogers, A.R. 1995. Genetic evidence for a Pleistocene population explosion. *Evolution* 49: 608–615.

Rogers, A.R. & Harpending, H. 1992. Population growth makes waves in the distribution of pairwise genetic differences. *Mol. Biol. Evol.* 9: 552–569.

Roman, J. & Palumbi, S.R. 2004. A global invader at home: population structure of the green crab, *Carcinus maenas*, in Europe. *Mol. Ecol.* 13: 2891–2898.

Rousset, F. 2004. *Genetic Structure and Selection in Subdivided Populations*. Princeton, NJ: Princeton University Press.

Ryman, N. 2006. CHIFISH: a computer program testing for genetic heterogeneity at multiple loci using chi-square and Fisher's exact test. *Mol. Ecol. Notes* 6: 285–287.

Schneider, S. & Excoffier, L. 1999. Estimation of past demographic parameters from the distribution of pairwise differences when the mutation rates vary among sites: application to human mitochondrial DNA. *Genetics* 152: 1079–1089.

Schubart, C.D., Diesel, R. & Hedges, S.B. 1998. Rapid evolution to terrestrial life in Jamaican crabs. *Nature* 393: 363–365.

Schubart, C.D. & Huber, M.G.J. 2006. Genetic comparisons of German populations of the stone crayfish, *Austropotamobius torrentium* (Crustacea: Astacidae). *Bull. Fr. Pêche Piscic.* 380–381: 1019–1028.

Shaw, P.W., Turan, C., Wright, J.M., OConnell, M & Carvalho, G.R. 1999. Microsatellite DNA analysis of population structure in Atlantic herring (*Clupea harengus*), with direct comparison to allozyme and mtDNA RFLP analyses. *Heredity* 83: 490–499.

Silva, I.C., Mesquita, N., Schubart, C.D., Alves, M.J. & Paula, J. 2009. Genetic patchiness of the shore crab *Pachygrapsus marmoratus* along the Portuguese coast. *J. Exp. Mar. Biol. Ecol.* 378: 50–57.

Simon, C., Frati, F., Beckenbach, A., Crespi, B., Liu, H. & Flook, P. 1994. Evolution, weighting, and phylogenetic utility of mitochondrial gene sequences and a compilation of conserved polymerase chain reaction primers. *Ann. Entomol. Soc. Am.* 87: 651–701.

Skliris, N., Sofianos, S. & Lascaratos, A. 2007. Hydrological changes in the Mediterranean Sea in relation to changes in the freshwater budget: a numerical modelling study. *J. Mar. Sys.* 65: 400–416.

Tajima, F. 1989. Statistical method for testing the neutral mutation hypothesis by DNA polymorphism. *Genetics* 123: 585–595.

Tajima, F. & Nei, M., 1984. Estimation of evolutionary distance between nucleotide sequences. *Mol. Biol. Evol.* 1: 269–285.

Tautz, D. 1989. Hypervariability of simple sequences as a general source for polymorphic DNA markers. *Nucleic Acids Res.* 17: 6463–6471.

Tessier, N., Bernatchez, L., Presa, P. & Angers, B. 1995. Gene diversity analysis of mitochondrial DNA, microsatellites, and allozymes in landlocked Atlantic salmon. *J. Fish Biol.* 47: 156–163.

Tóth, G., Gáspári, Z. & Jurka, J. 2000. Microsatellites in different eukaryotic genomes: survey and analysis. *Genome Res.* 10: 967–981.

van Oosterhout, C., Hutchinson, W.F., Wills, D.P.M. & Shipley, P. 2004. MICRO-CHECKER: software for identifying and correcting genotyping errors in microsatellite data. *Mol. Ecol. Notes* 4: 535–538.

Weersing, K. & Toonen, R.J. 2009. Population genetics, larval dispersal, and connectivity in marine systems. *Mar. Ecol. Progr. Ser.* 393: 1–12.

Zane, L., Bargelloni, L. & Patarnello, T. 2002. Strategies for microsatellites isolation: a review. *Mol. Ecol.* 11: 1–16.

Zariquiey Álvarez, R. 1968. Crustáceos decápodos Ibéricos. *Inv. Pesq.* 32: 1–510.

Zaykin, D.V. & Pudovkin, A.I. 1993. Two programs to estimate significance of χ^2 values using pseudo-probability tests. *J. Hered.* 84: 152.

III

Population genetics and phylogeography of limnic crustaceans

The history of the *Daphnia pulex* complex: asexuality, hybridization, and polyploidy

FRANCE DUFRESNE

Département de Biologie, Centre d'Études Nordiques, Université du Québec à Rimouski, Rimouski, Canada

ABSTRACT

Water fleas of the genus *Daphnia* (Crustacea: Cladocera) are key components of freshwater ecosystems, being primary consumers of algae and predated by fish and as such, have been extensively studied in the fields of limnology (Peters & de Bernardi 1987) and general ecology (Kalff 2002). In addition to their important ecological role, their wide geographic distribution, phenotypic variation, and peculiar reproductive mode (facultative parthenogenesis) have rendered them important models in evolutionary studies of speciation and adaptation (Schwenk et al. 2004). Extensive genetic surveys have revealed that the once thought cosmopolitan *Daphnia* species represent complexes of cryptic species with limited geographical distribution (Colbourne et al. 1998; Taylor et al. 1998). Fifteen different complexes were identified upon first sequencing 32 members of the genus *Daphnia* at the 16S mitochondrial gene (Colbourne & Hebert 1996). In this chapter, I will focus on one of these complexes: the *Daphnia pulex* complex that includes two major clades (the *Daphnia pulicaria* and the *Daphnia tenebrosa* clades) divided into ten divergent mitochondrial lineages (Colbourne et al. 1998). In North America, diploid *D. pulex* show a longitudinal gradient in their mode of reproduction with obligately asexual populations in eastern (temperate, subarctic, and arctic) Canada, mixed populations in Ontario, and cyclically parthenogenetic populations in western Canada (Hebert & Crease 1983; Hebert et al. 1993). Transitions to obligate parthenogenesis are thought to incur due to meiosis-suppressor genes that disrupt meiosis in females but not in males (Innes & Hebert 1988). As a result, males carrying these genes spread the "asexuality genes" in a contagious fashion. Meiosis-suppressor genes are thought to have arisen in an eastern *D. pulex* lineage and, hence, may explain the geographical distribution of obligate parthenogenesis in this species (Paland et al. 2005). In addition, glacial advances and retreat have provoked, through isolation in refugia and secondary contacts, hybridization and polyploidizations of numerous lineages of the complex in subarctic and arctic areas (Dufresne & Hebert 1997). Sharp phylogeographical patterns have been found in these young arctic lineages (Weider et al. 1999a; Marková et al. 2007) suggesting that historical processes have contributed significantly to the current distribution of these lineages. The recent sequencing of the *Daphnia pulex* genome will provide additional tools that will greatly contribute to our understanding of speciation.

1 GENERAL CHARACTERISTICS OF THE GENUS *DAPHNIA*

Daphnia are planktonic crustaceans that belong to the class Branchiopoda characterized by attened leaf-like legs used to produce a water current for the filtering apparatus (Figure 1). Branchiopoda comes from the Greek: branchia for gills and pous for foot. The presence of gills on many of the animal's appendages, including some of the mouthparts is responsible for the name of the group.

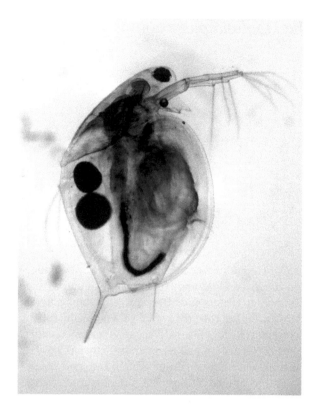

Figure 1. Photograph of *Daphnia tenebrosa* (lateral view).

Daphnia occupy most types of standing freshwater habitats ranging from small ephemeral pools to large lakes. They produce resistant diapausing stages that act as effective agents for long-term dispersal and hence are widely distributed on all continents. The cladoceran *Daphnia* reproduce by cyclic parthenogenesis, the ancestral breeding system of the genus *Daphnia* (Figure 2). This reproductive system is unique in that it consists of alternations between clonal and sexual reproduction (cyclic is not the best term to characterize this reproductive mode as clonal reproduction can be interrupted at any time by sexual reproduction). Cyclic parthenogens are thought to make the best of both worlds in that they avoid the costs of sexual reproduction and they exhibit enhanced efficiency of selection as compared to obligate parthenogens (Lynch & Gabriel 1983; Lynch 1984a). During the summer, females make direct-developing (subitaneaous) apomictic eggs that develop into miniature adults. This enables *Daphnia* to rapidly exploit the water column. In the fall, females produce haploid, diapausing eggs that require fertilization in order for eggs to be released into the developing ephippium. Male production is triggered upon a complex set of stimuli: increased competition, reduced food availability, and decreased day length. There are no sex chromosomes and sex is determined by complex genotype and environmental interactions (Innes & Dunbrack 1993; Innes & Singleton 2000). Populations are re-established from diapausing eggs in the spring. These diapausing eggs are capable of surviving harsh conditions and remain viable for a long time, forming resting egg banks in the sediments (Cáceres 1999). These resting stages are easily dispersed through migrating vectors such as birds since they are either attached or survive gut passage (Proctor 1964). This high dispersal potential is reflected in the rapid colonization of new habitats (Louette & De Meester 2004).

In some populations, females release diploid unreduced diapausing eggs into the ephippium (Crease & Hebert 1983). Male production is no longer required but some of these obligate parthenogens have retained the ability to produce males and these can mate with cyclical parthenogenetic females and generate new obligately parthenogenetic clones (Innes & Hebert 1988).

2   VARIATION IN PLOIDY LEVEL

Genetic variation within species is typically measured using molecular markers based on the variation of DNA sequence among homologous fragments of DNA in different specimens. Considerable insight on population genetic structure and speciation has been gained using this information. One less common measure of genetic variation is variation in chromosome number that arises as a result of whole genome duplications and/or hybridization. These changes in the number of chromosome sets (polyploidy) are often accompanied by changes in phenotypic characters and hence represent variation upon which selection can act. Furthermore, transitions to obligate parthenogenesis are frequently associated with these changes in ploidy level. Despite the predominance of polyploidy in nature, we still do not know very much about its evolutionary significance. In this chapter, I will focus on this type of genetic variation using the *Daphnia pulex* complex as a model system.

3   GENERAL VIEWS ON SPECIATION IN *DAPHNIA*

Early views on speciation in the plankton predicted that allopatric speciation would be rare in this group due to the high potential for long-distance dispersal (Mayr 1963). Previous work on the genus *Daphnia*, based solely on morphological characters, concluded that it consisted of a small number of taxa with wide distribution (Brooks 1957). The advent of molecular markers has radically changed our views on speciation in this genus showing that cosmopolitan species were in fact assemblages of genetically divergent taxa with narrow distributions (Frey 1987; Hebert & Wilson 1994; Hebert & Finston 1996). Therefore the long-standing concept of speciation shifted from cosmopolitanism

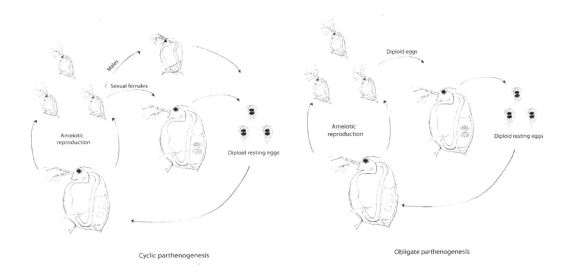

Figure 2. Reproduction in *Daphnia*. In cyclic parthenogenesis, the resting eggs are produced sexually whereas in obligate parthenogenesis, resting eggs are produced clonally.

to provincialism. The most common speciation mode in Cladocera is vicariant speciation where a physical barrier impedes gene flow and the separated populations evolve independently (Hebert & Wilson 1994). A recent study estimated that allopatric speciation was responsible for 42% of clado-genetic events among *Daphnia* species (Adamowicz et al. 2009). A second important speciation model developed by Lynch (1985) is the habitat shift model where lake *Daphnia* invade ponds and are selected for a pond life style. Concomitantly, drift operates since effective population size is reduced in ponds thus facilitating the speciation process. Many closely related species pairs occupying pond/lake habitats are known and might have originated following this model, e.g., *Daphnia pulex/Daphnia pulicaria*, and *Daphnia minnehaha/Daphnia catawba*. Hybrid speciation is also an important speciation mode in *Daphnia* with the formation of stabilized hybrid recombinants reproductively isolated from their parents. This is best exemplified by the zooplankter *Daphnia galeata* that appears to be a natural introgressant species with *Daphnia dentifera* throughout its North America and Japan range (Taylor et al. 2005). An extension of this model is the formation of polyploid species through interactions between asexual and sexual taxa and/or hybridization (ex. *D. pulex* complex see below).

4   GENETIC DIVERGENCE AT LOCAL, REGIONAL, AND CONTINENTAL SCALES

The first molecular studies using allozyme electrophoresis revealed large gene frequency differences (F_{ST} often greater than 0.2 and up to 0.7) in *Daphnia* from neighboring populations (Hebert 1974a) as a result of residual founder events (Boileau et al. 1992). By contrast, little macrogeographical differentiation of allozyme frequencies was found among geographical regions (Hebert et al. 1993). Similar findings were revealed using mtDNA and microsatellite markers despite large differences in mutation rates, dispersal probabilities, and modes of inheritance of these markers (Hebert et al. 1993; Vanoverbeke & De Meester 1997; Pálsson 2000; Ishida & Taylor 2007; Thielsch et al. 2009). No significant associations between genetic differentiation and geographical distances were detected (Pálsson 2000; Thielsch et al. 2009). This lack of differentiation on a macrogeographical scale has been attributed to the long-distance dispersal by birds.

De Meester (1996a) pointed to an apparent paradox in *Daphnia* that display extensive local adaptation despite evidence for high dispersal capability and rapid colonization of new habitats. This apparent paradox can be explained by combining the persistent founder effects proposed by Boileau et al. (1992) with a relative advantage of the resident population in the so-called Monopolization Hypothesis. This hypothesis takes into account the high population growth rates and the large resting egg banks of cyclic parthenogens. The first colonizers rapidly monopolize the habitat, due to these characteristics, preventing secondary colonizers from establishing themselves in the ponds. As a result, gene flow is restricted and local adaptation of the first settlers reinforced. Some *Daphnia* inhabiting small rock pools on subarctic Finnish islands exhibit a metapopulation structure characterized by high rates of extinction/recolonization. Strong genetic differentiation has been found in *D. magna* and *Daphnia longispina* populations within and among these small islands (Haag et al. 2006). Frequent bottlenecks and inbreeding during colonization coupled with a rapid size increase have been invoked for this pattern of genetic differentiation.

Vicariant events have acted as barriers to gene flow. For example, the Appalachian mountain range in North America has reduced gene flow between populations of *Daphnia laevis* (Taylor et al. 1998), *Daphnia ambigua* (Hebert et al. 2003), and *Sida crystallina* (Cox & Hebert 2001). Pleistocene glaciations have had dramatic effects on the genetic architecture of many North American and European species. Some species show stronger genetic signatures than others perhaps as a result of a complex interplay between gene flow, priority effects, and niche requirements. Very high levels of genetic divergence (4–7% at cytochrome c oxidase subunit I) were found among four phylogeographic assemblages corresponding to known refugia (Atlantic, Pacific, Mississippian, and

Beringian) in the cladoceran *Sida crystallina americana* (Cox & Hebert 2001). Genetic divergences were slightly more modest among the three phylogroups of *Daphnia ambigua* (2–4%) (Hebert et al. 2003) and the four phylogroups of *Daphnia obtusa* (1–2%) (Penton et al. 2004), and the three phylogroups of *Daphnia laevis* (Taylor et al. 1998). A survey of 75 European populations of *D. magna* revealed the existence of four phylogroups with very shallow genetic structure (De Gelas & De Meester 2005).

Daphnia history includes many intercontinental dispersal events and studies using molecular markers have helped clarify both the direction and timing of these events. Studies on the cladocerans *Holopedium*, *Polyphemus*, and *Leptodora* have revealed enough genetic divergence between populations from temperate North America and Europe to suggest that they have not exchanged genes for at least 5 million years (Hebert & Cristescu 2002). Populations of *Sida crystallina* from North America and Europe show even deeper genetic divergence suggesting an isolation time of 10–11 million years (Cox & Hebert 2001). A large-scale phylogenetic analysis has revealed that there have been at least 15 recent intercontinental dispersal events among 12 *Daphnia* species with multicontinental distribution (Adamowicz et al. 2009). Most of these intercontinental dispersion events are associated with divergence of less than 5% at the cytochrome c oxidase subunit I (COI) gene thus indicating that they are not the result of old vicariance but to more recent dispersal (less than 2.5 million years). Curiously, whereas some polar copepods have failed to recolonize large areas of the Canadian arctic in the 7000 years since deglaciation (Hebert & Hann 1986) other polar taxa, such as members of the *Daphnia pulex* complex have experienced recent exchanges as signalled by the low genetic divergence among some European and North American lineages (Colbourne et al. 1998; Weider et al. 1999a; Marková et al. 2007).

Finally, intercontinental transfers mediated by humans have allowed the rapid and successful spread of African *D. lumholtzi* in North America following co-introduction with the Nile perch (Havel & Hebert 1993). North American *D. pulex* has recently been found in Kenya (Mergeay et al. 2006). The low clonal diversity at microsatellite loci coupled with the high genetic similarities at the mitochondrial gene ND5 (NADH dehydrogenase) strongly suggest a recent human introduction.

5 EFFECTS OF CYCLIC PARTHENOGENESIS ON PATTERNS OF GENETIC VARIATION

Reproductive mode is typically deciphered using molecular markers as the sole presence of males is inconclusive. Hebert et al. (1993) used three parameters to determine if *Daphnia* populations reproduce by obligate or by cyclic parthenogenesis: the number of multilocus genotypes identified, the log-transformed probability of the observed genotypic distribution at polymorphic loci being in the Hardy-Weinberg equilibrium, and the genotypic diversity ratio (GDR—the observed number of multilocus genotypes divided by the expected number in a sample of the same size and with the same gene frequencies from a sexual population). These earlier studies were based on allozyme loci. The advent of DNA-based markers, such as microsatellites, now enable the genotyping of the two eggs enclosed in the egg case (provided that these eggs have been produced in the field or in the laboratory) and help confirm in a direct manner whether the dormant egg stages have been produced sexually.

It was first predicted that cyclic parthenogenesis, due to extended opportunities for clonal erosion during the growing season would lead to deviations from the Hardy-Weinberg equilibrium (H-W equilibrium). Pioneer studies of Hebert (1974a, b) using allozyme electrophoresis have revealed that populations of *Daphnia magna* from intermittent habitats were characterized by stable allele frequencies and genotype frequencies in the H-W equilibrium but those from permanent populations showed severe shifts in allele frequencies within and between growing seasons. Furthermore, these permanent populations commonly deviated from the H-W equilibrium exhibiting heterozygote excess. The differences in genetic structure might reflect a different amount of clonal erosion through interclonal selection (Hebert 1974a, b). By contrast, lake populations typically showed deviations from the H-W equilibrium because the stable conditions in these environments decreased the frequencies of sexual reproduction (Černý & Hebert 1993; Thielsch et al. 2009). The incidence of sexual reproduction can vary greatly among lakes often spanning a 30-fold difference between high- and low-sex populations (Cáceres & Tessier 2004). Few empirical studies have examined the population genetics consequences of cyclic parthenogenesis at nuclear gene-coding loci. The intervening period of asexual reproduction between sexual episodes enables benecial combinations of genes to undergo selection, resulting in a buildup of genetic disequilibrium. Periodic sex breaks up these disequilibria, leading to changes in expressed genetic variance. Values of nucleotide diversity (1–2% at synonymous sites) at gene-coding loci in *Daphnia* reproducing by cyclic parthenogenesis appear equivalent or slightly lower than in other invertebrates (Haag et al. 2009; Omilian & Lynch 2009). Recombination rates were similar to that found in other sexually reproducing species thus suggesting that even a small amount of sexual reproduction can compensate for the long period of clonal reproduction in cyclic parthenogens (Omilian & Lynch 2009).

6   EFFECTS OF CYCLIC PARTHENOGENESIS ON QUANTITATIVE TRAITS

Daphnia offers an ideal model to examine the variation of fitness-related characters due to their clonal mode of reproduction, their short generation time, and the fact that they can be easily grown in the laboratory. The genetic and environmental components of the phenotypic variance can be easily separated since measures are taken on multiple genetically identical individuals. Variation in phenotypic characters between individuals of the same clone corresponds to environmental variance and variation among clones represents the genetic component of phenotypic variance. Studies that have examined patterns of genetic differentiation in ecologically relevant traits have shown evidence of local adaptation (neckteeth induction: Parejko & Dodson 1991; body size: Leibold & Tessier 1991; phototactic behavior: De Meester 1996b). Genetic diversity at molecular markers is often taken as a surrogate for genetic variation for quantitative characters (O'Brien et al. 1985). This variation is presumed important since the ability of a population to respond to evolutionary changes is directly related to the additive genetic variance (Falconer & Mackay 1996). A comparison of genetic variance for life history characters to variation at allozymes and microsatellite loci in 14 populations of *D. pulicaria* revealed a significant correlation between both types of measure (such that pairs of populations that were most distant at the molecular levels were the ones exhibiting the most differences in quantitative traits (Morgan et al. 2001). Genetic variance in quantitative traits is also affected by cyclic parthenogenesis. Indeed the long periods of clonal reproduction in the summer lead to intense interclonal selection, gametic disequilibrium, and low levels of expressed genetic variance. The production of the resting eggs by meiosis in the fall breaks the genetic disequilibria. Lynch (1984b) found that the expressed variance for size and fitness-related traits was high at the beginning of the clonal phase, diminished after a period of clonal reproduction, and was restored in the hatchlings produced after the episode of sexual reproduction. By contrast, Pfrender & Lynch (2000) found the expressed genetic variance for size at birth, juvenile growth rate, and age at first reproduction was decreased after the episodes of sexual reproduction. Thus episodic bouts of sexual reproduction can act to either enhance or retard the advance of mean phenotype gained during the clonal phase. Differences may incur due to differential selection regimes in different populations.

7   TRANSITIONS TO OBLIGATE PARTHENOGENESIS

The switch to obligate parthenogenesis in *Daphnia* is thought to result from a dominant mutation that suppresses meiosis during ephippial egg formation in females but fails to suppress spermatogenesis in males such that males carrying the mutations can mate with females and the resulting progeny will be predominantly asexual (Innes & Hebert 1988). About half of the clones that reproduce by obligate parthenogenesis retain the ability to produce males and thus will continue to transmit these alleles (Innes & Hebert 1988; Lynch et al. 2008). An experimental test in the laboratory clearly showed the transmission of paternal allozyme markers in the F1 progeny (Innes & Hebert 1988). Meiosis suppressor alleles are known to have spread in the population in a contagious fashion and to have generated numerous apomictic clones with largely disjunct geographic distributions (Hebert et al. 1993). Phylogenies of cyclical and obligate parthenogens from 72 ponds along a northeastern transect revealed that obligate parthenogens do not form monophyletic groups but are distributed more or less randomly among cyclical parthenogenetic clones as expected from a contagious mode of asexuality (Paland et al. 2005). These obligate parthenogens are young with an estimated age of 172,000 years. Two genetically distinct clades including clones with both reproductive modes were identified. Analyses of genetic subdivision indicated that one of these northeastern groups has experienced a rapid range expansion into the midwestern part of the North American continent. As a result, northeastern populations of *D. pulex* are obligate asexuals, central populations are mixed, and northwestern and midwestern populations are sexuals (Hebert & Finston 2001). The alternative scenario of this cline in asexuality responding to a cline in environmental factors rather than to the contagious spread of the meiosis-suppressor alleles has not been tested but is less likely. The application of associative mapping to the same clones used in Paland et al. (2005) provided evidence for the inheritance of at least four unlinked loci with one entire chromosome inherited through males in a nonrecombining fashion (Lynch et al. 2008) rather than a single dominant locus as concluded from crosses performed by Innes and Hebert (1988).

7.1   Genetic diversity in obligate parthenogens

Allozyme analyses have shown that clonal diversity is very high as a result of polyphyletic origins of obligate parthenogenesis, with thousands of clones occurring over the species' distribution. Thus, as a whole, obligate parthenogens have captured much of the genetic variation found in the cyclic parthenogens. Single habitats typically harbour an average of three clones for *D. pulex* in both temperate and low arctic sites (Hebert & Crease 1983; Weider & Hebert 1987; Wilson & Hebert 1992) and an average of 4.5 clones for *D. middendorffiana* at a high arctic site (Weider et al. 1987). This is much lower than what is typically found in cyclic parthenogens where recombination generates new gene combinations and clonal diversity is very high (Hebert et al. 1993). Considerable ecological and morphological divergence can be found among obligate parthenogens. Some subarctic and arctic clones produce melanin as a protection against UV radiation. The melanic clones inhabit ponds with clear water whereas unpigmented clones are found in ponds with high humic contents (Weider & Hebert 1987). Laboratory studies have confirmed the existence of variation for salinity, temperature, and pH tolerance (Weider & Hebert 1987; Jose & Dufresne 2010). It is quite clear that the origins of loss of sex have been accompanied by tremendous genetic diversity. Despite this, obligate parthenogens carry a higher load of deleterious mutations than cyclic parthenogens due to the inefficiency of purifying selection in nonrecombining genomes (Paland & Lynch 2006). A comparison of mutation accumulations in mitochondrial genes of 14 obligate and 14 cyclic parthenogens of *D. pulex* revealed an excess of nonsynonymous mutations in the obligate parthenogens. Similar results have been found in the snail *Campeloma* where the ratio of nonsynonymous to synonymous substitutions (K_a/K_s ratios) are six times higher in asexual lineages as compared to sexuals (Johnson & Howard 2007) indicating that these lineages are not as good as the sexual ones at removing deleterious mutations from their genomes.

7.2   The *Daphnia pulex* complex

The *Daphnia pulex* complex is the sole North American species complex in the genus *Daphnia* (that includes some 150 species) that has made transitions to obligate parthenogenesis. This complex includes 10 distinct mitochondrial lineages (characterized into 7 species) that can be distinguished on the basis of morphological, ecological, and genetic data (Hebert 1995; Colbourne et al. 1998; Marková et al. 2007). A portion of the ND5 gene has proven very reliable in distinguishing the various species of the *D. pulex* complex (Colbourne et al. 1998). Three major clades can be found in this complex: 1) the *D. pulicaria* clade, 2) *D. tenebrosa* clade, and 3) European *D. pulex* clade.

The *D. pulicaria* clade is composed of Panarctic *D. pulex*, *D. arenata*, *D. melanica*, eastern, western, and polar *D. pulicaria*, and *D. middendorffiana*. The *D. tenebrosa* clade includes *D. tenebrosa* and European *D. pulicaria* (Figure 3). The two dominant species in North America (*D. pulex* and *D. pulicaria*) are morphologically similar and in the midst of speciation. Both species have populations with mixed breeding systems (Hebert et al. 1993; Černý & Hebert 1993). The allozyme locus Lactate dehydrogenase has also been used in the past to distinguish *D. pulex* from *D. pulicaria*. *Daphnia pulex* is homozygous for the slow allele whereas *D. pulicaria* is homozygous for a faster allele. Speciation in this group has apparently been mediated by habitat shifts; *D. pulex* is always found in ponds and *D. pulicaria* in lakes. These two species readily hybridize and hybrids typically harbour a heterozygous phenotype at LDH and reproduce by apomictic parthenogenesis in nature (Hebert et al. 1993; Hebert & Finston 2001). Hybrids produced in the laboratory reproduce by cyclic parthenogenesis (Heier & Dudycha 2009). In temperate regions, hybridization is highly asymmetrical and diploid hybrids always have *D. pulex* as the maternal parent (Crease et al. 1989). Triploid hybrids are restricted to subarctic and arctic sites (to the exception of a single triploid clone found in southern Ontario, Canada). Genetic analyses have shown that *D. pulicaria* is most often the maternal parent of the hybrids (Dufresne & Hebert 1994) and that *D. pulex* is the paternal parent (Vergilino et al. 2011). Polyploid clones genetically similar to *D. middendorffiana* have been found in Argentina and in the tropical Andes in Bolivia (Adamowicz et al. 2002; Mergeay et al. 2008). Recent molecular analyses have revealed that North American and European *D. pulicaria* are genetically divergent taxa that have been confused under the same name (Marková et al. 2007). Similar confusion exists concerning North American and European *D. pulex* since the latter has invaded Europe and now coexists in sympatry with its misnomer! In the *D. tenebrosa* clade, transitions to obligate parthenogenesis have occurred in some populations of both *D. tenebrosa* and European *D. pulicaria* (Dufresne & Hebert 1995; Marková et al. 2007).

7.3   Distributions of obligate parthenogens

The prevalence of asexuals at high latitudes and altitudes and in extreme environments has long been recognized and referred to as geographical parthenogenesis (Vandel 1928). Many hypotheses (not mutually exclusive) have been postulated to account for this pattern in nature. The relaxation of biotic pressures (fewer pathogens, competitors, predators) in extreme environments would allow asexuals to persist in these environments (Bell 1982). Demographic hypotheses stipulate that asexuals are better colonizers than sexuals since a single individual can found a population (Cuellar 1994). Hence, asexuals would preferentially colonize areas where sexuals are limited by their ability to find mates such as the geographic limit of species ranges (Peck et al. 1998). Asexuals would be better able to compete against sexuals in areas where the latter are in low density and inbred to due repeated bottlenecks (Haag & Ebert 2004). The high reproductive rate of asexuals relative to sexuals is thought to allow them to colonize new areas faster than sexuals (Law & Crespi 2002). Other hypotheses have singled out heterosis provided by the hybrid origins of many asexuals as the most important factor enabling them to invade extreme environments (Kearney 2005). Since polyploidy

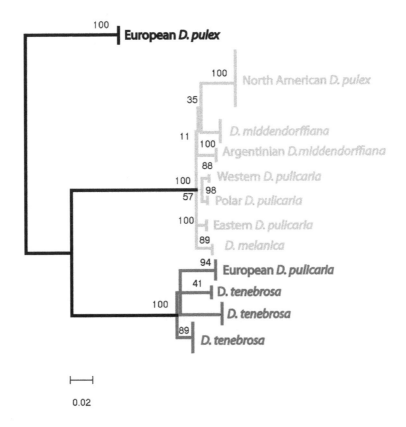

Figure 3. Genetic relationships among members of the *Daphnia pulex* complex as illustrated by a neighbor-joining tree (with branches collapsed) obtained from ND5 sequences. Members of the *D. tenebrosa* clade are shown in dark gray and members of the *D. pulicaria* clade are shown in light gray.

is often associated with apomixis, it is possible that elevated ploidy levels per se are advantageous at high latitude and altitude. In addition to showing intraspecific variation in reproductive mode, *D. pulex* exhibit a pattern of geographical parthenogenesis. Arctic and subarctic populations (starting at 54°N) of *D. pulex* and *D. pulicaria* reproduce by obligate parthenogenesis whereas temperate populations reproduce either by obligate or cyclic parthenogenesis (Hebert et al. 1993; Weider et al. 1999a). Obligate parthenogenesis would be advantageous in *Daphnia* from arctic areas since time available for reproduction is very short. Lakes in high arctic areas are often only free of ice for less than two weeks in the summer. Hence, bypassing male production certainly ensures that individuals will leave dormant propagules.

8 VARIATION IN PLOIDY LEVEL IN THE *DAPHNIA PULEX* COMPLEX

Polyploidy has arisen numerous times in the *D. pulex* complex (Beaton & Hebert 1988; Dufresne & Hebert 1994) and it is always associated with hybridization and apomixis. There is a geographical pattern to cytotypic distribution in *Daphnia* with polyploid populations found at high latitudes (53°N and higher) and altitude (Bolivian Andes) and diploid clones being prevalent in temperate regions (Beaton & Hebert 1988; Dufresne & Hebert 1995; Aguilera et al. 2007). A sole exception to this pattern is the discovery of two polyploid clones, closely related to the North American arctic clones, in Patagonia (Adamowicz et al. 2002). Members of polyploid populations of the *D. pulex* complex are obligate apomicts that develop from unfertilized egg by a mitosis-like cell division (Dufresne & Hebert 1995, 1997). Two species of the complex, *D. middendorffiana* and *D. tenebrosa*, are composed entirely or partially of polyploid populations (Dufresne & Hebert 1994, 1995, 1997) and have a circumarctic distribution (Weider et al. 1999a, b). Previous work based on feulgen microdensitometry had estimated that *D. middendorffiana* clones had 1.7 times the haploid genome size of diploid *D. pulex* clones and concluded that these clones were tetraploids that had lost DNA through genome rearrangments (Beaton & Hebert 1988). Recent estimates based on flow cytometry, revealed that *D. middendorffiana* clones had exactly 1.5 times the haploid value of diploid clones and hence were triploids (Vergilino et al. 2009). Microsatellite analyses further confirmed the presence of no more than three alleles at 9 loci. These findings shed new light on the mode of origins of these clones. These triploid clones likely arose following the union of unreduced ovule from obligate parthenogens and reduced sperm from cyclic parthenogens rather than through hybridization. By contrast, *D. tenebrosa* clones included both triploid and tetraploid clones and it is still not clear how polyploidy arose in this clade.

8.1 Origins of polyploids

The restriction of polyploid clones to subarctic and arctic sites can be explained by both proximate/causal and ultimate/functional hypotheses. Polyploidy might arise more frequently in cold climates because unreduced gametes are produced at a higher rate at low temperature (Mable 2004). Historical factors may also explain the higher incidence of polyploids at high latitudes/altitude. The secondary contact of divergent refugial races following range expansion has led to the formation of hybrid zones in many organisms (Hewitt 2004). Since hybridization is a frequent route to polyploidy, it is possible that the presence of many allopolyploids in arctic areas could be due to Pleistocene glaciations (Dufresne & Hebert 1997).

8.2 Impact of cytotype on life history traits and ecophysiological characteristics

Numerous phenotypic changes accompany polypoidy, the most universal change being an increase in cell size (Mirsky & Ris 1951). Other changes mediated by cell size are also noted in some species: increased developmental time, larger eggs, and larger body size (Dufresne & Hebert 1998). Life history experiments on sympatric diploid and polyploid clones from a low arctic site revealed that polyploid clones had larger eggs and neonates but lower fecundity than diploid clones at 10 °C, 17 °C, and 24 °C (Dufresne & Hebert 1998). Polyploid clones took more time to mature than diploid clones at 24 °C but less time to mature at 10 °C, suggesting that polyploid clones may be better adapted to low temperatures. One frequently invoked hypothesis to explain the prevalence of polyploid clones at high latitudes is that polyploidy confers greater tolerance to extreme environmental conditions because of higher heterozygosity and metabolic flexibility (Otto & Whitton 2000). Jose & Dufresne (2010) examined the tolerance of 12 clones with different ploidy levels and geographic origin to a suite of environmental factors. There was a single factor (low salinity) under which polyploid clones did better than diploid clones, therefore, their geographic distribution cannot be explained by an increased tolerance to more extreme environmental conditions.

9   GENOMICS

The *Daphnia* Genomics Consortium was launched in 2002 to "develop the *Daphnia* system to the same depth of molecular, cell, and developmental biological understanding as other model systems, but with the added advantage of interpretability of observations in the context of natural ecological communities." As it plays a pivotal role in aquatic ecosystems, *Daphnia* promises to be an ideal species for environmental genomics. *Daphnia pulex* has a genome of approximately 200 *Mb* with 31,000 genes. Many recent studies report the presence of numerous gene duplications in the *D. pulex* genome (e.g., Baldwin et al. 2009). Rho et al. (2010) have identified 333 intact long terminal repeats (LTR) in the *D. pulex* genome clustered into 142 families. These retroelements constitute 7.9% of the *D. pulex* genome, which is much higher than what is found in *Drosophila melanogaster*. As a first step towards the understanding of obligate parthenogenesis, Schurko et al. (2009) have inventoried 40 genes with diverse roles in meiosis based on knowledge from *Drosophila*. The expression of these genes was not different between parthenogenetic and meiotic reproduction in cyclically parthenogenetic females but the fact that many of these genes were present in multiple copies suggest that this may facilitate transitions to obligate parthenogenesis in this species. One exciting research avenue arising from the availability of genomic data is the identification of genes involved in the production of defensive structures. *Daphnia* represents the model case for elucidating the ecological and developmental mechanisms underlying polyphenism. These predator-induced defences involve morphological changes (production of neckteeth, helmets), behavioral, and life history changes upon exposure to invertebrate and fish predators (Tollrian & Harvell 1999). Miyakawa et al. (2010) compared the expression profiles in the presence/absence of predator kairomone in *D. pulex* embryos. Transcriptional profiling by real-time quantitative PCR revealed the up-regulation of 29 genes from the juvenile hormone and the insulin pathways in the exposed embryos. A whole-genome scan technique revealed that a single gene with no homology in other species but likely involved in signal perception and/or transduction was up-regulated in the embryonic stages. Another important field of research is the identification of genes involved in *Daphnia* immune system (McTaggart et al. 2009). As genome sequencing costs continue to drop, the genomes of additional clones of the *D. pulex* complex will be sequenced providing increased knowledge of the various forces at play during speciation and adaptive radiations.

10   CONCLUSIONS

The cladoceran *Daphnia* has served as a model species in ecological and evolutionary studies for the past century due to its important role in aquatic ecosystems and ease of culturing. Prior to the development of molecular markers, *Daphnia* studies were limited by the paucity of morphological characters arising from morphological stasis and constrained body plan. The advent of molecular markers has revealed the existence of numerous species complexes in *Daphnia*. The *Daphnia pulex* complex has been extensively studied since it is widely distributed in North America and shows transitions to obligate parthenogenesis often associated with polyploidy. Increases in ploidy level were found to be correlated with phenotypic changes in egg size and body size in *Daphnia* (Dufresne & Hebert 1998). The hypothesis that heterosis is associated with increased tolerance to various environmental conditions was not supported by Jose & Dufresne (2010). It is still not clear why polyploid *Daphnia* are restricted to high latitude environments. The recent sequencing of the *D. pulex* genome will help shed light on mechanisms of genome rearrangements and gene expression changes that accompany increases in ploidy, thus creating unprecedented opportunities to understand this important evolutionary process.

REFERENCES

Adamowicz, S.J., Gregory, T.R., Marinone, M.C. & Hebert, P.D.N. 2002. New insights into the distribution of polyploid *Daphnia*: the Holarctic revisited and Argentina explored. *Mol. Ecol.* 11: 1209–1217.

Adamowicz, S.J., Petrusek, A., Colbourne, J.K., Hebert, P.D.N. & Witt, J.D.S. 2009. The scale of divergence: a phylogenetic appraisal of intercontinental allopatric speciation in a passively dispersed freshwater zooplankton genus. *Mol. Phylogenet. Evol.* 50: 423–436.

Aguilera, X., Mergeay, J., Wollebrants, A., Declerck, S. & De Meester L. 2007. Asexuality and polyploidy in *Daphnia* from the tropical Andes. *Limnol. Oceanogr.* 52: 2079–2088.

Baldwin, W.S., Marko, P.B. & Nelson, D.R. 2009. The cytochrome P450 (CYP) gene superfamily in *Daphnia pulex*. *BMC Genomics* 10: 169.

Beaton, M.J. & Hebert, P.D.N. 1988. Geographical parthenogenesis and polyploidy in *Daphnia pulex*. *Amer. Naturalist* 132: 837–845.

Bell, G. 1982. *The Masterpiece of Nature: The Evolution and Genetics of Sexuality*. Berkeley, CA: University of California Press.

Boileau, M.G., Hebert, P.D.N. & Schwartz, S.S. 1992. Non-equilibrium gene frequency divergence: persistent founder effects in natural populations. *J. Evol. Biol.* 5: 25–39.

Brooks, J.L. 1957. The systematics of North American *Daphnia*. *Mem. Connecticut Acad. Arts Sci.* 13: 1–180.

Cáceres, C.E. 1999. Interspecific variation in the abundance, production, and emergence of *Daphnia* diapausing eggs. *Ecology* 79: 1699–1710.

Cáceres, C.E. & Tessier, A.J. 2004. Incidence of diapause varies among populations of *Daphnia pulicaria*. *Oecologia* 141: 425–431.

Černý, M. & Hebert, P.D.N. 1993. Genetic diversity and breeding system variation in *Daphnia pulicaria* from North American lakes. *Heredity* 71: 497–507.

Colbourne, J.K & Hebert, P.D.N. 1996. Phylogenetics and evolution of the *Daphnia longispina* group (Crustacea) based on 12S rDNA sequence and allozyme variation. *Mol. Phylogenet. Evol.* 5: 495–510.

Colbourne, J.K., Crease, T.J., Weider, L.J., Hebert, P.D.N., Dufresne, F. & Hobaek, A. 1998. Phylogenetics and evolution of a circumarctic species complex (Cladocera: *Daphnia pulex*). *Biol. J. Linn. Soc.* 65: 347–365.

Cox, A.J. & Hebert, P.D.N. 2001. Colonization, extinction, and phylogeographic patterning in a freshwater crustacean. *Mol. Ecol.* 10: 371–386.

Crease, T.J. & Hebert, P.D.N. 1983. A test for the production of sexual pheromones by *Daphnia magna* (Crustacea: Cladocera). *Freshw. Biol.* 13: 491–496.

Crease, T.J., Stanton, D.J. & Hebert, P.D.N. 1989. Polyphyletic origins of asexuality in *Daphnia pulex*. II. Mitochondrial DNA variation. *Evolution* 43: 1016–1026.

Cuellar, O. 1994. Biogeography of parthenogenetic animals. *Biogeographica* 70: 1–13.

De Gelas, K. & De Meester, L. 2005. Phylogeography of *Daphnia magna* in Europe. *Mol. Ecol.* 14: 753–764.

De Meester, L. 1996a. Local genetic differentiation and adaptation in freshwater zooplankton populations: patterns and processes. *Ecoscience* 3: 385–399.

De Meester, L. 1996b. Evolutionary potential and local genetic differentiation in a phenotypically plastic trait of a cyclical parthenogen, *Daphnia magna*. *Evolution* 50: 1293–1298.

Dufresne, F. & Hebert, P.D.N. 1994. Hybridization and origins of polyploidy. *Proc. R. Soc. Lond B* 258: 141–146.

Dufresne, F. & Hebert, P.D.N. 1995. Polyploidy and clonal diversity in an Arctic cladoceran. *Heredity* 75: 45–53.

Dufresne, F. & Hebert, P.D.N. 1997. Pleistocene glaciations and polyphyletic origins of polyploidy in an arctic cladoceran. *Proc. R. Soc. Lond. B* 264: 201–206.

Dufresne, F. & Hebert, P.D.N. 1998. Temperature-related differences in life-history characteristics between diploid and polyploid clones of the *Daphnia pulex* complex. *Ecoscience* 5: 433–437.

Falconer, D.S. & Mackay, T.F.C. (eds.) 1996. *Introduction to Quantitative Genetics*. 4th edition.

Harlow: Longman.

Frey, D.G. 1987. The taxonomy and biogegraphy of the Cladocera. *Hydrobiologia* 137: 97–115.

Haag, C.R. & Ebert, D. 2004. A new hypothesis to explain geographic parthenogenesis. *Ann. Zool. Fennici* 41: 539–544.

Haag, C.R., Riek, M.J., Hottinger, W., Pajunen, I. & Ebert, D. 2006. Founder events as determinants of within-island and among-island genetic structure of *Daphnia* metapopulations. *Heredity* 96: 150–158.

Haag, C.R., McTaggart, S.J., Didier, A., Little, T.J. & Charlesworth,D. 2009. Nucleotide polymorphism and within-gene recombination in *Daphnia magna* and *Daphnia pulex*, two cyclic parthenogens. *Genetics* 182: 313–323.

Havel, J.E. & Hebert, P.D.N. 1993. *Daphnia lumholtzi* in North America: another exotic zooplankter. *Limnol. Oceanogr.* 38: 1823–1827.

Hebert, P.D.N. 1974a. Enzyme variability in natural population of *Daphnia magna*. II. Genotypic frequencies in permanent populations. *Genetics* 77: 323–334.

Hebert, P.D.N. 1974b. Enzyme variability in natural populations of *Daphnia magna*. III. Genotypic frequencies in intermetting populations. *Genetics* 77: 335–341.

Hebert, P.D.N. 1995. The *Daphnia* of North America: an illustrated fauna. CD-ROM, University of Guelph.

Hebert, P.D.N & Crease, T.J.. 1983. Clonal diversity in populations of *Daphnia pulex* reproducing by obligate parthenogenesis. *Heredity* 51: 353–369.

Hebert, P.D.N. & Cristescu, M.E.A. 2002. Genetic perspectives on invasions: the case of the Cladocera. *Can. J. Fish. Aquat. Sci.* 62:1229–1234.

Hebert, P.D.N. & Finston, T.L. 1996. Genetic differentiation in *Daphnia obtusa*: A continental perspective. *Freshw. Biol.* 35: 311–321.

Hebert, P.D.N. & Finston, T.L. 2001. Macrogeographic patterns of breeding system diversity in the *Daphnia pulex* group from the United States and Mexico. *Heredity* 87: 153–161.

Hebert, P.D.N. & Hann, B.1986. Patterns in the composition of Arctic tundra pond microcrustacean communities. *Can. J. Fish. Aquat. Sci.* 43: 1416–1425.

Hebert, P.D.N., Swartz, S.S., Ward, R.D. & Finston, T.L. 1993. Macrogeographic patterns of breeding system diversity in the *Daphnia pulex* group. I. Breeding systems of Canadian populations. *Heredity* 70: 148–161.

Hebert, P.D.N. & Wilson, C.C. 1994. Provincialism in plankton-endemism and allopatric speciation in Australian *Daphnia*. *Evolution* 48:1339–1349.

Hebert, P.D.N., Witt, J.D.S. & Adamowicz, S.J. 2003. Phylogeographical patterning in *Daphnia ambigua*: regional divergence and intercontinental cohesion. *Limnol. Oceanogr.* 48: 261–268.

Heier, C.R. & Dudycha, J.L. 2009. Ecological speciation in a cyclic parthenogen: sexual capability of experimental hybrids between *Daphnia pulex* and *Daphnia pulicaria*. *Limnol. Oceanogr.* 54: 492–502.

Hewitt, G.M. 2004. Genetic consequences of climatic oscillations in the Quaternary. *Phil. Trans, R. Soc. Lond. B* 359: 183–195.

Innes, D.J. & Dunbrack, R.L. 1993. Sex allocation variation in *Daphnia pulex*. *J. Evol. Biol.* 6: 559–575.

Innes, D.J. & Hebert, P.D.N. 1988. The origin and genetic basis of obligate parthenogenesis in *Daphnia pulex*. *Evolution* 42: 1024–1035.

Innes, D.J. & Singleton, D.R. 2000. Variation in allocation to sexual and asexual reproduction among clones of cyclically parthenogenetic *Daphnia pulex* (Crustacea: Cladocera). *Biol. J. Linn. Soc.* 71: 771–787.

Ishida, S. & Taylor, D.J. 2007. Mature habitats associated with genetic divergence despite strong dispersal ability in an arthropod. *BMC Evol. Biol.* 7: 52.

Johnson, S.G. & Howard, R.S. 2007. Contrasting patterns of synonymous and nonsynonymous sequence evolution in asexual and sexual freshwater snail lineages. *Evolution* 61: 2728–2735.

Jose, C. & Dufresne, F. 2010. Differential survival among genotypes of *Daphnia pulex* differing in reproductive mode, ploidy level, and geographic origin. *Evol. Ecol.* 24: 413–421.

Kalff, J. 2002. *Limnology. Inland Water Ecosystems.* Upper Saddle River, NJ: Prentice Hall.

Kearney, M. 2005. Hybridization, glaciation and geographical parthenogenesis. *Trends Ecol. Evol.* 20: 495–502.

Law, J.H. & Crespi, B.J. 2002. The evolution of geographic parthenogenesis in *Timema* walking-sticks. *Mol. Ecol.* 11: 1471–1489.

Leibold, M. & Tessier, A.J. 1991. Contrasting patterns of body size for *Daphnia* species that segregate by habitat. *Oecologia* 86: 342–348.

Louette, G. & De Meester, L. 2004. Rapid colonization of a newly created habitat by cladocerans and the initial build-up of a *Daphnia*-dominated community. *Hydrobiologia* 513: 245–249.

Lynch, M. 1984a. The genetic structure of a cyclic parthenogen. *Evolution* 38: 186–203.

Lynch, M. 1984b. The limits to life history evolution in *Daphnia. Evolution* 38: 465–482.

Lynch, M. 1985. Speciation in the cladocera. *Verh. Internat. Verein. Theor. Angew. Limnol.* 22: 3116–3123.

Lynch, M. & Gabriel, W. 1983. Phenotypic evolution and parthenogenesis. *Amer. Naturalist* 122: 745–764.

Lynch, M., Seyfert, A., Eads, B. & Williams, E. 2008. Localization of the genetic determinants of meiosis suppression in *Daphnia pulex. Genetics* 180: 317–327.

Mable, B.K. 2004. Why polyploidy is rarer in animals than in plants: myths and mechanisms. *Biol. J. Linn. Soc.* 82: 453–466.

Marková, S., Dufresne, F., Rees, D.J., Černý, M. & Kotlík, P. 2007. Cryptic intercontinental colonization in water fleas *Daphnia pulicaria* inferred from phylogenetic analysis of mitochondrial DNA variation. *Mol. Phylogenet. Evol.* 44: 42–52.

Mayr, E. 1963. *Animal Species and Evolution.* Cambridge, MA: Harvard University Press.

McTaggart, S.J., Conlon, C., Colbourne, J.K., Blaxter, M.L & Little, T.J. 2009. The components of the *Daphnia pulex* immune system as revealed by complete genome sequencing. *BMC Genomics* 10: 175.

Mergeay, J., Aguilera, X., Declerck, S., Petrusek, A., Huyse, T. & De Meester, L. 2008. The genetic legacy of polyploid Bolivian *Daphnia*: the tropical Andes as a source for the North and South American *D. pulicaria* complex. *Mol. Ecol.* 17: 1789–1800.

Mergeay, J., Verschuren, D. & De Meester, L. 2006. Invasion of an asexual American water flea clone throughout Africa and rapid displacement of a native sibling species. *Proc. R. Soc. Lond. B* 273: 2839–2844.

Mirsky, A.E. & Ris, H. 1951. The deoxyriboneucleic acid content of animal cells and its evolutionary significance. *J. Genet. Physiol.* 34: 451–462.

Miyakawa, H., Imai, M., Sugimoto, N., Ishikawa, Y., Ishikawa, A., Ishigaki, H., Okada, Y., Miyazaki, S., Koshikawa, S., Cornette, R. & Miura, T. 2010. Gene up-regulation in response to predator kairomones in the water flea, *Daphnia pulex. BMC Dev. Biol.* 10: 45.

Morgan, K.K., Hicks, J., Spitze, K., Latta, L., Pfrender, M.E., Weaver, C.S., Ottone, M. & Lynch, M. 2001. Patterns of genetic architecture for life-history traits and molecular markers in a subdivided species. *Evolution* 55: 1753–1761.

O'Brien, S.J., Roelke, M.E., Marker, L., Newman, A., Winkler, C.A., Meltze, D., Colly, L., Evermann, J.F., Bush, M. & Wildt, D.E. 1985. Genetic basis for species vulnerability in the cheetah. *Science* 227: 1428–1434.

Omilian, A. R. & Lynch, M. 2009. Patterns of intraspecific DNA variation in the *Daphnia* nuclear genome. *Genetics* 182: 325–336.

Otto, S.P. & Whitton, J. 2000. Polyploid incidence and evolution. *Annu. Rev. Genet.* 34: 401–437.

Paland, S., Colbourne, J. K. & Lynch, M 2005. Evolutionary history of contagious asexuality in *Daphnia pulex*. *Evolution* 59: 800–813.

Paland, S. & Lynch, M. 2006. Transitions to asexuality result in excess amino acid substitutions. *Science* 311: 990–992.

Pálsson, S. 2000. Microsatellite variation in *Daphnia pulex* from both sides of the Baltic Sea. *Mol. Ecol.* 9: 1075–1088.

Parejko, K. & Dodson, S.I. 1991. The evolutionary ecology of an antipredator reaction norm: *Daphnia pulex* and *Chaoborus americanus*. *Evolution* 45: 1665–1674.

Peck, J.R., Yearsley, J.M. & Waxman, D. 1998. Explaining the geographic distributions of sexual and asexual populations. *Nature* 391: 889–892.

Penton, E.H., Hebert, P.D.N. & Crease, T.J. 2004. Mitochondrial DNA variation in North American populations of *Daphnia obtusa*: continentalism or cryptic endemism? *Mol. Ecol.* 13: 97–107.

Peters, R.H. & de Bernardi, R. (eds.), 1987. *Daphnia*. Verbania Pallanza: Consiglio Nazionale delle Ricerche, InstitutoIitalitano di Idrobiologia.

Pfrender, M.E. & Lynch, M. 2000. Quantitative genetic variation in *Daphnia*: temporal changes in genetic architecture. *Evolution* 54: 1502–1509.

Proctor, V.W. 1964. Viability of crustacean eggs recovered from ducks. *Ecology* 45: 656–658.

Rho, M., Schaack, S., Gao, X., Kim, S., Lynch, M. & Tang, H. 2010. LTR retroelements in the genome of *Daphnia pulex*. *BMC Genomics* 11: 425.

Schurko, A.M., Logsdon, J.M. & Eads, B. 2009. Meiosis genes in *Daphnia pulex* and the role of parthenogenesis in genome evolution. *BMC Evol. Biol.* 9: 78.

Schwenk, K., Junttila, P., Rautio, M., Bastiansen, F., Knapp, A., Dove, O., Billiones, R. & Streit, B. 2004. Ecological, morphological, and genetic differentiation of *Daphnia* (*Hyalodaphnia*) from the Finnish and Russian subarctic. *Limnol. Oceanogr.* 49: 532–539.

Taylor, D.J., Finston, T.L. & Hebert, P.D.N. 1998. Biogeography of a widespread freshwater crustacean: pseudocongruence and cryptic endemism in the North American *Daphnia laevis* complex. *Evolution* 52: 1648–1670.

Taylor, D.J., Sprenger, H.L. & Ishida, S. 2005. Geographic and phylogenetic evidence for dispersed nuclear introgression in a daphniid with sexual propagules. *Mol. Ecol.* 14: 525–537.

Thielsch, A., Brede, N., Petrusek, A., De Meester, L. & Schwenk, K. 2009. Contribution of cyclic parthenogenesis and colonization history to population structure in *Daphnia*. *Mol. Ecol.* 18: 1616–1628.

Tollrian, R. & Harvell, C.D. (eds.) 1999. *The Ecology and Evolution of Inducible Defenses*. Princeton, NJ: Princeton University Press.

Vandel, A. 1928. La parthénogénèse géographique. Contribution à l'étude biologique et cytologique de la parthénogénèse naturelle. *Bull. Biol. Franç. Belg.* 62: 164–182.

Vanoverbeke, J. & De Meester, L. 1997. Among-populational genetic differentiation in the cyclical parthenogen *Daphnia magna* (Crustacea, Anomopoda) and its relation to geographic distance and clonal diversity. *Hydrobiologia* 360: 135–142.

Vergilino, R., Belzile, C. & Dufresne, F. 2009. Genome size evolution and polyploidy in the *Daphnia pulex* complex (Cladocera: Daphniidae). *Biol. J. Linn. Soc.* 97: 68–79.

Vergilino, R., Markova, S., Ventura, M., Manca, M. & Dufresne, F. 2011. Reticulate evolution of the *Daphnia pulex* complex as revealed by nuclear markers. *Mol. Ecol.* 20: 1191–1207.

Weider, L.J. & Hebert, P.D.N. 1987. Ecological and physiological differentiation among low- Arctic clones of *Daphnia pulex*. *Ecology* 68: 188–198.

Weider, L. J., Beaton, M. J. & Hebert, P.D.N. 1987. Clonal diversity in high Arctic populations of *Daphnia pulex*, a polyploid, apomictic complex. *Evolution* 41: 1346–1355.

Weider, L.J., Hobæk, A., Colbourne, J.K., Crease, T.J., Dufresne, F. & Hebert, P.D.N. 1999a. Holarctic phylogeography of an asexual species complex I. Mitochondrial DNA variation in arctic *Daphnia*. *Evolution* 53: 777–792.

Weider, L.J., Hobæk, A., Hebert, P.D.N. & Crease, T.J. 1999b. Holarctic phylogeography of an asexual species complex – II. Allozymic variation and clonal structure in Arctic *Daphnia*. *Mol. Ecol.* 8: 1–13.

Wilson, C.C. & Hebert, P.D.N. 1992. The maintenance of taxon diversity in an asexual assemblage: an experimental analysis. *Ecology* 73: 1462–1472.

Phylogeographic patterns in *Artemia*: a model organism for hypersaline crustaceans

ILIAS KAPPAS, ATHANASIOS D. BAXEVANIS & THEODORE J. ABATZOPOULOS

Department of Genetics, Development & Molecular Biology, School of Biology, Aristotle University of Thessaloniki, 54124 Thessaloniki, Greece

ABSTRACT

Crustacea are an enormously diverse natural group. Although a few lineages have adapted to life on land, the vast majority of them (more than 52,000 described species) are aquatic, living in marine or freshwater environments, where they are as abundant as insects are on land. The aquatic realm also includes a proportion of settings where the modest $\approx 3.5\%$ of earth's oceans salt concentration can be vastly exceeded, at times approaching levels close to halite precipitation. Until recently, these aquatic ecosystems were largely considered devoid of life and trivial. The later discovery of rich microbial communities overturned such ideas and sparked interest. The list of eukaryotic salty-survivors though is very limited. Among metazoans, admittedly the most accomplished survivors of hypersaline settings come from Crustacea and are the brine shrimps of the genus *Artemia*. This ancient anostracan lineage is a close relative to other freshwater branchiopod representatives, which collectively comprise the main component of continental zooplankton. *Artemia* has developed remarkable adaptations that enable it to cope with the extreme challenges of brine life. Freshwater and hypersaline ecosystems share many attributes, but they also differ in many ways. We know more about phylogeographic patterns in the former type, presumably due to higher relative occurrence and associated lineage diversity. Similar research on the latter type of ecosystems has only recently been accumulating and a reasonable query is whether genealogical patterns in hypersaline habitats are in any way particular. In this chapter, an overview of current phylogeographic knowledge in brine settings is given using *Artemia* as a model-system. Phylogeographic investigations in this crustacean, coupled with or aided by phylogenetic and systematic assessments, have revealed patterns, some of which bear an idiosyncratic signature. The chapter concludes by highlighting the potential of *Artemia* in modern evolutionary research as well as in surveys dealing with genomic responses to extreme environments.

1 A PHYLOGEOGRAPHIC PARADOX

Phylogeography is the study of the geographic distributions of phylogenetic lineages. Its empirical roots can be traced back to the early 1970s when important developments in the study of mitochondrial DNA variation provided the first insights on the spatial and temporal context of gene genealogies. The explosive growth of phylogeographic appraisals was catalyzed by the seminal paper of Avise and colleagues (Avise et al. 1987) who coined the term, conceptualized the field, and introduced an array of testable hypotheses. Since then, phylogeographic investigations, coupled with certain analytical advances, have united the previously disparate fields of population genetics and systematics.

Although a great variety of taxa have been surveyed for geographic patterns of genetic variation, organisms inhabiting the continental aquatic realm are particularly suited for such approaches
as a result of their distinctive biology and the nature of their environments. The diversity of biological attributes in continental zooplankters, including sexuality, parthenogenesis (either obligate
or cyclic), dormant life stages, polyploidy and others, is equally matched by a remarkable habitat
diversity spanning the whole spectrum of water chemistry (freshwater to brackish to hypersaline),
physical structure (ponds and pools to large lakes), and predictability (temporal to permanent). Despite the island-like nature and geographic isolation of their habitats (Gajardo et al. 2006) continental zooplanktonic organisms exhibit typical cosmopolitan distributions, presumably due to passive
transport of their diapausing propagules through wind or waterfowl (Bilton et al. 2001; Figuerola
et al. 2005). Consequently, for several decades it was thought that their biogeography would resemble that of microbes and the high capacity for passive dispersal, conducive to homogenization
of gene pools, would eventually translate at the genetic level to effective panmixia (Mayr 1963).
However, over the last twenty years or so, molecular data have provided strong evidence for marked
phylogeographic structuring, both at a regional and a global scale (Boileau et al. 1992; De Meester
1996; Gómez et al. 2000; Lee 2000; Baxevanis et al. 2006; Ishida & Taylor 2007; Mills et al. 2007;
Ketmaier et al. 2008; Xu et al. 2009).

In an effort to explain this obvious high dispersal-low gene flow paradox, De Meester et al.
(2002) proposed the monopolization hypothesis. According to this, the structure of genetic diversity
in continental zooplankton is a function of priority effects and strong local adaptation that act to
severely reduce effective gene flow among resident populations. In particular, a combination of
stochastic and selection-driven processes have shaped to a large extent current phylogeographic
patterns of passive dispersers, including cladocerans, copepods, anostracans as well as rotifers,
bryozoans and others. In these taxa, colonization of new habitats by a limited number of propagules
leads to rapid population growth and a subsequent build-up of a large resting propagule bank. The
succeeding monopolization of resources is further enhanced by rapid local adaptation of the resident
population. Therefore, the effects of new immigrants are "diluted" and a self-reinforcing process
kicks off, further constraining gene flow and eventually registering strong signatures of priority
effects and prolonged impacts of founder events.

A large amount of empirical data seems to fit the above scenario. In genealogical terms, the
genetic diversity spectrum ranges from sharply demarcated phylogroups (Burton & Lee 1994; De
Gelas & De Meester 2005) to cryptic species assemblages (Taylor et al. 1998; Lee 2000; Witt
& Hebert 2000; Gómez et al. 2002; Hebert et al. 2003; Penton et al. 2004), frequently embedded in deeper phylogenetic patterns owing to vicariance or allopatric divergence within continents
(Adamowicz et al. 2009). Consequently, current diversity estimates in aquatic invertebrates are
tested against a default biogeographic hypothesis of provincialism (Hebert & Wilson 1994) in the
broader sense.

De Meester et al. (2002) have made specific predictions, based on the monopolization hypothesis, regarding the nature and degree of intraspecific genetic population structure. They mainly
considered cyclical parthenogens, obligate parthenogens, and obligate sexuals, for which most data
are available. These taxa are dominant components of terrestrial freshwater zooplankton. However,
a comparable proportion of earth's inland waters constitute hypersaline settings (Williams 2002), in
which organisms with similar life strategies are also found. For decades, these aquatic ecosystems
were considered as unimportant due to their limited species richness and diversity as well as various
misapprehensions regarding their distribution and global volume. Given the requirement for specific
adaptations by halophiles and the distinctive physico-chemical milieu of hypersaline settings, it is
interesting to examine whether genealogical patterns in respective taxa resemble those seen in their
freshwater relatives. Before that, however, it would be wise to briefly highlight the characteristics
of saline waters in general.

2 THE SALINE REALM

Conventionally, the salinity barrier separating fresh from saline waters is 3 g/l (3 ppt or 3 ‰). This value corresponds to the sum total of dissolved salts. Saline waters are categorized as hyposaline (3–20 ppt), mesosaline (20–50 ppt), and hypersaline (> 50 and up to 350 ppt) (Hammer 1986). They include permanent or temporary bodies of water (i.e., ponds, pools, lagoons, lakes) with a global distribution in dry regions where evaporation exceeds precipitation (Figure 1A). In these regions, saline waters are often more abundant than fresh waters. On a global basis, saline and fresh waters do not differ significantly in terms of total volume (85 versus 105×10^3 km^3; Shiklomanov 1990). The former, however, are significantly more diverse in many physico-chemical features than the latter, with salinity and other, often associated, parameters (e.g., ionic composition, oxygen concentration, pH, temperature) varying widely on a temporal basis (Williams 1998; Van Stappen 2002). These differences are reflected in a sharp decline in the composition and nature of the biota in saline waters. This marked effect on biodiversity is probably the most prominent attribute of saline ecosystems. The biodiversity spectrum includes several bacterial groups (archaeobacteria, eubacteria, cyanobacteria) and fewer representatives of algae, nonalgal macrophytes, crustaceans (anostracans, cladocerans, copepods, ostracodes), rotifers, insects, and fish. However, as salinity approaches the hypersaline boundary, a sharp decrease in species richness and diversity occurs (Hammer 1986). Thus, at salinities above 50 ppt most taxa are excluded and very few lineages (mostly microbes) continue to persist. In these environments and up to the point of saturation, the physiological demands imposed on organisms are severe and require a very specific adaptation kit. Among metazoans, the most renowned inhabitants of highly saline waters are the brine shrimps of the genus *Artemia*.

3 THE BRINE SHRIMP *ARTEMIA* AS A MODEL SYSTEM IN PHYLOGEOGRAPHY

Artemia (Crustacea, Anostraca) is a typical zooplankter of inland salt lakes and coastal lagoons. Although other variables (e.g., temperature, light intensity, primary food production) may have an influence on quantitative aspects of *Artemia* populations, salinity is without any doubt the principal abiotic factor determining its global distribution (Triantaphyllidis et al. 1998; Van Stappen 2002) (Figure 1B).

Due to the importance of *Artemia* in aquaculture and solar salt production, both permanent and temporal feral populations also exist worldwide as a result of anthropogenic introductions (Kappas et al. 2004). In the hypersaline biotopes where the brine shrimp thrives, predation pressure is very limited as potential predators cannot cross the salinity barrier, with the exception of course of waterfowl.

The genus *Artemia* is an ancient branchiopod lineage (Fryer 1987; Braband et al. 2002; Richter et al. 2007) comprising a complex of sexual and obligate parthenogenetic forms. Its biogeography bears a distinct pattern and is customarily described in terms of a New versus Old World divide. Typically, this split reflects the segregation of the two gene pools within the genus (sexual and asexual) and the fact that the Old World is considered as the radiation center of *Artemia* (Baxevanis et al. 2006). Currently, six bisexual species are recognized showing sharply delimited distributions and occasional endemicity (Abatzopoulos et al. 2002a). The Americas are inhabited by two species, *A. franciscana*, widely distributed in North and Central America and recently expanded over certain parts of South America (Amat et al. 2004), and *A. persimilis*, confined to Chile and Argentina (Gajardo et al. 2004; Ruiz et al. 2008). Waterbird-mediated dispersal of *A. franciscana* populations has been documented also in western Mediterranean following primary human introductions for aquaculture operations (Green et al. 2005).

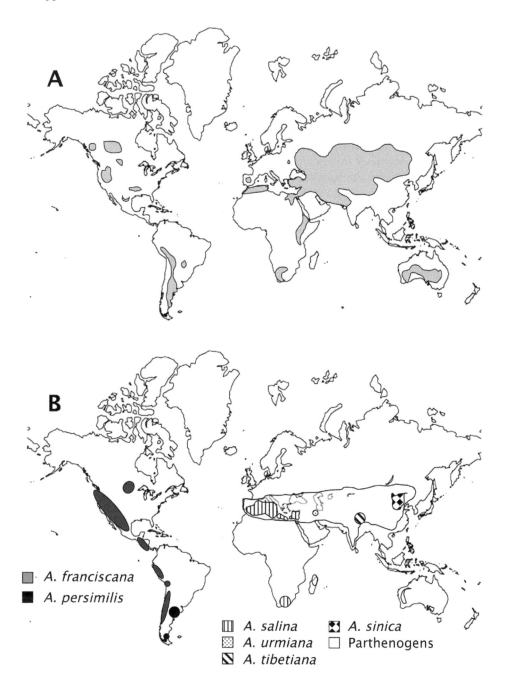

Figure 1. (A) World distribution of arid and semiarid regions where saline water bodies are mainly found (based on data from Hammer 1986). (B) Global distribution of the genus *Artemia*. The significant overlap is evident.

The Old World harbors a mixture of bisexual species and numerous parthenogenetic forms. *Artemia salina* is mainly distributed in the Mediterranean Basin, although recent reports also document the presence of a highly divergent South African lineage (Kaiser et al. 2006; Muñoz et al. 2008; Kappas et al. 2009). *Artemia sinica* is broadly distributed in north-eastern China and Inner Mongolia, while *A. tibetiana* and *A. urmiana* are typical endemics of Tibetan plateau lakes (Abatzopoulos et al. 1998, 2002a) and Urmia Lake, Iran (Abatzopoulos et al. 2006a; but see also Abatzopoulos et al. 2009), respectively. Parthenogenetic populations are characteristically restricted to the Old World where they comprise the majority and are occasionally sympatric with bisexual species (e.g., Spain, Italy, Central and North China, Iran; see Amat et al. 1995a; Van Stappen 2002; Abatzopoulos et al. 2006a).

Bisexuals are diploid with $2n = 42$ (except *A. persimilis* with $2n = 44$), while parthenogens, which can be either automictic or apomictic, range in ploidy from $2n$ to $5n$ (Abatzopoulos et al. 1986). Morphometric and/or morphological, life history, and genetic divergences are widely partitioned both within and between the different reproductive modes, and along with differences in parturition (encysted embryos or live birth) and ploidy, they present valuable opportunities for testing ecological and phylogeographic patterns (see Abatzopoulos et al. 2002b).

Cladistic analyses of mitochondrial and nuclear DNA sequence data (Baxevanis et al. 2006; Kappas et al. 2009) have provided strong evidence for a vicariance-based split of *A. persimilis* from Old World lineages following severance of mid-Cretaceous landmasses. Subsequent major radiations involving *A. salina* and *A. franciscana* sprang from a Mediterranean centre of origin, trailing the sequence of events leading to early Tertiary (emergence of the Mediterranean Sea, broad land connection between North America and Europe). The diversification of Asian species (*A. sinica*, *A. urmiana*, *A. tibetiana*) has been largely interweaved with that of parthenogens for which at least four independent origins have been deduced. Thus, the early phylogeography of *Artemia* bears the signature of deep phylogenetic splits due to continental fragmentation, a process that has also affected historically other terrestrial zooplankters, like *Streptocephalus* (Daniels et al. 2004) and *Daphnia* (Colbourne et al. 2006; Adamowicz et al. 2009).

4 GENETIC ARCHITECTURE AND GENEALOGY OF PARTHENOGENS

Parthenogenesis in *Artemia* is obligate; cyclic parthenogens have never been reported. The parthenogenetic types exhibit different ploidy levels (diploid to pentaploid) often confounded within the same population (Abatzopoulos et al. 2002b). The type of reproduction found in diploid parthenogens is automictic (normal meiosis occurs and diploidy is restored through fusion of meiotic division products), whereas apomixis (meiosis is thoroughly suppressed and only mitotic division takes place separating sister chromatids) is prevalent among polyploids (Barigozzi 1974; Maniatsi et al. 2011). Many parthenogenetic populations contain a small proportion (usually less than 1%) of males whose asexual origin has been confirmed through allozyme and mitochondrial DNA markers (Browne et al. 1991). The role of these "rare," although fertile, males remains obscure.

A strong positive correlation has been reported between gene diversity and ploidy level ($r = 0.89$, $p < 0.001$; Abreu-Grobois & Beardmore 1982) as expected if polyploids are better buffered against deleterious mutations than diploids. Due to apomixis, polyploids are more often monoclonal with very similar or identical genotypes found in distant localities (Abreu-Grobois & Beardmore 1982). However, this is less evident for triploids showing relatively large genetic distances among themselves and from other parthenogens. Diploids on the other hand are typically polyclonal. An outstanding example concerns the population at Salin de Giraud (France), harboring more than 60 genotypes and showing interclonal distances as high as 0.25 (Browne & Hoopes 1990). These genotypes appear to be of local and recent origin, also exhibiting a striking distribution along salinity gradients at the site as well as significant differences in lifespan and reproductive characteristics.

Phylogenetic inferences based on Internal Transcribed Spacer 1 (ITS-1) sequence and length data as well as 16S rRNA Restriction Fragment Length Polymorphism (RFLP) patterns (Baxevanis et al. 2006) have shown that the parthenogenetic gene pool is a heterogeneous amalgam of clones, including both narrow endemics and widespread lineages of distinctive spatial and temporal origins and genetic diversity. It is worth noting that diversity estimates for parthenogens as a whole match or even exceed those of their likely sexual ancestors (*A. sinica, A. urmiana*, and possibly *A. tibetiana*), including interspecific comparisons of the latter. This is also supported by additional preliminary 16S and cytochrome c oxidase subunit I (COI) sequence data (mean d for parthenogens = 7.6%, mean d for Asian sexuals = 0.75%). For ITS-1 in particular, the sequence diversity value of 2.0% within parthenogenetic *Artemia* is striking when compared with the ostracode *Darwinula stevensoni*, which is thought to be reproducing asexually for millions of years, yet being effectively invariant in a sampling range extending from Finland to South Africa (Schön et al. 1998).

The inferred evidence for polyphyletic origin of asexuality in *Artemia* (Baxevanis et al. 2006) in conjunction with sufficient screening of additional genomic regions may provide valuable information as to whether their substantial diversity is mutationally generated or recurrently captured from sexual ancestors (Crease et al. 1989; Chaplin & Hebert 1997). The lack of association ($r = 0.56$, $P > 0.05$) between nuclear and mitochondrial DNA divergences for pairs of clones points to the fact that the source of clonal diversity are independent transitions to asexuality from sexual relatives. However, it is surprising that the observed diversity of parthenogens does not mirror that of their nearest Asian sexuals, phylogenetically implicated in transitions to unisexuality (*A. urmiana, A. sinica*). As there is no evidence that *Wolbachia* infections may induce parthenogenesis in *Artemia* (Maniatsi et al. 2010), it remains that obligate parthenogenetic lineages may have been derived either spontaneously or through hybridization between bisexual species. Another possibility, the contagious origin via parthenogenetically-produced males also exists, however it is preconditioned on the pre-existence of parthenogenetic populations as well as the reproductive functionality of rare males that has not been verified in the brine shrimp (Browne 1992).

Closely linked to parthenogenesis is the issue of polyploidy, with a general trend of ploidy level increase, with an increase in latitude or altitude often being associated with a shift to asexuality (Otto & Whitton 2000). For example, in the North American *Daphnia pulex* group a striking geographical pattern in ploidy level exists with high Arctic sites dominated by polyploids, sub-Arctic ones by both diploids and polyploids, and temperate zones by diploids (Beaton & Hebert 1988). Similar patterns are also present in *D. pulex* populations from the United Kingdom and Scandinavia (Ward et al. 1994). On the other hand, polyploid lineages of the *D. pulicaria* group are abundant in the alpine lakes of the Bolivian Andes (Mergeay et al. 2008). Analogous inferences on the genealogical history and distribution of polyploids have been made in bosminid cladocerans (Little et al. 1997) and ostracodes (Turgeon & Hebert 1994).

One of the most intriguing ecological patterns is that of geographical parthenogenesis. It describes the tendency for parthenogenetic organisms to display biased distributions towards particular environmental settings, such as high latitudes, high altitudes, deserts, islands, and otherwise marginal or disturbed environments when compared with their close sexual relatives (Vandel 1928; see also Kearney 2005). Although the issue has attracted early attention in *Artemia*, no clear-cut patterns have been identified. Partly, this is probably due to the fact that additional factors like the polyphyletic origin of parthenogenesis, the mode of parthenogenetic reproduction (apomixis versus automixis), the extreme variation in ploidy as well as the primarily restricted distribution of the brine shrimp to hypersaline settings only, may yield a considerably complicated picture, overshadowing underlying trends. For example, if parthenogenetic lineages result from hybridization between bisexuals, then, initially, their occurrence should, to a lesser or a greater extent, be associated with that of their sexual parentals. However, parthenogenetic genotypes can be extremely widespread and of variable diversity (Baxevanis et al. 2006; Maniatsi et al. 2011). In addition, locally synthesized diploids in Salin de Giraud (France) display identical mitochondrial DNA RFLP profiles, suggestive of recent origin, yet genotypes of other *Artemia* populations are absent from the same site (Browne 1992). In the long term, ecological tolerances and genetic architecture of asexuals should jointly determine niche breadth. A number of studies have highlighted marked ecological diversification and interesting tendencies (Browne & Hoopes 1990; Zhang & Lefcort 1991; Zhang & King 1992, 1993; Barata et al. 1995, 1996a, b; Baxevanis & Abatzopoulos 2004), like the higher frequency of polyploids at high latitudes or contrasting environmental optima for life history traits between diploids and polyploids. These results, however, have not been verified by field data which should also critically account for the updated biogeography of the genus (Van Stappen 2002). Evidence for the primary determinant of ecological success of asexual lineages in *Artemia* is still elusive since potential patterns of geographical polyploidy may be embedded within broader patterns of geographical parthenogenesis or hybridity (see Kearney 2005; Adolfsson et al. 2010).

5 PHYLOGEOGRAPHIC PATTERNS IN SEXUAL *ARTEMIA*

The biogeography of *Artemia* bisexuals bears strong signatures of vicariance for deep phylogenetic splits, of intracontinental allopatric divergence for mesoscale separations, whereas typical endemism or provincialism is found for more recent radiations or intraspecifically (Gajardo et al. 2002; Baxevanis et al. 2006; Muñoz et al. 2008; Kappas et al. 2009). Excluding anthropogenic introductions of strains for aquaculture purposes, a sharp pattern of intercontinental genetic discontinuities is evident indicating ancient speciation events for major groups (*A. persimilis*, *A. salina*, *A. franciscana*). Intercontinental affinities of < 5% in mitochondrial DNA markers have been reported between North and South America and between North America and Europe for several *Daphnia* species (Schwenk et al. 2000; Hebert et al. 2003; Ishida & Taylor 2007; Mergeay et al. 2008). The same is true for species of *Bosmina* and *Moina* (Taylor et al. 2002; Petrusek et al. 2004). Given the remarkable capacity for dispersal of continental zooplanktonic groups (Brendonck & Riddoch 1999; Havel et al. 2000), it is of special interest why the observed intercontinental associations are not eroded by ongoing gene flow. This question has been recently investigated, in part, by Adamowicz et al. (2009) using a compilation of mitochondrial DNA data (COI, 12S, 16S) in 92 species from all *Daphnia* subgenera. The authors concluded that at least 30% of cladogenetic events are attributable to vicariance and in particular to the break-up of Pangaea and subsequent continental shifts following the fragmentation of Gondwanaland. Another 12% of total diversification is explained by intracontinental allopatric speciation. Within these proportions, dispersal events, both at large and regional scales, are responsible for several divergences between sister species. Three different explanations were advanced to account for the intriguing persistence of these ancient biogeographic patterns: i) shifts in zooplankton dispersal rates related to changes in bird migration pathways, ii) changing probabilities of establishment versus extinction of intercontinental migrants over time, and iii) stable background extinction of members of different intercontinentally dispersed species pairs. The above study is an example of the type of research needed in continental aquatic biogeography, trying to resolve long-recognized paradoxes and to combine the basic rationale behind the monopolization hypothesis (De Meester et al. 2002) with known mechanisms of geographic speciation operating in zooplankters.

To this end, comparative data are lacking especially in halophiles of similar evolutionary age. The best source of our knowledge comes from *Artemia*, in which, at first glance, patterns of speciation and genetic differentiation resemble those seen in freshwater relatives. All else being equal (life cycle, dispersal capacity, reproductive modes, etc.), vicariance and pronounced ecological diversification seem to have shaped current biogeographic patterns to a great extent. However, although saline (excluding marine) and fresh waters are more or less similar in global volume, their distributions are markedly distinct, showing little overlap. Moreover, given the extreme physico-chemical diversity of hypersaline environments, the considerable effects registered by Pleistocene glaciations (including desertification; Kearney 2005), and the increasing salinization of modern inland aquatic ecosystems (Williams 2001, 2002), notable particularities in halophile phylogeographic patterns may emerge from comparative analyses with freshwater taxa.

5.1   New World *Artemia*

5.1.1   Artemia persimilis: *species interactions and hybridization in South America*

New World endemics include representatives of *A. franciscana* and *A. persimilis*. Based on ITS-1 and 16S RFLP and sequence data (Baxevanis et al. 2006; Kappas et al. 2009), it has been inferred that the *A. persimilis* clade diverged from the ancestral stock of *Artemia* populations between 80 and 90 mya, at the time of separation of Africa from South America. The diversity of climates, hydrobiological, and geological conditions, even on a short distance, in Argentina and Chile provided opportunities for pronounced differentiation among *A. persimilis* populations, which is evident in genetic, cytogenetic, and biochemical characters (see Gajardo et al. 1995, 2004; Baxevanis et al. 2006; Ruiz et al. 2007, 2008; Papeschi et al. 2008). Preliminary analysis of 16S rRNA sequence data (see also Kappas et al. 2009) provide strong indications for a marked splitting between populations of the Chilean Patagonia (far south) and those of the Argentine Patagonia (further north) (see Figure 2). Using the same marker, Ruiz et al. (2008) reported F_{ST} values of 0.81 ($P < 0.05$) as well as significant heterogeneity in haplotype frequencies, yet in the absence of isolation by distance.

The region mapped out by Argentina and Chile is probably the only area in the world where *Artemia* bisexuals interact so intensively. An intriguing case is the Chilean population of Pichilemu at $\approx 34°S$. This latitude demarcates the distributions of the two New World endemics, with *A. franciscana* occurring north and *A. persimilis* south of this point (Amat et al. 2004; Gajardo et al. 2004). Over the years, a number of studies have provided varied evidence for the species status of *Artemia* from Pichilemu. Allozymes and morphometric data have assigned this population to *A. franciscana* (Gajardo et al. 1995; Zuñiga et al. 1999), whereas diploid and chromocenter numbers, 16S RFLP patterns, and ITS-1 sequence data have ascribed it to *A. persimilis* (Gajardo et al. 2004; Baxevanis et al. 2006). Based on cladistic and network methods, Kappas et al. (2009) provided evidence that hybridization between *A. franciscana* and *A. persimilis* occurs in Pichilemu. They also detected mitochondrial introgression from *A. persimilis* to *A. franciscana*, thus pointing to fertility of F_1 hybrids. Given additional genetic features in *Artemia* (e.g., $2n = 42$ versus $2n = 44$, female sex heterogamety), it is impressive that species separated by tens of millions of years are able to hybridize following secondary contact. The case of *Artemia* from Pichilemu highlights a delay of reproductive isolation relative to genetic isolation that is also shared by other branchiopods, like *Streptocephalus* (Wiman 1979), *Simocephalus* (Hann 1987), *Branchinecta* (Maeda-Martinez et al. 1992), and *Daphnia* (Schwenk et al. 2000). It also underlines the role of contemporary hybridization in determining population structure and marks out possible particularities in genomic evolution and patterns of reproductive isolation in terrestrial zooplankters.

5.1.2   Artemia franciscana: *incipient speciation and ecological isolation*

In the rest of the continent, *A. franciscana* displays an array of populations stretching up north through Central America and the western interior of North America as far as Canada (Van Stappen 2002). Compared with other *Artemia* species, *A. franciscana* is more euryhaline, and eurythermal performing better in terms of life history and reproductive characteristics (see Kappas et al. 2004 and references therein). Based on allozymes, considerable differentiation (mean $d = 12.6\%$) and substructuring (mean $F_{ST} = 0.24$) (see Gajardo et al. 2002) has been reported between geographic populations of *A. franciscana*. For South American populations, mean F_{ST} values range from 0.38 (allozymes, Gajardo et al. 1995) to 0.91 (16S RFLPs, Gajardo et al. 2004).

Marked genetic distances have also been reported between Mexican and North American *A. franciscana* (Tizol-Correa et al. 2009). Owing to its extreme radiation in the American continent and the pronounced differentiation of populations, *A. franciscana* is considered as a "superspecies"

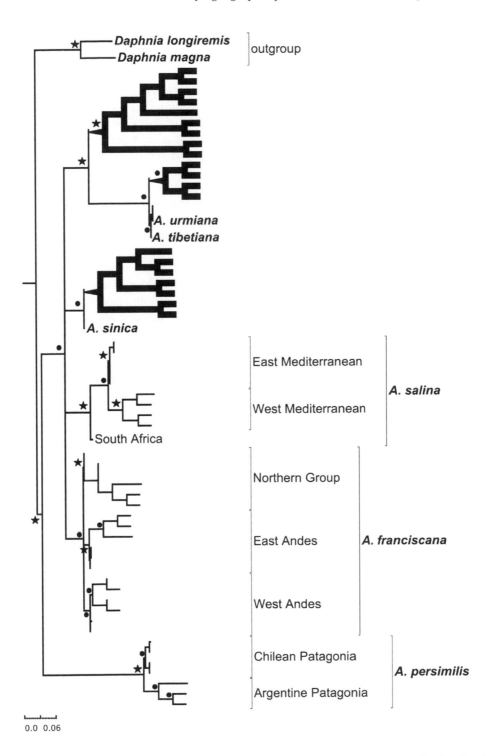

Figure 2. Maximum likelihood phylogeny of *Artemia* based on 16S rRNA sequences (for details of data analysis, see Kappas et al. 2009; *Daphnia longiremis*: AY921454, *D. magna*: AY921452). Bootstrap values (filled circles: 60–80%, stars: 80–100%) are shown at nodes. Bold branches indicate parthenogenetic strains.

in the process of incipient speciation. Recently, Maniatsi et al. (2009a, b) analyzed patterns of genealogical concordance using mitochondrial (16S, COI) and nuclear (ITS-1, intron-2 of p26) DNA sequence data in 14 American populations. Sharp phylogeographic breaks were identified in mitochondrial DNA (Figure 3), while a signal of advanced lineage sorting was evident for the nuclear loci. The results of the above work nicely portray the genealogical phase between gene splitting and population divergence, the evolutionary interval where speciation occurs. The substantial amount of genetic differentiation between mitochondrial phylogroups (2.2% for 16S, 4.2% for COI), the increased support for reciprocal monophyly, and the absence of co-distributed haplotypes provide evidence for a pattern of regional endemism of lineages and severe restriction of effective gene flow (Maniatsi et al. 2009a). Using reported molecular rates for COI sequences of snapping shrimp (Knowlton & Weigt 1998), the divergence of mitochondrial clades coincides with the final uplift of the Andes in late Pliocene (Gregory-Wodzicki 2000; Ramírez et al. 2008). This implies that populations evolved from gene pools that persisted through the more recent periods of repeated glaciations in isolated refugia on both slopes of the Andes. The spatial structuring of *A. persimilis* lineages may have been similarly affected by the Patagonian ice cap.

In a demonstration of an extreme example of ecological isolation within the *A. franciscana* group, Bowen et al. (1985) studied the *Artemia* population from Mono Lake, California. *Artemia franciscana* populations arguably display the widest variability with respect to tolerance for waters of different ionic composition, with reproductive isolation occurring among populations due to intolerance to each others chemical habitat. The San Francisco Bay strain of *A. franciscana* can hatch in the carbonate water of Mono Lake, but the nauplii die within a few days. Conversely, Mono Lake *Artemia* cannot be raised in other high-chloride media. The closest relatives of *Artemia* from Mono Lake are the North American strains from San Francisco Bay and Great Salt Lake (mean $d = 9.8\%$). High heterozygosity levels have been reported for the Mono Lake population ($H_e = 0.18$), following a more general and significant trend for higher genetic variability in non-chloride populations (mean $H_e = 0.138$) as opposed to chloride ones ($H_e = 0.077$) (Abreu-Grobois & Beardmore 1982). Thus, certain environmental factors may have a bearing on the genetic makeup of strains. Alternatively, the retention of high heterozygosity in the Mono Lake population may be due to input of genes from other populations via rare pre-adapted genotypes or passage through intermediate ecosystems, in order to overcome the problem imposed by the toxicity of Mono Lake water. Surprisingly, the population of the soda lake Fallon, Nevada, can survive either in carbonate or in chloride water, and is interfertile with Mono Lake shrimps, thereby potentially preventing genetic differentiation. On the other hand, although Mono Lake hosts a rich avifauna and lies on the route of migration flyways, its isolation may be more intense compared with similar-type habitats as a result of the specific cyst behavior (cysts sinking in the water column) of Mono Lake *Artemia*. These findings demonstrate that sharp genetic discontinuities and cryptic lineages may be fruitfully sought in patterns of ecological segregation in halophiles.

5.2   Old World *Artemia*

5.2.1   Artemia salina: *Mediterranean phylogroups and outliers*

Species interactions in the Old World mainly involve sexual strains and parthenogens (in terms of co-occurrence and possible sympatry). The four Old World bisexual species (*A. salina, A. urmiana, A. sinica, A. tibetiana*) show largely disjunct distributions with a high degree of endemicity. The evolutionary radiation of *A. salina* seems to have been largely determined by the geological history of the Mediterranean Basin. The first sequence-based (ITS-1) estimates for *A. salina* (Baxevanis et al. 2006) have provided solid evidence for substantial differentiation into eastern and western phylogroups (mean $d = 2.4\%$), confirming earlier Amplified Fragment Length Polymorphism (AFLP)

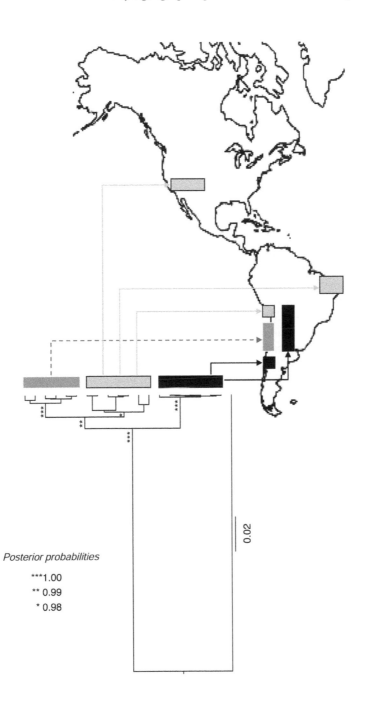

Figure 3. Bayesian analysis of *Artemia franciscana* based on COI haplotypes (re-analyzed data from Maniatsi et al. 2009a). Analysis parameters using BEAST 1.5.3 (Drummond & Rambaut 2007); HKY + I + G, 4 gamma rate categories, two codon partitions (1 + 2, 3), two independent MCMC analyses, 10^7 generations, sampling every 10^3, first 2×10^6 generations discarded as burnin. Basal branch denotes the outgroup *A. salina*. Three reciprocally monophyletic clades are recovered. Posterior probabilities for major phylogroups are shown above branches. The geographic distributions of lineages are shown on the map (see text for details).

analyses (Triantaphyllidis et al. 1997). Molecular clock calibrations place the initial eastern-western split at the beginning of the Messinian salinity crisis, mapping the land exposure between southern Turkey and Egypt to the eastern clade and that between the Gulf of Sidra and the Strait of Sicily to the western clade. Subsequent studies (Kappas et al. 2009) have shown that 16S sequence diversity within *A. salina* may be as high as 3.61%. More importantly though, yet another example of natural hybridization between *A. franciscana* and *A. salina* from South Africa was inferred on the basis of combined nuclear and mitochondrial DNA data. Previous morphometric and mating data (Amat et al. 1995b) had produced strong indications for the presence of *A. salina* in South Africa. In their review of the African distribution of *Artemia*, Kaiser et al. (2006) estimated that among a total of 127 records, a fraction of about 50% likely belong to *A. salina*.

The divergent South African lineage of *A. salina* was also confirmed by Muñoz et al. (2008) using COI sequence data. In their analysis of 23 *A. salina* populations from the Mediterranean Basin, the authors found evidence for a pattern of extensive regional endemism and a possible Pleistocene range expansion. A number of Iberian refugia were inferred, echoing patterns in other invertebrates (Gómez et al. 2000; Gómez & Lunt 2007), as well as long-distance colonizations involving populations from the east (Cyprus, Egypt), southwest Spain, Sardinia, and Sicily. Four South African haplotypes were also identified, forming a monophyletic cluster and showing an average distance from Mediterranean groups of 10.4%. This estimate is within the range of interspecific COI divergences for crustaceans (Costa et al. 2007) and well above those between Asian bisexuals (*A. urmiana*, *A. sinica*, *A. tibetiana*). However, the nested clade analysis of Muñoz et al. (2008) was unable to recover the historical and phylogeographic signal present in South African *A. salina*. In fact, the authors even excluded this lineage due to the presence of several substitutions separating it from Mediterranean *A. salina*. They were thus confronted with the problems of the actual origin of these mutations and of distinguishing between a hypothesis of ancestry or one of migration. In this respect, testing for associations between geography and patterns of gene genealogy for inferring historical processes should be performed in more rigorous, model-based statistical frameworks, which account for the stochasticity in lineage sorting (see Knowles & Maddison 2002; Hickerson et al. 2010). The results of Muñoz et al. (2008) combined with those of Baxevanis et al. (2006) and Kappas et al. (2009) on the global phylogeny of the genus imply that the ancestral stock of *Artemia* populations had already diverged to some extent prior to the split of Gondwanaland. To this end, recent improvements of dispersal-vicariance analysis (see Yu et al. 2010) may be particularly helpful as they explicitly correct for uncertainty in phylogenetic inference. Future work should exploit such approaches for extracting phylogeographic signal and assessing different biogeographic scenarios and geographic lineage associations.

5.2.2   Artemia urmiana: *endemicity and niche partitioning*

From the remaining Asian bisexuals, *A. urmiana* from Lake Urmia (Iran) presents a very interesting case due to i) the characteristics of its habitat, ii) its large endemicity, and iii) its genetic associations and close proximity with certain parthenogenetic strains. Lake Urmia, in north-western Iran, is among the largest hypersaline permanent water bodies in western Asia. It is situated at an altitude of 1250 m above sea level, with a total surface area between 4750 and 6100 km^2, and average and maximum depths of 6 and 16 m, respectively (Agh et al. 2007). In the late 1990s, a combination of severe drought and inflow diversions caused a dramatic increase in salinity (reaching over 300 ppt) that subsequently affected the density of the brine shrimp population. From 2003 to the present, favorable climatic conditions resulted in a decrease of water salinity at $\approx$ 200 ppt. Not surprisingly (see below), contrasting results have been obtained for the species status of the endemic *Artemia* population over the years (see Agh et al. 2007 and references therein). This endemicity, however, had never been challenged until recently, when Abatzopoulos et al. (2009) using a multidisciplinary approach (discriminant analysis of morphometry, scanning electron microscopy, ITS-1 sequence data) reported the presence of *A. urmiana* in Lake Koyashskoe, a hypersaline lake on the Black Sea coast of the Crimean Peninsula. This was the first time *A. urmiana* had been reported outside Lake Urmia. Based on limited sequence divergence (0.70%) between Lake Koyashskoe and Lake Urmia populations, the authors considered either bird-mediated dispersal or human transfer through salt trade routes in recent historical times as alternative hypotheses for the presence of *A. urmiana* in the Crimea. Data from several other lakes in the area are currently limited. Yet, the region is of special interest as most localities represent newly opened niches, now gradually transformed into salt lakes following recent abandonment by humans. It is therefore possible that priority effects are at work in these unoccupied habitats. Although Lake Urmia is home to many bird species, which can be potential dispersers, the distinctive structure of *A. urmiana* cysts causing them to sink (similarly to Mono Lake cysts) may drastically reduce migration. Using buoyancy tests and transmission electron microscopy, Abatzopoulos et al. (2006b) confirmed that different cyst chorion characteristics are responsible for the reduced buoyancy of Urmia cysts (over 60% sank compared to only 10% of the Great Salt Lake strain). On the contrary, the fate (sink or float) of *Daphnia* ephippia (chitinous shells enclosing diapausing eggs) is determined by the moulting behavior of females (Slusarczyk & Pietrzak 2008). In the same study of Abatzopoulos et al. (2006b), investigation of reproductive and life span characteristics in salinities ranging from 35 to 180 ppt, revealed a preference for high salinity (140–180 ppt) in *A. urmiana* individuals. The striking similarities between Mono Lake and Urmia Lake demonstrate that ecological factors may strongly affect population dynamics and interfere with colonization capacity and lineage distributions in halophilic taxa. In one of the most detailed studies of Urmia Lake, Agh et al. (2007) investigated survival in the laboratory as well as population composition in earthen ponds (constructed in the vicinity of the lake) over a period of two years. Results indicated that while the lake itself is dominated by sexual *Artemia* (*A. urmiana*), asexual populations are also found, although exclusively restricted to the shores or several lagoons adjacent to the lake, where no *A. urmiana* is found. This niche partitioning is largely determined by salinity as parthenogens mature and reproduce at very low salinities (15–33 ppt), whereas higher salinities (> 50 ppt) are required by *A. urmiana* to attain sexual maturity. The ephemeral lagoons are periodically restocked from a relatively stable, mixed-source population (Urmia Lake). In this way, they resemble ecological models, where extinction and recolonization determine intraspecific genetic differentiation (see Avise 2000 and references therein). However, they are largely atypical since they involve both environmental periodicity and absence of gene flow. Appropriate molecular surveys should provide valuable insights on whether parthenogens have indeed achieved some degree of regional radiation or are invariably locked in a narrow zone of dynamic coexistence with *A. urmiana*. Additional work (Abatzopoulos et al. 2006a) has documented the exclusive presence of parthenogens in 17 different geographical locations, including salt lakes, lagoons, and salty rivers throughout Iran. Initial screening of a number of these populations has revealed genetic uniformity of 16S RFLP patterns (including *A. urmiana*), yet also morphometric segregation in two clusters (one sexual and one parthenogenetic) (Agh et al. 2009). Integration of physiological, ecological, and genetic assays is needed to elucidate critical phylogeographic parameters of *Artemia* in Iran and the crucial role of *A. urmiana* in the spread of parthenogenetic strains.

5.2.3   Artemia sinica: *habitat type and size*

The vast arid-semiarid expanses of Asia Interior, extending east of the Caspian Sea and through Kazakhstan and southern Siberia to Mongolia and north-eastern China (Inner Mongolia), are limnologically very rich regions, yet largely unexplored. China itself is characterized by the highest concentration of salt lakes in the world. Considering taxon sampling issues, Baxevanis et al. (2006) demonstrated that *A. sinica* is the closest sexual relative to a group of Chinese parthenogens. This relationship is recurrently recovered with an increased support both for nuclear and mitochondrial markers (see also Figure 2). Thus, *A. sinica* is one of the lineages implicated in the polyphyletic origin of asexual *Artemia*. The split between *A. sinica* and Chinese parthenogens has been estimated at ≈ 3.5 mya. It has been suggested that coastal sites in China are dominated by parthenogens, while inland locations mainly host bisexual strains (Xin et al. 1994). This habitat segregation is even more pronounced at the microscale; in provinces with both coastal and inland sites, parthenogens are generally found at the coast while bisexuals are restricted at inland localities. It has been hypothesized that distribution patterns caused by the prevalence of a particular, habitat-dependent mode of reproduction, are the result of specific ionic requirements of local populations, yet no detailed data exist. Equally interesting is the fact that the above pattern also displays phylogenetic partitioning. Using AFLP markers, Sun et al. (1999) recovered three distinct phylogroups among Chinese populations and other reference strains: i) inland parthenogens, grouped with *A. urmiana*, ii) coastal parthenogens, and iii) Chinese bisexuals (*A. sinica*) and *A. tibetiana*, as a sister clade to coastal parthenogens. There is thus the possibility that inland parthenogens, concentrated in the remote north-west provinces of Xinjiang and Qinghai, where the largest salt lakes exist, bear specific affinities to *A. urmiana*. All previous results indicate that additional data and taxon sampling are needed on which well-formulated hypotheses regarding the evolutionary radiation of Asian bisexuals and parthenogens can be tested.

The degree of genetic differentiation among *A. sinica* populations has been used as evidence for reduced founder effects in large compared to small populations. Using variation at nine allozyme loci, Naihong et al. (2000) reported a clear tendency for an increase in heterozygosity with increasing habitat size ($R^2 = 0.62$, $P = 0.04$) for nine bisexual populations from China. All studied populations inhabit lakes ranging in surface area from 3.5 to $64 \, \text{km}^2$. Although a positive relationship between genetic differentiation and geographical distance was evident, overall F_{ST} values ranged from 0.05–0.1. For populations separated by as much as 1200 km, F_{ST} values were between 0.02 and 0.25. Admittedly, one would expect higher levels of population differentiation. The authors interpreted the observed homogeneity by considering (albeit not explicitly analyzing) an effect due to habitat size. Accordingly, large populations reduce the speed with which populations differentiate from each other through genetic drift. In addition, propagule sizes are large, since colonization success of immigrants remains relatively high for a long time, as the size of the habitat increases and saturation delays. Consequently, a more representative sample of the regional gene pool can be expected, leading to founder effects of reduced impact (see De Meester et al. 2002). Although the above explanation is sufficient, it does not consider the possibility of subpopulation structure which is higher in large habitats. Larger lakes are expected to have greater environmental heterogeneity and the correlation between heterozygosity and surface area of the water body might also be a consequence of that, apart from differences in population size. In their investigation, Naihong et al. (2000) used a low number of loci and did not account for the above possibility and the associated Wahlund effect that might have been strong as gauged by the heterozygote deficits observed in seven out of nine loci. It remains to be seen, pending additional and more informative data, if this is a localized effect or a more general attribute of *Artemia*. So far the literature does not support such a conclusion, and differentiation among populations of comparable sizes is without any doubt substantial (see Abatzopoulos et al. 2002b and references therein).

5.2.4   Artemia tibetiana: *the geology of the Tibetan Plateau*

The most recently discovered species of the genus is *A. tibetiana* (Abatzopoulos et al. 1998). As the name implies, *A. tibetiana* is found in the high elevations of the Tibetan Plateau. This region has an average altitude exceeding 4500 m and, according to Zheng et al. (1993), it hosts more than 350 geologically and chemically diverse saline lakes of various sizes and salinities reaching up to 390 ppt (see also Zheng 1997). The type locality of the species is Lagkor Co, a carbonate lake situated at 4490 m above sea level in the arid-temperate plateau zone of Tibet. This salt lake has a salinity of ±60 ppt, with alkaline water (pH 8.8) and the temperature varies from a maximum of ±24 °C to a minimum of ± − 26 °C, with average annual air temperature of ±1.6 °C (see Zheng 1997).

Since its first characterization (Abatzopoulos et al. 1998, 2002a), *A. tibetiana* has been studied using an impressive array of approaches, including biometrics of cysts and nauplii, cytogenetics, allozymes, cross-breeding tests, Random Amplified Polymorphic DNA (RAPD), AFLPs, morphometry, 16S RFLPs, thermal tolerance of cysts, temperature and media preferences, highly unsaturated fatty acids (HUFA) content, and others (see Baxevanis et al. 2005 and references therein). Data so far strongly indicate that *A. tibetiana* represents a distinct lineage with a varying propinquity to *A. sinica* and *A. urmiana*. Marked phenotypic and genetic segregation has been documented based on the biometry of cysts and nauplii, morphometry of adults, and genetic markers. For example, *A. tibetiana* has the largest ever recorded values for both bisexual and parthenogenetic strains in cyst diameter, length of instar-I nauplii, and total body length of adults (Abatzopoulos et al. 1998). Similar divergence is evident for ITS-1 (Baxevanis et al. 2006) and 16S sequence data (see Figure 2). Laboratory cross-breeding tests (Abatzopoulos et al. 2002a) have shown complete infertility between *A. tibetiana* and all other species, although isolating barriers were less strong in crosses to *A. urmiana* and *A. sinica* indicating a recent separation.

Unfortunately, the inaccessibility of the Tibetan plateau and the shortage of samples have constrained detailed inferences on the population structure and lineage distribution of *A. tibetiana*. Nevertheless, the close genetic affinity of *A. tibetiana* and certain parthenogenetic strains is currently accepted both on empirical data (Baxevanis et al. 2006) and on grounds of adjacent distributions. In addition, its divergence from *A. sinica* has been estimated to have occurred around 8 mya, within a period for which geological and thermochronological evidence (Clark et al. 2005) demarcate the elevation of the Tibetan plateau and subsequent separation of eastern and western landscapes. Thus, the particular characteristics of this region including effects by glaciations and aridity as well as the plethora and diversity of saline lakes, provide unique settings for investigating patterns of lineage diversification. This can be aided further by consideration of special adaptations required to cope with cold tolerance and ultraviolet (UV) exposure at high altitudes and their possible effects on molecular rates (see Clegg et al. 2001; Tanguay et al. 2004).

6 ARE THERE HYPERSALINE-SPECIFIC PHYLOGEOGRAPHIC PATTERNS?

The preceding treatment of *Artemia*, the most renowned inhabitant of brine settings, was deliberately organized in such a way as to accentuate two important aspects. The first, is the abundance and heterogeneity of mechanisms, both within and between *Artemia* species, that bring about substantial genetic differentiation and lineage radiation. Using the brine shrimp as a representative and model organism of hypersaline ecosystems, extensive similarities with freshwater organisms are all too clear, owing to common life history strategies and evolutionary trajectories. This congruence in phylogeographic patterns covers all levels of genealogical structure, from intraspecific lineage separations of various depths and distributions (allopatry versus sympatry) (De Gelas & De Meester 2005; Mills et al. 2007; Muñoz et al. 2008) to extended genealogical concordance among multiple loci, codistributed species, or even between molecular data and historical biogeographic information (Daniels et al. 2004; Baxevanis et al. 2006; Adamowicz et al. 2009; Maniatsi et al. 2009a, 2011).

The second aspect has been hopefully emphasized or intuitively acknowledged by the unconventional species-by-species treatment of *Artemia*. Notwithstanding the initial intention, it should be more than obvious that *Artemia* is a depauperate genus. Only six bisexual species have been described, using multidisciplinary approaches and critically cross-breeding tests (e.g., Pilla & Beardmore 1994; Abatzopoulos et al. 2002a). Thus, there is currently little disagreement regarding species boundaries and phylogenetic/evolutionary partitions. Even sidestepping the most widely accepted yardstick of empirically delimiting species, i.e., the criterion of reproductive isolation, the number of distinct phenotypic and genotypic clusters in *Artemia* probably would remain more or less the same (see also Suatoni et al. 2006). Current phylogenetic evidence, despite some genome porosity between lineages (Kappas et al. 2009), supports that detouring the biological species concept for demarcating species would mainly have an effect on the parthenogenetic gene pool by partitioning it to distinct units. Although the biological relevance of splitting asexual taxa on phylogenetic grounds can be highly questionable (e.g., Barraclough et al. 2003), the resulting novel assemblages would insignificantly increase species numbers.

Assuming for a moment that the above observation of a comparatively species-poor lineage holds true, a number of additional factors make it even more problematic and paradoxical. First, the prolific cladogenesis in *Daphnia* (Adamowicz et al. 2009), a freshwater relative of comparable age, is by all standards qualitatively similar to *Artemia*, yet species numbers between these two lineages differ by more than 10-fold. Second, even within hypersaline taxa and accounting for the extent of geographic distribution, great differences exist. For example, the Australian endemic brine shrimp *Parartemia* has diversified into probably more than 13 species (Remigio et al. 2001; Timms & Hudson 2009; Timms et al. 2009). Third, large-scale comparative analyses of various groups have demonstrated a highly consistent and significant positive association between ecological divergence and reproductive isolation across taxa (Funk et al. 2006). In the previous sections, several examples of ecological divergence have shown that ecology holds a prominent role in promoting biological diversification in *Artemia*. Moreover, the hypersaline niche does not have less diversity or temporal-spatial variance (probably the opposite) compared with freshwaters. Fourth, recent molecular work (Hebert et al. 2002; Colbourne et al. 2006) has provided strong indications for habitat-specific rates of evolution. Halophilic crustaceans show consistent acceleration in rates of molecular evolution, presumably due to the effects of ionic strength and UV exposure on the fidelity of DNA replication. For phylads of comparable age, increased substitution rates should translate to increased cladogenesis if molecular evolution is clock-like and birth-death rates of diversification are invariable. Yet, most halophilic lineages are characteristically destitute in that respect.

The above considerations, as exemplified by *Artemia*, highlight the potential presence of idiosyncratic patterns in genealogical structure and speciation in eukaryotic halophiles. These particularities should be registered in both the micro- and macroevolutionary scales. As Avise (1994) rightly put it, many authors have viewed the speciation process (though not necessarily its products) as a rather unexceptional continuation of the microevolutionary processes generating geographic population structure, with the added factor of the evolutionary acquisition of intrinsic reproductive isolation. Intraspecific population differentiation and interspecific genetic divergence are essentially the outcomes of the very same deterministic and stochastic forces operating in different time scales.

Our knowledge on the evolutionary mechanisms operating in the continental aquatic realm will benefit from future comparative approaches based on an integration of molecular, biogeographic, and ecological data aided by modern analytical tools. Do the salinity constraints on biodiversity also extend to restraints on cladogenesis? Do differences in depths of phylogeographic structure among terrestrial zooplankters covary with habitat size and type, predation pressure, and size of cyst banks? This chapter started with a paradox and ends with one more. In that respect, the use of *Artemia* as a model system will certainly contribute to a more unified theory and interpretive framework of inland waters.

ACKNOWLEDGEMENTS

We would like to thank Prof. Dr. Stefan Koenemann and the editorial team for kindly inviting us to contribute to this book. Also, two anonymous reviewers are thoughtfully acknowledged for their constructive comments.

REFERENCES

Abatzopoulos, T.J., Kastritsis, C.D. & Triantaphyllidis, C.D. 1986. A study of karyotypes and heterochromatic associations in *Artemia*, with special reference to two N. Greek populations. *Genetica* 71: 3–10.

Abatzopoulos, T.J., Zhang, B. & Sorgeloos, P. 1998. *Artemia tibetiana*: preliminary characterization of a new *Artemia* species found in Tibet (People's Republic of China). International Study on *Artemia*. LIX. *Int. J. Salt Lake Res.* 7: 41–44.

Abatzopoulos, T.J., Kappas, I., Bossier, P., Sorgeloos, P. & Beardmore, J.A. 2002a. Genetic characterization of *Artemia tibetiana* (Crustacea: Anostraca). *Biol. J. Linn. Soc.* 75: 333–344.

Abatzopoulos, T.J., Beardmore, J.A., Clegg, J.S. & Sorgeloos, P. 2002b. *Artemia: Basic and Applied Biology*. Dordrecht: Kluwer Academic Publishers.

Abatzopoulos, T.J., Agh, N., Van Stappen, G., Razavi Rouhani, S.M. & Sorgeloos, P. 2006a. *Artemia* sites in Iran. *J. Mar. Biol. Ass. UK* 86: 299–307.

Abatzopoulos, T.J., Baxevanis, A.D., Triantaphyllidis, G.V., Criel, G., Pador, E.L., Van Stappen, G. & Sorgeloos, P. 2006b. Quality evaluation of *Artemia urmiana* Günther (Urmia Lake, Iran) with special emphasis on its particular cyst characteristics (International Study on *Artemia* LXIX). *Aquaculture* 254: 442–454.

Abatzopoulos, T.J., Amat, F., Baxevanis, A.D., Belmonte, G., Hontoria, F., Maniatsi, S., Moscatello, S., Mura, G. & Shadrin, N.V. 2009. Updating geographic distribution of *Artemia urmiana* Günther, 1890 (Branchiopoda, Anostraca) in Europe: an integrated and interdisciplinary approach. *Internat. Rev. Hydrobiol.* 94: 560–579.

Abreu-Grobois, F.A. & Beardmore, J.A. 1982. Genetic differentiation and speciation in the brine shrimp *Artemia*. In: Barigozzi, C. (ed.), *Mechanisms of Speciation*: 345–376. New York, NY: Alan R. Liss, Inc.

Adamowicz, S.J., Petrusek, A., Colbourne, J.K., Hebert, P.D.N. & Witt, J.D.S. 2009. The scale of divergence: a phylogenetic appraisal of intercontinental allopatric speciation in a passively dispersed freshwater zooplankton genus. *Mol. Phylogenet. Evol.* 50: 423–436.

Adolfsson, S., Michalakis, Y., Paczesniak, D., Bode, S.N.S., Butlin, R.K., Lamatsch, D.K., Martins, M.J.F., Schmit, O., Vandekerkhove, J. & Jokela, J. 2010. Evaluation of elevated ploidy and asexual reproduction as alternative explanations for geographic parthenogenesis in *Eucypris virens* ostracods. *Evolution* 64: 986–997.

Agh, N., Abatzopoulos, T.J., Kappas, I., Van Stappen, G., Rouhani, S.M.R. & Sorgeloos, P. 2007. Coexistence of sexual and parthenogenetic *Artemia* populations in Lake Urmia and neighbouring lagoons. *Internat. Rev. Hydrobiol.* 92: 48–60.

Agh, N., Bossier, P., Abatzopoulos, T.J., Beardmore, J.A., Van Stappen, G., Mohammadyari, A., Rahimian H. & Sorgeloos, P. 2009. Morphometric and preliminary genetic characteristics of *Artemia* populations from Iran. *Internat. Rev. Hydrobiol.* 94: 194–207.

Amat, F., Barata, C., Hontoria, F., Navarro, J.C. & Varó, I. 1995a. Biogeography of the genus *Artemia* (Crustacea, Branchiopoda, Anostraca) in Spain. *Int. J. Salt Lake Res.* 3: 175–190.

Amat, F., Barata, C. & Hontoria, F. 1995b. A Mediterranean origin for the Veldrif (South Africa) *Artemia* Leach population. *J. Biogeogr.* 22: 49–59.

Amat, F., Cohen, R.G., Hontoria, F. & Navarro, J.C. 2004. Further evidence and characterization of *Artemia franciscana* (Kellogg, 1906) populations in Argentina. *J. Biogeogr.* 31: 1735–1749.

Avise, J.C. 1994. *Molecular Markers, Natural History and Evolution*. New York, NY: Chapman and Hall.

Avise, J.C. 2000. *Phylogeography: The History and Formation of Species*. Cambridge, MA: Harvard University Press.

Avise, J.C., Arnold, J., Ball, R.M., Bermingham, E., Lamb, T., Neigel, J.E., Reeb, C.A. & Saunders,

N.C. 1987. Intraspecific phylogeography: the mitochondrial DNA bridge between population genetics and systematics. *Ann. Rev. Ecol. Syst.* 18: 489–522.

Barata, C., Hontoria, F. & Amat, F. 1995. Life history, resting egg formation, and hatching may explain the temporal-geographical distribution of *Artemia* strains in the Mediterranean basin. *Hydrobiologia* 298: 295–305.

Barata, C., Hontoria, F., Amat, F. & Browne, R. 1996a. Demographic parameters of sexual and parthenogenetic *Artemia*: Temperature and strain effects. *J. Exp. Mar. Biol. Ecol.* 196: 329–340.

Barata, C., Hontoria, F., Amat, F. & Browne, R. 1996b. Competition between sexual and partheno-genetic *Artemia*: Temperature and strain effects. *J. Exp. Mar. Biol. Ecol.* 196: 313–328.

Barigozzi, C. 1974. Artemia: a survey of its significance in genetic problems. *Evol. Biol.* 7: 221–252.

Barraclough, T.G., Birky, W. Jr. & Burt, A. 2003. Diversification in sexual and asexual organisms. *Evolution* 57: 2166–2172.

Baxevanis, A.D. & Abatzopoulos, T.J. 2004. The phenotypic response of ME_2 (M. Embolon, Greece) *Artemia* clone to salinity and temperature. *J. Biol. Res.-Thessalon.* 1: 107–114.

Baxevanis, A.D., Triantaphyllidis, G.V., Kappas, I., Triantafyllidis, A., Triantaphyllidis, C.D. & Abatzopoulos, T.J. 2005. Evolutionary assessment of *Artemia tibetiana* (Crustacea, Anostraca) based on morphometry and 16S rRNA RFLP analysis. *J. Zool. Syst. Evol. Res.* 43: 189–198.

Baxevanis, A.D., Kappas, I. & Abatzopoulos, T.J. 2006. Molecular phylogenetics and asexuality in the brine shrimp *Artemia. Mol. Phylogenet. Evol.* 40: 724–738.

Beaton, M.J. & Hebert, P.D.N. 1988. Geographical parthenogenesis and polyploidy in *Daphnia pulex. Am. Nat.* 132: 837–845.

Bilton, D.T., Freeland, J.R. & Okamura, B. 2001. Dispersal in aquatic invertebrates. *Annu. Rev. Ecol. Syst.* 32: 159–181.

Boileau, M.G., Hebert, P.D.N. & Schwartz, S.S. 1992. Nonequilibrium gene frequency divergence: persistent founder effects in natural populations. *J. Evol. Biol.* 5: 25–39.

Bowen, S.T., Fogarino, E.A., Hitchner, K.N., Dana, G.L., Chow, V.H.S., Buoncristiani, M.R. & Carl, J.R. 1985. Ecological isolation in *Artemia*: population differences in tolerance of anion concentrations. *J. Crust. Biol.* 5: 106–129.

Braband, A., Richter, S., Hiesel, R. & Scholtz, G. 2002. Phylogenetic relationships within the Phyllopoda (Crustacea, Branchiopoda) based on mitochondrial and nuclear markers. *Mol. Phylogenet. Evol.* 25: 229–244.

Brendonck, L. & Riddoch, B.J. 1999. Wind-borne short-range egg dispersal in anostracans (Crustacea: Branchiopoda). *Biol. J. Linn. Soc.* 67: 87–95.

Browne, R.A. 1992. Population genetics and ecology of *Artemia*: insights into parthenogenetic reproduction. *Trends Ecol. Evol.* 7: 232–237.

Browne, R.A. & Hoopes, C.W. 1990. Genotypic diversity and selection in asexual brine shrimp (*Artemia*). *Evolution* 44: 1035–1051.

Browne, R.A., Li, M., Wanigasekera, G., Brownlee, S., Eiband, G. & Cowan, J. 1991. Ecological, physiological and genetic divergence of sexual and asexual (diploid and polyploid) brine shrimp (*Artemia*). *Adv. Ecol.* 1: 41–52.

Burton, R.S. & Lee, B.-N. 1994. Nuclear and mitochondrial gene genealogies and allozyme polymorphism across a major phylogeographic break in the copepod *Tigriopus californicus. Proc. Natl. Acad. Sci. USA* 91: 5197–5201.

Chaplin, J.A. & Hebert, P.D.N. 1997. *Cyprinotus incongruens* (Ostracoda): an ancient asexual? *Mol. Ecol.* 6: 155–168.

Clark, M.K., House, M.A., Royden, L.H., Whipple, K.X., Burchfiel, B.C., Zhang, X. & Tang, W. 2005. Late Cenozoic uplift of southeastern Tibet. *Geology* 33: 525–528.

Clegg, J.S., Van Hoa, N. & Sorgeloos, P. 2001. Thermal tolerance and heat shock proteins in encysted embryos of *Artemia* from widely different thermal habitats. *Hydrobiologia* 466: 221–229.

Colbourne, J.K., Wilson, C.C. & Hebert, P.D.N. 2006. The systematics of Australian *Daphnia* and *Daphniopsis* (Crustacea: Cladocera): a shared phylogenetic history transformed by habitat-specific rates of evolution. *Biol. J. Linn. Soc.* 89: 469–488.

Costa, F.O., deWaard, J.R., Boutillier, J., Ratnasingham, S., Dooh, R.T., Hajibabaei, M. & Hebert, P.D.N. 2007. Biological identification through DNA barcodes: the case of the Crustacea. *Can. J. Fish. Aquat. Sci.* 64: 272–295.

Crease, T.J., Stanton, D.J. & Hebert, P.D.N. 1989. Polyphyletic origins of asexuality in *Daphnia pulex*. II. Mitochondrial-DNA variation. *Evolution* 43: 1016–1026.

Daniels, S.R., Hamer, M. & Rogers, C. 2004. Molecular evidence suggests an ancient radiation for the fairy shrimp genus *Streptocephalus* (Branchiopoda: Anostraca). *Biol. J. Linn. Soc.* 82: 313–327.

De Gelas, K. & De Meester, L. 2005. Phylogeography of *Daphnia magna* in Europe. *Mol. Ecol.* 14: 753–764.

De Meester, L. 1996. Local genetic differentiation and adaptation in freshwater zooplankton: patterns and processes. *Ecoscience* 3: 385–399.

De Meester, L., Gómez, A., Okamura, B. & Schwenk, K. 2002. The Monopolization Hypothesis and the dispersal-gene flow paradox in aquatic organisms. *Acta Oecol.* 23: 121–135.

Drummond, A.J. & Rambaut, A. 2007. BEAST: Bayesian evolutionary analysis by sampling trees. *BMC Evol. Biol.* 7: 214.

Figuerola, J., Green, A.J. & Michot, T.C. 2005. Invertebrate eggs can fly: evidence of waterfowl-mediated gene flow in aquatic invertebrates. *Am. Nat.* 165: 274–280.

Fryer, G. 1987. A new classification of the branchiopod Crustacea. *Zool. J. Linn. Soc. Lond.* 91: 357–383.

Funk, D.J., Nosil, P. & Etges, W.J. 2006. Ecological divergence exhibits consistently positive associations with reproductive isolation across disparate taxa. *Proc. Natl. Acad. Sci. USA* 103: 3209–3213.

Gajardo, G., Da Conceicao, M., Weber, L. & Beardmore, J.A. 1995. Genetic variability and interpopulational differentiation of *Artemia* strains from South America. *Hydrobiologia* 302: 21–29.

Gajardo, G., Abatzopoulos, T.J., Kappas, I. & Beardmore, J.A. 2002. Evolution and speciation. In: Abatzopoulos, T.J., Beardmore, J.A., Clegg, J.S. & Sorgeloos, P. (eds), *Artemia: Basic and Applied Biology*: 225–250. Dordrecht: Kluwer Academic Publishers.

Gajardo, G., Crespo, J., Triantafyllidis, A., Tzika, A., Baxevanis, A.D., Kappas, I. & Abatzopoulos, T.J. 2004. Species identification of Chilean *Artemia* populations based on mitochondrial DNA RFLP analysis. *J. Biogeogr.* 31: 547–555.

Gajardo, G., Sorgeloos, P. & Beardmore, J.A. 2006. Inland hypersaline lakes and the brine shrimp *Artemia* as simple models for biodiversity analysis at the population level. *Saline Systems* 2: 14.

Gómez, A. & Lunt, D.H. 2007. Refugia within refugia: patterns of phylogeographic concordance in the Iberian Peninsula. In: Weiss, S. & Ferrand, N. (eds), *Phylogeography of Southern European Refugia*: 155–188. Dordrecht: Springer-Verlag.

Gómez, A., Carvalho, G.R. & Lunt, D.H. 2000. Phylogeography and regional endemism of a passively dispersing zooplankter: mitochondrial DNA variation in rotifer resting egg banks. *Proc. R. Soc. Lond. B* 267: 2189–2197.

Gómez, A., Serra, M., Carvalho, G.R. & Lunt, D.H. 2002. Speciation in ancient cryptic complexes: evidence from the molecular phylogeny of *Brachionus plicatilis* (Rotifera). *Evolution*

56: 1431–1444.

Green, A.J., Sanchez, M.I., Amat, F., Figuerola, J., Hontoria, F., Ruiz, O. & Hortas, F. 2005. Dispersal of invasive and native brine shrimps *Artemia* (Anostraca) via waterbirds. *Limnol. Oceanogr.* 50: 737–742.

Gregory-Wodzicki, K.M. 2000. Uplift history of the Central and Northern Andes: a review. *Geol. Soc. Am. Bull.* 112: 1091–1105.

Hammer, U.T. 1986. *Saline Lake Ecosystems of the World.* Dordrecht: Dr. W. Junk Publishers.

Hann, B.J. 1987. Naturally occurring interspecific hybridization in *Simocephalus* (Cladocera, Daphniidae): its potential significance. *Hydrobiologia* 145: 219–224.

Havel, J.E., Colbourne, J.K. & Hebert, P.D.N. 2000. Reconstructing the history of intercontinental dispersal in *Daphnia lumholtzi* by use of genetic markers. *Limnol. Oceanogr.* 45: 1414–1419.

Hebert, P.D.N. & Wilson, C.C. 1994. Provincialism in plankton: endemism and allopatric speciation in Australian *Daphnia. Evolution* 48: 1333–1349.

Hebert, P.D.N., Remigio, E.A., Colbourne, J.K., Taylor, D.J. & Wilson, C.C. 2002. Accelerated molecular evolution in halophilic crustaceans. *Evolution* 56: 909–926.

Hebert, P.D.N., Witt, J.D.S. & Adamowicz, S.J. 2003. Phylogeographic patterning in *Daphnia ambigua*: regional divergence and intercontinental cohesion. *Limnol. Oceanogr.* 48: 261–268.

Hickerson, M.J., Carstens, B.C., Cavender-Bares, J., Crandall, K.A., Graham, C.H., Johnson, J.B., Rissler, L., Victoriano, P.F. & Yoder, A.D. 2010. Phylogeographys past, present, and future: 10 years after Avise, 2000. *Mol. Phylogenet. Evol.* 54: 291–301.

Ishida, S. & Taylor, D.J. 2007. Quaternary diversification in a sexual Holarctic zooplankter, *Daphnia galeata. Mol. Ecol.* 16: 569–582.

Kaiser, H., Gordon, A.K. & Paulet, T.G. 2006. Review of the African distribution of the brine shrimp genus *Artemia. Water SA* 32: 597–604.

Kappas, I., Abatzopoulos, T.J., Van Hoa, N., Sorgeloos, P. & Beardmore, J.A. 2004. Genetic and reproductive differentiation of *Artemia franciscana* in a new environment. *Mar. Biol.* 146: 103–117.

Kappas, I., Baxevanis, A.D., Maniatsi, S. & Abatzopoulos, T.J. 2009. Porous genomes and species integrity in the branchiopod *Artemia. Mol. Phylogenet. Evol.* 52: 192–204.

Kearney, M. 2005. Hybridization, glaciation and geographical parthenogenesis. *Trends Ecol. Evol.* 20: 495–502.

Ketmaier, V., Pirollo, D., De Matthaeis, E., Tiedemann, R. & Mura, G. 2008. Large-scale mitochondrial phylogeography in the halophilic fairy shrimp *Phallocryptus spinosa* (Milne-Edwards, 1840) (Branchiopoda: Anostraca). *Aquat. Sci.* 70: 65–76.

Knowles, L.L. & Maddison, W.P. 2002. Statistical phylogeography. *Mol. Ecol.* 11: 2623–2635.

Knowlton, N. & Weigt, L.A. 1998. New dates and new rates for divergence across the Isthmus of Panama. *Proc. R. Soc. Lond. B* 265: 2257–2263.

Lee, C.E. 2000. Global phylogeography of a cryptic copepod species complex and reproductive isolation between genetically proximate "populations." *Evolution* 54: 2014–2027.

Little, T.J., Demelo, R., Taylor, D.J. & Hebert, P.D.N. 1997. Genetic characterization of an arctic zooplankter: insights into geographic polyploidy. *Proc. R. Soc. Lond. B* 264: 1363–1370.

Maeda-Martinez, A.M., Obregon-Barboza, H. & Dumont, H.J. 1992. *Branchinecta belki* n. sp. (Branchiopoda: Anostraca), a new fairy shrimp from Mexico, hybridizing with *B. packardi* Pearse under laboratory conditions. *Hydrobiologia* 239: 151–162.

Maniatsi, S., Kappas, I., Baxevanis, A.D., Farmaki, T. & Abatzopoulos, T.J. 2009a. Sharp phylogeographic breaks and patterns of genealogical concordance in the brine shrimp *Artemia franciscana. Int. J. Mol. Sci.* 10: 5455–5470.

Maniatsi, S., Baxevanis, A.D. & Abatzopoulos, T.J. 2009b. The intron 2 of p26 gene: a novel genetic marker for discriminating the two most commercially important *Artemia franciscana* subspecies. *J. Biol. Res.-Thessalon.* 11: 73–82.

Maniatsi, S., Bourtzis, K. & Abatzopoulos, T.J. 2010. May parthenogenesis in *Artemia* be attributed to *Wolbachia*? *Hydrobiologia* 651: 317–322.

Maniatsi, S., Baxevanis, A.D., Kappas, I., Deligiannidis, P., Triantafyllidis, A., Papakostas, S., Bougiouklis, D. & Abatzopoulos, T.J. 2011. Is polyploidy a persevering accident or an adaptive evolutionary pattern? The case of the brine shrimp *Artemia*. *Mol. Phylogenet. Evol.* 58: 353–364.

Mayr, E. 1963. *Animal Species and Evolution.* Cambridge, MA: Belknap Press of Harvard University Press.

Mergeay, J., Aguilera, X., Declerck, S., Petrusek, A., Huyse, T. & De Meester, L. 2008. The genetic legacy of polyploid Bolivian *Daphnia*: the tropical Andes as a source for the North and South American *D. pulicaria* complex. *Mol. Ecol.* 17: 1789–1800.

Mills, S., Lunt, D.H. & Gómez, A. 2007. Global isolation by distance despite strong regional phylogeography in a small metazoan. *BMC Evol. Biol.* 7: 225.

Muñoz, J., Gómez, A., Green, A.J., Figuerola, J., Amat, F. & Rico, C. 2008. Phylogeography and local endemism of the native Mediterranean brine shrimp *Artemia salina* (Branchiopoda: Anostraca). *Mol. Ecol.* 17: 3160–3177.

Naihong, X., Audenaert, E., Vanoverbeke, J., Brendonck, L., Sorgeloos, P. & De Meester, L. 2000. Low among-population genetic differentiation in Chinese bisexual *Artemia* populations. *Heredity* 84: 238–243.

Otto, S.P. & Whitton, J. 2000. Polyploid incidence and evolution. *Annu. Rev. Genet.* 34: 401–437.

Papeschi, A., Lipko, P., Amat, F. & Cohen, R.G. 2008. Heterochromatin variation in *Artemia* populations. *Caryologia* 61: 53–59.

Pilla, E.J.S. & Beardmore, J.A. 1994. Genetic and morphometric differentiation in Old World bisexual species of *Artemia* (the brine shrimp). *Heredity* 73: 47–56.

Penton, E.H., Hebert, P.D.N. & Crease, T.J. 2004. Mitochondrial DNA variation in North American populations of *Daphnia obtusa*: continentalism or cryptic endemism? *Mol. Ecol.* 13: 97–107.

Petrusek, A., Černý, M. & Audenaert, E. 2004. Large intercontinental differentiation of *Moina micrura* (Crustacea: Anomopoda): one less cosmopolitan cladoceran? *Hydrobiologia* 526: 73–81.

Ramírez, C.C., Salazar, M., Palma, R.E., Cordero, C. & Mezabasso, L. 2008. Phylogeographical analysis of neotropical *Rhagoletis* (Diptera: Tephritidae): Did the Andes uplift contribute to current morphological differences? *Neotrop. Entomol.* 37: 651–661.

Remigio, E.A., Hebert, P.D.N. & Savage, A. 2001. Phylogenetic relationships and remarkable radiation in *Parartemia* (Crustacea: Anostraca), the endemic brine shrimp of Australia: evidence from mitochondrial DNA sequences. *Biol. J. Linn. Soc.* 74: 59–71.

Richter, S., Olesen, J. & Wheeler, W.C. 2007. Phylogeny of Branchiopoda (Crustacea) based on a combined analysis of morphological data and six molecular loci. *Cladistics* 23: 1–36.

Ruiz, O., Medina, G.R., Cohen, G., Amat, F. & Navarro, J.C. 2007. Diversity of the fatty acid composition of *Artemia* sp. cysts from Argentinean populations. *Mar. Ecol. Prog. Ser.* 335: 155–165.

Ruiz, O., Amat, F., Saavedra, C., Papeschi, A., Cohen, R.G., Baxevanis, A.D., Kappas, I., Abatzopoulos, T.J. & Navarro, J.C. 2008. Genetic characterization of Argentinean *Artemia* species with different fatty acid profiles. *Hydrobiologia* 610: 223–234.

Schön, I., Butlin, R.K., Griffiths, H.I. & Martens, K. 1998. Slow molecular evolution in an ancient asexual ostracod. *Proc. R. Soc. Lond. B* 265: 235–242.

Schwenk, K., Posada, D. & Hebert, P.D.N. 2000. Molecular systematics of European *Hyalodaphnia*: the role of contemporary hybridization in ancient species. *Proc. R. Soc. Lond. B* 267: 1833–1842.

Shiklomanov, I.A. 1990. Global water resources. *Nat. Resour.* 26: 34–43.

Slusarczyk, M. & Pietrzak, B. 2008. To sink or float: the fate of dormant offspring is determined by maternal behaviour in *Daphnia. Freshwater Biol.* 53: 569–576.

Suatoni, E., Vicario, S., Rice, S., Snell, T. & Caccone, A. 2006. An analysis of species boundaries and biogeographic patterns in a cryptic species complex: the rotifer—*Brachionus plicatilis. Mol. Phylogenet. Evol.* 41: 86–98.

Sun, Y., Song, W., Zhong, Y., Zhang, R., Abatzopoulos, T.J. & Chen, R. 1999. Diversity of genetic differentiation in *Artemia* species and populations detected by AFLP markers. *Int. J. Salt Lake Res.* 8: 314–350.

Tanguay, J.A., Reyes, R.C. & Clegg, J.S. 2004. Habitat diversity and adaptation to environmental stress in encysted embryos of the crustacean *Artemia. J. Biosci.* 29: 489–501.

Taylor, D.J., Finston, T.L. & Hebert, P.D.N. 1998. Biogeography of a widespread freshwater crustacean: pseudocongruence and cryptic endemism in the North American *Daphnia leavis* complex. *Evolution* 52: 1648–1670.

Taylor, D.J., Ishikane, C.R. & Haney, R.A. 2002. The systematics of Holarctic bosminids and a revision that reconciles molecular and morphological evolution. *Limnol. Oceanogr.* 47: 1486–1495.

Timms, B.V. & Hudson, P. 2009. The brine shrimps (*Artemia* and *Parartemia*) of South Australia, including descriptions of four new species of *Parartemia* (Crustacea: Anostraca: Artemiina). *Zootaxa* 2248: 47-68.

Timms, B.V., Pinder, A.M. & Campagna, V.S. 2009. The biogeography and conservation status of the Australian endemic brine shrimp *Parartemia* (Crustacea, Anostraca, Parartemiidae). *Conserv. Sci. W. Aust.* 7: 413–427.

Tizol-Correa, R., Maeda-Martínez, A.M., Weekers, P.H.H., Torrentera, L. & Murugan, G. 2009. Biodiversity of the brine shrimp *Artemia* from tropical salterns in southern México and Cuba. *Curr. Sci.* 96: 81–87.

Triantaphyllidis, G.V., Criel, C.R.J., Abatzopoulos, T.J., Thomas, K.M., Peleman, J., Beardmore, J.A. & Sorgeloos, P. 1997. International Study on *Artemia.* LVII. Morphological and molecular characters suggest conspecificity of all bisexual European and North African *Artemia* populations. *Mar. Biol.* 129: 477–487.

Triantaphyllidis, G.V., Abatzopoulos, T.J. & Sorgeloos, P. 1998. Review of the biogeography of the genus *Artemia* (Crustacea, Anostraca). *J. Biogeogr.* 25: 213–226.

Turgeon, J. & Hebert, P.D.N. 1994. Evolutionary interactions between sexual and all-female taxa of *Cyprinotus* (Ostracoda, Cyprinidae). *Evolution* 48: 1855–1865.

Vandel, A. 1928. La parthénogenèse geographique. Contribution à létude biologique et cytologique de la parthénogenèse naturelle. *Bull. Biol. France Belg.* 62: 164–281.

Van Stappen, G. 2002. Zoogeography. In: Abatzopoulos, T.J., Beardmore, J.A., Clegg, J.S. and Sorgeloos, P. (eds), *Artemia: Basic and Applied Biology*: 171–224. Dordrecht: Kluwer Academic Publishers.

Ward, R.D., Bickerton, M.A., Finston, T. & Hebert, P.D.N. 1994. Geographical cline in breeding systems and ploidy levels in European populations of *Daphnia pulex. Heredity* 73: 532–543.

Williams, W.D. 1998. Salinity as a determinant of the structure of biological communities in salt lakes. *Hydrobiologia* 381: 191–201.

Williams, W.D. 2001. Anthropogenic salinisation of inland waters. *Hydrobiologia* 466: 329–337.

Williams, W.D. 2002. Environmental threats to salt lakes and the likely status of inland saline ecosystems in 2025. *Environ. Conserv.* 29: 154–167.

Wiman, F.H. 1979. Mating patterns and speciation in the fairy shrimp genus *Streptocephalus. Evolution* 33: 172–181.

Witt, J.D.S. & Hebert, P.D.N. 2000. Cryptic species diversity and evolution in the amphipod genus

Hyalella within central glaciated North America: a molecular phylogenetic approach. *Can. J. Fish. Aquat. Sci.* 57: 687–698.

Xin, N., Sun, J., Zhang, B., Triantaphyllidis, G.V., Van Stappen, G. & Sorgeloos, P. 1994. International Study on *Artemia*. LI. New survey of *Artemia* resources in the People's Republic of China. *Int. J. Salt Lake Res.* 3: 105–112.

Xu, S., Hebert, P.D.N., Kotov, A.A. & Cristescu, M.E. 2009. The noncosmopolitanism paradigm of freshwater zooplankton: insights from the global phylogeography of the predatory cladoceran *Polyphemus pediculus* (Linnaeus, 1761) (Crustacea, Onychopoda). *Mol. Ecol.* 18: 5161–5179.

Yu, Y., Harris, A.J. and He, X. 2010. S-DIVA (statistical dispersal-vicariance analysis): a tool for inferring biogeographic histories. *Mol. Phylogenet. Evol.* 56: 848–850.

Zhang, L. & Lefcort, H. 1991. The effects of ploidy level on the thermal distributions of brine shrimp *Artemia parthenogenetica* and its ecological implications. *Heredity* 66: 445–452.

Zhang, L. & King, C.E. 1992. Genetic variation in sympatric populations of diploid and polyploid brine shrimp (*Artemia parthenogenetica*). *Genetica* 85: 211–221.

Zhang, L. & King, C.E. 1993. Life-history divergence of sympatric diploid and polyploid populations of brine shrimp *Artemia parthenogenetica*. *Oecologia* 93: 177–183.

Zheng, M. 1997. *An Introduction to Saline Lakes on the Qinghai-Tibet Plateau*. Dordrecht: Kluwer Academic Publishers.

Zheng, M., Tang, J., Liu, J. & Zhang, F. 1993. Chinese saline lakes. *Hydrobiologia* 267: 23–36.

Zuñiga, O., Wilson, R., Amat, F. & Hontoria, F. 1999. Distribution and characterization of Chilean populations of the brine shrimp *Artemia* (Crustacea, Branchiopoda, Anostraca). *Int. J. Salt Lake Res.* 8: 23–40.

Intraspecific geographic differentiation and patterns of endemism in freshwater shrimp species flocks in ancient lakes of Sulawesi

KRISTINA VON RINTELEN

Museum für Naturkunde, Leibniz-Institut für Evolutions- und Biodiversitätsforschung an der Humboldt-Universität zu Berlin, Invalidenstr. 43, 10115 Berlin, Germany

ABSTRACT

Freshwater shrimps of the genus *Caridina* are the most speciose group within the family Atyidae. This diversity can be witnessed in two ancient lake systems of the Indonesian island Sulawesi, where 21 currently described species, reflect the largest radiations of freshwater shrimps known worldwide. Both lake systems provide similar environmental conditions with locally heterogeneous habitats: Lake Poso is a single lake with less pronounced geographic structure than the Malili lake system with five lakes forming part of the same drainage system. The spatial subdivision in the Malili lakes and the higher number of species compared to Lake Poso (15 vs. 6) renders this system special interest for the study of geographic differentiation.

Using mitochondrial DNA data, the phylogeographic patterns of the majority of the lacustrine species in the radiations are assessed. Both lake systems have been sampled comprehensively over several years, i.e. 137 localities altogether. A molecular phylogeny based on two standard genes (16S and COI) was reconstructed. A mismatch between molecules and morphology hints towards cases of introgression or incomplete lineage sorting in some species. To analyse geographic differentiation, geographic data for each species was mapped onto the phylogeny. In addition, haplotype networks were constructed to assess intralacustrine geographic patterns. In general, the molecular phylogeny revealed that all species, with one exception, are endemic to the ancient lakes, always showing local endemism within one system or even single lakes. Species from the Malili lakes revealed more geographic differentiation than species from the solitary Lake Poso. In several species from the Malili lakes, geographically separated populations form distinct subclades, suggesting limited gene flow between these populations. In other cases, allopatrically occurring populations within species were found in separate clades that were not even sister groups to each other in the overall phylogeny. This suggests not only the existence of cryptic species but also of unrecorded endemism.

1 Introduction

Freshwater caridean shrimps occur in all biogeographic regions bar Antarctica (De Grave et al. 2008), but are in general among the less studied groups of decapod crustaceans. This might not be surprising regarding the fact that the majority of shrimp-like decapods are found in marine environments. Freshwater taxa only account for approximately a quarter of all described Caridea and are numerically dominated by the two families Atyidae and Palaemonidae (see De Grave et al. 2008). At present, the Atyidae contain 42 extant genera (De Grave et al. 2009; Cai 2010). The vast majority of species are described within the genus *Caridina* H. Milne Edwards 1837 (around 200 species were estimated by De Grave et al. 2008), which is widely distributed throughout the Indo-West Pacific. These small shrimps (2–3 cm) are often abundant in various freshwater habitats (including cave systems). Regardless of its doubtful monophyly (Page et al. 2007), *Caridina* is generally interesting for phylogeographic studies: it comprises widespread species with high dispersal abilities (catadromous marine larval development) and land-locked species with low dispersal abilities (direct larval development) only occurring in true freshwater environments (Lai & Shy 2009).

Typical land-locked freshwater habitats are, for example, highland rivers and lakes. In ancient lakes, several crustacean groups have produced extensive species flocks with a high degree of endemicity (Martens & Schön 1999). For biologists, an ancient lake is generally an extant lake that exists for at least 100,000 years (Gorthner 1994; Martens 1997). The best known example certainly is the amphipod radiation in Lake Baikal (Macdonald et al. 2005). Among the few cases of ancient lake radiations in decapod crustaceans are the endemic species flocks of freshwater crabs and shrimps in Lake Tanganyika (Fryer 2006; Marijnissen et al. 2008) and of crabs and shrimps in the two ancient lake systems of the Indonesian island Sulawesi, the former Celebes (Schubart & Ng 2008; von Rintelen & Cai 2009).

The two Sulawesi lake systems, the focus of this chapter, are located in the central mountains of the island (Figure 1A). Lake Poso (Figure 1B) is a deep, solitary and trough-like lake that is drained northwards via Poso River into the sea. The Malili lake system comprises five lakes sharing a common drainage (Figure 1C): the three larger lakes, Matano, Mahalona, and Towuti, are directly connected via Petea and Tominanga River, whereas the two smaller satellite lakes, Lontoa and Masapi, are not directly connected. The Malili system is drained westwards via Larona River into the sea (Figure 1C). The age of Lake Matano (with 590 m, the eighth deepest lake in the world) was estimated at 1–2 my (Geoffroy Hope, pers. comm.); Lake Towuti may be ca. 600,000 years old based on seismic data (James Russell, pers.comm.). Both lake systems provide similar environmental conditions (oligotrophy, very low nutrient and organic content, high transparency; Giesen 1994; Haffner et al. 2001) with locally heterogeneous habitats ranging from different types of soft (sand, mud, macrophytes) to hard substrates, such as wood or rocky drop-offs (von Rintelen et al., in press). In general, the solitary Lake Poso has a less pronounced geographic structure than the five connected Malili lakes and a less heterogeneous habitat fragmentation. The spatial subdivison makes the Malili system especially interesting for the study of geographic differentiation.

Four major groups of freshwater organisms have radiated in the central lakes of Sulawesi: fishes, snails, crabs, and shrimps (for details see von Rintelen et al., in press). Both lake systems harbour endemic species flocks of *Caridina* first described by Woltereck (1937) for the Malili lakes and by Schenkel (1902) for Lake Poso. Today, 21 endemic species are known (including the river systems), six from Lake Poso and 15 from the Malili lake system (von Rintelen & Cai 2009). The Malili flock represents the largest radiation within the genus *Caridina* and even within the Atyidae (von Rintelen & Cai 2009). A similar atyid radiation is so far only known from Lake Tanganyika. However, the Tanganyikan species flock comprises three different genera of other atyid shrimps with a total of 11 species altogether (Fryer 2006).

A recent study inferred from mitochondrial DNA sequences revealed three highly supported lake clades, one in Lake Poso and two in the Malili lake system (Figure 1D), that were interpreted as three independent lake colonizations (von Rintelen et al. 2010). In both species flocks, morphological, genetic, and ecological data were further clearly indicative of adaptive radiations, involving habitat-specific diversification of feeding appendages (von Rintelen et al. 2007, 2010). In case of the Malili lakes, von Rintelen et al. (2010) and Roy et al. (2006) already found hints for a geographic differentiation within the entire species flock and within single species, e.g. genetic differences between allopatrically occurring populations. Thus, it was suggested that inter-lacustrine allopatry plays an important role in species diversification and speciation in the Malili system (von Rintelen et al. 2010). In Lake Poso, results showed no significant intraspecific genetic structuring so far, which was accounted for by the lack of geographic subdivisions in this trough-shaped lake (von Rintelen et al. 2007).

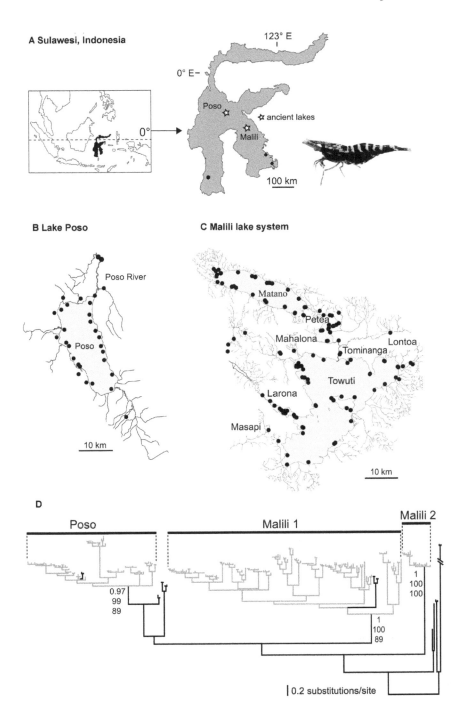

Figure 1. Sulawesi (Indonesia) and the ancient lakes. (A) Geographic location of the ancient lakes. (B) Lake Poso with collecting sites. (C) Malili lake system with collecting sites. (D) Bayesian Inference phylogram (1332 basepairs of combined 16S and COI mtDNA) showing three ancient lake clades (grey lines) in Sulawesi. The black lines indicate species from rivers outside of the lakes. Numbers next to each lake clade are, from top, Bayesian posterior probabilities, ML and MP bootstrap values (tree modified from von Rintelen et al. 2010).

Table 1. Maximum intraspecific genetic divergences (p-distance in %) within the three ancient lake clades. Source: *: von Rintelen et al. (2007); **: von Rintelen et al. (2010).

Species of *Caridina*	COI	16S
Poso (entire clade)*	**7.7**	**3.7**
*C. acutirostris***	1.2	0.7
*C. caerulea***	0.8	0.6
*C. ensifera***	2.3	0.9
*C. longidigita***	1.3	0.4
*C. sarasinorum***	0.9	0.2
*C. schenkeli***	4.9	2.1
Malili 1 (entire clade)**	**11.5**	**6.4**
*C. lingkonae****	0.8	0.4
*C. dennerli****	0.8	0.4
*C. glaubrechti****	9.7	3.1
*C. holthuisi****	5.2	2.4
*C. loehae****	2.2	0.7
*C. mahalona****	8.3	2.7
*C. masapi****	2.5	1.8
*C. parvula****	5.0	2.0
*C. profundicola****	1.8	0.4
*C. spinata****	0.5	0.0
*C. spongicola****	1.8	1.1
*C. striata****	6.8	2.0
*C. tenuirostris****	3.3	1.8
*C. woltereckae****	5.0	1.5
Malili 2**		
*C. lanceolata****	2.8	0.4

On the basis of previous studies (von Rintelen & Cai 2009; von Rintelen et al. 2007, 2010), this chapter compares results from both lake systems; it summarizes and reviews some general ideas on radiation and species diversity in both ancient lake systems, whereas in earlier molecular studies results were discussed separately and mostly at the interspecific level (Lake Poso: von Rintelen et al. 2007; Malili lake system: von Rintelen et al. 2010). As a new methodological approach in this chapter, phylogeography is studied by utilizing haplotype networks and mapping geographic distribution of haplotypes for each of the species. Hereby, new hypotheses could be tested such as populations from the Malili lakes showing higher genetic and geographic structure than populations from Lake Poso due to the higher geographic division of the Malili lake system. The so far rather superficial phylogeographic approach to the Sulawesi lake radiations of *Caridina* is therefore taken further into detail with the aim to reveal new insights into intraspecific geographic differentiation and patterns of endemism of atyid shrimps from the islands ancient lakes.

2   MATERIAL AND METHODS

Specimens were collected from both ancient lake systems and surrounding rivers between 2002 and 2007 from 137 sites (107 from the Malili lakes, 30 from Lake Poso). Sampling was either done with hand nets or with plastic containers while snorkeling and scuba diving. Sampling sites were chosen to cover all geographic subdivisions of the lakes as well as all major river systems. Basic ecological data such as substrate or depth were recorded. The dense coverage of sampling points is visualized in Figures 1A and 1C (the lack of sampling points in rivers can be attributed to the absence of shrimps, e. g. in the rivers southeast of Lake Towuti in Figure 1C). Specimens were fixed and preserved in 70–95% ethanol, vouchers are deposited in the crustacean departments of the Museum of Natural History of Berlin, Germany (ZMB), and the Museum Zoologicum Bogoriense, Indonesia (MZB); for details on voucher numbers see von Rintelen et al. (2010). Species were generally identified using a suite of morphological characters commonly employed in atyid taxonomy as described by von Rintelen & Cai (2009).

The three major lakes of the Malili lake system are drained and/or connected by the larger rivers Petea, Tominanga, and Larona (compare Figure 1C). As these can be considered extensions of the lakes themselves (albeit with currents), specimens from these rivers have been treated here as lacustrine and not as riverine samples. In consequence, these localities have been assigned to the lake of which the respective river is draining: Petea = Lake Matano; Tominanga = Lake Mahalona; Larona = Lake Towuti. The tree topology of the molecular phylogeny shown in Figures 1–3 and the maximum intraspecific genetic divergences shown in Table 1 of species from both lake systems were taken and modified from the studies by von Rintelen et al. (2007, 2010), in which two mitochondrial gene fragments, approximately 560 basepairs (bp) of the large ribosomal subunit (16S) and 861 bp of the cytochrome c oxidase subunit I (COI) were sequenced; for details on molecular methods please see von Rintelen et al. (2010). To illustrate intraspecific geographic differentiation of single species occurring in different lakes of the Malili lake system, geographic data for each species were mapped onto the phylogeny in Figure 2 (this was not done for Lake Poso species due to the single lake situation).

The dataset for haplotype networks were taken from von Rintelen et al. (2010) for both genes separately: 141 specimens of 15 species from the Malili lake system and 70 specimens of six species from Lake Poso. DAMBE v. 5.0.66 (Xia & Xie 2001) was used for reducing the dataset to unique haplotypes. Network reconstruction was done using TCS v. 1.21 (Clement et al. 2000) with a connection limit (parsimony criterion) of 95 and occasionally 90% for connection of more distant networks (as in *C. mahalona*, *C. tenuirostris*, *C. woltereckae*, *C. lanceolata*, and *C. glaubrechti*). For the analysis of intraspecific geographic differentiation between and within lakes, the geographic distribution of haplotypes was then mapped onto images of both lake systems.

3   RESULTS AND DISCUSSION

3.1   Radiation and species diversity

The three ancient lakes' clades (Figure 1D) are shown in detail for species of the Malili lake system (Figure 2) and Lake Poso (Figure 3A). The majority of species from the Malili lakes (Figure 2) form a single clade (Malili 1). The second clade (Malili 2) consists of only one species, *Caridina lanceolata*. In Lake Poso (Figure 3A), all species can be found in only one clade. These three clades were interpreted by von Rintelen et al. (2010) as three independent lake colonization events with subsequent adaptive radiations in two cases (Malili 1 and Poso).

A mainly morphology- and ecology-based revision of all ancient lake species revealed a total number of 21 species for both lake systems and, with 15 (Malili) vs. 6 (Poso) species, a higher diversity in the Malili lakes (von Rintelen & Cai 2009). This difference in taxonomic diversity is also reflected in the maximum intraspecific genetic distances found in each clade (Poso: 7.7%; Malili 1: 11.5%; Malili 2: 2.8%; COI, p-distance; Table 1). The additional molecular analyses based on mitochondrial DNA (von Rintelen et al. 2007, 2010) in part corroborated the results of

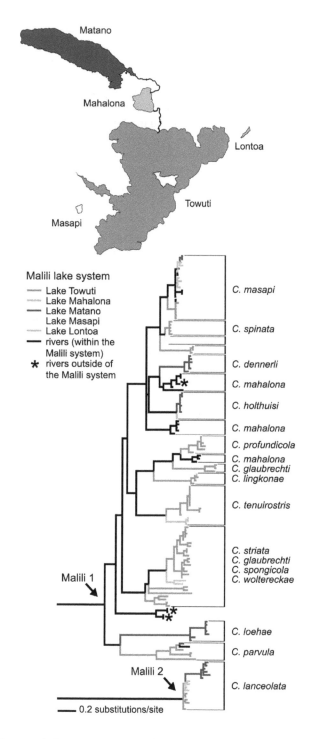

Figure 2. Bayesian Inference phylogram (1332 basepairs of combined 16S and COI mtDNA) of *Caridina* from the Malili lake system. Detail topology of the two Malili clades with its 15 species (for the entire topology, compare Figure 1D). The occurrence of sequenced specimens in single lakes and surrounding rivers are color-coded (see Figure 6 in Color insert; modified from von Rintelen et al. 2010).

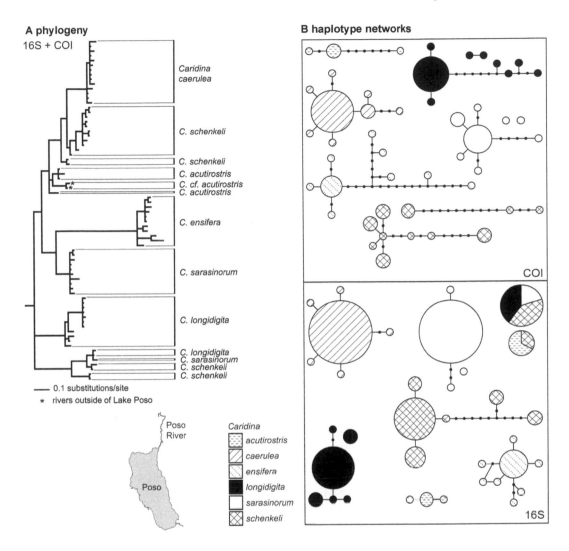

Figure 3. Lake Poso *Caridina*. (A) Bayesian Inference phylogram (mtDNA, 16S and COI) of *Caridina* from Lake Poso and surrounding rivers. Detail topology of the Poso clade with its six species (modified from von Rintelen et al. 2010). (B) Haplotype networks for the six Poso species from two mitochondrial genes, COI (upper panel) and 16S (lower panel).

the morphological species delimitation: several species from both systems appear monophyletic in the molecular phylogeny, for example *Caridina spinata*, *C. dennerli*, *C. lingkonae* from the Malili lakes (Figure 2) and *C. caerulea* , *C. ensifera* from Lake Poso (Figure 3A). However, several other species could not be recovered as monophyletic in the molecular tree, for example the *Caridina striata/glaubrechti/spongicola/woltereckae* clade (Figures 2 and 6) and *C. schenkeli*, *C. longidigita*, *C. sarasinorum* from Lake Poso (Figure 3A). Table 1 further illustrates the generally higher maximum genetic divergences (p-distance in %) in these species compared to those that appear monophyletic in the molecular tree: good examples are *Caridina striata/C. glaubrechti/C. woltereckae* (6.8%/9.7%/5.0%) vs. *C. spinata/C. dennerli/C. lingkonae* (0.5%/0.8%/0.8%) from the Malili lakes (Figure 2) and *C. caerulea* (0.8%) vs. *C. longidigita* (1.3%) from Lake Poso (Figure 3A). Haplotype networks (without geographic context) are shown for the six Poso species in Figure 3B. The COI-based networks (upper panel) show a good resolution at species level. All haplotypes appear in separated and species-specific networks, but haplotypes of three species, viz. *Caridina schenkeli*, *C. sarasinorum* and *C. longidigita*, are not always connected to their main network. In the 16S-based network (lower panel), representatives of these three species share a common haplotype which is different from their respective species-specific network (see also shared haplotype by *C. acutirostris* and *C. schenkeli*). Corrsepondingly, in the phylogeny, the shared haplotypes of these species also cluster together in a single clade (Figure 3A), which is distinct from the 'main' clades comprising the vast majority of haplotypes of each species. This is an indication for additional cryptic species which appear to hybridize with these genetically heterogeneous species. Less likely is the explanation of introgressive hybridization or incomplete lineage sorting as suggested in von Rintelen et al. (2007), as these represent unrelated haplogroups.

In general, all 21 former morpho-species in both lake systems were regarded and described as valid species (von Rintelen & Cai 2009). However, previous studies using molecular methods gave evidence for mismatch between morphology and molecules, which could be explained by introgressive hybridization, incomplete lineage sorting (persistence of ancestral polymorphism after recent speciation events), or the prevalence of cryptic species (von Rintelen et al. 2007, 2010). In Lake Poso, two morphologically almost indistinguishable (except for body coloration of living specimens in the field) species, *Caridina caerulea* and *C. ensifera* (compare Figure 3A), were considered one species, before genetic data revealed the existence of a "cryptic" species (von Rintelen et al. 2007): sympatrically occurring populations appeared in two separate clades that were not sister groups to each other. All members of each clade were characterized by different color morphs ("blue" or "red") as living animals in the field (not after alcohol bleaching). Later, other morphological and behavioural differences between the two species were found. Based on these results, von Rintelen and Cai (2009) described the former unrecognized species as a new taxon. For the Malili species flock, similar cases of cryptic species were found, although with a geographic background (von Rintelen et al. 2010). In three cases (*Caridina masapi*, *C. holthuisi*, and *C. mahalona*), allopatric populations (from different lakes) of single species appeared in separate clades that are not sister groups to each other (Figure 2), e.g. a Lake Towuti clade and a Lake Matano/Mahalona clade in *Caridina holthuisi* and a separate Lake Masapi clade in *C. masapi*. Based on these results, the real species diversity might even be higher than described.

3.2   Patterns of endemism

In general, all ancient lake species are endemic to their respective lake system (von Rintelen & Cai 2009; von Rintelen et al. 2010). *Caridina mahalona* is the only species of which one specimen was also collected from a different drainage system just north of Lake Matano (asterisk within the *C. mahalona* clade in Figure 2). In both lake systems, lacustrine species mainly occurring within the lakes can be distinguished from riverine species mainly or exclusively occurring in the surrounding river systems (compare maps in Figures 4–6). In the Malili lake system, only *Caridina mahalona* is an exclusive river dweller (Figure 4). The majority (almost 80%) of the 15 Malili species are endemic to one ($n = 6$) or two lakes ($n = 5$) within the system, e.g. *Caridina profundicola* in Lake Towuti (Figures 2 and 4) or *C. dennerli* in Lake Matano (Figures 2 and 5). Only a few species are widely distributed, e.g. *Caridina lanceolata* (Figures 2 and 4). In Lake Poso, four species are exclusively lacustrine (Figure 6), i. e. they are endemic to the lake itself. The other two species only occur in the Poso river system or at the outlet of the lake but not within the lake proper (*Caridina schenkeli* and *C. acutirostris* in Figure 7).

In at least three cases from the Malili lakes, the existence of cryptic species was suggested (see above). In these species, the genetically divergent populations are restricted to different geographic

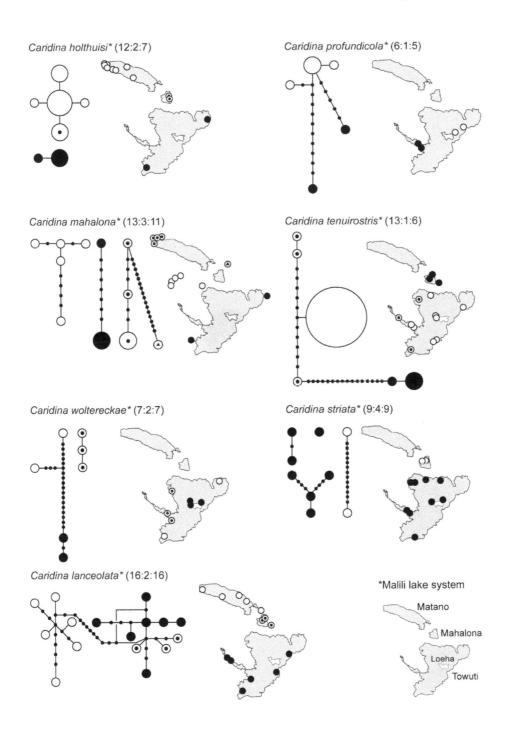

Figure 4. Phylogeography of Malili species. The COI haplotype network is shown for each species. The geographic distribution of the respective haplotypes is shown on a simplified map next to each network. The numbers in brackets indicate the species-specific haplotype information for: number of specimens examined: number of unconnected networks/haplotypes: number of haplotypes.

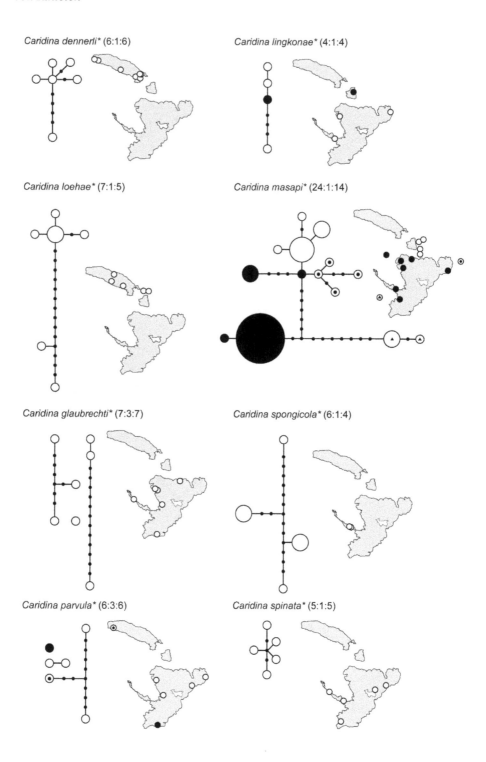

Caridina dennerli* (6:1:6)

Caridina lingkonae* (4:1:4)

Caridina loehae* (7:1:5)

Caridina masapi* (24:1:14)

Caridina glaubrechti* (7:3:7)

Caridina spongicola* (6:1:4)

Caridina parvula* (6:3:6)

Caridina spinata* (5:1:5)

Figure 5. Phylogeography of Malili species continued. The COI haplotype network is shown for each species. The geographic distribution of the respective haplotypes is shown on a simplified map next to each network. Lake names and numbers in brackets as in Figure 4.

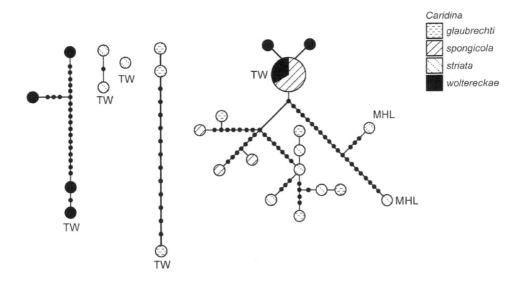

Figure 6. Phylogeography of Malili species continued. The COI haplotype network is shown for the unresolved *C. glaubrechti/spongicola/striata/woltereckae* clade. Lake names and numbers in brackets as in Figure 4.

areas of the Malili system and appear as clades in a non-sister relationship in the overall phylogeny (Figure 2). A good example is provided by the populations of *Caridina holthuisi* that are different in lakes Matano/Mahalona compared to Lake Towuti (Figures 2 and 4). In riverine *Caridina mahalona*, cryptic endemism is less obvious (Figure 4), although the distribution patterns of populations of the four clades (Figure 2) do not overlap and are restricted to different catchment areas within the Malili system (i.e. to Lake Matano catchment, Lake Mahalona catchment, to an area west of Lake Towuti and to Lake Towuti catchment; Figure 4). Another case of cryptic endemism may be present in *Caridina masapi*: the populations from the type locality, Lake Masapi, are genetically distinct and form the sister group of a *C. spinata* clade (Figure 2), making it likely that the other, geographically separated populations of this species are not conspecific (Figure 5). In Lake Poso, no cases of cryptic endemism have been recorded so far.

3.3 Geography and intraspecific differentiation

The Malili lakes and likewise the two Malili clades are strongly structured geographically. Strong intraspecific genetic differences were not only found among clades (as defined in Figure 2) and postulated above to represent cryptic species, but likewise within clades (see also von Rintelen et al. 2010). In several species, populations from different lakes form distinct subclades within clades (Figure 2) such as in *Caridina lanceolata* (Towuti/Mahalona/Matano subclades) and *C. tenuirostris* (Towuti/Mahalona subclades). These species with intraspecific allopatric distribution show higher genetic divergence as species without subclades (COI p–distance > 2%, Table 1). For this type of differentiation, rivers between the respective lakes (Tominanga River and Petea River; Figure 1C) were suggested as barriers to gene flow leading to small-scale vicariance (Roy et al. 2006; von Rintelen et al. 2010). However, neither these rivers nor other potential barriers within the Malili system can be regarded as universal barriers, because in other species populations from the respective lakes are genetically only marginally distinct (von Rintelen et al. 2010), as in populations of *Caridina holthuisi* from lakes Matano and Mahalona (Figures 2 and 4). So far, no consistent morphological differences were found between different populations in any of the ancient lake species.

Intraspecific geographic differentiation becomes even more obvious when haplotype networks for all ancient lake species are compared (Figures 4–7). In several cases, allopatric distribution patterns were verified by unconnected networks at the 90% parsimony level (e.g. *Caridina holthuisi* in Figure 4) or if haplotypes were only connected by long branches with many missing haplotypes (e.g. the Lake Masapi populations of *C. masapi* in Figure 5). Surprisingly, and unexpected at least for Lake Poso (see von Rintelen et al. 2007), populations previously considered as uniform entities (von Rintelen & Cai 2009; von Rintelen et al. 2010) were found to be geographically subdivided. In *Caridina profundicola*, a species restricted to Lake Towuti, populations are divided into one at the eastern part of the lake and one at the outlet bay in the western part (Figure 4). These populations are separated by a comparatively high number of mutational steps (5 and 10, respectively; see Figure 4). Furthermore, the two haplotypes (single specimen for each haplotype) occurring in the western part are more distinct from each other than both from the eastern haplotypes (16 mutational steps within the western population in comparison to 5 and 10 steps between western and eastern populations). Similar examples of geographically separated populations are found in *C. woltereckae* from Lake Towuti (Figures 4 and 6) and *C. parvula* from lakes Towuti and Matano (Figure 5). In Lake Poso highest geographic differentiation is found in *Caridina ensifera* (Figure 7), which might explain its relatively high intraspecific genetic divergence (COI p–distance 2.3%; Table 1). In this species, populations from the central eastern shore of Lake Poso are genetically distinct from other populations. Other species from Lake Poso show different geographic patterns (*C. longidigita, C. sarasinorum, C. schenkeli*; Figure 7). Nevertheless, regarding the rather small number of specimens for all studied species, further sampling of the respective populations from both lake systems would be needed to test any differentiation hypothesis more thoroughly and with the corresponding F–statistics. In only few cases, no intraspecific pattern was found, e.g. in *Caridina dennerli* from Lake Matano (Figure 5), in *C. glaubrechti* and *C. spinata* from Lake Towuti (Figures 5 and 6), or in *C. caerulea* from Lake Poso (Figure 7).

4   CONCLUSIONS

The potential of geographic subdivision, past and present, to shape the evolution of species has long been known (e.g. Mayr 1963) and the resulting genetic signature in the fast-evolving mtDNA was the initially preferred tool for the comparatively new field of phylogeography (Avise et al. 1987). Ideally, the geographic placement and inferred age of a genetic subdivision will allow one to draw conclusions on the geographic causation. Here, genetic subdivisions are shown to be present in most species in the ancient lake radiations of *Caridina* of Sulawesi. While details are shown above, three more general points should be made. First of all, genetic subdivisions almost always coincide with a geographic subdivision. This may seem trivial in a strongly structured system like the Malili lakes, but is less obvious in Lake Poso due to the lack of geographic subdivision. However, the possibility of past allopatric settings cannot be ruled out as the palaeohydrology of Lake Poso is not known (von Rintelen et al. 2007). Secondly, presumed geographic barriers between the different Malili lakes (rivers Petea and Tominanga) are apparently not universal in their effect on species with a distribution across the barrier, e.g. different patterns found in *Caridina lanceolata* and *C. holthuisi*. This might indicate either a difference in the potential of individual species to cross this barrier, or a varying effectiveness of the barrier through time. Thirdly, in a few cases, such as in *C. holthuisi*, intraspecific differentiation with respect to geography is indicative of cryptic species and thus unrecorded endemism. While ecological factors have been suggested to play a major role in the Sulawesi radiations of *Caridina* (see von Rintelen et al. 2010), geographic subdivision seems to be another driving force in differentiation by limiting gene flow and dispersal between allopatric

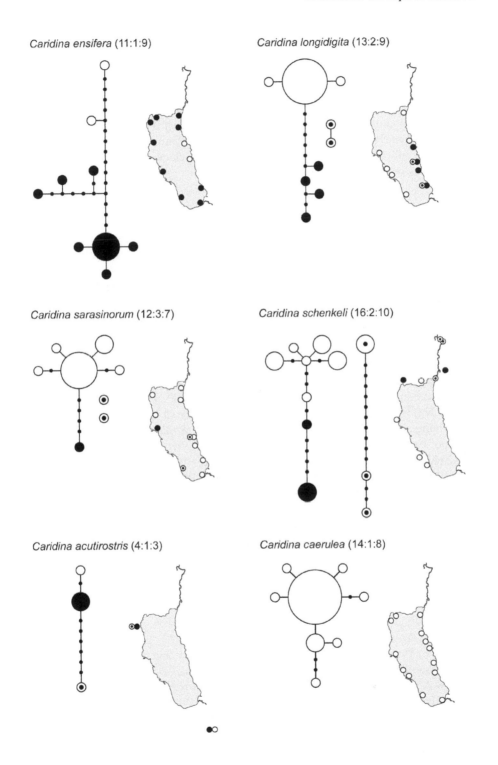

Figure 7. Phylogeography of Lake Poso species. The COI haplotype network is shown for each species. The geographic distribution of the respective haplotypes is shown on a simplified map next to each network. Numbers in brackets as in Figure 4.

populations. Corrspondingly, the higher species diversity in the Malili lake system may be explained by the pronounced geographic structure of this system.

ACKNOWLEDGEMENT

The following people are acknowledged for logistic, financial and other support relevant for this book chapter: Ristiyanti Marwoto and Daisy Wowor (MZB); Thomas von Rintelen, Matthias Glaubrecht, and Björn Stelbrink (ZMB); Christoph Schubart (University of Regensburg, Germany). I further thank the editors for the invitation to write this book chapter and two anonymous reviewers for their comments on the manuscript.

REFERENCES

Avise J.C., Arnold, J., Ball, R.M., Bermingham, E., Lamb, T., Neigel, J.E., Reeb, C.A. & Saunders, N.C. 1987. Intraspecific phylogeography: the mitochondrial DNA bridge between population genetics and systematics. *Annu. Rev. Ecol. Syst.* 18: 489–522.

Cai, Y. 2010. *Atydina*, a new genus for *Caridina atyoides* Nobili, 1900, from Indonesia (Crustacea: Decapoda: Atyidae). *Zootaxa* 2372: 75–79.

Clement, M., Posada, D. & Crandall, E. 2000. TCS: a computer program to estimate gene genealogies. *Mol. Ecol.* 9: 1657–1659.

De Grave, S., Cai, Y. & Anker, A. 2008. Global diversity of shrimps (Crustacea: Decapoda: Caridea) in freshwater. *Hydrobiologia* 595: 287–293.

De Grave, S., Pentcheff, N.D., Ahyong, S.T., Chan, T.-Y., Crandall, K.A., Dworschak, P.C., Felder, D.L., Feldmann, R.M., Fransen, C.H.J.M., Goulding, L.Y.D., Lemaitre, R., Low, M.E.Y., Martin, J.W., Ng, P.K.L., Schweitzer, C.E., Tan, S.H., Tshudy, D. & Wetzer, R. 2009. A classification of living and fossil genera of decapod crustaceans. *Raffles Bull. Zool.* 21: 1–109.

Fryer, G. 2006. Evolution in ancient lakes: radiation of Tanganyikan atyid prawns and speciation of pelagic cichlid fishes in Lake Malawi. *Hydrobiologia* 568: 131–142.

Giesen, W. 1994. Indonesia's major freshwater lakes: a review of current knowledge, development processes and threats. *Mitt. int. Vereinig. Limnol.* 24: 115–128.

Gorthner, A. 1994. What is an ancient lake? In: Martens, K., Goddeeris, B. & Coulter, G. (eds.), *Speciation in Ancient Lakes*: 97–100. Stuttgart: Schweizerbart'sche Verlagsbuchhandlung.

Haffner, G.D., Hehanussa, P.E. & Hartoto, D. 2001. The biology and physical processes of large lakes of Indonesia: Lakes Matano and Towuti. In: Munawar, K. & Heck, R.E. (eds.), *The Great Lakes of the World (GLOW): Food-web, Health and Integrity*: 183–192. Leiden: Backhuys Publishers.

Lai, H.-T. & Shy, J.-Y. 2009. The larval development of *Caridina pseudodenticulata* (Crustacea: Decapoda: Atyidae) reared in the laboratory, with a discussion of larval metamorphosis types. *Raffles Bull. Zool.* 20: 97–107.

Macdonald, K.S., Yampolsky, L. & Duffy, J.E. 2005. Molecular and morphological evolution of the amphipod radiation in Lake Baikal. *Mol. Phylogenet. Evol.* 35: 323–343.

Marijnissen, S.A.E., Michel, E. & Kamermans, M., Olaya-Bosch, K., Kars, M., Cleary, D.F.R., van Loon, E.E., Rachello Dolmen, P.G. & Menken, S.B.J. 2008. Ecological correlates of species differences in the Lake Tanganyika crab radiation. *Hydrobiologia* 615: 81–94.

Martens, K. 1997. Speciation in ancient lakes. *Trends Ecol. Evol.* 12: 177–182.

Martens, K. & Schön, I. 1999. Crustacean biodiversity in ancient lakes: a review. *Crustaceana* 72: 899–910.

Mayr, E. 1963. *Animal Species and Evolution.* Cambridge, MA: Belknap Press of Harvard University.

Page, T.J., von Rintelen, K. & Hughes J.M. 2007. Phylogenetic and biogeographic relationships of subterranean and surface genera of Australian Atyidae (Crustacea: Decapoda: Caridea) inferred with mitochondrial DNA. *Invertebr. Syst.* 21: 137–145.

Roy, D., Kelly, D.W., Fransen, C.H.J.M., Heath, D.D. & Haffner, G.D. 2006. Evidence of small-scale vicariance in *Caridina lanceolata* (Decapoda: Atyidae) from the Malili lakes, Sulawesi. *J. Evol. Biol.* 20: 1126–1137.

Schenkel, E. 1902. Beitrag zur Kenntnis der Dekapodenfauna von Celebes. *Verh. Naturf. Ges. Basel* 13: 485–585.

Schubart, C.D. & Ng, P.K.L. 2008. A new molluscivore crab from Lake Poso confirms multiple colonization of ancient lakes in Sulawesi by freshwater crabs (Decapoda: Brachyura). *Zool. J. Linn. Soc.* 154: 211–221.

von Rintelen, K. & Cai, Y. 2009. Radiation of endemic species flocks in ancient lakes: systematic revision of the freshwater shrimp *Caridina* H. Milne Edwards, 1837 (Crustacea: Decapoda: Atyidae) from the ancient lakes of Sulawesi, Indonesia, with the description of eight new species. *Raffles Bull. Zool.* 57: 343–452.

von Rintelen, K., von Rintelen, T. & Glaubrecht, M. 2007. Molecular phylogeny and diversification of freshwater shrimps (Decapoda, Atyidae, *Caridina*) from ancient Lake Poso (Sulawesi, Indonesia)—the importance of being colourful. *Mol. Phylogenet. Evol.* 45: 1033–1041.

von Rintelen, K., Glaubrecht, M., Schubart, C.D., Wessel, A. & von Rintelen, T. 2010. Adaptive radiation and ecological diversification of Sulawesi's ancient lake shrimps. *Evolution* 64: 3287–3299.

von Rintelen, T., von Rintelen, K. Glaubrecht, M., Schubart, C.D. & Herder, F. (in press). Aquatic biodiversity hotspots in Wallacea—the species flocks in the ancient lakes of Sulawesi, Indonesia. In: Gower, D.J., Johnson, K.G., Richardson, J.E., Rosen, B.R., Rüber, L. & Williams, S.T. (eds.), *Biotic Evolution and environmental Change in Southeast Asia*. Cambridge: Cambridge University Press.

Woltereck, E. 1937. Systematisch-variationsanalytische Untersuchungen über die Rassen- und Artbildung bei Süßwassergarnelen aus der Gattung *Caridina* (Decapoda, Atyidae). *Int. Revue ges. Hydrobiol. Hydrogr.* 34: 208–262.

Xia, X. & Xie, Z. 2001. DAMBE: data analysis in molecular biology and evolution. *J. Hered.* 92: 371–373.

Molecular and conservation biogeography of freshwater caridean shrimps in north-western Australia

BENJAMIN D. COOK[1,2], TIMOTHY J. PAGE[2] & JANE M. HUGHES[1,2]

[1] *Tropical Rivers and Coastal Knowledge Commonwealth Environmental Research Facility*
[2] *Australian Rivers Institute, Griffith University, Nathan, Queensland, Australia, 4111*

ABSTRACT

Rivers in north-western Australia have been long isolated systems and several of them, including those in the central north Kimberley (CNK) and Pilbara (PIL) regions, are known for their endemism in freshwater fishes. Furthermore, some species of freshwater fish are shared between rivers of the south-western Kimberley (SWK) and rivers of the western Arafura-Carpentarian (WAC) regions to the exclusion of CNK, despite CNK being geographically intermediate to these regions. This suggests that the freshwater biota of the CNK may have an independent evolutionary history with respect to the SWK and the WAC, and that the latter two may have relatively strong biogeographic affinities. This pattern warrants further examination using phylogeographic analyses of freshwater biota, such as freshwater caridean shrimp. In this study we used mtDNA data and conservation biogeographic approaches to assess molecular patterns of biodiversity within freshwater shrimp, genus *Caridina* (Atyidae), in north-western Australia using reported biogeographic patterns in freshwater fishes as a hypothesis. Specifically, we expected CNK and PIL to have higher endemism and phylogenetic diversity (PD) within the *Caridina* than SWK and WAC, and shrimps of SWK and WAC to have stronger biogeographic affinities with one another than either one has with CNK. Results showed high endemism within genus *Caridina* from the CNK and PIL, although PD and species richness was lowest for PIL and highest for CNK. Two lineages within *Caridina* were shared between the SWK and the WAC to the exclusion of the CNK. The results of this study for *Caridina* shrimp are thus strikingly similar to previous analyses of freshwater fishes, and support earlier studies that define PIL as a bioregion. We suggest that the incorporation of phylogeographic data for both fishes and freshwater macroinvertebrates (i.e., shrimps and molluscs) in future analyses may also identify CNK as a bioregion.

1 INTRODUCTION

Rivers are highly isolated ecological systems, somewhat akin to oceanic islands (MacArthur & Wilson 1967); thus they have long interested biogeographers (Rauchenberger 1988; Banarescu 1990) and phylogeographers (Bermingham & Avise 1986). Island biogeography theory predicts that highly isolated habitats (e.g., islands, rivers, lakes, mountain tops) contain a high proportion of endemic species because opportunities for biotic exchange between unconnected habitats are very limited (MacArthur & Wilson 1967; Whittaker et al. 2008). Examples of isolated freshwater systems with high endemism include various lake systems around the world (e.g., Africa's Rift Valley lakes, Marijnissen et al. 2009; Lake Poso, Indonesia, von Rintelen et al. 2007), desert springs of the

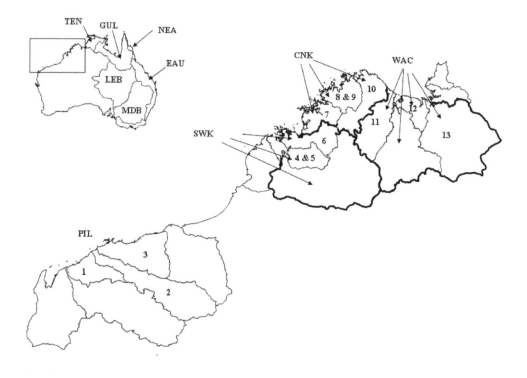

Figure 1. Map of north-western Australia showing the rivers and regions considered in this study. The north-western regions depicted are as follows: PIL: Pilbara; SWK: south-western Kimberly; CNK: central-northern Kimberley; WAC: western Arafura-Carpentaria. Numbers refer to rivers as presented in Table 1. Regions from elsewhere in Australia are shown on the inset, as follows: TEN: Top End of Australia; GUL: rivers draining to the Gulf of Carpentaria; NEA: North-eastern Australia; EAU: Eastern Australia; MDB: Murray-Darling Basin; LEB: Lake Eyre Basin.

United States of America, Mexico, and central Australia (Colgan & Ponder 2000; Kodric-Brown & Brown 2007), and various stygobiont ecosystems, globally (e.g., Edward & Harvey 2008).

Endemic taxa may also be relicts, i.e., geographically restricted taxa that are phylogenetically distinct from their closest relatives (Erwin 1991). Thus, the places they occur have high endemism and high phylogenetic diversity (PD, Faith 1992). Endemism and PD are therefore complementary metrics that can be used to assess the conservation value of places and the historical biogeography of biota in isolated habitats. Habitats with a high proportion of relict species would have high conservation value and indicate that the long-term isolation of "place" is a key driver of concerted and strong biogeographic patterns across the species, including the long-term stability of species assemblages (Zink 2002; Lapointe & Rissler 2005). Use of these metrics in molecular conservation biogeographic analyses may therefore identify key patterns of freshwater biodiversity for freshwater bioregionalization and conservation.

North-western Australia (Figure 1) is renowned for its ancient (i.e., Proterozoic) geology and geographic isolation, particularly the many sandstone escarpments and gorges (Ollier et al. 1988; Bowman et al. 2010). Consequently, the rivers of the Kimberley and Pilbara regions are highly isolated ecological systems and most remained discrete basins during historical periods of lowered sea levels when many other rivers of northern Australia in other regions were repeatedly connected by ancient confluences (Voris 2000; Harris et al. 2005). The freshwater fish fauna of the Kimberley and Pilbara regions therefore contain a high proportion of endemic species (Unmack 2001; Allen et al. 2002; Morgan et al. 2009) and both are recognized freshwater bioregions on account of their unique freshwater fish faunas (Unmack 2001; Abell et al. 2008). Patterns of endemism in freshwater biota in the Kimberley region are especially pronounced for rivers draining the sandstone plateau country of the central and northern Kimberley (CNK) (e.g., Morgan et al. 2009; Figure 1). Indeed, riverine biota from the south-western Kimberley (SWK) region may have stronger biogeographic affinities with the biota of the western-most rivers in the Arafura-Carpentaria bioregion (Western Arafura-Carpentaria, WAC; Figure 1) than with the CNK. For example, Unmack (2001) reports that several species of freshwater fish are shared between the SWK and WAC, and are absent from CNK. The biogeographic distinctiveness of the CNK warrants further examination using phylogeographic analyses of key freshwater groups.

One group that is widespread in northern Australia is the freshwater caridean shrimp, genus *Caridina* (Decapoda: Atyidae), that has been used to demonstrate strong freshwater biogeographic and phylogeographic patterns elsewhere in Australia (Page & Hughes 2007; Cook et al. 2008). In the present study we used mtDNA data and conservation biogeographic approaches to assess molecular patterns of biodiversity within the freshwater shrimp, genus *Caridina* (Atyidae), in north-western Australia using reported biogeographic patterns in freshwater fishes as a hypothesis. Specifically, we expected CNK and PIL to have higher endemism and phylogenetic diversity within *Caridina* than SWK and WAC, and shrimps of SWK and WAC to have stronger biogeographic affinities with one another than either one has with CNK. Support for these predictions would suggest that CNK should be considered a distinct freshwater bioregion.

2 MATERIALS AND METHODS

2.1 Samples used and genotyping method

A total of 58 individual shrimps from 13 rivers (23 sites) from the four regions were sequenced for a 453 basepair fragment of the cytochrome c oxidase subunit I (COI) mtDNA gene (Table 1; Appendix, Tables A1 and A2). Total genomic DNA was extracted from each individual using the CTAB/phenol chloroform method and a fragment of the COI mtDNA gene was amplified using polymerase chain reaction (PCR) and the CDC0.La (5'-CCN GGG TTY GGR ATA ATT TCT C-3'; Page et al. 2005a) and COIa (5'- AGT ATA AGC GTC TGG GTA GTC-3'; Palumbi et al. 1991) primers. PCR cycling conditions were as follows: 3 min at 94 °C; 15 cycles of 30 s at 94 °C, 30 s at 40 °C, 60 s at 72 °C; then 25 cycles of 30 s at 94 °C, 30 s at 55 °C, 60 s at 72 °C; 7 min at 72 °C, and finally held at 4 °C. PCR product was purified with the exonucleoase I-shrimp alkaline phosphatase method, using 2.5 μl PCR product, 2.0 μl shrimp alkaline phosphatase (Promega) and 0.5 μl exonucleoase I (Fermentas), and a two-step thermocycling profile: 35 min at 37 °C, 20 min at 80 °C. Ten micro-liter sequencing reactions contained 0.5 μl purified product, 0.32 μl forward primer (3.5 pmol/μl), 2 μl BigDye v1.1 (Applied Biosystems) and 2 μl BigDye, 5× sequencing buffer (Applied Biosystems), and standard thermal cycling conditions were used: 1 min at 96 °C; 30 cycles of 10 s at 96 °C, 5 s at 50 °C, 4 min at 60 °C; and a hold period of 4 °C. Two exemplars of each lineage were sequenced using the reverse primer to verify basepair composition. Sequencing was conducted on a 3130xl Capillary Electrophoresis Genetic Analyser (Applied Biosystems) and sequences were aligned and edited using SEQUENCHER version 4.1.2 (Gene Codes), yielding 453 unambiguous basepairs, and deposited in GenBank. To test for the possibility of nuclear copies of mitochondrial DNA (NUMTs) in the data and other nucleotide anomalies in the COI mtDNA data,

Table 1. Origin of freshwater shrimp of *Caridina* used for this study with regions and rivers sampled and sample sizes for each river (number of sites and number of individuals). River numbers correspond to numbers used in Figure 1.

Region	River	No. of sites	No. of individuals
PIL	1. Onslow Coast	3	5
PIL	2. Fortescue River	3	4
PIL	3. Port Headland Coast	1	3
SWK	4. May River	2	8
SWK	5. Robinson River	1	5
SWK	6. Isdell River	1	1
CNK	7. Prince Regent River	1	1
CNK	8. Mitchell River	3	7
CNK	9. King Edward River	3	9
CNK	10. Drysdale River	2	5
WAC	11. Pentecost River	1	4
WAC	12. Keep River	1	3
WAC	13. Victoria River	1	3

which may artificially produce false genetic lineages in phylogenetic analyses (Song et al. 2008), nucleotide sequences were translated to amino acid sequences using the invertebrate mtDNA code as implemented in GENEDOC version 2.7.000 (Nicholas & Nicholas 1997). The presence of one or more stop codons in the amino acid sequence would suggest the possibility of NUMTs in the data. Exemplars of each COI lineage were sequenced for a 455 base-pair fragment of the 16S mtDNA gene (methods described in von Rintelen et al. 2007; primer sequences: 16S-F-Car: 5'-TGC CTG TTT ATC AAA AAC ATG TC-3', 16S-R-Car: 5'-AGA TAG AAA CCA ACC TGG CTC-3') and aligned with published 16S sequences of other Australian *Caridina* from GenBank (Appendix, Table A1). All new sequences (COI and 16S) were submitted to Genbank and can be retrieved under accession numbers JN012514–JN012565.

2.2   Data analysis

Phylogenetic analyses for both the COI and 16S datasets were performed using 1,000 bootstrap pseudoreplicates of the Maximum Likelihood (ML) method incorporating the GTR model as implemented in PHYML (Guindon & Gascuel 2003). We used *Caridina serratirostris* (GenBank Accession number DQ478515) from north-eastern Australia as an outgroup for the COI phylogeny, and *Paratya australiensis* (GenBank Accession number DQ78566) as an outgroup for the 16S phylogeny. The gene trees with best log likelihood scores were used in analysis of endemism and phylogenetic diversity after nodes with less than 50% bootstrap support were collapsed using MESQUITE (Maddison & Maddison 2000).

Endemism was calculated for each river basin and each region using the Site Endemism Index (SEI, Rebelo & Siegfried 1992) and the identified COI lineages as taxa, with $SEI = \sum k/a_i$, where k is the total number of sites (i.e., river basin or region) and a_i is the number of sites at which species i occurs. SEI ranges from one, where all taxa at a location have broad geographical ranges, through to infinity, with larger values indicating the presence of an increasing number of geographically restricted taxa. SEI therefore reflects the richness of endemic species. Phylogenetic diversity (PD, Faith 1992) was then calculated for each river basin and each region for the COI

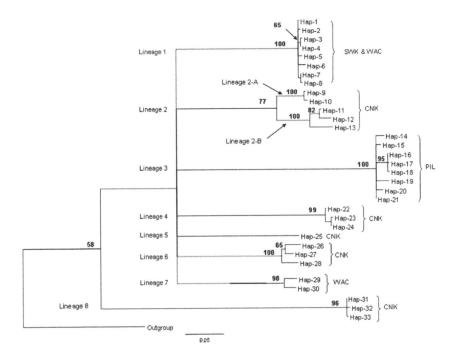

Figure 2. Maximum likelihood COI mtDNA gene tree (ln -2715.56) for *Caridina* spp. sampled from north-western Australia, showing bootstrap values (after 1,000 pseudorelilcates) along branches. Nodes with less than 50% bootstrap support were collapsed and the scale bar indicates five percent difference in basepair composition. The distributions of the lineages with respect to the regions are indicated on the right-hand side of the tree.

mtDNA ML phylogeny, with branch lengths calculated using MESQUITE and summed for each region and river, respectively. PD is a measure that integrates phylogeny and complementarity (Faith 1992) and ranges from zero, where a location has low endemism and complementarity with respect to the total phylogeny of the taxa, through to infinity, with larger values indicating that the location contains taxa that incorporate increasing proportions of the phylogeny that are not present elsewhere in the landscape. Because some regions had relatively few sites sampled, we tested for correlations between the number of sites and the diversity metrics (i.e., endemism and PD) using Pearson's correlation coefficient in SPSS.

Finally, we used the 16S phylogeny and regional-scale analyses of SEI and PD to compare patterns of molecular biodiversity within *Caridina* in north-western Australia with other regions throughout Australia (see Figure 1 and Appendix, Table A1 for description of these regions). Whereas lineages were used to calculate COI SEI and PD, several of the 16S lineages were comprised of sublineages, many relating to described or previously distinguished taxa (see Results). Thus, taxa (rather than lineages) were used in analyses of 16S endemism and PD.

3   RESULTS

Stop codons and double peaks were not detected in the COI data, reducing the possibility of including NUMTs in our analyses. Eight divergent COI mtDNA lineages were recovered by ML analyses (Figure 2), six of which received high bootstrap support (i.e., Lineages 1, 3, 4, 6, 7, and 8; all boot-

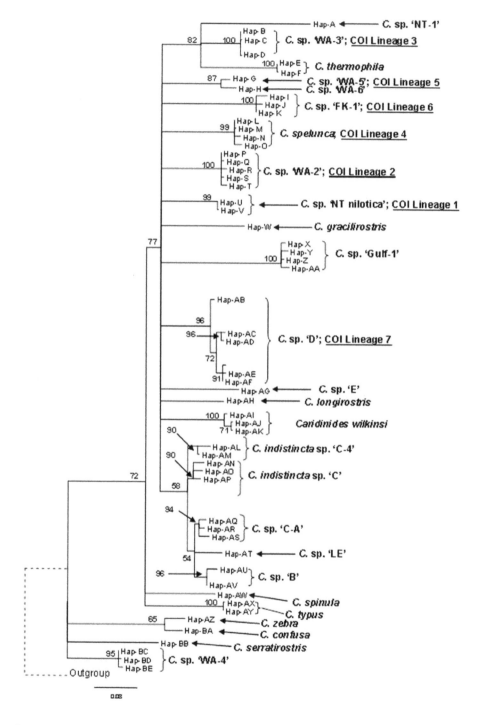

Figure 3. Maximum likelihood 16S mtDNA gene tree (ln -3801.86) for *Caridina* spp. sampled from throughout Australia, showing bootstrap values (after 1,000 pseudoreplicates) along branches. Nodes with less than 50% bootstrap support were collapsed and the scale bar indicates eight percent difference in basepair composition. The COI lineages corresponding to the 16S lineages are indicated on the right-hand side of the tree in underlined font. The distributions of 16S taxa from throughout Australia are presented in Appendix, Table A1.

strap values > 95%), one (i.e., Lineage 2), which had good bootstrap support (i.e., 77%), and one (i.e., Lineage 5) which was unable to be assessed by bootstrapping because it was represented by only a single haplotype. Lineage 2 contained two strongly supported sublineages (i.e., 2-A and 2-B), and Lineage 8 appears to be monophyletic to all other lineages, although this lineage had relatively low bootstrap support. The phylogenetic relationships between Lineages 1 to 7 could not be determined due to poor bootstrap support at several of the deeper nodes, which were collapsed into a soft polytomy.

ML analysis of the 16S data recovered 20 lineages (Figure 3), although several of these were comprised of sublineages, many relating to described or previously distinguished taxa. Five of the COI lineages related to previously reported 16S taxa, as follows: Lineage 1 relates to *Caridina* sp. "NT *nilotica*" (Page et al. 2007a); Lineage 2 relates to *Caridina* sp. "WA-2" (Page et al. 2007a); Lineage 3 relates to *Caridina* sp. "WA-3" (Page et al. 2007a); Lineage 4 relates to *Caridina spelunca*; and Lineage 7 relates to *Caridina* sp. "D" (Page et al. 2005a). Two COI lineages (i.e., Lineages 5 and 6) do not relate to previously reported 16S taxa and COI Lineage 8 was not represented in the 16S phylogeny because only a very short 16S fragment (i.e., < 300 bp) could be amplified for this taxon. We note here that this short fragment does not align with previously reported 16S taxa, suggesting that these three COI lineages (i.e., Lineages 5, 6, and 8) represent new taxa. We included an additional 16S lineage shared between SWK and WAC for which we did not have COI data (i.e., WA-4; Figure 3); thus, two 16S lineages were shared between SWK and WAC to the exclusion of CNK.

The geographic distribution of the COI mtDNA lineages varied, with five lineages restricted to CNK (i.e., Lineages 2, 4, 5, 6, and 8), one lineage restricted to PIL (i.e., Lineage 3), one lineage restricted to the WAC (i.e., Lineage 7), and one lineage shared among regions (i.e. Lineage 1 shared by WAC and SWK). These striking patterns of distribution resulted in similarly impressive patterns of endemism at both river and region scales, with the King Edward River and CNK regions having the highest levels of endemism, respectively (Figure 4). Endemism was lowest for the May, Robinson, Pentecost and Keep Rivers, and the SWK region, respectively. Endemism was not correlated with the number of sample sites in a drainage basin (Pearson's correlation coefficient = 0.418, $P = 0.177$).

Phylogenetic Diversity (PD, Faith 1992) as assessed for the COI mtDNA ML phylogeny was highest in the King Edward River and CNK region, respectively (Figure 5). The Mitchell and Drysdale Rivers also had high PD at the river basin scale. Rivers of the SWK and WAC regions had the lowest PD at the river basin scale, whereas PIL had the lowest PD at the region scale. PD was correlated with the number of sampled sites in a drainage basin (Pearson's correlation coefficient = 0.772, $P = 0.003$).

The 16S data indicated that north-eastern Australia is the region with the greatest endemism within *Caridina*, and that CNK was ranked as sharing the second-highest levels of endemism with eastern Australia, closely followed by the Top End region (Figure 6). PIL had a low score for SEI, but the single 16S lineage detected from this region is endemic. SWK, WAC, and Gulf of Carpentaria region (GUL) all have relatively low levels of endemism as most taxa from these regions are common to at least one other region. North-eastern Australia also received the highest score for PD, with CNK being ranked as equal second highest with the Top End region. Eastern Australia had third-highest PD, although unlike endemism this value was not much greater than for most other regions.

4   DISCUSSION

The molecular biogeographic analyses of *Caridina* shrimp from north-western Australia indicate striking phylogeographic structuring and endemism, including three previously unreported taxa.

A

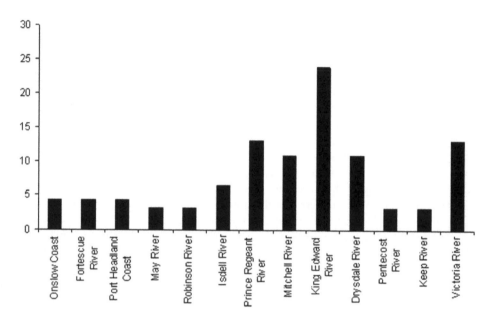

B

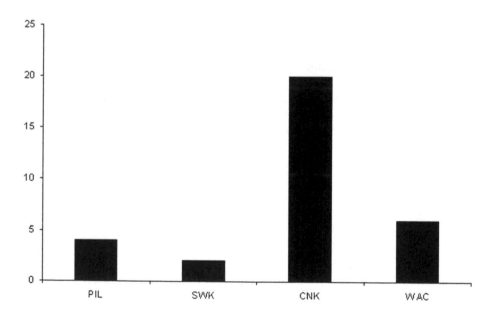

Figure 4. Histogram of COI mtDNA endemism as assessed using the site endemism index (SEI) within the genus *Caridina* from north-western Australia for A: rivers, and B: regions.

A

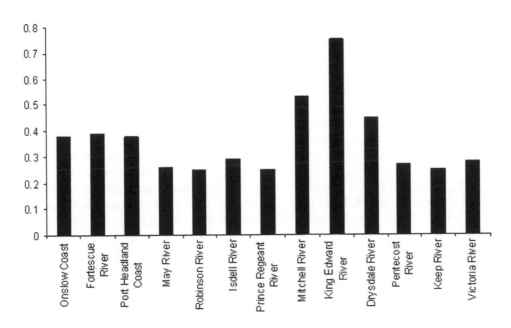

B

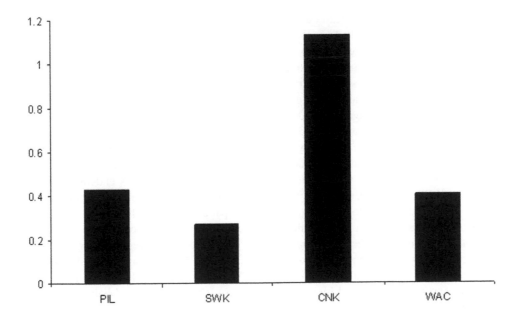

Figure 5. Histogram of COI mtDNA phylogenetic diversity (PD) within the genus *Caridina* from north-western Australia for A: rivers, and B: regions.

Biogeographic patterns of high species richness and endemism in freshwater fishes from the CNK (Allen et al. 2002; Unmack 2001; Morgan et al. 2009) were also found within *Caridina* (this study). The region has remained geographically isolated for both freshwater fishes and shrimps over evolutionary time scales and the many gorges in the region have acted as refugia, enabling the long-term persistence of freshwater species (Unmack 2001). Interestingly, freshwater crayfish are absent from this region, although they are common in most other regions of Australia (Whiting et al. 2000). Overall, these data indicate biogeographic distinctiveness of the freshwater biota of this region, similar to that shown for freshwater biodiversity of South Africa's Cape floristic region (Wishart & Day 2002) and patterns of diversity within *Caridina* from other isolated habitats (e.g., Lake Poso, von Rintelen et al. 2007). Whilst the phylogenetic relationships between most taxa within the Australian *Caridina* could not be established, it is clear from the 16S data that at least some taxa from north-western Australia are more closely related to species in more eastern regions of Australia than to other taxa from the west. A very similar pattern was found for species within the freshwater fish genus *Craterocephalus* (i.e., *C. helenae*, endemic to the CNK, and *C. lentigenosus*, distributed across SWK, CNK, and WAC), which were more closely related to geographically distant species than to each other (Unmack & Dowling 2010). Indeed, some Australian species of *Caridina* are known to be more closely related to species from Asia than to other species from Australia (Page et al. 2007a).

Endemism and phylogenetic diversity are complimentary approaches for assessing diversity and bioregional patterns in biota. For the COI analyses, both metrics identified the same river (i.e., King Edward River) and region (i.e., CNK) as having the highest diversity. These results indicate the presence of a high proportion of endemic taxa, which contrasts patterns found in the shrimp genus *Paratya* from eastern Australia in which high diversity at the river basin scale was due to the co-occurrence of occasionally detected endemic taxa with widespread taxa (Cook et al. 2008). We note, however, that PD was correlated with the number of sites sampled in a river basin; thus, our results for PD in this study must be considered a prediction which we hope to test in the near future using more comprehensive sampling. For the continent-wide 16S analyses, SEI and PD both identified north-eastern Australia as the region with the highest diversity in Australia within *Caridina*, and both metrics ranked CNK as having equal second highest diversity, indicating this region has significant freshwater shrimp biodiversity at the national scale. However, CNK was equal second with eastern Australia for endemism, and equal second with the Top End region for PD. These contrasting patterns demonstrate the utility of applying both these metrics in biodiversity assessments; eastern Australia had high endemism due to a large number of cryptic taxa within the *C. indistincta* complex (Page et al. 2005a) but had lower PD because these species were very closely related relative to the fewer but more distantly related taxa present in the Top End. Similarly, PIL had low SEI within *Cardinia* relative to PD. However, as SEI increases within increasing richness of endemics, places with low species richness will have low SEI, even if the few species are endemic. The single taxon we detected within *Caridina* from PIL was endemic (i.e., PIL has 100% endemism for *Caridina*), suggesting perhaps that both SEI and percent endemism should be assessed in conservation biogeographic analyses. The pattern of low richness and high endemism in freshwater biodiversity from PIL is reflected in freshwater fishes at both species (e.g., *Craterocephalus cuneips* and *Leiopotherapon aheneus*) and genus (i.e., *Milyeringa veritas*) levels, as well as in caridean shrimp, including species within *Caridina* (i.e., COI Lineage 3) and for the endemic shrimp genus *Stygiocaris* (Atyidae) (Page et al. 2008a). Low species richness and high endemism is a classical expectation of island biogeographic theory for highly isolated habits (MacArthur & Wilson 1967).

Previous biogeographic analyses report the shared distribution of several species of freshwater fish (e.g., *Anodontiglanis dahli*, *Arius midgleyi*, and *Craterocephalus stramineus*) between the SWK and WAC, with these species being absent from CNK despite the CNK being geographically

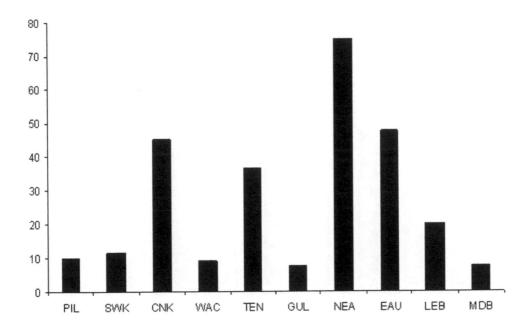

Figure 6. Histogram of regional levels of 16S endemism within the genus *Caridina* in Australia. The regions considered are: TEN: Top End of Australia; GUL: rivers draining into the Gulf of Carpentaria; NEA: North-eastern Australia; EAU: Eastern Australia; MDB: Murray-Darling Basin; LEB: Lake Eyre Basin.

intermediate to SWK and WAC (Unmack 2001). Our analyses of *Caridina* show a similar pattern, with two 16S lineages shared between these two regions, to the exclusion of CNK, suggesting close biogeographic affinities between these two regions, probably reflecting past inland drainage connections. Past offshore riverine connections mediated by palaeoeustatic processes likely had an important influence on freshwater biogeographic patterns in other regions of northern Australia, such as the Gulf of Carpentaria (e.g., de Bruyn et al. 2004; Cook & Hughes 2010), although drainages of north-western Australia remained discrete basins during glacial phases (see Harris et al. 2005). The offshore and extensive palaeo-hydrosystems to the west of the Kimberly and Pilbara predicted by Harris et al. (2005) were therefore extremely unlikely conduits for past connectivity for *Caridina* shrimp between WAC and SWK, to the exclusion of CNK. We are presently investigating the biogeographic affinities of WAC and SWK, to the exclusion of the distinctive and diverse CNK, using phylogeographic methods in both freshwater fishes and other invertebrates (e.g., other genera of caridean shrimp and molluscs).

We note the need for more comprehensive sampling of *Caridina* from throughout the study area, particularly considering that PD was correlated with the number of sampled sites in a river basin, to demarcate freshwater bioregional units in north-western Australia. Phylogeographic patterns in other freshwater species, including fish and other invertebrate groups, would also contribute greatly to developing a freshwater bioregionalization for north-western Australia. Whilst unrelated taxa may not have shared evolutionary histories and may therefore not share the same bioregional boundaries (e.g., Growns 2009), the striking concordance in diversity patterns within freshwater fish and *Caridina* shrimp suggests a strong degree of congruence in biogeographic history. Abell et al. (2008) suggest that periodic reviews of the bioregional boundaries they propose should incorpo-

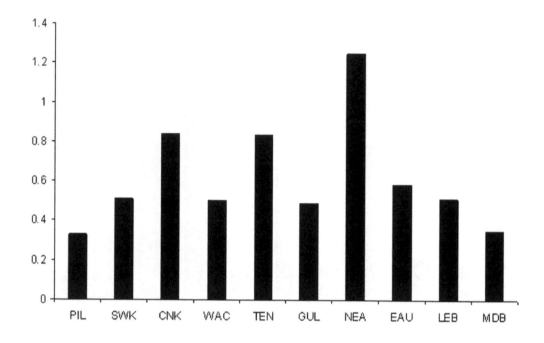

Figure 7. Histogram of regional levels of 16S phylogenetic diversity within genus *Caridina* in Australia. The regions considered are: TEN: Top End of Australia; GUL: rivers draining to the Gulf of Carpentaria; NEA: North-eastern Australia; EAU: Eastern Australia; MDB: Murray-Darling Basin; LEB: Lake Eyre Basin.

rate updated data, including fish and taxa other than fish. They also state that obligate freshwater macroinvertebrates respond to localized ecological and evolutionary factors that are too small to be meaningful for ecoregion delineation. However, conservation biogeographic analyses of caridean shrimp in eastern Australia were very similar to cryptic biodiversity patterns in freshwater fishes (Cook et al. 2008), as shown for fishes and shrimps in this study. It would be interesting to assess phylogeographic patterns in other obligate freshwater macroinvertebrates (e.g., molluscs) throughout north-western Australia, many of which do have very localised population structures (e.g., Carini & Hughes 2006; Colgan & Ponder 2000), to determine if similar biogeographic patterns are found in these taxa.

The delineation of boundaries between bioregional units is a debated issue (e.g., Unmack 2001; Filipe et al. 2009), with discrepancies concerning the use of single versus multiple taxonomic groups (as noted in the previous paragraph) and criteria for boundaries, such as the use of alpha or beta components of biodiversity. Alpha diversity criteria include species richness or percent endemism, although the percentage of endemism that should be used to separate biogeographic units is not established in biogeographic theory (Unmack 2001). Beta diversity criteria may include geographical points of turnover of multiple species' distributions and turnover points of divergent genetic lineages within species, although no rule of thumb exists concerning the proportion of taxa needed to exhibit turnover at either species or genetic levels to constitute a "unit of biodiversity with a shared evolutionary history" which is a commonly used definition of "bioregion" and its synonyms (e.g., "province" and "ecoregion", Unmack 2001; Spalding et al. 2007; Abell et al. 2008). In practice, however, the criteria used to delineate bioregional boundaries will be determined by data availability (Richardson & Whittaker 2010). Our analysis of endemism and PD, which are both components of alpha biodiversity, showed markedly different patterns among the four regions of north-western Australia, notably with both metrics indicating higher within-region diversity for CNK. High endemism also suggests substantial differences in the composition of biodiversity between biogeographic units (i.e., beta diversity), although we did not explicitly test biotic dissimilarity within *Caridina* among the regions. A conservation biogeographic study of caridean shrimp in eastern Australia showed that the composition of cryptic biodiversity within shrimp of the genera *Caridina* and *Paratya* can be significant among river basins (Cook et al. 2008), and we suggest that such analyses based on more comprehensive sampling would likely indicate that CNK and PIL are both distinct bioregional units. In contrast, it is probable that SWK and WAC would be shown to have relatively high similarity in alpha and beta components of biodiversity within *Caridina*. Thus, our analyses of *Caridina* support previous analysis of freshwater fishes that define PIL as a freshwater bioregion (Unmack 2001; Abell et al. 2008) and indicate that CNK could also be considered a distinct bioregion. This study demonstrates that molecular-based bioregional studies of freshwater shrimps can indicate strong patterns of freshwater biodiversity and identify units for freshwater conservation management, as also shown in other regions of Australia and for other freshwater crustacean groups (e.g., Whiting et al. 2000; Cook et al. 2008; Bentley et al. 2010).

ACKNOWLEDGEMENTS

The Tropical Rivers and Coastal Knowledge Commonwealth Environmental Research Facility (TRaCK) receives major funding for its research through the Australian Government's Commonwealth Environment Research Facilities initiative, the Australian Government's Raising National Water Standards Program, Land and Water Australia, the Fisheries Research and Development Corporation, and the Queensland Government's Smart State Innovation Fund. We are grateful to the Traditional Owners of the land and Western Australian and Northern Territory State Governments for permission to conduct sampling, and thank Adrian Pinder and Mike Scanlon (CALM) for providing additional material. Thanks to Augie Unghango, Mark Unghango, William Maraltadj, Sylvester Mangolamara, Raphael Karadada, Edmund Ngerdu, Lindsey, Le Roy, Joseph, Mildred Mungula, Teresa Numunduma, Phil Palmer, Tom Vigilante, Steve Sharpe, James Fawcett, and Kate Masci for helping out in the field. We also thank two anonymous reviewers for constructive criticism.

REFERENCES

Abell, R., Thieme, M.L., Revenga, C., Bryer, M. Kottelat, M., Bogutskaya, N., Coad, B., Mandrak, N. Balderas, S.C., Bussing, W., Stiassny, M.L.J., Skelton, P., Allen, G.R., Unmack, P., Naseka, A., Ng, R., Sindorf, N., Roberston, J., Armijo, E., Higgins, J.V., Heibel, T.J., Wikramanayake, E., Olson, D., López, H.L., Reis, R.E., Lundberg, J.G., Sabaj Pérez, M. H, & Petry, P. 2008. Freshwater ecoregions of the world: a new map of biogeographic units for freshwater biodiversity conservation. *BioScience* 58: 403–414.

Allen, G.R., Midgley, S.H. & Allen, M. 2002. *Field Guide to the Freshwater Fishes of Australia.* Victoria, CSIRO Publishing.

Banarescu, P. 1990. *Zoogeography of Freshwaters. Volume 1. General Distribution and Dispersal of Freshwater Animals.* Wiesbaden: Aula-Verlag.

Bentley, A.I., Schmidt, D.J. & Hughes, J.M. 2010. Extensive intraspecific genetic diversity of a freshwater crayfish in a biodiversity hotspot. *Freshw. Biol.* 55: 1861–1873.

Bermingham, E. & Avise, J.C. 1986. Molecular zoogeography of freshwater fishes in the southeastern United States. *Genetics* 113: 939–965.

Bowman, D.M.J.S., Brown, G.K., Braby, M.F., Brown, J.R., Cook, L.G., Crisp, M.D., Ford, F., Haberle, S., Hughes J., Isagi, Y., Joseph, L., McBride, J., Nelson, G. & Ladiges, P.Y. 2010.

Biogeography of the Australian monsoon tropics. *J. Biogeogr.* 37: 201–216.

Carini, G. & Hughes, J.M. 2006. Subdivided population structure and phylogeography of an endangered freshwater snail, *Notopala sublineata* (Conrad, 1850) (Gastropoda: Viviparidae), in Western Queensland, Australia. *Biol. J. Linn. Soc.* 88: 1–16.

Colgan, D.J. & Ponder, W.F. 2000. Incipient speciation in aquatic snails in an arid-zone spring complex. *Biol. J. Lin. Soc.* 71: 625–641.

Cook, B.D., Page, T.J. & Hughes, J.M. 2008. Importance of cryptic species for identifying "representative" units of biodiversity for freshwater conservation. *Biol. Conserv.* 141: 2821–2831.

Cook, B.D. & Hughes J.M. 2010. Historical population connectivity and fragmentation in a tropical freshwater fish with a disjunct distribution (pennyfish, *Denariusa bandata*). *J. N. Amer. Benthol. Soc.* 29: 1119–1131.

de Bruyn, M., Wilson, J.C. & Mather, P.B. 2004. Reconciling geography and genealogy: phylogeography of giant freshwater prawns from the Lake Carpentaria region. *Mol. Ecol.* 13: 3515–3526.

Edward, K.L. & Harvey, M.S. 2008. Short-range endemism in hypogean environments: the pseudoscorpion genera *Tyrannochthonius* and *Lagynochthonius* (Pseudoscorpiones: Chthoniidae) in the semiarid zone of Western Australia. *Invertebr. Syst.* 22: 259–293.

Erwin, T.L. 1991. An evolutionary basis for conservation strategies. *Science* 253: 750–752.

Faith, D.P. 1992. Conservation evaluation and phylogenetic diversity. *Biol. Conserv.* 61: 1–10.

Filipe, A.F., Araújo, M. B., Doadrio, I., Angermeier, P.L. & Collares-Pereiral, M.J. 2009. Biogeography of Iberian freshwater fishes revisited: the roles of historical versus contemporary constraints. *J. Biogeogr.* 36: 2096–2110.

Growns, I. 2009. Differences in bioregional classifications among four aquatic biotic groups: implications for conservation reserve design and monitoring programs. *J. Environ. Manage.* 90: 2652–2658.

Guindon, S. & Gascuel, O. 2003. A simple, fast, and accurate algorithm to estimate large phylogenies by maximum likelihood. *Syst. Biol.* 52: 696–704.

Harris, P., Heap, A., Passlow, V., Sbaffi, L. Fellows, M., Porter-Smith, R., Buchanan, C. & Daniell, J. 2005. *Geomorphic Features of the Continental Margin of Australia. Geoscience Australia. Record 2003/30.* Canberra: Geoscience Australia.

Kodric-Brown, A. & Brown, J.H. 2007. Native fishes, exotic mammals, and the conservation of desert springs. *Front. Ecol. Environ.* 5: 549–553.

Lapointe, F.-J. & Rissler L.J. 2005. Congruence, concensus, and the comparative phylogeography of codistributed species in California. *Amer. Nat.* 166: 290–299.

MacArthur, R.H. & Wilson, E.O. 1967. *The Theory of Island Biogeography.* Princeton, NJ: Princeton University Press.

Maddison, W.P. & Maddison, D.R. 2000. MESQUITE: a modular system for evolutionary analysis. Version 2.72. http://mesquiteproject.org.

Marijnissen, S.A.E., Michel, E., Cleary, D.F.R. & McIntyre, P.B. 2009. Ecology and conservation status of endemic freshwater crabs in Lake Tanganyika, Africa. *Biodivers. Conserv.* 18: 1555–1573.

Morgan, D., Cheinmora, D., Charles, A., Nulgit, P. & Kimberley Language Resources Centre. 2009. *Fishes of the King Edward and Carson Rivers with their Belaa and Ngarinyin Names. Report for Land & Water Australia Project No. UMU22.* Canberra: Land and Water Australia.

Nicholas, K.B. & Nicholas, H.B. Jr. 1997. GeneDoc: a tool for editing and annotating multiple sequence alignments. Distributed by author.

Ollier, C.D., Armidale, G., Gaunt, F.M. & Jurkowski, I. 1988. The Kimberley Plateau, Western Australia: a Precambrian erosion surface. *Zeitschr. Geomorph.* 32: 239–246.

Page, T.J., Choy, S.C. & Hughes, J.M. 2005a. The taxonomic feedback loop: symbiosis of morphology and molecules. *Biol. Lett.* 1: 139–142.

Page, T.J., Baker, A.M., Cook, B.D. & Hughes, J.M. 2005b. Historic transoceanic dispersal of a freshwater shrimp: colonisation of the South Pacific by genus *Paratya*. *J. Biogeogr.* 32: 581–593.

Page, T.J. & Hughes, J.M. 2007. Radically different scales of phylogeographic structuring within cryptic species of freshwater shrimp (Atyidae: Caridina). *Limnol. Oceanogr.* 52: 1055–1066.

Page, T.J., von Rintelen, K. & Hughes, J.M. 2007a. An island in the stream: Australia's place in the cosmopolitan world of Indo-West Pacific freshwater shrimp (Decapoda: Atyidae: *Caridina*). *Mol. Phylogenet. Evol.* 43: 645–659.

Page, T.J., von Rintelen, K. & Hughes, J.M. 2007b. Phylogenetic and biogeographic relationships of subterranean and surface genera of Australian Atyidae (Crustacea: Decapoda: Caridea) inferred with mitochondrial DNA. *Invertebr. Syst.* 21: 137–145.

Page, T.J., Humphries, W.F. & Hughes, J.M. 2008a. Shrimps down under: evolutionary relationships of subterranean crustaceans from Western Australia (Decapoda: Atyidae: *Stygiocaris*). *PLoS One* 3: e1618.

Page, T.J., Cook, B.D., von Rintelen, K, von Rintelen, T. & Hughes J.M. 2008b. Evolutionary relationships of atyid shrimps imply both ancient Caribbean radiations and common marine dispersals. *J. N. Amer. Benthol. Soc.* 27: 68–83.

Palumbi, S.R., Martin, A., Romano, S., McMillan, W.O., Stice, L. & Grabowski, G. 1991. *A Simple Fool's Guide to PCR*. Honolulu, HI: University of Hawaii Press.

Rauchenberger, M. 1988. Historical biogeography of poeciliid fishes in the Caribbean. *Syst. Zool.* 37: 356–365.

Rebelo, A.G. & Siegfried, W.R. 1992. Where should nature reserves be located in the Cape Floristic Region, South Africa? Models for the spatial configuration of a reserve network aimed at maximising the protection of floral diversity. *Conserv. Biol.* 6: 243–252.

Richardson, D. M. & Whittaker, R.J. 2010. Conservation biogeography—foundations, concepts and challenges. *Divers. Distrib.* 16: 313–320.

Song, H., Buhay, J.E., Whiting, M.F. & Crandall, K.A. 2008. Many species in one: DNA barcoding overestimates the number of species when nuclear mitochondrial pseudogenes are coamplified. *Proc. National Acad. Sci. USA* 105: 13486–13491.

Spalding, M.D., Fox, H.E., Allen, G.R., Davidson, N., Ferdaña, Z.A., Finlayson, M., Halpern, B.S., Jorge, M.A., Lombana, A., Lourie, S.A., Martin, K.D., McManus, E., Molnar, J., Recchia, C.H., & Robertson, J. 2007. Marine ecoregions of the world: a bioregionalisation of coastal and shelf areas. *BioScience* 57: 573–583.

Unmack, P.J. 2001. Biogeography of Australian freshwater fishes. *J. Biogeogr.* 28: 1053–1098.

Unmack, P.J. & Dowling, T.E. 2010. Biogeography of the genus *Craterocephalus* (Teleostei: Atherinidae) in Australia. *Mol. Phylogenet. Evol.* 55: 968–984.

von Rintelen, K., von Rintelen, T. & Glaubrecht, M. 2007. Molecular phylogeny and diversification of freshwater shrimp (Decapoda: Atyidae: Caridina) from ancient Lake Poso (Sulawesi, Indonesia)—the importance of being colourful. *Mol. Phylogenet. Evol.* 45: 1033–1041.

Voris, H.K. 2000. Maps of Pleistocene sea levels in South East Asia: shorelines, river systems and time durations. *J. Biogeogr.* 27: 1153–1167.

Wishart, M.J. & Day, J.A. 2002. Endemism in the freshwater fauna of the south-western Cape, South Africa. *Verh. Internat. Vereinig. theor. angew. Limnol.* 28: 1–5.

Whittaker, R.J., Triantis, K.A. & Ladle, R.J. 2008. A general dynamic theory of oceanic island biogeography. *J. Biogeogr.* 35: 977–994.

Whiting, A.S., Lawler, S.H., Horwitz, P. & Crandall, K.A. (2000). Biogeographic regionalization of Australia: assigning conservation priorities based on endemic freshwater crayfish phylogenetics. *Anim. Conserv.* 3: 155–163.

Zink, R.M. 2002. Methods in comparative phylogeography, and their application to studying evolution in the North American arid lands. *Integr. Comp. Biol.* 42: 953–959.

APPENDIX

Table A1. 16S mtDNA data used, with Genbank numbers, source publication and geographic distribution represented by the sequence. TEN: Top End of Australia; GUL: rivers draining to the Gulf of Carpentaria; NEA: North-eastern Australia; EAU: Eastern Australia; MDB: Murray-Darling Basin; LEB: Lake Eyre Basin.

Species	Distribution	GenBank number	Reference
Caridina confusa	NEA	DQ478495	Page et al. (2007a)
Caridina gracilirostris	NEA	DQ478452	Page et al. (2007a)
Caridina longirostris	NEA	DQ478507	Page et al. (2007a)
Caridina serratirostris	NEA	DQ478515	Page et al. (2007a)
Caridina sp. D	GUL, EAU, MDB	AY795052	Page et al. (2005a)
Caridina sp. D	GUL, EAU, MDB	DQ478523	Page et al. (2007a)
Caridina sp. D	WAC		this study
Caridina sp. DG	GUL	DQ478519	Page et al. (2007a)
Caridina sp. DG	GUL	DQ478520	Page et al. (2007a)
Caridina sp. E	EAU	AY795051	Page et al. (2005a)
Caridina sp. FK1	CNK		this study
Caridina sp. FK1	CNK		this study
Caridina sp. FK1	CNK		this study
Caridina sp. Gulf 1	TEN, GUL		this study
Caridina sp. Gulf 1	TEN, GUL		this study
Caridina sp. Gulf 1	TEN, GUL	DQ478531	Page et al. (2007a)
Caridina sp. Gulf 1	TEN, GUL	DQ478533	Page et al. (2007a)
Caridina sp. *indistincta* A	EAU		Page et al. (2007a)
Caridina sp. *indistincta* A	EAU	DQ478499	Page et al. (2007a)
Caridina sp. *indistincta* A	EAU	AY795039	Page et al. (2005a)
Caridina sp. *indistincta* B	EAU, MDB	AY795040	Page et al. (2005a)
Caridina sp. *indistincta* B	EAU, MDB	AY795043	Page et al. (2005a)
Caridina sp. *indistincta* C	EAU	AY795046	Page et al. (2005a)
Caridina sp. *indistincta* C	EAU	AY795045	Page et al. (2005a)
Caridina sp. *indistincta* C	EAU	DQ478503	Page et al. (2007a)
Caridina sp. *indistincta* C4	EAU	AY795048	Page et al. (2005a)
Caridina sp. *indistincta* C4	EAU	AY795049	Page et al. (2005a)
Caridina sp. LE	LEB	DQ478534	Page et al. (2007a)
Caridina sp. NT 1	TEN	DQ478537	Page et al. (2007a)
Caridina sp. NT *nilotica*	SWK, WAC, NT		this study
Caridina sp. NT *nilotica*	SWK, WAC, NT	DQ478510	Page et al. (2007a)
Caridina sp. WA 2	CNK		this study
Caridina sp. WA 2	CNK		this study
Caridina sp. WA 2	CNK		this study
Caridina sp. WA 2	CNK	DQ478550	Page et al. (2007a)
Caridina sp. WA 2	CNK	DQ478551	Page et al. (2007a)
Caridina sp. WA 3	PIL	DQ478552	Page et al. (2007a)
Caridina sp. WA 3	PIL		this study
Caridina sp. WA 3	PIL		this study
Caridina sp. WA 4	SWK, WAC, TEN		this study
Caridina sp. WA 4	SWK, WAC, TEN	DQ478554	Page et al. (2007a)
Caridina sp. WA 4	SWK, WAC, TEN	DQ478555	Page et al. (2007a)

Table A1. Continuation.

Species	Distribution	GenBank number	Reference
Caridina sp. WA 5	CNK		this study
Caridina sp. WA 6	CNK		this study
Caridina spelunca	CNK, SWK		this study
Caridina spelunca	CNK, SWK	EU123845	Page et al. (2008a)
Caridina spelunca	CNK, SWK	DQ478548	Page et al. (2007a)
Caridina spelunca	CNK, SWK	DQ478549	Page et al. (2007a)
Caridina spinula	NEA	DQ478527	Page et al. (2007a)
Caridina thermophila	LEB	EU123846	Page et al. (2008a)
Caridina thermophila	LEB	DQ478556	Page et al. (2007a)
Caridina typus	NEA	DQ478561	Page et al. (2007a)
Caridina typus	NEA	DQ478562	Page et al. (2007a)
Caridina zebra	NEA	AY661486	Page et al. (2005b)
Caridinides wilkinsi	TEN, NEA		this study
Caridinides wilkinsi	TEN, NEA	DQ681272	Page et al. (2008b)
Caridinides wilkinsi	TEN, NEA	DQ681273	Page et al. (2007b)

Table A2. Collection accession numbers of specimens and species from this study. AM: Australian Museum; GU: Griffith University; QM: Queensland Museum; VM: Museum Victoria; WAM: West Australian Museum; ZMB: Museum fr Naturkunde Berlin.

Species	Institution	Specimen number
Specimens from this study		
Caridinides wilkinsi	GU	CAR99
Caridina sp. FK1	GU	CAR54-6, CAR128, TP777
Caridina sp. FK2	GU	CAR121-6
Curidina sp. Gulf 1	GU	CAR68, CAR103
Caridina sp. NT *nilotica*	GU	CAR36-7, CAR39, CAR79-96
Caridina sp. D	GU	CAR70, CAR73, CAR75, TP540
Caridina spelunca/sp. WA1	GU	CAR62-3, TP308
Caridina sp. WA 2	GU	CAR24-5, CAR61, CAR64, CAR132, CAR156, TP533
Caridina sp. WA 3	GU	TP306-7, TP309, TP335-6, TP539, TP563, TP763-4, TP833, TP862
Caridina sp. WA 4	GU	TP1257
Caridina sp. WA 5	GU	TP1196
Caridina sp. WA 6	GU	TP1198
Same taxa in other institutions		
Caridinides wilkinsi	QM	W22083
Caridina sp. Gulf 1	ZMB	29.24
Caridina sp. NT *nilotica*	ZMB	29.191
Caridina sp. D	VM	J 53098
Caridina spelunca	AM	P38512
Caridina sp. WA 5	WAM	C38998
Caridina sp. WA 6	WAM	C39000

Comparing phylogeographic patterns across the Patagonian Andes in two freshwater crabs of the genus *Aegla* (Decapoda: Aeglidae)

MARCOS PÉREZ-LOSADA[1], JIAWU XU[2], CARLOS G. JARA[3] & KEITH A. CRANDALL[2]

[1] *CIBIO, Centro de Investigação em Biodiversidade e Recursos Genéticos, Universidade do Porto, Campus Agrário de Vairão, 4485-661 Vairão, Portugal*

[2] *Department of Biology & Monte L. Bean Life Science Museum, Brigham Young University, Provo, UT 84602, U. S. A.*

[3] *Instituto de Zoología, Casilla 567, Universidad Austral de Chile, Valdivia, Chile*

ABSTRACT

Andean orogeny and Quaternary glacial cycles have played a major role in shaping the Patagonian freshwater fauna. However, few studies have compared the phylogeographic patterns across the Andes in this region using freshwater taxa. Here we used mitochondrial DNA sequence data to study the population structure of two Patagonian freshwater crabs of the genus *Aegla* and time their divergence. We reanalyzed previously published *A. alacalufi* cytochrome c oxidase subunit I (COI) and subunit II (COII) as well as 16S data from Chile (12 localities; 88 individuals) and newly collected *A. neuqensis* COI sequences from Argentina (7 localities; 28 individuals). Our phylogenetic and population genetic analyses showed that glaciated continental and nonglaciated insular *A. alacalufi* samples have been separated during the Last Glacial Maximum (LGM), with the insular samples showing deeper genetic and time divergence, and the continental ones showing sudden population expansion. The same analyses indicated isolation and dispersion of *A. neuquensis* populations along the Negro River Basin into the Chico-Chubut River Basin, with the Negro samples showing deeper genetic and time divergence and the Chico-Chubut samples showing demographic expansion. Evolution of *A. alacalufi* (21–104 ky) was greatly influenced by Late Pleistocene glaciations, while *A. neuquensis* differentiation (21–164 ky), although facilitated by the LGM, was mainly driven by drainage divides established long before the LGM. Similar phylogeographic patterns have been described in freshwater fishes on each side of the Patagonian Andes.

1 INTRODUCTION

Patagonian biodiversity has been severely affected by the last Andean orogeny since the early Miocene (23–16 my) to the Pleistocene glaciations (1.8–0.012 my) (Vuilleumier 1971; Antonelli et al. 2009). The uplift of the Andes changed forever the South American drainage pattern, generating multiple episodes of vicariance plus drainage system coalescence and extension into new watersheds of emergent land (Lundberg et al. 1998). Pleistocene glacial cycles remodeled some of those environments, especially on the western slopes of the Andes. Glaciers changed drainage patterns and lake distributions, resulting in displacement of plant and animal populations. Ice sheets covered broad areas of Patagonia from ≈ 36°S to 56°S (McCulloch et al. 2000; Hulton et al. 2002), but extensive parts of the Argentinean steppe in the east and stretches of the coastal Chile in the west remained ice free (Rabassa & Clapperton 1990; Clapperton 1993). Over the last 200 ky (thousand years) three significant glaciations have been recorded and labeled according to Oxygen Isotope

Stages (OIS), these are OIS 6 (180–140 ky), OIS 4 (70–60 ky) and OIS 2—Last Glacial Maximum (LGM) (25–23 ky) (see Ruzzante et al. 2008 and references therein).

Patagonian freshwater fauna is likely to show phylogeographic patterns that reflect the major climatic and geologic changes, such as glaciation and orogeny (Xu et al. 2009). Yet only a few fishes (e.g., Ruzzante et al. 2006, 2008; Zemlak et al. 2008, 2010; Unmack et al. 2009) and one crab (Xu et al. 2009) have been examined in Patagonian phylogeographic studies. To better test the effects of the Late Pleistocene glaciations in this region, we need to increase taxon sampling and include species with different biological characteristics (desirably with limited dispersal ability). *Aegla* freshwater crabs are excellent models for phylogeographic studies at both macro- and microgeographical scales: several species are widely distributed across Patagonia (both Chile and Argentina), they have limited dispersal ability compared to fishes, they are easy to collect and their phylogenetic relationships are well known (Pérez-Losada et al. 2009).

Here we focus on two *Aegla* species, *A. alacalufi* Jara and López 1981 and *A. neuquensis* Schmitt 1942. *Aegla alacalufi* occurs in drainages across southern Chile, including glaciated and nonglaciated areas during the LGM (Figure 1). *Aegla neuquensis* has been described in two large rivers in the Argentinean steppe, the Negro (including Neuquen and Limay rivers) and the Chico-Chubut rivers (Figure 1). These two river basins were never covered by ice and are currently isolated from each other, although their headwaters may have been connected during the LGM (Rabassa & Clapperton 1990; Clapperton 1993). In the present study, we use previously published mtDNA and newly collected COI sequence data to compare the phylogeography of *A. alacalufi* (western Andes) and *A. neuquensis* (eastern Andes). Considering how differently these two areas have been affected by Pleistocene glaciations since the uplift of the Andes, we expect both species to show different phylogeographic patterns, too.

2   MATERIALS AND METHODS

2.1   Samples of *Aegla*

A total of 88 *A. alacalufi* freshwater crabs corresponding to 12 locations across southern Chile (Figure 1), covering the distributional range of the species, were chosen from Xu et al. (2009) to represent the species genetic diversity. The five insular samples (Linao, Coinco (Chiloe Island) and Guafo, Ipun, and Guamblin islands) in the west were not covered by ice (nonglaciated samples) within glacial periods (Clapperton 1993). While the seven continental samples in the east (Ralun, Cochamo, Chaica, Contao, Santa Barbara, El Amarillo, and Rosselot, from north to south) were once covered by glacial ice (glaciated samples) (Rabassa & Clapperton 1990; Clapperton 1993). *Aegla hueicollensis*, *A. manni*, and *A. denticulata* were used as outgroups for the phylogenetic analyses.

A total of 28 *A. neuquensis* crabs were collected during 2006–2009 from seven locations in Patagonia covering the distributional range of the species: four sampling sites in the Negro River [Ingeniero Ballester, Covunco, and Alamito (Neuquen) and Comallo (Limay)], and three in the Chico-Chubut River (Chubut and Altares (Chubut) and Sarmiento (Chico)) (Figure 1). *Aegla jujuyana* and *A. sanlorenzo* were used as outgroups. Specimens were collected by dipnet or hand and stored in the crustacean collections at the Monte L. Bean Life Sciences Museum at Brigham Young University (KACa1040 to KACa1067). Gill tissue was removed for DNA extraction.

2.2   Sequence data

Previously collected mtDNA sequence data for partial regions of the cytochrome c oxidase subunit I (COI), subunit II (COII) and 16S ribosomal RNA (16S) genes in *A. alacalufi* were used. For *A.*

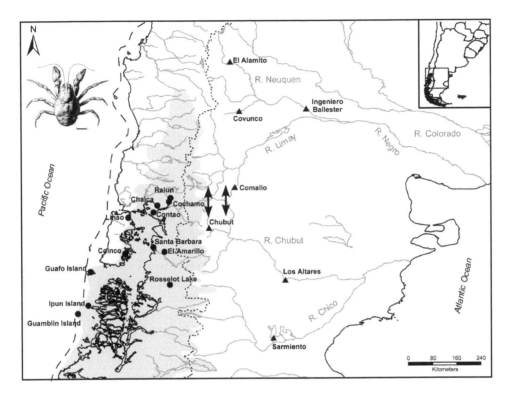

Figure 1. Map of the sampled locations for *Aegla alacalufi* (circles; Chile) and *A. neuquensis* (triangles; Argentina). The dashed line indicates exposed coastal regions during the time of lower sea level (-150 m) in Chile. The gray areas indicate the ice sheet during the Last Glacial Maximum (see Xu et al. 2009 and reference therein). Potential connections between Limay and Chubut rivers are indicated (double arrows). Drawing corresponds to *A. alacalufi*.

neuquensis, genomic DNA was extracted using the DNeasy Blood & Tissue Kit (Qiagen). Partial fragments of the COI gene (659 bp) were amplified and sequenced using primers and conditions described in Xu et al. (2009).

2.3 Genetic analysis

Sequences were aligned with MAFFT 6.0 (Katoh et al. 2005). Model fit was assessed with JModel-Test 0.1.1 (Posada 2009). The GTR + G + I model was chosen for both *Aegla* datasets. Maximum-likelihood (ML) trees were built in RAxML 7.2.6 (Stamatakis 2006). We carried out 1000 bootstrap runs and searched for the best-scoring ML tree.

Population structure was investigated using Analysis of Molecular Variance (AMOVA) (Excoffier et al. 1992) and Φ_{ST} statistics in ARLEQUIN 3.5 (Excoffier & Lischer 2010). Significance was assessed by 1000 permutations. Both tests of population structure were performed using Tamura-Nei genetic distance with gamma correction.

Population demographics of *Aegla* were examined using Tajima's D (Tajima 1989) and Fu's F_S (Fu 1997) neutrality tests and mismatch distribution analysis as implemented in DNASP 5 (Librado

& Rozas 2009). The validity of the expansion model was tested using parametric bootstrap (1000 pseudoreplicates) in ARLEQUIN. The fit to the expected mismatch distribution was quantified by the sum of squared deviations (SSD) between the observed and simulated distributions on one hand and the expected distribution on the other. Current (θ_π) and historical (θ_W) genetic diversities were also estimated in DNASP. θ_π is based on pairwise differences between sequences, while θ_W is based on the number of segregating sites among the sequences. As a consequence, the θ_π estimator reflects current diversity by virtue of incorporating frequency information for different genotypes. On the other hand, θ_W is a genealogical estimator that ignores frequencies of genotypes and estimates diversity based on the differences over a coalescent process thereby reflecting a historical perspective of genetic diversity. Therefore, comparing these two estimates provides insight into population dynamics over recent evolutionary history; differences between them are indicative of recent bottlenecks (if $\theta_\pi < \theta_W$) or recent population growth (if $\theta_\pi > \theta_W$) (Templeton 1993; Pearse & Crandall 2004).

We used *A. alacalufi* divergence times (i.e., time to the most recent common ancestor—TMRCA) and their High Posterior Density (HPD) intervals from Xu et al. (2009). TMRCA and HPD of *A. neuquensis* for selected clades on the ML tree were estimated using the coalescent Bayesian skyline plot model as implemented in BEAST 1.5.3 (Drummond & Rambaut 2007) and visualized in Tracer 1.5 (Rambaut & Drummond 2009). We applied a lognormal relax-clock model. The LGM calibration used in Xu et al. (2009) was also implemented here under a normal prior (mean = 21; SD = 2.1 ky). We performed four independent runs with a Markov chain Monte Carlo (MCMC) chain length of 20 million and sampling every 1000 generations.

3 RESULTS

3.1 *Aegla alacalufi*—Chile

The ML tree (Figure 2) revealed four main ingroup clades supported by high bootstrap values (> 85%). The three most basal clades consisted of all nonglaciated samples (Ipun, Guamblin, Linao (clade 1), Coinco (clade 2), and Guafo (clade 3)), while the most derived clade included all the glaciated populations (Ralun, Cochamo, Chaica, Contao, Santa Barbara, El Amarillo, and Rosselot). For the most part, the glaciated populations and their associated drainages did not form monophyletic clades. Uncorrected genetic distances (current genetic diversity, θ_π; Table 1) among nonglaciated samples ranged from 0.0 to 0.0171 (mean = 0.0095; median = 0.0112) and were significantly higher ($P < 0.001$; AMOVA test) than those observed among glaciated samples, which ranged from 0 to 0.0071 (mean = 0.0029; median = 0.0030). Hierarchical AMOVA revealed that 71% of the total genetic variation could be explained by differences between populations from glaciated and nonglaciated areas, while only 29% of the variation could be explained by differences within populations. The Φ_{ST} index was also high (0.71; $P < 0.001$).

The mismatch distribution analysis showed a unimodal distribution ($P_{SSD} = 0.56$) for the glaciated samples (Figure 3A) and a multimodal distribution for the nonglaciated ones (Figure 3B). Tajima's D values were significant and negative for the glaciated samples, but positive and nonsignificant for the nonglaciated ones (Table 1). Fu's F_S values were significant and negative for both non- and glaciated samples, although estimates were 3.6 times higher for the latter than the former (Table 1). All this evidence combined suggests sudden demographic expansion for the glaciated samples and demographic equilibrium for the nonglaciated samples.

Genetic diversity estimates were lower for the glaciated samples than for the nonglaciated ones (Table 1), as expected in suddenly expanded populations (Avise 2000). Historical diversity (θ_W) was higher than current diversity (θ_π) for the glaciated samples and the opposite for the nonglaciated ones, implying the occurrence of a recent bottleneck in the glaciated samples and recent population growth for the nonglaciated samples (Templeton 1993).

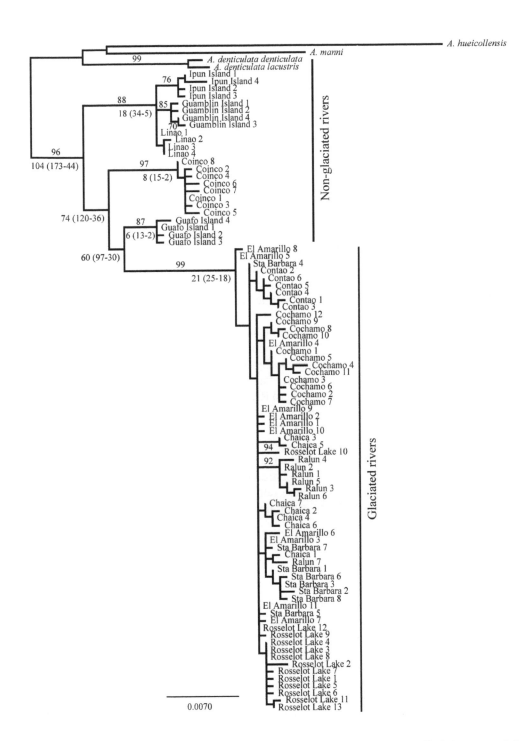

Figure 2. ML tree of *Aegla alacalufi* mtDNA (COI, COII and 16S) sequences. Clade bootstrap (after 1000 reiterations) proportions (if ≥ 70%) and mean divergence times (ky (95% High Density Posterior intervals)) are indicated above and below branches, respectively.

3.2   *Aegla neuquensis*—Argentina

The ML phylogeny (Figure 4) had a well-supported ($\geq$ 78%) pectinate tree shape, where the Neuquen and Limay (Negro) River samples formed four consecutive basal clades and the Chubut and Chico River samples formed a shallower derived clade. Uncorrected genetic distances (θ_π; Table 1) among the Negro River samples ranged from 0.0015 to 0.0455 (mean = 0.0252; median = 0.0288) and were significantly higher ($P < 0.001$; ANOVA test) than those observed among Chico-Chubut River samples, which ranged from 0.0015 to 0.0106 (mean = 0.0056; median = 0.0061). Hierarchical AMOVA revealed that 59% of the total genetic variation could be explained by differences between populations from the Negro and Chico-Chubut rivers, while only 41% of the variation could be explained by differences within populations. The F_{ST} index was also very high (0.59; $P < 0.001$).

The mismatch distribution analysis showed a unimodal distribution ($P_{SSD} = 0.31$) for the Chico-Chubut River samples (Figure 3C) and a multimodal distribution for the Negro River ones (Figure 3D). Tajima's D values were negative for the Chico-Chubut River samples, but positive for the Negro River samples, and all nonsignificant (Table 1). Fu's F_S values were significant and negative for both river samples, although estimates were almost three times higher for the Chico-Chubut than the Negro River samples (Table 1). All these analyses combined suggest a sudden demographic expansion for the Chico-Chubut River samples and demographic equilibrium for the Negro River samples.

Genetic diversity estimates were lower for Chico-Chubut River samples than for the Negro River ones (Table 1), as expected for suddenly expanded populations (Avise 2000). Historical diversity (θ_W) was higher than current diversity (θ_π) for the Chico-Chubut River samples and the opposite for the Negro River ones, implying the occurrence of a recent bottleneck in the Chico-Chubut River samples and recent population growth for the Negro River samples.

The TMRCA (95% HPD) for *A. neuquensis* was estimated at 164 ky (292–59 ky; Figure 4). The Negro River samples diverged 164 (292–59) to 41 (68–21) ky ago (kya) and the Chico-Chubut River diverged 21 (25–16) kya. Based on our LGM calibration, the estimated mean substitution rate for COI was 48.6% per my, almost 25 times higher than the normally applied rate of 2% per my (Avise, 2000) and 5 times higher than those reported in intraspecific data (Emerson 2007), but similar to that estimated for *A. alacalufi*. Nonetheless these rate estimates must be interpreted with caution considering the wide time range of the LGM calibration of the only calibration used.

4   DISCUSSION

4.1   Phylogeography of *Aegla alacalufi*—Chile

The population structure of *A. alacalufi* in the western Andes appears to be driven by the Late Pleistocene glacial cycles. Our mtDNA phylogeny (Figure 2) and AMOVA test revealed high population structuring in *A. alacalufi*, and nonglaciated samples showed deeper genetic and time divergences than glaciated ones—as expected under glacial cycles and also has been described in freshwater fishes (Bernatchez & Wilson 1998; Ruzzante et al. 2006).

A single recent colonizing event of the glaciated areas is indicated by our phylogenetic analysis (Figure 2) with the Amarillo as the potential source (refuge). Amarillo is a hot spring that could be left uncovered by the ice of the LGM. Glacial refugia, both within and beyond the glacial ice shield, have been proposed for Patagonian fauna and flora in the coastal Chilean Cordillera and the Andes (Kim et al. 1998; Smith et al. 2001; Muellner et al. 2005; Palma et al. 2005). Comparative phylogeographic analysis of other species within this region may confirm Amarillo as a refugium in the western Andes during the LGM.

Table 1. Genetic diversity parameters for *Aegla alacalufi* (COI, 16S and 12S) and *A. neuquensis* (COI). Sample size (N), historical (θ_W) and current (θ_π) genetic diversity, and D and F_S neutrality tests.

Species/region	N	θ_W	θ_π	F_S	D
A. alacalufi					
Nonglaciated rivers	24	0.0084	0.0095	−9.246∗∗∗	0.517
Glaciated rivers	64	0.0073	0.0029	−33.818∗∗∗	−2.02∗
A. neuquensis					
Negro River	14	0.0239	0.0252	−4.401∗∗∗	0.531
Chico-Chubut River	14	0.0076	0.0056	−12.874∗∗∗	−0.867

Significant evidence of sudden demographic expansion was observed in the glaciated samples, but demographic stability was observed in the nonglaciated ones (Table 1 and Figure 3). Our phylogenetic analysis does not reveal the direction of this expansion, but previous cladogram and nested clade analysis in Xu et al. (2009) seem to indicate a downward altitudinal dispersion following the flood of melting ice from the Amarillo (264 m) to and across all the other lower altitude areas (0–71 m). Northward and southward expansions have also been proposed for other fauna and flora in the Chilean Patagonia after the LGM (Morrone & Lopretto 1994; Premoli et al. 2000; Smith et al. 2001; Muellner et al. 2005; Palma et al. 2005).

Last Pleistocene glaciation had less or no direct impact on the nonglaciated (islands) *A. alacalufi* populations (Figure 1). Their deep genetic and time divergence (Figure 2) indicate long-term isolation (≥ 60 ky), before the LGM (25–23 ky) and within the OIS 4 (70–60 ky). During this period a large coastal ocean landmass connecting all the western islands was exposed (Figure 1), allowing crabs to reach these islands via watersheds and facilitated by ice melting floods. Colonization (or expansion) across islands in freshwater species usually occurred in glacial stages with the formation of land bridges following the sea level drops (e.g., Okazaki et al. 1999). After colonization, insular populations may have experienced fragmentation from their continental relatives during interglacial periods in which ocean water levels rose again. Moreover, drainage isolation may have also contributed to increase genetic differences among samples, as described in other freshwater crustaceans and fishes with limited dispersal capacity (e.g., Daniels et al. 2006; Gouws et al. 2006; Ruzzante et al. 2006; Zemlak et al. 2010).

Comparisons of current and historical genetic diversities (Table 1) within glaciated and nonglaciated samples indicate recent population increase for the latter and drastic decrease for the former. The continental habitats of *A. alacalufi* have been deeply altered due to deforestation and drainage sedimentation caused by large fires and logging (Pérez-Losada et al. 2002; Habit et al. 2010). Habitat fragmentation and population isolation have been observed in this and many other *Aegla* species (Pérez-Losada et al. 2009).

4.2 Phylogeography of *Aegla neuquensis*—Argentina

The population structure of *A. neuquensis* in the eastern Andes appears to be driven by the drainage divides of the Negro and Chico-Chubut rivers. Our COI phylogeny (Figure 4) and AMOVA test revealed high population structuring, and the Limay-Neuquen River samples showed deeper genetic and time divergences than the Chico-Chubut River ones. The ML tree also suggests a longitudinal dispersion first along the Neuquen into the Limay basins, then across river drainages into and along the Chubut River, and finally to the Chico River. The riverheads of the Limay and Chubut are only a few kilometers apart in two sections (see Figure 1), and at least the most western drainage was

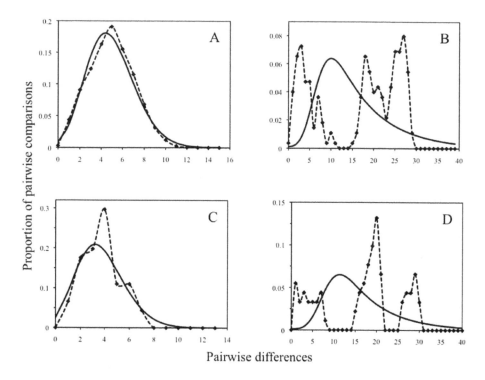

Figure 3. Mismatch distribution analyses of *Aegla alacalufi* glaciated (A) and nonglaciated samples (B) and *A. neuquensis* samples from the Chico-Chubut (C) and Negro (D) Rivers. Observed and expected distributions are depicted with dashed and solid lines, respectively.

under ice during the LGM. One could hence postulate that after the LGM, the ice melting flood could have connected the headwaters of both rivers for a short period of time and facilitated *A. neuquensis* dispersal along the Chubut River. Moreover, large Pleistocene palaeolakes have been also proposed east of the Andes during periods of glacial retreat, and one such palaeolake, Lake Cari Laufquen, was presumably located at ≈ 41°S (Clapperton 1993) and may have connected the headwaters of the two rivers. Similar dispersal routes have been also proposed in several fish groups (Zattara & Premoli 2005; Ruzzante et al. 2006, 2008; Zemlak et al. 2008, 2010) and other decapods (Morrone & Lopretto 1994).

The last Pleistocene glaciation did not reach *A. neuquensis* populations in the Negro River (Figure 1). Moreover, their deep genetic and time divergence (Figure 4) indicate longer isolation (164–41 ky). During that period the Neuquen-Limay basins were already established (Lundberg et al. 1998), hence despite the potential for population connectivity, some other environmental (e.g., water temperature, turbidity), ecological (e.g., habitat requirements), and/or biological (e.g., limited dispersal ability) factors, have precluded animal migration (gene flow) across the sampled areas and caused population structuring (Table 1). Unraveling *A. neuquensis* population structuring in the Negro River will require further sampling.

Comparisons of current and historical genetic diversities (Table 1) within Negro and Chico-Chubut River samples indicate recent population increase for the former and decrease for the latter. The Chico-Chubut Basin is particularly arid and has been suffering severe water reductions due to crop irrigation. This could have caused recent population size reduction and erosion of genetic diversity in these *Aegla* populations, as seen in other *Aegla* species (Pérez-Losada et al. 2009).

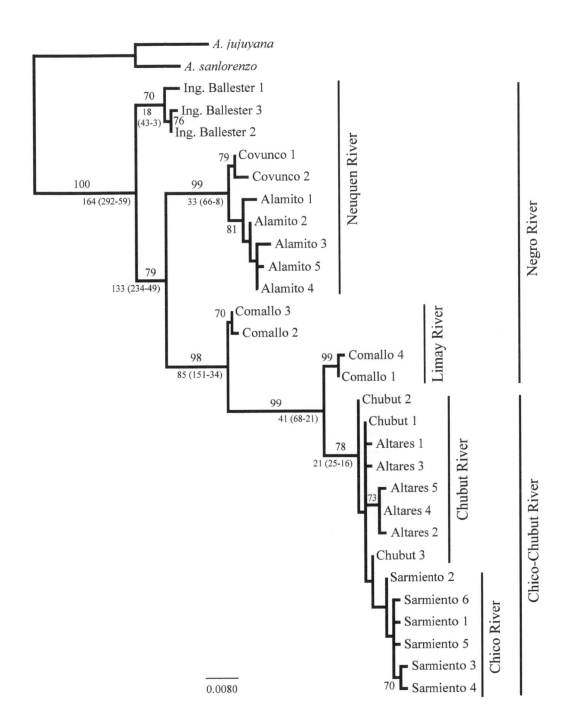

Figure 4. ML tree of *Aegla neuquensis* COI sequences. Clade bootstrap (after 1000 reiterations) proportions (if ≥ 70%) and mean divergence times (ky (95% High Posterior Density intervals)) are indicated above and below branches, respectively.

4.3   Comparing phylogeographic patterns across the Patagonian Andes

The uplift of the Andes changed drastically the Patagonian river systems (Lundberg et al. 1998), which, subsequently, were altered by the Pleistocene glaciations (Clapperton 1993). Chilean rivers (western Andes) are less numerous, short, and fast flowing; they were severely affected by glacial cycles which changed their shape, slope, flowing direction, and connections both at their headwaters (e.g., palaeolakes) and mouths (e.g., sea level changes). Argentinean rivers (eastern Andes) are less numerous, long and slow flowing; they were well established before Pleistocene glaciations (Lundberg et al. 1998) and for the most part remained free of ice. Only the headwaters of the southern drainages were possibly affected by the last glacial cycles, which nonetheless created opportunities for drainage connectivity through palaeolakes and river capture (Rabassa & Clapperton 1990; Clapperton 1993). Hence, considering their different hydrological history, one would expect that freshwater species occurring at both sides of the Andes might show different phylogeographic patterns. Surprisingly, very few phylogeographic studies have compared freshwater fauna across the Andean slopes.

Our data and phylogenetic (Figures 2 and 4) and time estimates indicate that geological and climatic events occurring during the Late Pleistocene glaciations have differently affected species of *Aegla* occurring at both sides of the Andes. For *A. alacalufi* population structuring was primarily the result of the OIS 4 (insular samples) and LGM glaciations (continental samples) and subsequent drainage isolation. For *A. neuquensis* population structuring was mainly driven by pre-Pleistocene river divides. Pleistocene glaciations did not seem to play a major role at shaping *A. neuquensis* diversity, although we think that the LGM has facilitated its southward dispersion by connecting the headwaters of the Limay and Chubut rivers. Similar biogeographical patterns have been already proposed in the Patagonian Andes for other freshwater fauna (Morrone & Lopretto 1994; Zattara & Premoli 2005; Ruzzante et al. 2006, 2008; Zemlak et al. 2008, 2010).

Our analyses also revealed higher genetic diversity, deeper phylogenetic structure and older divergence times in the Argentinean *A. neuquensis* samples than in the Chilean *A. alacalufi* samples. Comparative studies on freshwater fishes, however, suggest greater homogeneity in the eastern Andes populations than in the western ones (Ruzzante et al. 2006; Zemlak et al. 2008, 2010; Unmack et al. 2009). According to these authors, eastern populations may have experienced larger population admixture and dispersal during the Pleistocene through palaeolakes formed in the interglacial periods that connected the headwaters of the Argentinean drainages. Western (Chilean) fish populations, however, may have remained more isolated through the same period. Hence, our *Aegla* data suggest that Pleistocene glacial cycles have affected freshwater species differently. Biological differences (i.e., dispersal ability, reproductive strategies, population sizes) between crabs and fishes may be the key to explaining such differences. Our conclusions highlight the importance of using taxa with different biological characteristics when comparing phylogeographic patterns (Bernatchez & Wilson 1998; Ruzzante et al. 2008; Xu et al. 2009).

ACKNOWLEDGEMENTS

We thank Peter Unmack for his assistance with the map drawing. We also thank the two referees and the editor for their constructive comments. This work was supported by grants from the U.S. National Science Foundation (0530267; 0520978), the Monte L. Bean Life Science Museum at Brigham Young University, and the Fundação para a Ciência ea Tecnologia (PTDC/BIA-BEC/098553/2008) to MP-L, and benefited from a NESCent catalysis group.

REFERENCES

Antonelli, A., Nylander, J.A.A., Persson, C. & Sanmartin, I. 2009. Tracing the impact of the Andean uplift on Neotropical plant evolution. *Proc. Natl. Acad. Sci. USA* 106: 9749–9754.

Avise, J. 2000. *Phylogeography*. Cambridge, MA: Harvard University Press.

Bernatchez, L. & Wilson, C.C. 1998. Comparative phylogeography of Nearctic and Palearctic fishes. *Mol. Ecol.* 7: 431–452.

Clapperton, C. 1993. *Quaternary Geology and Geomorphology of South America*. Amsterdam: Elsevier.

Daniels, S.R., Gouws, G. & Crandall, K.A. 2006. Phylogeographic patterning in a freshwater crab species (Decapoda: Potamonautidae: *Potamonautes*) reveals the signature of historical climatic oscillations. *J. Biogeogr.* 33: 1538–1549.

Drummond, A.J. & Rambaut, A. 2007. BEAST: Bayesian evolutionary analysis by sampling trees. *BMC Evol. Biol.* 7: 214.

Emerson, B.C. 2007. Alarm bells for the molecular clock? No support for Ho et al.'s model of time-dependent molecular rate estimates. *Syst. Biol.* 56: 337–345.

Excoffier, L. & Lischer, H.E.L. 2010. Arlequin suite ver 3.5: a new series of programs to perform population genetics analyses under Linux and Windows. *Mol. Ecol. Resourc.* 10: 564–567.

Excoffier, L., Smouse, P.E. & Quattro, J.M. 1992. Analysis of molecular variance inferred from metric distances among DNA haplotypes: application to human mitochondrial DNA restriction data. *Genetics* 131: 479–491.

Fu, Y.X. 1997. Statistical tests of neutrality of mutations against population growth, hitchhiking and background selection. *Genetics* 147: 915–925.

Gouws, G., Stewart, B.A. & Daniels, S.R. 2006. Phylogeographic structure of a freshwater crayfish (Decapoda: Parastacidae: *Cherax preissii*) in south-western Australia. *Mar. Freshw. Res.* 57: 837–848.

Habit, E., Piedra, P., Ruzzante, D.E., Walde, S.J., Belk, M.C., Cússac, V.E., González, J. & Colin, N. 2010. Changes in the distribution of native fishes in response to introduced species and other anthropogenic effects. *Global Ecol. Biogeogr.* 19: 697–710.

Hulton, N.R.J., Purves, R.S., McCulloch, R.D., Sugden, D.E. & Bentley, M.J. 2002. The last glacial maximum and deglaciation in southern South America. *Quaternary Sci. Rev.* 21: 233–241.

Jara, C.G. & López, M.T. 1981. A new species of freshwater crab (Crustacea: Anomura: Aeglidae) from insular Chile. *Proc. Biol. Soc. Wash.* 94: 88–93.

Katoh, K., Kuma, K., Toh, H. & Miyata, T. 2005. MAFFT version 5: improvement in accuracy of multiple sequence alignment. *Nucleic Acids Res.* 33: 511–518.

Kim, I., Phillips, C.J., Monjeau, J.A., Birney, E.C., Noack, K., Pumo, D.E., Sikes, R.S. & Dole, J.A. 1998. Habitat islands, genetic diversity, and gene flow in a Patagonian rodent. *Mol. Ecol.* 7: 667–678.

Librado, P. & Rozas, J. 2009. DnaSP v5: a software for comprehensive analysis of DNA polymorphism data. *Bioinformatics* 25: 1451–1452.

Lundberg, J.G., Marshall, L. G., Guerrero, J., Horton, B., Malabarba, M.C.S.L. & Wesselingh, F. 1998. The stage for Neotropical fish diversification: a history of tropical South American rivers. In: Malabarba, L.R., Reis, R.E., Vari, R.P., Lucena, Z.M.S. & Lucena, C.A.S. (eds.) *Phylogeny and Classification of Neotropical Fishes*: 13–48. Porto Alegre, Brazil: EDIPUCRS.

McCulloch, R.D., Bentley, M.J., Purves, R.S., Hulton, N.R.J., Sugden, D.E. & Clapperton, C.M. 2000. Climatic inferences from glacial and paleoecological evidence at the last glacial termination, southern South America. *J. Quaternary Sci.* 15: 409–417.

Morrone, J.J. & Lopretto, E.C. 1994. Distributional patterns of fresh-water Decapoda (Crustacea, Malacostraca) in southern South-America—a panbiogeographic approach. *J. Biogeogr.* 21:

97–109.

Muellner, A.N., Tremetsberger, K., Stuessy, T. & Baeza, C.M. 2005. Pleistocene refugia and re-colonization routes in the southern Andes: insights from *Hypochaeris palustris* (Asteraceae, Lactuceae). *Mol. Ecol.* 14: 203–212.

Okazaki, T., Jeon, S.R., Watanabe, M. & Kitagawa, T. 1999. Genetic relationships of Japanese and Korean bagrid catfishes inferred from mitochondrial DNA analysis. *Zool. Sci.* 16: 363–373.

Palma, R.E., Rivera-Milla, E., Salazar-Bravo, J., Torres-Pérez, F., Pardiñas, U.F.J., Marquet, P.A., Spotorno, A.E., Meynard, A.P. & Yates, T.L. 2005. Phylogeography of *Oligoryzomys longicaudatus* (Rodentia: Sigmodontinae) in temperate South America. *J. Mammal.* 86: 191–200.

Pearse, D.E. & Crandall, K. 2004. Beyond F_{ST}: analysis of population genetic data for conservation. *Conserv. Genet.* 5: 585–602.

Pérez-Losada, M., Bond-Buckup, G., Jara, C.G. & Crandall, K.A. 2009. Conservation assessment of southern South American freshwater ecoregions on the basis of the distribution and genetic diversity of crabs from the genus *Aegla. Conserv. Biol.* 23: 692–702.

Pérez-Losada, M., Jara, C.G., Bond-Buckup, G. & Crandall, K.A. 2002. Conservation phylogenetics of Chilean freshwater crabs *Aegla* (Anomura, Aeglidae): assigning priorities for aquatic habitat protection. *Biol. Conserv.* 105: 345–353.

Posada, D. 2009. Selection of models of DNA evolution with JModelTest. *Methods Mol. Biol.* 537: 93–112.

Premoli, A.C., Kitzberger, T. & Veblen, T.T. 2000. Isozyme variation and recent biogeographical history of the long-lived conifer *Fitzroya cupressoides. J. Biogeogr.* 27: 251–260.

Rabassa, J. & Clapperton, C. 1990. Quaternary glaciations of the southern Andes. *Quaternary Sci. Rev.* 9: 153–174.

Rambaut, A. & Drummond, A.J. 2009. *Tracer: MCMC trace analysis tool 1.5.* Institute of Evolutionary Biology: Edinburgh.

Ruzzante, D.E., Walde, S.J., Cussac, V.E., Dalebout, M.L., Seibert, J., Ortubay, S. & Habit, E. 2006. Phylogeography of the Percichthyidae (Pisces) in Patagonia: roles of orogeny, glaciation, and volcanism. *Mol. Ecol.* 15: 2949–2968.

Ruzzante, D.E., Walde, S.J., Gosse, J.C., Cussac, V.E., Habit, E., Zemlak, T.S. & Adams, E.D. 2008. Climate control on ancestral population dynamics: insight from Patagonian fish phylogeography. *Mol. Ecol.* 17: 2234–2244.

Schmitt, W.L. 1942. The species of *Aegla*, endemic South American fresh-water crustaceans. *Proc. U. S. Nat. Mus.* 91: 431–520.

Smith, M.F., Kelt, D.A. & Patton, J.L. 2001. Testing models of diversification in mice in the *Abrothrix olivaceus/xanthorhinus* complex in Chile and Argentina. *Mol. Ecol.* 10: 397–405.

Stamatakis, A. 2006. RAxML-VI-HPC: maximum likelihood-based phylogenetic analyses with thousands of taxa and mixed models. *Bioinformatics* 22: 2688–2690.

Tajima, F. 1989. Statistical method for testing the neutral mutation hypothesis by DNA polymorphism. *Genetics* 123: 585–595.

Templeton, A.R. 1993. The "Eve" hypothesis: a genetic critique and reanalysis. *Amer. Anthropol.* 95: 51–72.

Unmack, P.J., Bennin, A.P., Habit, E.M., Victoriano, P.F. & Johnson, J.B. 2009. Impact of ocean barriers, topography, and glaciation on the phylogeography of the catfish *Trichomycterus areolatus* (Teleostei: Trichomycteridae) in Chile. *Biol. J. Linn. Soc.* 97: 876–892.

Vuilleumier, B.S. 1971. Pleistocene changes in the fauna and flora of South America. *Science* 173: 771–780.

Xu, J., Pérez-Losada, M., Jara, C.G. & Crandall, K.A. 2009. Pleistocene glaciation leaves deep signature on the freshwater crab *Aegla alacalufi* in Chilean Patagonia. *Mol. Ecol.* 18: 904–918.

Zattara, E.E. & Premoli, A.C. 2005. Genetic structuring in Andean landlocked populations of *Galaxias maculatus*: effects of biogeographic history. *J. Biogeogr.* 32: 5–14.

Zemlak, T.S., Habit, E.M., Walde, S.J., Battini, M.A., Adams, E.D. & Ruzzante, D.E. 2008. Across the southern Andes on fin: glacial refugia, drainage reversals and a secondary contact zone revealed by the phylogeographical signal of *Galaxias platei* in Patagonia. *Mol. Ecol.* 17: 5049–5061.

Zemlak, T.S., Habit, E.M., Walde, S.J., Carrea, C. & Ruzzante, D.E. 2010. Surviving historical Patagonian landscapes and climate: molecular insights from *Galaxias maculatus*. *BMC Evol. Biol.* 10: 67.

Molecular diversity of river versus lake freshwater anomurans in southern Chile (Decapoda: Aeglidae) and morphometric differentiation between species and sexes

HEATHER D. BRACKEN-GRISSOM[1], TIFFANY ENDERS[1], CARLOS G. JARA[2] & KEITH A. CRANDALL[1,3]

[1] Department of Biology, Brigham Young University, Provo, UT 84602, U. S. A.
[2] Instituto de Zoología, Casilla 567, Universidad Austral de Chile, Valdivia, Chile
[3] Monte L. Bean Life Science Museum, Brigham Young University, Provo, UT 84602, U. S. A.

ABSTRACT

The family Aeglidae consists of a single genus (*Aegla*) of anomuran crabs restricted to the Neotropical region of South America. Much of the present-day distribution of freshwater species in southern Chile has been impacted by Pleistocene glacial cycles. The melting of ice sheets created elaborate lake and river systems throughout this area and played an important role in the speciation of *Aegla*. In this study, we sampled phylogenetically closely related species of *Aegla* from one river (*A. cholchol*) and one lake (*A. rostrata*), to examine the molecular divergence across three genetic loci (elongation factor 1 (EF1) intron, 16S, cytochrome c oxidase subunit I (COI)) and how genetic variation differs between the two habitats. We estimate the relative timing of divergence using Bayesian molecular dating methods and the associated molecular data. We then examine whether morphometrics allow differentiation between the two species and sexes.

1 INTRODUCTION

The family Aeglidae consists of a single genus (*Aegla*) of anomuran crabs restricted to the Neotropical region of South America. Aeglid crabs are morphologically distinct, sharing several external characters including branchial morphology and carapace structure (Martin & Abele 1988). They are the only taxon of the Anomura found entirely in freshwater habitats, making them ecologically unique. These crabs occupy a variety of habitats such as streams, lakes, and caves, from depths that plunge 320 m (Jara 1977) to heights that reach 3,500 m (Bond-Buckup & Buckup 1994).

Much of the present-day distribution of freshwater species in southern Chile has been impacted by Pleistocene glacial cycles (e.g., Xu et al. 2009). The melting of ice sheets created elaborate lake and river systems throughout this area and played an important role in the speciation of *Aegla*. There are 19 species of aeglid crabs in Chile, 16 of which are endemic to the country, and many of which may be in grave danger due to the deterioration of the Chilean stream environments (Pérez-Losada et al. 2002a). Past studies have suggested conservation strategies to protect these species, and a recent assessment of the Chilean species lists three as critically endangered, two as extinct in the wild, one as near threatened, and six as vulnerable (Pérez-Losada et al. 2002a). In particular, *A. cholchol* has undergone a population reduction of at least 30% (reported in Pérez-Losada et al. 2002a), and studies have shown that their habitats are subject to contamination, alteration, and

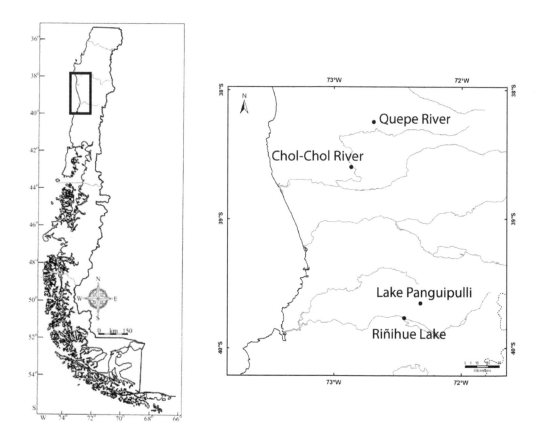

Figure 1. Sampling localities of *Aegla* in southern Chile.

pollution from agricultural runoff, cattle farms, urbanization, and farming (Jara 1996; Bahamonde et al. 1998). The implication for conservation efforts within aeglid crabs, in combination with their morphological and ecological uniqueness, emphasizes the need for increased attention and research on the group.

Aegla cholchol is a river species found within the Chol-Chol, Cautín, and Quepe rivers; all tributaries of the Imperial River that drain into the Pacific Ocean. *Aegla cholchol* is morphologically similar to *A. rostrata* but differs in pereopodal characters and attributes of the carapace (Jara & Palacios 1999). *Aegla rostrata* is found primarily in lakes (Riñihue, Panguipulli, and Calafquén), but occasionally in river environments (Calle Calle and Huanehue) (Jara 1977). Jara (1977) suggested that the current distribution of *A. rostrata* is the result of melting ice sheets following the Pleistocene glaciation.

In this study, we examine the genetic relationships and divergence times of a pair of closely related (Pérez-Losada et al. 2002b, 2004) river (*A. cholchol*) and lake species (*A. rostrata*) in southern Chile, and correlate species diversification and origins with the Last Glacial Maximum in the region. It is presumed that river basins located near or at the Coastal Range served as refugial areas for the freshwater fauna during glacial maxima, subsequently serving as source areas for the colonization of pre-Andean lakes once they became deglaciated (see Jara 1977, 1989). Furthermore, previous research suggests that there is a tendency towards a reduction in rostrum length, width of the forehead, and ornamentation as an adaptation to river or stream environments (Ringuelet 1949). On the contrary, the reverse tendency can be considered when riverine populations of *Aegla* are considered as ancestors (plesiomorph) to the lacustrine populations, so that a magnified ornamentation conveys a derived (apomorph) condition (see Jara 1989, 1996). Using a combination of morphometrics and genetics, we compare a river species and lake species to test hypotheses associated with the morphological adaptations between ecotypes (lacustrine vs. riverine) and provide possible explanations for the morphological trends suggested by the morphological and genetic data.

2 MATERIAL AND METHODS

2.1 Taxon and locality sampling

Freshwater crabs of *A. cholchol* and *A. rostrata* were sampled from two localities (one river and one lake) across southern Chile from February through December 2009 (Figure 1). Individuals of *A. cholchol* were collected from the Quepe River (38° 15′ 06″ S, 72° 41′ 31″ W) approximately 45 m above sea level, and individuals of *A. rostrata* were collected from Lake Panguipulli (39° 39′ 06″ S, 72° 18′ 77″W), approximately 130 m above sea level. Lake Panguipulli is a pre-Andean lake interconnected by a series of short rivers with at least two other lakes, i.e., Calafquen and Riñihue, which are part of a seven lakes chain in the upper Valdivia River Basin, associated with the temperate rain forests of southern Chile. Studies looking at the lake's bathymetric profile and morphometric parameters have concluded it to be of glacial origins (Campos et al. 1981). The Quepe River originates in Lake Quepe, located in the Araucanía region of southern Chile. This river joins the Cautín River 112 km downstream, eventually forming the Imperial River where the Cautín and Chol-Chol rivers merge. All crabs were collected under rocks with dip nets in the river (*A. cholchol*) or by placing bag nets baited with fish heads at 45 m depth on the lake's sandy bottom (*A. rostrata*). Specimens were preserved in ethanol and deposited in the Monte L. Bean Life Science Museum at Brigham Young University (see Appendix, Table A1).

To broaden our sampling efforts across the geographic distribution of these two species, Gen-Bank sequences of *A. rostrata* and *A. cholchol* were incorporated in the molecular analyses (Appendix, Table A1). These included sequences from four individuals of *A. cholchol* collected from the Chol-Chol River (38° 36′ S, 72° 52′ W) and two individuals of *A. rostrata* collected from Riñihue Lake (39° 46′ S, 72° 27′ W) (Figure 1; Pérez-Losada et al. 2004).

We selected two species of *Aegla* (*A. abtao* and *A. alacalufi*) for inclusion in the molecular analysis based on recent genetic studies and available sequence data (Pérez-Losada et al. 2002b, 2004). All outgroup sequences were obtained from GenBank (Appendix, Table A1). All individuals used in the molecular analysis (excluding GenBank taxa) were included in the morphometric analysis.

2.2 DNA extraction, PCR, and sequencing

Total genomic DNA was extracted from the gills or muscle using the Qiagen DNeasy® Blood and Tissue Kit (Cat. No. 69582). Targeted gene regions were amplified by means of the polymerase chain reaction (PCR) using one or more sets of primers: 16S, large ribosomal RNA subunit (≈ 450 basepairs (bp), Pérez-Losada et al. 2002b); COI, cytochrome c oxidase subunit I (≈ 870 bp, Xu et al. 2009) (both mtDNA); EF1, elongation factor 1 intron (≈ 350 bp, Xu et al. 2009) (nuclear DNA).

Reactions were performed in 25 μl volumes containing 10 μM forward and reverse primer for each gene, 200 μmol each dNTP, 10× PCR buffer with 25 mM Mg^{2+}, 1 unit HotMasterTaq polymerase (5 PRIME), and 30–100 ng extracted DNA. The thermal cycling profile conformed to the following parameters: initial denaturation for 1 min at 94 °C followed by 30–40 cycles of 1 min at 94 °C, 1 min at 48–50 °C (depending on gene region), 1 min at 72 °C, and a final extension of 10 min at 72 °C. PCR products were purified using filters (PrepEase™ PCR Purification 96-well Plate Kit, USB Corporation) and sequenced with ABI BigDye® terminator mix (Applied Biosystems, Foster City, CA, USA). An Applied Biosystems 9800 Fast Thermal Cycler (Applied Biosystems, Foster City, CA, USA) was used in PCR and cycle sequencing reactions, and sequencing products were run (forward and reverse) on an ABI 3730xl DNA Analyzer 96-capillary automated sequencer.

2.3   Phylogenetic analysis

Sequences were assembled, cleaned, and edited using the computer program Sequencher 4.7 (GeneCodes, Ann Arbor, MI, USA). Alignments were created using MUSCLE (multiple sequence comparison by log-expectation) or MAFFT, which have been found to be more accurate and faster than other alignment algorithms (Edgar 2004; Katoh et al. 2005). Alignments were concatenated into two separate datasets, one consisting of mitochondrial (mt) genes (16S, COI; 1328 bp), and a second consisting of all 3 gene regions (16S, COI, EF1 intron; 1694 bp). To check for pseudogenes within our dataset, we followed suggestions outlined by Song et al. (2008). This included translating COI sequences into amino acids to check for stop codons, using species-specific primers and generating individual gene trees.

The model of evolution that best fit the individual datasets (16S, COI, EF1) was determined by MODELTEST 3.7 (Posada & Crandall 1998). The Maximum likelihood (ML) analysis was conducted using RAxML (Randomized Axelerated Maximum Likelihood) (Stamatakis et al. 2005, 2007, 2008) with computations performed on the computer cluster of the Cyberinfrastructure for Phylogenetic Research Project (CIPRES) at the San Diego Supercomputer Center. Likelihood settings followed the General time reversible model (GTR) with a gamma distribution and invariable sites and RAxML estimated all free parameters following a partitioned dataset. Confidence in the resulting topology was assessed using nonparametric bootstrap estimates (Felsenstein 1985) with 1,000 pseudoreplicates; values > 50% are presented on the resulting phylogeny.

The Bayesian analysis (BA) was conducted in MrBayes v3.1.2b4 (Huelsenbeck & Ronquist 2001) on the Marylou5 Dell PowerEdge M610 computing cluster at Brigham Young University. Three independent BA analyses (each consisting of four chains) were performed using parameters selected by MODELTEST. All Markov chain Monte Carlo (MCMC) algorithms ran for 10,000,000 generations, sampling one tree every 1,000 generations. To ensure that independent analyses converged on similar values, we graphically compared all likelihood parameters and scores (means and variances) using the program Tracer v1.4 (Rambaut & Drummond 2007). Observation of the likelihood (-Ln L) scores and split frequencies allowed us to determine burn-ins ($\approx$1,000,000 generations) and stationary distributions for the data. Once the values reached a plateau, a 50% majority-rule consensus tree was obtained from the remaining saved trees. Posterior probabilities (pP) for clades were compared for congruence and post-burn-in-trees were combined between individual runs. Values > 0.5 are presented on the BA phylogram (presented as percentages).

2.4   Divergence time analyses

To estimate the relative timing of divergence between *A. cholchol* and *A. rostrata* we used Bayesian molecular dating methods implemented in BEAST v1.5.2 (Bayesian evolutionary analysis by sampling trees) (Drummond & Rambaut 2007). We chose a strict molecular clock, calibrated by specifying a substitution rate, to determine the rates among branches. Using a set substitution rate, we can estimate the divergence dates for any particular clade. A recent study, examining a closely related aeglid species (*A. alacalufi*) and similar gene set (16S, COI, COII), estimated the substitution rate for mtDNA to be $\approx$ 0.118 substitutions per site per million years (Xu et al. 2009). This study used the Last Glacial Maximum between glaciated and nonglaciated populations in Chile as a calibration

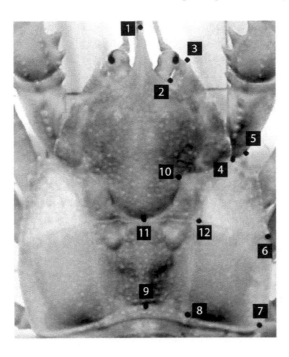

Figure 2. Description of locations and landmarks for geometric morphometric analysis. 1: tip of the rostrum; 2: orbital spine; 3: tip of the anterolateral spine; 4: union between the third hepatic lobe and the epibranchial tooth; 5: tip of the epibranchial tooth; 6: union between the branchial line and the posterior of the linea aeglica lateralis; 7: posterior vertices of the cephalothorax; 8: posterior extreme of the dorsal longitudinal linea; 9: center-posterior extremes of the cephalothorax; 10: cervical groove; 11: midpoint of the transverse dorsal linea; 12: extreme of the bar line.

point in their divergence time analysis. To calculate a new rate for our two-gene (16S, COI) dataset, we reran the Xu et al. (2009) analysis following the author's instructions and estimated the mean rates for each individual gene. We applied these new rates to our study. To ensure that our data were clocklike, we computed the log likelihood values (-ln L) for trees with and without the molecular clock enforced (in PAUP*, Swofford 2002), and a likelihood ratio test (LTR) was calculated to test the null hypothesis that the data follow a molecular clock. Three independent runs with MCMC chain length of 10 millions were performed, sampling every 1,000 generations. To ensure that analyses converged on similar values, we graphically compared all likelihood parameters and scores using the program Tracer v1.4 (Rambaut & Drummond 2007). Estimates of the mean divergence times with 95% highest posterior density regions (HPD) and posterior probabilities (represented as percentages) were noted on the chronogram.

2.5   Geometric morphometric analyses

Data were recorded for 104 individuals of *A. cholchol* (65 males and 39 females) and 82 individuals of *A. rostrata* (46 males and 36 females). A subset of individuals from the morphometric study was included in the phylogenetic analysis. Digital pictures of all specimens were taken with a Nikon Coolpix PS100 camera. Distances were standardized between the specimens and the camera to ensure similar orientation and scaling for all photographs. The crabs were placed in a Petri dish with ethanol and positioned over a grid for scaling and numbering purposes. Twelve primary landmarks were chosen to capture the fundamental shape of the carapace (Figure 2). To reduce redundancy due to the symmetrical nature of the carapace, landmarks were only placed on the right side of the crab. The landmarks were digitized using the tpsDig2 program v2.12 and tpsUtil v1.44 (Rohlf 2004a, b).

The configurations of the specimens were superimposed to remove nonshape variation through Generalization Procrustes Analysis (GPA, Rohlf & Slice 1990, Rohlf 1999) in the tpsRelw program v1.46 (Rohlf 2005). Nonshape variation included translation, rotation and scaling. The GPA process worked in three steps to align the landmarks involved with each specimen (Buchanan & Collard 2010). First, the set of landmark coordinates was centered at the origin, or centroid, and scaled all configurations to the unit centroid size (centroid size is a measurement of overall size of a specimen computed as the square root of the sum of the squared distances from all the landmarks to the centroid). Secondly, the consensus configuration was determined. Lastly, each landmark configuration was rotated to minimize the sum-of-squared residuals. We used the tpsRelw program to compute relative warps. The relative warps were generated from the weight matrix, which is derived from the differences during superimposition. The relative warps first described the major patterns of shape variation within and between the groups. Using the extremes of the relative warps, hypothetical specimens were created and compared.

A MANOVA (multivariate analysis of variance) was run on shape variables and centroid size using SAS statistical analysis software. This tested size and shape variation within and between species to determine, if differences in and between species were statistically significant.

3   RESULTS

3.1   Phylogenetic analysis

In total, we included 85 sequences for 16S and COI, and 77 sequences for EF1 (Appendix, Table A1). Missing data were designated as a "?" in the alignment. Three individuals (KACa1963, KACa1981, KACa2002) were excluded from the combined (nuclear/mitochondrial) analysis because of missing data. The optimal models of evolution selected in MODELTEST were the Hasegawa, Kishino and Yano (HKY) model with invariant sites (16S), Tamura-Nei (TrN) model with gamma-distributed among-site rate heterogeneity and invariant sites (COI), and Kimura (K81uf) model with invariant sites (EF1 intron). Individual gene trees for all three genes showed very similar patterns, although the nuclear tree (EF1 intron) was less resolved. Topologies derived from the ML and Bayesian (BA) analyses on the concatenated dataset were strongly congruent. Here, we present the BA phylogram of the combined genes (Figure 3).

Our study recovered the river species, *A. cholchol* to be a paraphyletic group divided into several clades (Figure 3). Phylogenetic relationships were not correlated with locality as individuals collected from the Chol-Chol River and Quepe River did not form reciprocally monophyletic groups. The lake species, *A. rostrata* was recovered as a monophyletic group. As seen in *A. cholchol*, populations from different localities (Lake Panguipulli and Riñihue Lake) did not form genetically distinct clades (Figure 3).

The combined (16S, COI, EF1 intron) and mitochondrial (16S, COI) phylogenetic analyses (Figures 3 and 4) generated congruent topologies with the exception of three individuals (KACa1993, KACa1995, KACa1996). In the mitochondrial phylogeny these individuals grouped within the *A. rostrata* clade (Figure 4), however with the addition of the nuclear gene they formed a sister group to *A. rostrata* (Figure 3). Individual gene trees showed similar trends; 16S and COI trees retained these samples within *A. rostrata*, while EF1 intron placed them outside *A. rostrata*. A possible explanation for this finding could be incomplete lineage sorting or amplification of undetectable pseudogenes (see Discussion).

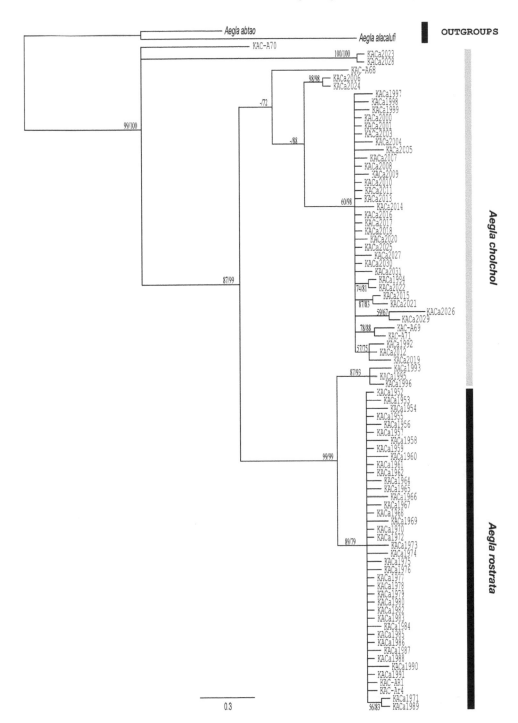

Figure 3. Bayesian (BA) phylogram for *Aegla cholchol* ($n = 43$), *Aegla rostrata* ($n = 40$), and outgroups ($n = 2$) based on a 16S (mtDNA), COI (mtDNA), and EF1 intron (nDNA) concatenated dataset. ML bootstrap values and BA posterior probabilities are represented as percentages and noted above or below the branches (ML/BA). Values < 50% are not shown.

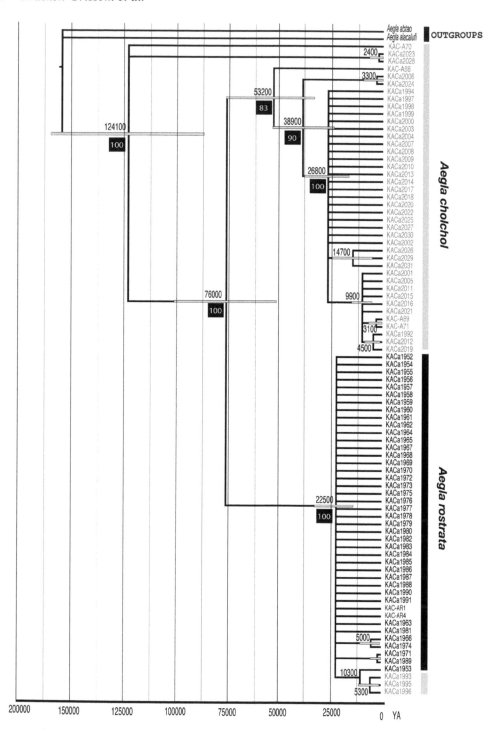

Figure 4. Divergence time chronogram based on a 16S (mtDNA) and COI (mtDNA) dataset. Vertical bars (black or gray) indicate individuals of *Aegla cholchol*, *Aegla rostrata*, or outgroup taxa. Divergence time estimates (YA) are noted adjacent to their respective nodes. Horizontal gray bars represent the 95% highest posterior density regions (HPD). Black boxes contain Bayesian posterior probabilities represented by percentages. Values < 50% are not shown.

3.2 Divergence time analysis

In total, we ran three separate divergence time analyses under a strict molecular clock model starting from different random seed values to ensure the program converged on similar mean ages and highest posterior density regions (HPD). The individual mean rates for each gene were estimated at 0.04 substitutions/site per million years for 16S and 0.187 substitutions/site per million years in COI (combined mt mean rate = 0.114 substitutions/site per million years). The time to most common recent ancestor (TMCRA) between *A. rostrata* and *A. cholchol* was estimated at 124,100 years ago (ya) (161,600–87,300) with the two lineages diverging around 76,000 (101,800–51,700) ya (Figure 4). Following this split, there was a radiation within *A. cholchol*, giving rise to three major clades in our phylogeny (Figure 4). Results suggest that the *A. rostrata* lineage radiated around 22,500 (33,000–14,100), which corresponds closely to the Last Glacial Maximum in southern South America (25,000–23,000).

We did explore a relaxed molecular clock and ran multiple trees under this assumption. We set a prior on the mean rates (16S, COI and combined mt) under a lognormal distribution (also implemented different distributions) and compared the results to our strict molecular clock analysis. We found that 1) all of the mean nodes ages of the strict clock were nested inside the 95% highest posterior density (HPD) regions of the relaxed molecular clock analysis, 2) the 95% HPD regions of the strict clock were much tighter when compared to the relaxed model, 3) the relaxed molecular clock analyses computed unrealistically high mean rates of evolution for the mitochondrial dataset (0.7-0.4 substitutions per site per million years) depending on the prior distribution we implemented (normal vs. uniform vs. lognormal). Lastly, to ensure that our data are clocklike, we computed the log likelihood values (-ln L) for trees with and without the molecular clock enforced (in PAUP*); a likelihood ratio test failed to reject the null hypothesis that the data followed a molecular clock ($P > 0.25$).

3.3 Morphometric analyses

3.3.1 *Interspecific variation*

Analysis of the shape variation revealed a clear separation between two clusters (Figure 5), suggesting a strong distinction between *A. rostrata* and *A. cholchol*. The cluster on the right side of the

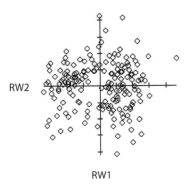

Figure 5. Interspecies comparison (*Aegla cholchol* vs. *Aegla rostrata*) of shape variation on relative warps 1 and 2.

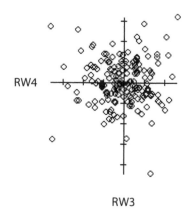

RW4

RW3

Figure 6. Intraspecies comparison (male vs. female) of shape variation on relative warps 3 and 4.

graph corresponds to the shape variation within *A. rostrata* and the cluster on the left corresponds to the shape variation within *A. cholchol*. Relative warp 1 (RW1) accounts for 26.07% and relative warp 2 (RW2) accounts for 17.59%. Together the cumulative variance of RW1 and RW2 is 46.66%. The multivariate test comparing the two species (*A. cholchol* vs. *A. rostrata*) was highly significant ($F = 11.71$, $P < 0.0001$). The distinction between *A. cholchol* and *A. rostrata* was especially evident on the first relative warp. Deformation grids corresponding to extremes of variation along this axis help visualize shape difference between the two taxa (Figure 7). *Aegla rostrata* has a longer rostrum (landmark 1) and ornamentation is more exaggerated (landmarks 4 and 5). They also have a more slender anterior and posterior cephalothroax region (to a lesser degree) and the cervical groove area is reduced (landmark 10). *Aegla cholchol* generally has a shorter rostrum and wider anterior cephalothorax region. The posterior cephalothorax is also wider, but to a lesser degree. The cervical groove area is slightly larger as well.

A multivariate test was run on centroid size between the two species with highly significant results ($F = 4.60$, $P < 0.0001$). This size difference between the two species is easy to see with the naked eye; *A. rostrata* is larger than *A. cholchol*. Thus, the two species show variation in both shape and size.

3.3.2   *Intraspecific variation*

Analysis of shape variation between sexes (male vs. female) did not show clear separated clusters (relative warps 3 and 4, Figure 6). However, the multivariable test shows highly significant difference in sex shape ($F = 2.08$, $P < 0.0042$). Dimorphism is moving in the same direction for both species. Females of *A. rostrata* have a slightly wider and lower cephalothorax area compared to the males. The same trend is seen in females of *A. cholchol*.

4   DISCUSSION

4.1   Molecular phylogenetic analysis

Our phylogenetic analysis suggests a high level of genetic diversity and structure within the river species, *A. cholchol* (Figure 3). This level of divergence was unexpected, especially since a majority

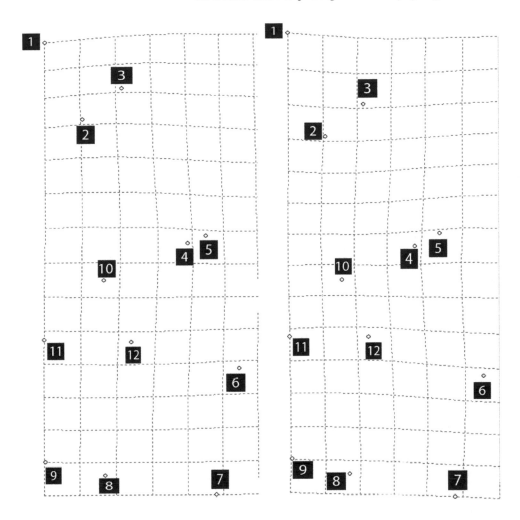

Figure 7. Deformation grids from relative warp 1. The right side is the extreme right, representing *Aegla rostrata*. The left side is the extreme left, representing *Aegla cholchol*.

of the samples were collected from one sampling locality (Quepe River) in the Araucanía region of southern Chile. We acknowledge that our sampling was limited, and with the exploration of more populations and potentially cryptic lineages the amount of diversity may increase. Past studies have found increased levels of population differentiation within freshwater aeglid crabs and accredit this phenomenon to a small population size, habitat fragmentation, and poor dispersal ability (Xu et al. 2009).

Additionally, if it is considered that Quaternary glaciation events occurred at least four times in southern Chile (Illies 1970), range restriction of *Aegla* populations toward the Coastal Cordillera should have occurred the same number of times. Each time the range reduction and the subsequent postglacial range expansion to the east could have caused population fragmentation and incipient differentiation, so that the high level of riverine population differentiation could be the result of those long-term recurrent processes superimposed over time on the same territory. The level of molecular divergence within *A. cholchol* exemplifies the need for species-level studies of aeglid crabs in southern Chile in an effort to identify and preserve the genetic diversity that exists within these environments. *Aegla cholchol* being a river species, listed as vulnerable and subject to the habitat deterioration and degradation, this study reinforces the fact that genetic diversity, especially in river or stream environments has yet to be discovered and protected.

Our findings suggesting that the river species, *A. cholchol*, is nonmonophyletic are in congruence with recent molecular studies (Pérez-Losada et al. 2002a, 2004; Bond-Buckup et al. 2008). More robust sampling throughout the species distribution may corroborate this conclusion and determine how many evolutionary significant units or cryptic species are involved. We are seeing this trend in all three individual gene trees (not shown), the mitochondrial gene tree (16S/COI combined, Figure 4), and the combined tree (16S/COI/EF1, Figure 3). One explanation for the nonmonophyly within *A. cholchol* could be attributed to incomplete lineage sorting, especially in recently diverged species. At shallow time depths, this phenomenon may occur when genetic drift has not have enough time to bring loci to fixation (Pamilo & Nei 1988). This may lead to misleading phylogenies that do not accurately portray the true species tree. More specifically, this explanation may apply to the three individuals of *A. cholchol* (KACa1993, KACa1996, KACa1995) that are grouping with *A. rostrata* in the mitochondrial phylogeny (Figure 4). With the addition of an unlinked nuclear loci (EF1 intron, Figure 3), these individuals formed a sister group to the *A. rostrata* clade. It is possible undetectable pseudogenes may be present, however we took precautionary measures to avoid the amplification of pseudogenes and sequenced these individuals multiple times to confirm our findings (see Methods). Although incomplete lineage sorting or pseudogenes may explain this particular case, more robust phylogenies exploring deep relationships among aeglid crabs and utilizing a suite of nuclear and mitochondrial genes continue to recover nonmonophyletic relationships within *A. cholchol* (see Pérez-Losada et al. 2004; Bond-Buckup et al. 2008). A second explanation may be that *A. cholchol* should be split into separate species based on genetic, morphological, and ecological observations. Two morphotypes have been observed for *A. cholchol* (Jara, personal observation) and past research has reported this species to occupy two types of benthic habitats in the Chol-Chol River (pebble/stony vs. sandy) (Jara 1996). In accordance with recent postulation, we agree that habitat partitioning in combination with genetic and morphological differentiation could promote sympatric speciation within *A. cholchol* (see Pérez-Losada et al. 2002a).

4.2 Divergence time estimates and South American glacial cycles

The Last Glacial Maximum (LGM) in southern South America occurred around ≈25,000–23,000 years ago (ya) with deglaciation occurring around 17,500–17,150 ya due to a response in climate warming (Hulton et al. 2002; Sugden et al. 2005, Ruzzante et al. 2008). Recent modeling studies have predicted that the ice sheet extended between 38 and 55 degrees S, with a western extension reaching the edge of the continental shelf south of 43 degrees S (Hulton et al. 2002). Upon initiation, deglaciation occurred rapidly in the north and glaciers began to retreat within 10 km of their source within ≈2,000 years (Hulton et al. 2002). Based on this information, it becomes evident that our sampling localities (Quepe River, Chol-Chol River, Lake Panguipulli, Riñihue Lake) fall just within the northern latitudinal limits of the LGM and would have been the first to experience deglaciation in the area. Our chronogram suggests that the LGM had little impact on the *A. cholchol* populations, which may have survived in alternative refugia during the glacial cycle or were positioned far enough north to escape the northern extent of the ice sheet. In either case, our results suggest the *A. cholchol* lineage originated around 124,100 ya with subsequent radiations around 53,200–26,800 ya. The origin of *A. rostrata* was estimated around 22,500 ya (33,000–14,100), which corresponds more closely with the LGM and the onset of deglaciation in the area. Previous research examining Lake Panguipulli's bathymetric profile and morphometric parameters have concluded it to be of glacial origins (Campos et al. 1981). It is plausible that *A. rostrata* speciated in this new habitat once the lake basin became free of ice, possibly from invading *A. cholchol*-like forebearers coming from western refuges. Other studies have suggested that the current distribution of *A. rostrata* is the result of melting ice sheets following the Pleistocene glaciation (Jara 1977).

4.3   Morphometric analyses

In this study, we performed a morphometric analysis to examine differences in morphological adaptations associated with living in a lacustrine vs. riverine ecotype. *Aegla cholchol* was chosen to represent the river species, whereas *A. rostrata* was chosen to represent the lake species. Despite the overlapping morphological characters of *A. cholchol* and *A. rostrata*, the geometric morphometric analysis revealed a clear distinction in shape. Previous research suggests that there is a tendency towards a reduction in rostrum length, width of the forehead, and ornamentation as an adaptation to rivers or stream environments (Ringuelet 1949). These adaptations allow an organism to be more streamline in a system where currents and flow are continuously changing. In the river species *A. cholchol*, the rostrum length is shortened and there is visibly less expressed ornamentation (not statistically shown), however the width in shape of the anterior cephalothorax region is larger. The lake species, *A. rostrata*, have a longer rostrum, thinner anterior cephalothorax, and more pronounced ornamentation. Although the size of the cephalothorax is larger in *A. cholchol* when compared to *A. rostrata*, other morphological features represent adaptations to a dynamic environment such as a river or stream.

In the past, there have been contrasting descriptions of size differences between lake and river species. Giri and Loy (2008) found river crabs to be larger, when comparing *Aegla riolimayana* and *A. neuquensis*, while Jara (1989) found lake crabs to be larger when describing *A. denticulata lacustris* and *A. denticulata denticulata*. Our findings are similar to Jara's results: size is a significant factor and *A. rostrata* are significantly larger than *A. cholchol*. The smaller overall body size may be another adaption to the river environment, allowing crabs to be more streamline and resistant to drag.

This study also found sexual dimorphism in both species. Significant differences were revealed in shape and size of the cephalothorax between males and females. Females of both *A. cholchol* and *A. rostrata* were found to have a wider posterior cephalothorax when compared to males. With this trait the number of eggs that females can potentially carry should be maximized (Jara 1994). Our study highlights the importance of adaptive morphological evolution in distinguishing populations, even when the neutral genetic markers have not had time to completely sort to species (Crandall et al. 2000).

ACKNOWLEDGEMENTS

We would like to thank J. Xu, M. Pérez-Losada and anonymous reviewers for advice on the manuscript and/or labwork. This work was supported by grants from the National Science Foundation (OISE-0530267, EF-0531762) and Brigham Young University.

REFERENCES

Bahamonde, N., Carvacho, A., Jara, C., López, M., Ponce, F., Retamal, M.A. & Rudolph, E. 1998. Categorías de conservación de decapodos nativos de aguas continentales de Chile. *Bol. Mus. Nacion. Hist. Nat.* 47: 91–100.

Bond-Buckup, G. & Buckup, L. 1994. A família Aeglidae (Crustacea, Decapoda, Anomura). *Arq. Zool.* 32: 159–347.

Bond-Buckup, G., Jara, C.G., Pérez-Losada, M., Buckup, L. & Crandall, K.A. 2008. Global diversity of crabs (Aeglidae: Anomura: Decapoda) in freshwater. *Hydrobiologia* 595: 267–273.

Buchanan, B. & Collard, M. 2010. A geometric morphometrics-based assessment of blade shape differences among Paleoindian projectile point types from western North America. *J. Archaeol. Sci.* 37: 350–359.

Campos, H., Arenas, J., Steffen, W. & Aquero, G. 1981. Morphometrical, physical and chemical limnology of Lake Panguipulli (Valdivia, Chile). *Neues Jahrb. Geol. Paläont. Mh.* 10: 603–625.

Crandall, K.A., Bininda-Emonds, O.R.P., Mace G.M. & Wayne, R.K. 2000. Considering evolutionary processes in conservation biology. *Trends Ecol. Evol.* 15: 290–295.

Dana, J.D. 1852. Crustacea. Part I. *U. S. Explor. Exped. 1838–1842* 13: 1–685.

Drummond, A.J. & Rambaut, A. 2007. BEAST: Bayesian evolutionary analysis by sampling trees. *BMC Evol. Bio.* 7: 214.

Edgar, R.C. 2004. MUSCLE: multiple sequence alignment with high accuracy and high throughput. *Nucleic Acids Res.* 32: 1792–1797.

Felsenstein, J. 1985. Confidence-limits on phylogenies with a molecular clock. *Syst. Zool.* 34: 152–161.

Giri, F. & Loy, A. 2008. Size and shape variation of two freshwater crabs in Argentinean Patagonia: the influence of sexual dimorphism, habitat, and species interactions. *J. Crust. Bio.* 28: 37–45.

Huelsenbeck, J.P. & Ronquist, F. 2001. MRBAYES: Bayesian inference of phylogeny. *Biometrics* 17: 754–755.

Hulton, N.R.J., Purves, R.S., McCulloch, R.D., Sugden, D.E. & Bentley, M.J. 2002. The Last Glacial Maximum and deglaciation in southern South America. *Quatern. Sci. Rev.* 21: 233–241.

Illies, H. 1970. *Geología de los alrededores de Valdivia y Volcanismo y Tectónica en márgenes del Pacífico en Chile meridional.* Valdivia: Universidad Austral de Chile, Instituto de Geología y Geografía.

Jara, C. 1977. *Aegla rostrata* n. sp., (Decapoda, Aeglidae), nuevo crustáceo dulceacuícola del Sur de Chile. *Stud. Neotrop. Fauna Environ.* 12: 165–176.

Jara, C. 1994. *Aegla pewenchae,* a new species of central Chilean freshwater decapods (Crustacea: Anomura: Aeglidae). *Proc. Biol. Soc. Wash.* 107: 325–339.

Jara, C. 1996. *Taxonomía, sistemática y zoogeográfica de las especies chilenas del género Aegla Leach (Crustacea: Decapoda: Anomura: Aeglidae).* Ph.D. dissertation, University of Concepción, Chile.

Jara, C.G. 1989. *Aegla denticulata lacustris,* new subspecies, from Lake Rupanco, Chile (Crustacea, Decapoda, Anomura, Aeglidae). *Proc. Biol. Soc. Wash.* 102: 385–393.

Jara, C.G. & López, M.T. 1981. A new species of freshwater crab (Crustacea: Anomura: Aeglidae) from insular south Chile. Proc. Biol. Soc. Wash. 94: 34–41.

Jara, C.G. & Palacios, V.L. 1999. Two new species of *Aegla* Leach (Crustacea: Decapoda: Anomura: Aeglidae) from southern Chile. *Proc. Biol. Soc. Wash.* 112: 106–119.

Katoh, K., Kuma, K., Toh, H. & Miyata, T. 2005. MAFFT version 5: improvement in accuracy of multiple sequence alignment. *Nucleic Acids Res.* 33: 511–518.

MacLeay, W.S. 1838. On the brachyurous decapod Crustacea. In: Smith, A. (ed.), *Illustrations of the Zoology of South Africa; consisting chiefly of figures and descriptions of the objects of natural history collected during an expedition into the interior of South Africa, in the years 1834, 1835, and 1836; fitted out by "The Cape of Good Hope Association for Exploring Central Africa": together with a summary of African Zoology, and an inquiry into the geographical ranges of species in that quarter of the globe*: 53–71. London: Smith, Elder and Co.

Martin, J.W. & Abele, L.G. 1988. External morphology of the genus *Aegla* (Crustacea: Anomura:

Aeglidae). *Smithson. Contrib. Zool.* 453: 1–46.

Pamilo, P. & Nei, M. 1988. Relationships between gene trees and species trees. *Mol. Biol. Evol.* 5: 568–583.

Pérez-Losada, M., Jara, C.G., Bond-Buckup, G. & Crandall, K.A. 2002a. Conservation phylogenetics of Chilean freshwater crabs *Aegla* (Anomura, Aeglidae): assigning priorities for aquatic habitat protection. *Biol. Conserv.* 105: 345–353.

Pérez-Losada, M., Jara, C.G., Bond-Buckup, G. & Crandall, K.A. 2002b. Phylogenetic relationships among the species of *Aegla* (Anomura: Aeglidae) freshwater crabs from Chile. *J. Crust. Biol.* 22: 304–313.

Pérez-Losada, M., Bond-Buckup, G., Jara, C.G. & Crandall, K.A. 2004. Molecular systematics and biogeography of the southern South American freshwater "crabs" *Aegla* (Decapoda: Anomura: Aeglidae) using multiple heuristic tree search approaches. *Syst. Biol.* 53: 767–780.

Posada, D. & Crandall, K.A. 1998. MODELTEST: Testing the model of DNA substitution. *Bioinformatics* 14: 817–818.

Rambaut, A. & Drummond, A.J. 2007. Tracer v1.4. http://beast.bio.ed.ac.uk/Tracer.

Ringuelet, R. 1949. Consideraciones sobre las relaciones filogenéticas entre las especies del género *Aegla* Leach (decápodos anomuros). *Not. Mus. La Plata, Zool.* 14: 111–118.

Rohlf, F.J. 1999. Shape statistics: procrustes superimpositions and tangent spaces. *J. Classif.* 16: 197–223.

Rohlf, F.J. 2004a. Tps-Dig,Version1.40. New York, NY: Department of Ecology and Evolution, State University of New York at Stony Brook.

Rohlf, F.J. 2004b. TpsUtil, Version 1.21.0.1. New York, NY: Department of Ecology and Evolution, State University of New York at Stony Brook.

Rohlf, F.J. 2005. TpsRelw, Relative Warp Analysis,Version 1.39. New York, NY: Department of Ecology and Evolution, State University of New York at Stony Brook.

Rohlf, F.J. & Slice, D. 1990. Extension of the procrustes method for the optimal superimposition of landmarks. *Syst. Zool.* 39: 40–59.

Ruzzante, D.E., Walde, S.J., Gosse, J.C., Cussac, V.E., Habit, E., Zemlak, T.S. & Adams, E.D. 2008. Climate control on ancestral population dynamics: insight from Patagonian fish phylogeography. *Mol. Ecol.* 17: 2234–2244.

Schmitt, W.L. 1942. Two new species of *Aegla* from Chile. *Rev. Chilena Hist. Nat.* 44: 25–31.

Song, H., Buhay, J.E., Whiting, M.F. & Crandall, K.A. 2008. Many species in one: DNA barcoding overestimates the number of species when nuclear mitochondrial pseudogenes are coamplified. *Proc. Natl. Acad. Sci. USA* 105: 13486–13491.

Stamatakis, A., Blagojevic, F., Nikolopoulos, D.S. & Antonopoulos, C.D. 2007. Exploring new search algorithms and hardware for phylogenetics: RAxML meets the IBM cell. *J. Signal. Process. Sys.* 48: 271–286.

Stamatakis, A., Hoover, P. & Rougemont, J. 2008. A rapid bootstrap algorithm for the RAxML web servers. *Syst. Biol.* 57: 758–771.

Stamatakis, A., Ludwig, T. & Meier, H. 2005. RAxML-III: a fast program for maximum likelihood-based inference of large phylogenetic trees. *Bioinformatics* 21: 456–463.

Sugden, D.E., Bentley, M.J., Fogwill, C.J., Hulton, N.R.J., McCulloch, R.D. & Purves, R.S. 2005. Late-glacial glacier events in southernmost South America: a blend of "northern" and "southern" hemispheric climatic signals? *Geogr. Ann.* A 87A: 273–288.

Swofford, D.L. 2002. PAUP*. Phylogenetic Analysis Using Parsimony (*and other methods). Sunderland, MA, Sinauer Associates.

Xu, J.W., Pérez-Losada, M., Jara, C.G. & Crandall, K.A. 2009. Pleistocene glaciation leaves deep signature on the freshwater crab *Aegla alacalufi* in Chilean Patagonia. *Mol. Ecol.* 18: 904–918.

APPENDIX

Table A1. Taxonomy, voucher catalog numbers, and GenBank accession numbers for gene sequences used in study. An "N/A" (not available) indicates missing sequence data. New sequences are indicated in bold.

Taxon	Locality	Voucher	GenBank accession numbers		
			16S	**COI**	**EF1 intron**
Outgroup taxa					
Anomura MacLeay 1838					
Aeglidae Dana 1852					
Aegla abtao Schmitt 1942		KAC-Aa5	AY050067	AY050113	N/A
Aegla alacalufi Jara & López 1981		KACa1142	FJ472205	FJ471839	FJ472271
Ingroup taxa					
Anomura MacLeay 1838					
Aeglidae Dana 1852					
Aegla cholchol Jara & Palacios 1999	Quepe	KACa1992	**HQ236179**	**HQ236258**	**HQ236337**
Aegla cholchol Jara & Palacios 1999	Quepe	KACa1993	**HQ236180**	**HQ236259**	**HQ236338**
Aegla cholchol Jara & Palacios 1999	Quepe	KACa1994	**HQ236181**	**HQ236260**	**HQ236339**
Aegla cholchol Jara & Palacios 1999	Quepe	KACa1995	**HQ236182**	**HQ236261**	**HQ236340**
Aegla cholchol Jara & Palacios 1999	Quepe	KACa1996	**HQ236183**	**HQ236262**	**HQ236341**
Aegla cholchol Jara & Palacios 1999	Quepe	KACa1997	**HQ236184**	**HQ236263**	**HQ236342**
Aegla cholchol Jara & Palacios 1999	Quepe	KACa1998	**HQ236185**	**HQ236264**	**HQ236343**
Aegla cholchol Jara & Palacios 1999	Quepe	KACa1999	**HQ236186**	**HQ236265**	**HQ236344**
Aegla cholchol Jara & Palacios 1999	Quepe	KACa2000	**HQ236187**	**HQ236266**	**HQ236345**
Aegla cholchol Jara & Palacios 1999	Quepe	KACa2001	**HQ236188**	**HQ236267**	**HQ236346**
Aegla cholchol Jara & Palacios 1999	Quepe	KACa2002	N/A	**HQ236268**	**HQ236347**
Aegla cholchol Jara & Palacios 1999	Quepe	KACa2003	**HQ236189**	**HQ236269**	**HQ236348**
Aegla cholchol Jara & Palacios 1999	Quepe	KACa2004	**HQ236190**	**HQ236270**	**HQ236349**
Aegla cholchol Jara & Palacios 1999	Quepe	KACa2005	**HQ236191**	**HQ236271**	**HQ236350**
Aegla cholchol Jara & Palacios 1999	Quepe	KACa2006	**HQ236192**	**HQ236272**	**HQ236351**
Aegla cholchol Jara & Palacios 1999	Quepe	KACa2007	**HQ236193**	**HQ236273**	**HQ236352**
Aegla cholchol Jara & Palacios 1999	Quepe	KACa2008	**HQ236194**	**HQ236274**	**HQ236353**
Aegla cholchol Jara & Palacios 1999	Quepe	KACa2009	**HQ236195**	**HQ236275**	**HQ236354**
Aegla cholchol Jara & Palacios 1999	Quepe	KACa2010	**HQ236196**	**HQ236276**	**HQ236355**
Aegla cholchol Jara & Palacios 1999	Quepe	KACa2011	**HQ236197**	**HQ236277**	**HQ236356**
Aegla cholchol Jara & Palacios 1999	Quepe	KACa2012	**HQ236198**	**HQ236278**	**HQ236357**
Aegla cholchol Jara & Palacios 1999	Quepe	KACa2013	**HQ236199**	**HQ236279**	**HQ236358**
Aegla cholchol Jara & Palacios 1999	Quepe	KACa2014	**HQ236200**	**HQ236280**	**HQ236359**
Aegla cholchol Jara & Palacios 1999	Quepe	KACa2015	**HQ236201**	**HQ236281**	**HQ236360**
Aegla cholchol Jara & Palacios 1999	Quepe	KACa2016	**HQ236202**	**HQ236282**	**HQ236361**
Aegla cholchol Jara & Palacios 1999	Quepe	KACa2017	**HQ236203**	**HQ236283**	**HQ236362**
Aegla cholchol Jara & Palacios 1999	Quepe	KACa2018	**HQ236204**	**HQ236284**	**HQ236363**
Aegla cholchol Jara & Palacios 1999	Quepe	KACa2019	**HQ236205**	**HQ236285**	**HQ236364**
Aegla cholchol Jara & Palacios 1999	Quepe	KACa2020	**HQ236206**	N/A	**HQ236365**
Aegla cholchol Jara & Palacios 1999	Quepe	KACa2021	**HQ236207**	**HQ236286**	**HQ236366**
Aegla cholchol Jara & Palacios 1999	Quepe	KACa2022	**HQ236208**	**HQ236287**	**HQ236367**
Aegla cholchol Jara & Palacios 1999	Quepe	KACa2023	**HQ236209**	**HQ236288**	**HQ236368**
Aegla cholchol Jara & Palacios 1999	Quepe	KACa2024	**HQ236210**	**HQ236289**	**HQ236369**
Aegla cholchol Jara & Palacios 1999	Quepe	KACa2025	**HQ236211**	**HQ236290**	**HQ236370**
Aegla cholchol Jara & Palacios 1999	Quepe	KACa2026	**HQ236212**	**HQ236291**	**HQ236371**
Aegla cholchol Jara & Palacios 1999	Quepe	KACa2027	**HQ236213**	**HQ236292**	N/A
Aegla cholchol Jara & Palacios 1999	Quepe	KACa2028	**HQ236214**	**HQ236293**	**HQ236372**

Table 1. Continuation.

Taxon	Locality	Voucher	GenBank accession numbers		
			16S	**COI**	**EF1 intron**
Aegla cholchol Jara & Palacios 1999	Quepe	KACa2029	**HQ236215**	**HQ236294**	**HQ236373**
Aegla cholchol Jara & Palacios 1999	Quepe	KACa2030	**HQ236216**	**HQ236295**	**HQ236374**
Aegla cholchol Jara & Palacios 1999	Quepe	KACa2031	**HQ236217**	**HQ236296**	**HQ236375**
Aegla cholchol Jara & Palacios 1999	CholChol	KAC-A70	AY050049	AY050095	N/A
Aegla cholchol Jara & Palacios 1999	CholChol	KAC-A69	AY050048	AY050094	N/A
Aegla cholchol Jara & Palacios 1999	CholChol	KAC-A71	AY050050	AY050096	N/A
Aegla cholchol Jara & Palacios 1999	CholChol	KAC-A68	AY050047	AY050093	N/A
Aegla rostrata Jara 1977	Panguipulli	KACa1952	**HQ236141**	**HQ236218**	**HQ236297**
Aegla rostrata Jara 1977	Panguipulli	KACa1953	**HQ236142**	**HQ236219**	**HQ236298**
Aegla rostrata Jara 1977	Panguipulli	KACa1954	**HQ236143**	**HQ236220**	**HQ236299**
Aegla rostrata Jara 1977	Panguipulli	KACa1955	**HQ236144**	**HQ236221**	**HQ236300**
Aegla rostrata Jara 1977	Panguipulli	KACa1956	**HQ236145**	**HQ236222**	**HQ236301**
Aegla rostrata Jara 1977	Panguipulli	KACa1957	**HQ236146**	**HQ236223**	**HQ236302**
Aegla rostrata Jara 1977	Panguipulli	KACa1958	**HQ236147**	**HQ236224**	**HQ236303**
Aegla rostrata Jara 1977	Panguipulli	KACa1959	**HQ236148**	**HQ236225**	**HQ236304**
Aegla rostrata Jara 1977	Panguipulli	KACa1960	**HQ236149**	**HQ236226**	**HQ236305**
Aegla rostrata Jara 1977	Panguipulli	KACa1961	**HQ236150**	**HQ236227**	**HQ236306**
Aegla rostrata Jara 1977	Panguipulli	KACa1962	**HQ236151**	**HQ236228**	**HQ236307**
Aegla rostrata Jara 1977	Panguipulli	KACa1963	N/A	**HQ236229**	**HQ236308**
Aegla rostrata Jara 1977	Panguipulli	KACa1964	**HQ236152**	**HQ236230**	**HQ236309**
Aegla rostrata Jara 1977	Panguipulli	KACa1965	**HQ236153**	**HQ236231**	**HQ236310**
Aegla rostrata Jara 1977	Panguipulli	KACa1966	**HQ236154**	**HQ236232**	**HQ236311**
Aegla rostrata Jara 1977	Panguipulli	KACa1967	**HQ236155**	**HQ236233**	**HQ236312**
Aegla rostrata Jara 1977	Panguipulli	KACa1968	**HQ236156**	**HQ236234**	**HQ236313**
Aegla rostrata Jara 1977	Panguipulli	KACa1969	**HQ236157**	**HQ236235**	**HQ236314**
Aegla rostrata Jara 1977	Panguipulli	KACa1970	**HQ236158**	**HQ236236**	**HQ236315**
Aegla rostrata Jara 1977	Panguipulli	KACa1971	**HQ236159**	**HQ236237**	**HQ236316**
Aegla rostrata Jara 1977	Panguipulli	KACa1972	**HQ236160**	**HQ236238**	**HQ236317**
Aegla rostrata Jara 1977	Panguipulli	KACa1973	**HQ236161**	**HQ236239**	**HQ236318**
Aegla rostrata Jara 1977	Panguipulli	KACa1974	**HQ236162**	**HQ236240**	**HQ236319**
Aegla rostrata Jara 1977	Panguipulli	KACa1975	**HQ236163**	**HQ236241**	**HQ236320**
Aegla rostrata Jara 1977	Panguipulli	KACa1976	**HQ236164**	**HQ236242**	**HQ236321**
Aegla rostrata Jara 1977	Panguipulli	KACa1977	**HQ236165**	**HQ236243**	**HQ236322**
Aegla rostrata Jara 1977	Panguipulli	KACa1978	**HQ236166**	**HQ236244**	**HQ236323**
Aegla rostrata Jara 1977	Panguipulli	KACa1979	**HQ236167**	**HQ236245**	**HQ236324**
Aegla rostrata Jara 1977	Panguipulli	KACa1980	**HQ236168**	**HQ236246**	**HQ236325**
Aegla rostrata Jara 1977	Panguipulli	KACa1981	N/A	**HQ236247**	**HQ236326**
Aegla rostrata Jara 1977	Panguipulli	KACa1982	**HQ236169**	**HQ236248**	**HQ236327**
Aegla rostrata Jara 1977	Panguipulli	KACa1983	**HQ236170**	**HQ236249**	**HQ236328**
Aegla rostrata Jara 1977	Panguipulli	KACa1984	**HQ236171**	**HQ236250**	**HQ236329**
Aegla rostrata Jara 1977	Panguipulli	KACa1985	**HQ236172**	**HQ236251**	**HQ236330**
Aegla rostrata Jara 1977	Panguipulli	KACa1986	**HQ236173**	**HQ236252**	**HQ236331**
Aegla rostrata Jara 1977	Panguipulli	KACa1987	**HQ236174**	**HQ236253**	**HQ236332**
Aegla rostrata Jara 1977	Panguipulli	KACa1988	**HQ236175**	**HQ236254**	**HQ236333**
Aegla rostrata Jara 1977	Panguipulli	KACa1989	**HQ236176**	**HQ236255**	**HQ236334**
Aegla rostrata Jara 1977	Panguipulli	KACa1990	**HQ236177**	**HQ236256**	**HQ236335**
Aegla rostrata Jara 1977	Panguipulli	KACa1991	**HQ236178**	**HQ236257**	**HQ236336**
Aegla rostrata Jara 1977	Riñihue	KAC-Ar4	AY050074	AY050120	N/A
Aegla rostrata Jara 1977	Riñihue	KAC-Ar1	AY050073	AY050119	N/A

Population structure of two crayfish with diverse physiological requirements

Jesse W. Breinholt[1], Paul E. Moler[2] & Keith A. Crandall[1,3]

[1] *Department of Biology, Brigham Young University, Provo, UT, U. S. A.*
[2] *Florida Fish and Wildlife Conservation Commission, Gainesville, FL, U. S. A.*
[3] *Monte L. Bean Life Science Museum, Brigham Young University, UT, Provo, U. S. A.*

ABSTRACT

Freshwater crayfish occur in a wide variety of habitats. Comparison of how these habitats shape population structure can lead to a better understanding of crayfish evolution and diversification. Both biotic and abiotic factors impact species distributions, with different species responding differently to changes in habitat/climate. Crayfish inhabit streams, caves, lakes, or burrows and have a variety of physiological requirements associated with these habitats. Yet few studies have examined how these different habitats and physiological requirements impact population structure. In this study, population connectivity of a stream crayfish *Procambarus spiculifer* is compared to that of a tertiary burrowing crayfish, *Procambarus paeninsulanus*, using two mitochondrial genes, 16S and COI. *Procambarus spiculifer* is dependent on habitat with moving water with high oxygen content. The habitat requirements for *Procambarus paeninsulanus* are less stringent, and it can be found nearly anywhere water has pooled or periodically pools within its native range. Not only is *P. spiculifer* dependent on moving water with higher oxygen content, it also has a lower heat tolerance than *P. paeninsulanus*. These physiological attributes of *P. spiculifer* limit the amount of habitat that can support this species. In contrast, *P. paeninsulanus* has higher heat tolerance and lower oxygen requirements, which opens a wider range of available habitat. Comparing population connectivity of these species across the same geographic region allows us to examine how physiological attributes have affected the population structure of each species. We hypothesize that the species with the more stringent physiological needs, *P. spiculifer*, would have more population substructure resulting in less genetic diversity within subpopulations compared to among subpopulations. On the other hand, *P. paeninsulanus*, with fewer physiological constraints, would have less population structure and, therefore, more similar levels of genetic diversity both within and among subpopulations. Our genetic data support the hypothesis that physiological attributes have affected the population structure of these species, with the more physiologically constrained species showing greater population structure.

1 INTRODUCTION

Many species exhibit habitat specificity and inhabit environments with a narrow range of biotic and abiotic characteristics such as temperature, oxygen levels, soil type, moisture, and vegetation communities. These required characteristics can shape the distribution of a species and geographically isolate demes due to a discontinuous distribution of the required habitat. The distribution of appropriate habitat can affect the genetic structure of a species by limiting the frequency and patterning of migration events. Likewise, physiological constraints may limit a species' ability to cope with habitat changes (e.g., Bernardo et al. 2007). For example, the Utah endemic plant *Astragalus ampullariodes* is soil specific to two clay rich soils, chinle and moenave, derived from two distinct geological formations. Due to the limited amount of available habitat and the physiological dependence on specific soil type, the seven remaining populations are severely fragmented and experience little gene flow (Breinholt et al. 2009b). Thus, the patchy distribution of the required habitat and the unsuitable soil surrounding these patches serve as barriers limiting the distribution of this species and contribute to the lack of gene flow. Soil dependence can also influence species boundaries. For example, *Gryllus pennsylvanicus* and *G. firmus* are crickets that inhabit different soil types (loamy or sandy soil, respectively) and their physiological needs or preference to soil type affect the distributions of these two species (Rand & Harrison 1989). As a result of physiological needs or preference to soil type the transition zone between these soils forms a geographic boundary between these two species. Physiological requirements such as minimum dissolved oxygen levels can also affect distributions of species within a watershed. A recent study of two streams known to have once supported a freshwater shrimp, *Syncaris pacifica*, reported that habitat degradation and low levels of dissolved oxygen that approached or dropped below concentrations known to be lethal in other shrimp species may have been the reason for not finding any shrimp in one of the two streams (Martin et al. 2009).

Physiological attributes can also affect population structure and species distribution. Bernardo et al. (2007) discussed the utility and infrequent use of data about species physiological requirements and the implications they have for the conservation of a species. They also point out that the effects of intraspecific physiological variation on population structure and geographic distribution have not been well studied. Bernardo et al. (2007) demonstrated with salamanders from a similar (to crayfish) riparian habitat that species with strict physiological requirements have overall lower within-population genetic diversity with more population substructure than salamanders with less strict physiological requirements, which in contrast have higher genetic diversity and less population structure. Therefore, by contrasting species with different physiological needs, it is possible to test if the physiological requirements have affected the population genetic structure.

Freshwater crayfish inhabit a wide variety of habitats such as streams, swamps, ponds, and even subterranean waters. The many diverse habitats occupied by crayfish make them an excellent system to study how physiological attributes can shape population structure and distribution. Although some crayfish require moving water or standing water, others are burrowers, in most cases occurring in areas with a permanent connection to the water table. Hobbs (1942, 1981) identified three types of burrowing crayfish (primary, secondary, or tertiary burrowers) distinguished by their utilization of and time spent in burrows. Primary burrowers spend a majority of their life in burrows and are thought to only occur in areas with a permanent connection to the water table (Hobbs 1942, 1981; Welch & Eversole 2006). Secondary burrowers inhabit open water during the wet season and burrow when there is no standing water, when bearing eggs or young, or during the day when there is little cover. Tertiary burrowers live in open water and tend to construct simple burrows in or connecting to permanent water. Burrowing and aquatic crayfish differ physiologically in their ability to use oxygen in and out of water (McMahon & Stuart 1999). In a study that compared these differences, McMahon (2001) noted that burrowing crayfish appear to be equally adapted to take up oxygen in and out of water, whereas more aquatic crayfish performed significantly worse out of water. McMahon & Stuart (1999) also showed that the secondary burrower *Procambarus clarkii* can survive up to 28 days out of water and that their ability to respire without water is better than that of aquatic crayfish, but not as efficient as that of a primary burrower. The ability to utilize oxygen without water allows some primary burrower species to live significantly further away from a water source. For example, this ability allows *Distocambarus crockeri* to live in nonhydric soils. Welch & Eversole (2006) found that the geographic distribution of *D. crockeri* was positively correlated with nonhydric soil and negatively correlated with hydric soil. The correlation was so strong that Welch and Eversole (2006) equated *D. crockeri* to the terrestrial burrowing crayfish in the Australian genus *Engaeus* (Horwitz & Richardson 1986; Schultz et al. 2009). Physiological dependence or strong preference for specific soil conditions affects the distribution of *D. crockeri* and may also strongly affect population genetic structure of this species. Although crayfish have different physiological requirements (e.g., hydric soil and required dissolved oxygen levels) and those requirements can shape their geographic distribution, the effects of these physiological requirements on population genetic structure have yet to be studied.

Comparing population structure and connectivity of crayfish with different physiological requirements across the same geographic region allows us to examine how these requirements have affected the population structure of each species and can lead to a better understanding of the evolution of crayfish (i.e., what determines their species boundaries and how they might be impacted by changing climates and habitats). In this study, population distribution and connectivity of a stream crayfish *Procambarus (Pennides) spiculifer* (LeConte 1856) is compared to a tertiary burrowing crayfish *Procambarus (Scapulicambarus) paeninsulanus* (Faxon 1914). *Procambarus spiculifer* is physiologically constrained to lotic waters with high oxygen content. Within its native range, *P. paeninsulanus* can be found nearly anywhere water has pooled or periodically pools and is not limited to moving water. *Procambarus paeninsulanus* and *P. clarkii* are closely related and are classified in the same subgenus, *Scapulicambarus*. Both species seem to be able to survive long periods (up to 28 days) without water. Hobbs (1942) noted that he had observed *P. paeninsulanus* wandering around in daylight ≈ 400 m from any water source. Physiological comparisons of these species revealed that *P. paeninsulanus* has a lower critical oxygen level requirement and is more robust to higher temperatures than *P. spiculifer* (Caine 1978). Both species tend to wander instead of burrowing when artificial laboratory habitat is drained (Caine 1978). This tendency to wander would allow *P. paeninsulanus* to find new habitat with water and may have contributed to the migration and distribution of *P. paeninsulanus* across Florida. Although *P. spiculifer* is reported to wander when there is not flowing water (Caine 1978), migration would need to be a very short distance and the new habitat would have to fulfill physiological requirements, making overland migration unlikely. For *P. spiculifer*, stream capture is the most likely means of migration, resulting in its large distribution. Thus, we have two crayfish species whose habitat requirements and physiological tolerances are distinct and we can, therefore, make predictions about their associated population structure and genetic diversity. Specifically, we predict that the species with strict physiological requirements (*P. spiculifer*) will have lower within-population genetic diversity and more population substructure than the species with less stringent physiological requirements (*P. paeninsulanus*).

Phylogenetic and population genetic methods are used to explore temporal and spatial genetic relationships within both species. We hypothesize that physiological attributes have affected the genetic structure of these two species. Therefore, we compare the population structure of these species to test for a contrasting genetic pattern that we would expect to be exhibited by species with differing physiological requirements. We use two mitochondrial genes, 16S and cytochrome c oxidase subunit I (COI), to examine and contrast population genetic structure, genetic diversity, and effective population sizes for these two species. Our null hypothesis is that levels of genetic diversity and population structure are the same in the two species, as they cover roughly the same geographic area and are phylogenetically and taxonomically close relatives. The alternative hypothesis is that *P. spiculifer* will show greater population subdivision, higher among population diversity, and lower within-population diversity compared to the more physiologically tolerant *P. paeninsulanus*.

2   MATERIALS AND METHODS

2.1   Taxon and locality sampling

Procambarus spiculifer and *Procambarus paeninsulanus* samples were collected from 54 localities (Appendix, Tables A1 and A2) throughout their distribution in northern Florida, USA (Figure 1). A total of 253 individuals were collected, with 190 *P. paeninsulanus* from 34 localities and 162 *P. spiculifer* from 21 localities (Figure 1). *Procambarus paeninsulanus* ranges from southern Georgia into Florida from the Choctawhatchee Basin (Walton County, Florida) east and south to Flagler, Marion, and Hillsborough counties in Florida. *Procambarus spiculifer* ranges from the upper Savannah Basin south to the Saint Johns and Suwannee river systems in Florida and westward to the Alabama-Mobile River basin. Despite sampling efforts, collections of the target species were sparse between longitudes 82° and 84° (Figure 1). Another species, *Procambarus fallax*, is also distributed in this area and many of the locations we expected to find *P. paeninsulanus* were occupied instead by *P. fallax*. The distribution of *P. spiculifer* is also limited in this region, and we have single sample locations from the Ochlockonee River (sample site 32) and Suwannee River (sample site 43). To be able to test for monophyly of each species, eight additional crayfish species were used as outgroups, including *Procambarus clarkii* and *Procambarus versutus*, hypothesized as close relatives of our target species (Appendix, Table A2).

2.2   DNA extraction, PCR and sequencing

Genomic DNA was extracted from gill or muscle tissue with Qiagen DNeasy® Blood and Tissue Kit (Cat. No. 69582). Polymerase chain reaction (PCR) products for two mitochondrial genes, the partial 16S ($\approx$ 460 bp; (Crandall & Fitzpatrick 1996)) and partial COI ($\approx$ 700 bp; (Folmer et al. 1994)), were amplified following protocols of Porter et al. (2005) and Crandall & Fitzpatrick (1996). Sequences were generated bidirectionally on an ABI Prism 3730XL capillary autosequencer using the ABI Big Dye Ready-Reaction kit following standard cycle sequencing protocols, except using 1/16th of the suggested reaction volume. The mitochondrial genes 16S and COI have appropriate amounts of variation and are commonly used for population level analyses in crayfish (e.g., Fetzner & Crandall 2003; Zaccara et al. 2005; Buhay et al. 2007). In order to ensure nuclear mitochondrial pseudogenes (numts) were avoided when amplifying COI, we followed Song et al. (2008) by extracting DNA from tissue with high amounts of mitochondria (gill tissue), checking PCR results with gel electrophoresis post-PCR, translating sequences to check for indels and stop codons, and finally comparing sequences to closely-related species.

2.3   Sequence analyses

Sequencher 4.8 (GeneCodes, Ann Arbor, MI, USA) was used to assemble and clean the bidirectional sequences. Mitochondrial data from 16S and COI were aligned separately using MAFFT (Katoh et al. 2001, 2005) implementing the G-INS-I alignment algorithm. The iterative algorithms used by MAFFT allow for fast and repeatable alignments. The two genes for all crayfish were then concatenated using MESQUITE version 2.72 (Maddison & Maddison 2009) and used for all phylogenetic analyses. Two other alignments were estimated for each species independently in MAFFT and used for all population level analyses. The best fit model of evolution was estimated with MODELTEST 3.7 (Posada & Crandall 1998) for each gene and each alignment using Akaike Information Criterion (AIC) (Akaike 1974). In the Bayesian analysis the best model of evolution estimated by

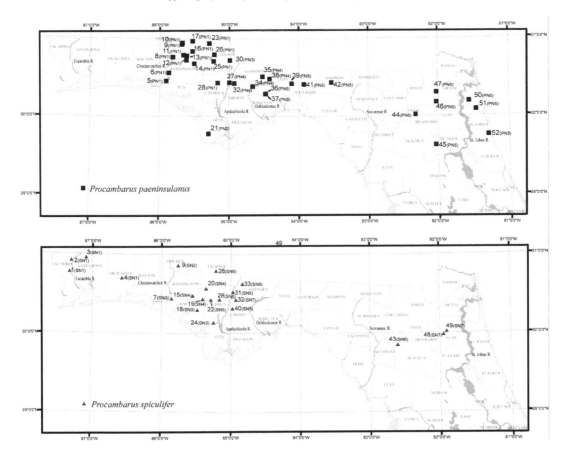

Figure 1. Map showing the sample distribution of *Procambarus paeninsulanus* (upper) and *P. spiculifer* (lower). Each location is labeled with sample site number followed by population in parentheses.

the AIC was used to test against simpler models with Bayes factors (Kass & Raftery 1995; Suchard et al. 2001) in the program Tracer v1.4 (Rambaut & Drummond 2007).

2.4   Phylogenetic and effective population size analyses

Phylogenies were estimated using two optimality criteria, Maximum Likelihood (ML) and Bayesian Analysis (BA). The program RAxML (Randomized Axelerated Maximum Likelihood) was used for ML estimation partitioned by gene and using only unique haplotypes defined by RAxML from the combined 16S and COI sequences (Stamatakis et al. 2005). We implemented 200 independent likelihood searches starting from random trees as well as 1000 bootstrap pseudoreplicates followed by a ML search. Divergence times are estimated in order to calibrate Bayesian skyline analysis (BSP) (Drummond et al. 2005), which estimates effective population size (N_e) over time (τ) ($N_e\tau$ = the product of the effective population size N_e and the generation length over time τ). BEAST v1.5.2 (Bayesian evolutionary analysis by sampling trees) was used to coestimate phylogeny and molecular divergence times for all samples (Drummond & Rambaut 2007). Our divergence time analysis mirrors Breinholt et al. (2009a) with the use of their fossil calibrations and taxa for Astacidea and differs in the number of genes and the use of BEAST instead of the program Multidivtime. In BEAST, we partitioned the dataset by gene and assigned independent models estimated from MOD-ELTEST 3.7 to each gene. We implemented a relaxed uncorrelated lognormal model using a Yule speciation tree prior. In order to use the same fossil calibrations, we fixed the topology to match Breinholt et al. (2009a) but did not fix relationships between or within *P. spiculifer*, *P. versutus*, *P. paeninsulanus*, and *P. clarkii*. Two dates based on fossil calibrations were used to date the origin of Astacidae (*Astacus licenti* Van Straelen 1928 and *Astacus spinirostris* Imaizumi 1938 both dated to Late Jurassic 144–159 Ma) and *Procambarus* (*Procambarus primaevus*: late Eocene 52.6–53.4 Ma Feldmann et al. (1981)). We used point estimations and standard deviation (SD) for a normal prior on the date of these nodes using a mean of 151.552 and SD of 1.698 for the root of the tree and mean of 54.461 and SD of 11.8 to calibrate the genus *Procambarus* (Breinholt et al. 2009a). We implemented three independent BEAST runs starting with random trees for 100 million generations recording every 10,000 samples. The three runs were checked for convergence to a similar posterior distribution of tree likelihood and time estimations graphically in Tracer v1.4. The Bayesian skyline plot (BSP) was estimated using a relaxed uncorrelated lognormal clock model and was calibrated at the root using a normal distribution prior with the mean estimated time to coalescence of the species included in analysis and a standard deviation large enough to include the 95% confidence interval (CI) of the time estimation. The Bayesian skyline tree model is a coalescent-based model that requires one to specify the number of allowable discrete changes in the population history. We set the allowable discrete population changes to 20 for each skyline analysis. The linear Bayesian skyline plot model was chosen to allow populations to grow or decline linearly between change-points instead of remaining constant between change-points. For each species, we ran three independent runs starting with a random tree for 100 million generations recording every 10,000 samples.

2.5   Mitochondrial evolution and genetic diversity

We tested the mitochondrial data for each species for the common assumption of neutral evolution by estimating Tajima's D (Tajima 1989) and Fu's F_s (Fu 1997) in ARLEQUIN 3.1 (Excoffier et al. 2005) using 10,000 simulated samples. Nucleotide diversity (π) and haplotype diversity (h_d) of each sample site and each of the resulting populations were estimated in the program DnaSP (Rozas et al. 2003). For each resulting population the total diversity, average intersampling site diversity, average intrasampling site diversity, and coefficient of diversification (G_{ST}) (Nei & Kumar 2000) were estimated in MEGA version 4 with the complete deletion option (Tamura et al. 2007). For each species the total diversity, average interpopulation diversity, average intrapopulation diversity, and coefficient of diversification (G_{ST}) (Nei & Kumar 2000) were also estimated in MEGA version 4 with the complete deletion option. The magnitude of increase for intersampling site and population to intrasampling site and population diversity was calculated by dividing the average intrasample site and population diversity by intersampling site and population diversity, respectively. ARLEQUIN 3.1 was also used to run an AMOVA (analysis of molecular variance) to examine the partitioning of genetic variation located within sample locations, between sample locations within resulting populations, and between populations for each species (Excoffier et al. 1992).

2.6   Population structure

Population structure of each species was measured using multiple methods to cross-validate findings, as these different approaches make different assumptions about the underlying populations. First, we calculated the ratio of the average percent sequence divergence (%SD) between sample locations over geographic distance in kilometers for each species and used this statistic as a phylogeographic structure parameter (%SD/Km) (Bernardo et al. 2007). Species with more genetic population structure have a higher (%SD/Km) than species with a lower genetic population structure. Population structure and diversity within versus between populations was estimated using F_{ST} (Weir & Cockerham 1984) and calculated in ARLEQUIN 3.1. F_{ST} is a measure of drift and gene flow equilibrium and assumes the population sizes are infinite and are equal-sized island populations that exchange migrants at a constant rate. While it has been shown that migration estimates, such as number of migrants (N_m) estimated from F_{ST}, can be inaccurate, F_{ST} still remains effective as a relative measure of population divergence (Neigel 2002; Pearse & Crandall 2004). The model based clustering method STRUCTURE version 2.2.3 of Falush et al. (2003) was used to estimate the number of natural genetic groups (K) using an admixture model with correlated allele frequencies. We ran STRUCTURE 20 times for each K ranging from 2 to 20 using a burnin of 50,000 and 10,000 Markov chain Monte Carlo (MCMC) reps after burnin. The online program STRUCTURE HARVESTER v0.56.4 (Earl 2009) was used to determine K using the largest ΔK (rate of change in the log probability of data between successive K) following Evanno et al. (2005). Coalescence based statistical parsimony with 95% connection limit (Templeton et al. 1992) was used to estimate networks based on the measurable signal of the coalescence process in each species using TCS (Clement et al. 2000). TCS networks that do not connect at the 95% limit are considered separate populations and multiple independent networks are indicative of heavily structured populations. The resulting TCS networks for each species were used to define populations. Lastly, we used a Mantel test (Mantel 1967) implemented in the program zt (Bonnet & Van de Peer 2002) to test if the current population structure is due to isolation by distance (IBD). We used the pairwise F_{ST} matrix and tested for correlation with a pairwise geographic distance matrix, in kilometers, estimated from the latitude and longitude of collection sites in GENALEX 6 (Peakall & Smouse 2006), using 100,000 randomizations. IBD is considered to be a result of relatively recent events, which can be confounded by historical events or distributions. In order to account for historical divergence and look at the most recent events, a partial Mantel test was used to account for the possible confounding effect, using a model based on deep divergence in the resulting phylogeny (Bonnet & Van de Peer 2002; Wolf et al. 2008).

3   RESULTS

3.1   Sequence analyses

For the full alignment, the general time reversible plus Gamma (GTR + G) model was estimated as the best-fit model of sequence evolution for both 16S and COI and was used for the ML estimation. The Bayes Factor supported the GTR + G model over the Hasegawa-Kishino-Yano with and without Gamma (HKY + G, HKY) models with substantial support (Log 10 Bayes Factor ≥ 4.872) (Kass & Raftery 1995). For the species alignments, the AIC indicated the HKY + G as the optimal model. Bayes Factor also strongly supported the HKY + G over the HKY model (Log 10 Bayes Factor ≥ 28.6844) for both species and for both genes and was used in the skyline estimation.

3.2   Phylogenetic and effective population size analyses

RAxML reduced the full dataset of 362 individuals to 117 unique combined 16S and COI sequences for ML phylogeny estimation. The tree with the best score (-4722.641148) includes 63 *P. paeninsulanus* and 46 *P. spiculifer* haplotypes (Figure 2). Both species of interest form highly supported monophyletic groups with the expected species *P. clarkii* and *P. versutus* being sister species to *P. paeninsulanus* and *P. spiculifer*, respectively. The three independent BA runs converged to a similar posterior distribution and the run with the highest effective sample size (ESS) for each parameter was used to estimate topology and time estimation from postburnin (burnin = 10,000,000) posterior

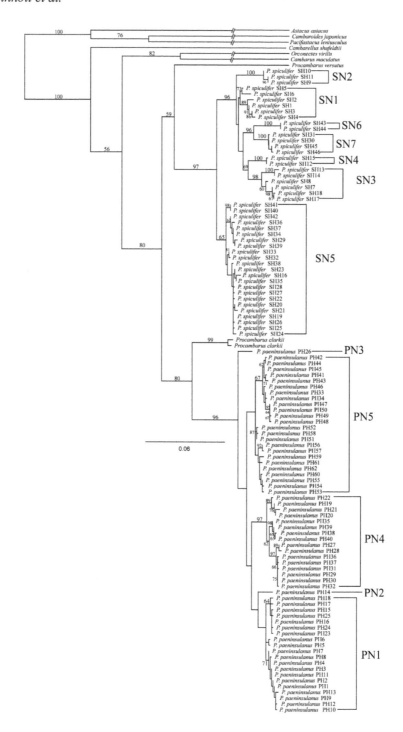

Figure 2. ML phylogentic hypothesis score -4722.641148 with 117 individuals estimated from COI and 16S mitochondrial gene regions with RAxML using a GTR + G model partitioned by gene with branch support estimated with 1000 pseudoreplicates and placed on branches leading to the node of support. Terminal tips are labeled with genus and species followed by the haplotype number. For each species populations are labeled with brackets.

Table 1. Data for each population: number of sample locations used (N_{SLU}), number of samples (N_{S}), total diversity (D_{total}), average intersample site diversity ($\overline{SSD}_{\mathrm{inter}}$), average intrasample site diversity ($\overline{SSD}_{\mathrm{intra}}$), magnitude of increase from inter- to intrasample site diversity (M_{inc}), G_{ST}, and average F_{ST} between sample sites in a population.

Population	N_{SLU}	N_{S}	D_{total}	$\overline{SSD}_{\mathrm{inter}}$	$\overline{SSD}_{\mathrm{intra}}$	M_{inc}	G_{ST}	F_{ST}
Procambarus spiculifer								
SN1	3	11	0.0075087	0.0010724	0.0064363	6.0018	0.85718	0.8215
SN2	1	11	0.00099	na	na	na	na	na
SN3	3	43	0.00696	0.00108	0.00589	5.4537	0.84514	0.9536
SN4	2	8	0.00023	0.00018	0.00005	0.2778	0.2	0.0000
SN5	5	66	0.00321	0.00098	0.00223	2.2755	0.69561	0.6912
SN6	1	5	0	na	na	na	na	na
SN7	4	15	0.00118	0.0057	0.00061	0.1070	0.51995	0.5182
Procambarus paeninsulanus								
PN1	7	45	0.009765	0.001242	0.0032523	2.6186	0.789103	0.5875
PN2	1	1	na	na	na	na	na	na
PN3	1	3	0	na	na	na	na	na
PN4	4	40	0.005626	0.0017266	0.0038997	2.2586	0.6931206	0.5529
PN5	9	76	0.0062915	0.0012803	0.0050113	3.9142	0.7965033	0.7978

distribution of tree and time estimations. The topology of the BA phylogeny is the same as the ML topology for *P. spiculifer*, whereas in *P. paeninsulanus* the same major clades are found, but relationships between the major clades that are weakly supported are different (e.g., BA estimation PN1 and PN2 are sister to PN5). The mean time estimations (not reported) for all concordant nodes fit within the 95% confidence interval estimated for each node by Breinholt et al. (2009a). Time to coalescence for the samples of both species were very similar and used for calibration of the BSP (*P. paeninsulanus* 74.3317 MY with 95% CI of 50.2204–84.7922, *P. spiculifer* 75.3935 with 95% CI 49.9304–85.7659). The three independent Bayesian skyline runs for each species converged and one run was used to construct each skyline plot. Current $N_e\tau$ for both species is $\approx$ 110 and they both have fairly stable effective population sizes through time (Figure 3). The skyline plot indicates that both species experienced a population decline starting at approximately 4–5 MYA and then a sharp increase in size beginning in the last million years for *P. paeninsulanus* and in the last half million years for *P. spiculifer*.

3.3 Mitochondrial evolution and genetic diversity

The test for neutral evolution of the combined COI and 16S mitochondrial genes for both species resulted in positive values with nonsignificant P values (Tajima's D $P \geq 0.43717$ and Fu's F_s $P \geq 0.40400$). Neutral evolution tests for each resulting population resulted in positive and negative values for both species and were also nonsignificant (Tajima's D $P \geq 0.2163$ and Fu's $F_s p \geq 0.06330$). Nucleotide diversity and haplotype diversity are reported in Appendix (Table A1) by species and by sample location. *Procambarus paeninsulanus* has lower average nucleotide diversity with $\pi = 0.01476$ compared to $\pi = 0.02036$ for *P. spiculifer*, but higher haplotype diversity (h_d). Each of the populations defined by resulting TCS networks for *P. paeninsulanus* had more

Table 2. Data for each species: number of sample locations used (N_{SLU}), number of samples (N_S), total diversity (D_{total}), average interpopulation diversity ($\overline{PD}_{inter}$), average intrapopulation diversity ($\overline{PD}_{intra}$), magnitude of increase from inter- to intrasample site diversity (M_{inc}), G_{ST}, and average F_{ST} between all population.

N_{SLU}	N_S	D_{total}	$\overline{PD}_{inter}$	$\overline{PD}_{intra}$	M_{inc}	G_{ST}	F_{ST}
Procambarus spiculifer							
7	162	0.02092	0.00279	0.01813	6.4982	0.86679	0.926402048
Procambarus paeninsulanus							
4	189	0.01503	0.0038	0.01123	2.955263158	0.74717	0.8442841

total nucleotide diversity than *P. spiculifer* populations (Table 1). Two populations of *P. spiculifer* (SN1 and SN3) had high G_{ST}, but the remaining three had values similar to or below those of *P. paeninsulanus*. Although *P. spiculifer* had higher total diversity and higher average intrapopulation diversity, it had lower average interpopulation diversity than *P. paeninsulanus* (Table 2). *Procambarus spiculifer* had more than double the magnitude of increase of diversity from inter- to intrapopulation diversity compared to *P. paeninsulanus* (Table 2). AMOVA analysis indicates that *P. paeninsulanus* has more of its genetic variation partitioned between populations and more variation partitioned within sample locations than *P. spiculifer* (Tables 3 and 4). *Procambarus spiculifer* had a higher F_{ST} among locations within a network and higher F_{CT} within sample locations than *P. paeninsulanus* (Tables 3 and 4).

3.4 Population structure

The phylogeographic structure parameter (%SD/Km) between sample sites was 0.0003212 for P. *spiculifer* and 0.0001512 for *P. paeninsulanus* indicating that *P. paeninsulanus* has less population structure. Statistical parsimony estimated with a 95% connection limit (14 steps for both species) estimated five independent networks for *P. paeninsulanus* (named PN1– PN5) and seven independent networks for *P. spiculifer* (named SN1–SN7). We do not report all F_{ST} values, but they are generally high for both species, with a majority of the values above 0.80. The average of the F_{ST} values for *P. spiculifer* (0.926402) is higher than that of *P. paeninsulanus* (0.844248) (Table 2). The average F_{ST} for *P. spiculifer* populations tends to be higher than those for *P. paeninsulanus* (Table 1). Over the 20 Structure runs for each K both species showed variation in likelihood score between runs yet the maximum ΔK was clearly identifiable and estimated at $K = 8$ for *P. paeninsulanus* and $K = 13$ for *P. spiculifer*. Although estimations indicated that *P. paeninsulanus* has less structure than *P. spiculifer*, the Mantel and partial Mantel tests indicate that both species exhibit IBD (Mantel, *P. paeninsulanus*, $r = 0.358243$, $P = 0.00001$; partial Mantel, *P. paeninsulanus* $r = 0.323967$, $P = 0.00001$; Mantel, *P. spiculifer* $r = 0.206286$, $P = 0.002630$; partial Mantel, *P. spiculifer* $r = 0.214154$, $P = 0.009570$).

4   DISCUSSION

The diversity and population structure predictions of this study were based on the distribution of habitat that can support the physiological needs of these two related species with similar geographic

Table 3. AMOVA statistics for *Procambarus paeninsulanus*.

Source of variation	d.f.	Sum of squares	Variance components	% of variation	Fixation indices
Among populations	4	1122.199	$8.37358V_a$	70.11	F_{SC}: 0.81845
Among locations within a population	29	496.962	$2.89182V_b$	24.50	F_{ST}: 0.94613
Within sample location	156	100.071	$0.64148V_c$	5.39	F_{CT}: 0.70326
Total	189	1692.232	11.84941		

All significant P values $= 0.00 \pm 0.00$

Table 4. AMOVA statistics for *Procambarus spiculifer*.

Source of variation	d.f.	Sum of squares	Variance components	% of variation	Fixation indices
Among populations	6	1501.386	$11.57526V_a$	78.28	F_{SC}: 0.79563
Among locations within a population	14	270.271	$2.55584V_b$	17.28	F_{ST}: 0.95600
Within sample location	141	92.570	$0.65653V_c$	4.44	F_{CT}: 0.78277
Total	161	1864.228	14.78753		

All significant P values $= 0.00 \pm 0.00$

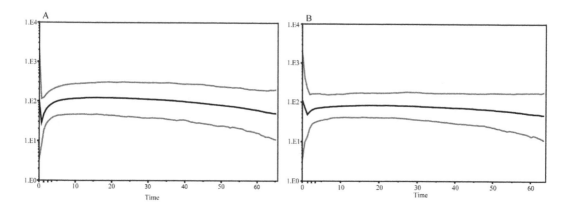

Figure 3. Bayesian skyline plots with time in millions of years on the x-axis and $N_e\tau$ the product of the effective population size and the generation length in millions years on the y-axis. A: *Procambarus spiculifer*; B: *P. paeninsulanus*.

distributions. We predicted that *Procambarus spiculifer* might show greater population structure as a result of physiological constraints compared to a less physiological constrained species, *P. paeninsulanus*. Although we find that both species have significant population structure, the more physiologically constrained species, *P. spiculifer*, clearly has more population structure than the less physiologically constrained species, *P. paeninsulanus*. All of the approaches used to estimate population structure give similar results indicating that *P. spiculifer* is more structured than *P. paeninsulanus*. The uses of multiple diverse methods, namely, the F_{ST} statistics, phylogeographic structure statistic %SD/km, model-based clustering method, and coalescence based gene genealogies networks, add overwhelming support that there is an actual difference in population structure between these two species with different physiological requirements.

We further predicted that *P. spiculifer* might show more total genetic diversity as a result of physiological constraints compared to a less physiologically constrained species, *P. paeninsulanus*. To account for possible effects of unequal population sizes on diversity, we estimated effective population history through time with a BSP. Effective population size estimations with the BSP indicate that the two species share similar current N_e and fairly similar demographic histories. Bottleneck events are predicted in both species in the BSP, yet it has been shown that spurious bottlenecks can be found in highly structured populations (Chikhi et al. 2010). A strategy to avoid spurious results is to include as many demes as possible (Chikhi et al. 2010). In our BSP, we used all of our samples, which represent at least 5 demes for *P. paeninsulanus* and 7 demes for *P. spiculifer*. Chikhi et al. (2010) suggests that when a bottleneck is detected in structured populations and samples for estimation come from multiple demes the results suggest that the whole metapopulation may have been subject to a population size change. Our estimates of Tajima's D also seem to suggest a possible population bottleneck with positive, yet nonsignificant values for both species. Tajima's D is also known to be affected by population structure. Stadler et al. (2009) showed that evaluating structured populations under a model of panmictic population can lead to errors, so they suggest estimating statistics for samples from populations as well as for samples pooled between populations to evaluate demographic history. For both species of interest, estimates of Tajima's D by population (more likely to fit a panmictic model) resulted in positive and negative values, but all failed to reject neutrality with nonsignificant P values. Our pooled Tajima's D estimations likely reflected these contrasting positive and negative values and represented an overall population decrease across the metapopulation as the BSP predicts. If the predicted bottleneck is a spurious result, both species have such a trend, and the factors that contribute to this result are likely to be common to both species. On the other hand, if the BSP predicted bottlenecks are real, then *P. paeninsulanus* would have had a longer time to recover from the bottleneck and, therefore, could have higher diversity than *P. spiculifer*. Despite the possibility of *P. spiculifer* having a more recent recovery from a predicted bottleneck, it is evident in the total diversity estimations for each species that *P. spiculifer* has more overall diversity than *P. paeninsulanus*.

Last, we predicted less diversity within a population and more diversity when including deeper levels of population structure for *P. spiculifer* when compared to *P. paeninsulanus*. We find that *P. spiculifer* has less average intersample site diversity within a population and less average interpopulation diversity than *P. paeninsulanus*. If our diversity predictions were true, we would find more increase in the diversity from the average intersample site diversity and average interpopulation diversity to the intrasample site and intrapopulation diversity in the most physiologically constrained species. This holds true for the average inter- to intrapopulation diversity for *P. spiculifer* when compared to *P. paeninsulanus*. For the average inter- to intrasample site diversity, it only holds true for two of the *P. spiculifer* populations (SN1 and SN4). In these populations the magnitude of change from inter- to intrapopulation diversity is much higher than values estimated for *P. paeninsulanus*. The populations of *P. spiculifer* that exhibit lower or similar magnitude of change from inter- to intrapopulation diversity than *P. paeninsulanus* either were sampled from a single drainage (SN4 and SN5) or may represent a more recent migration (SN7). This diversity prediction is also supported by *P. spiculifer* having less of its genetic variation found within sample location than *P. paeninsulanus*. By examining diversity in multiple levels of the population structure, we find evidence for our prediction that *P. spiculifer* would have higher among population diversity and lower within population diversity compared to *P. paeninsulanus*.

The null hypothesis for this study is that there would be no diversity or population structure differences found between the two species with similar geographic distributions. Not only do these species share a similar geographic distribution, they also have similar divergence times and effective population sizes and histories, yet we found different amounts of population structure and different partitioning and amounts of diversity. The data support the alternative hypothesis that population structure of *P. spiculifer* has been affected by its physiological requirements. While our data support the alternative hypothesis, there are many unknown factors that may have contributed to the underlying genetic structures of these two species. A broader study is needed to cross-validate these results using multiple species with differing physiological requirements that inhabit diverse habitats to investigate the broader role of physiological constraints on population structure.

Unfortunately, species with strict physiological requirements also tend to have reduced within-population genetic diversity, as exhibited by *P. spiculifer*, threatening a species' survival (Spielman et al. 2004). As Bernardo et al. (2007) suggested, physiological characteristics that have shaped a species distribution and population structure should be considered when estimating the effects of global warming and conservation decisions. Although neither of these species is currently listed as threatened, when considering the effects of global warming on freshwater habitats, the physiological requirements of *P. spiculifer* are of concern. The upper temperature for a typical *P. spiculifer* habitat is $\approx 23.5\,°C$ and the upper estimated temperature to be lethal is $33.3 \pm 0.9\,°C$ (Caine 1978). In contrast, the upper lethal temperature for *P. paeninsulanus* was estimated to be $37.1 \pm 1.2\,°C$ and the typical upper temperature for *P. paeninsulanus* habitat was estimated at $\approx 24.5\,°C$ (Caine 1978). In comparing the critical oxygen levels for these two species, Caine (1978) found that *P. spiculifer* requires approximately one and an half times more mg/l of oxygen than *P. paeninsulanus*. While global warming is not expected to raise freshwater habitats to lethal levels any time soon, the rise in temperature and resulting reduction in dissolved oxygen levels could be a deadly combination for *P. spiculifer*. As Bernardo et al. (2007) suggested, the results of this study reiterate the importance of using physiological data in combination with genetic data in making conservation predictions for a given species.

ACKNOWLEDGEMENTS

We thank the many BYU undergraduates, especially Jordan Fritzsche and Dohyup Kim, in the Crandall Lab who participated in various stages of collecting genetic data for this project. This work was supported by a grant from the Florida Fish and Wildlife Conservation Commission NG07-104 awarded to KAC as well as partial support from a grant from the U. S. NSF EF-0531762 awarded to KAC and Brigham Young University.

REFERENCES

Akaike, H. 1974. A new look at the statistical model indentification. *IEEE Trans. Autom. Contr.* 19: 716–723.

Bernardo, J., Ossola, R.J., Spotila, J. & Crandall, K.A. 2007. Interspecies physiological variation as a tool for cross-species assessments of global warming-induced endangerment: validation of an intrinsic determinant of macroecological and phylogeographic structure. *Biol. Lett.* 3: 695–699.

Bonnet, E. & Van de Peer, Y. 2002. Zt: A software tool for simple and partial mantel tests. *J. of Stat. Softw.* 7: 1–12.

Breinholt, J., Pérez-Losada, M. & Crandall, K.A. 2009a. The timing of the diversification of the freshwater crayfishes. In: Martin, J.W., Crandall, K.A., & Felder, D.L. (eds.), *Crustacean Issues 18. Decapod Crustacean Phylogenetics*: 305–318. Boca Raton, FL: CRC Press.

Breinholt, J.W., Van Buren, R., Kopp, O.R. & Stephen, C.L. 2009b. Population genetic structure of an endangered Utah endemic, *Astragalus ampullarioides* (Fabaceae). *Amer. J. Bot.* 96: 661–667.

Buhay, J.E., Moni, G., Mann, N. & Crandall, K.A. 2007. Molecular taxonomy in the dark: evolutionary history, phylogeography, and diversity of cave crayfish in the subgenus *Aviticambarus*, genus *Cambarus*. *Mol. Phyl. Evol.* 42: 435–488.

Caine, E.A. 1978. Comparative ecology of epigean and hypogean crayfish (Crustacea: Cambaridae) from northwestern Florida. *Amer. Midland Naturalist* 99: 315–329.

Chikhi, L., Sousa, V.C., Luisi, P., Goossens, B. & Beaumont, M.A. 2010. The confounding effects of population structure, genetic diversity and the sampling scheme on the detection and quantification of population size changes. *Genetics* 186: 983–995.

Clement, M., Posada, D. & Crandall, K.A. 2000. TCS: A computer program to estimate gene genealogies. *Mol. Ecol.* 9: 1657–1659.

Crandall, K.A. & Fitzpatrick, J.F., Jr. 1996. Crayfish molecular systematics: using a combination of procedures to estimate phylogeny. *Syst. Biol.* 45: 1–26.

Drummond, A. & Rambaut, A. 2007. Beast: Bayesian evolutionary analysis by sampling trees. *BMC Evol. Biol.* 7: 214.

Drummond, A.J., Rambaut, A., Shapiro, B. & Pybus, O.G. 2005. Bayesian coalescent inference of past population dynamics from molecular sequences. *Mol. Biol. Evol.* 22: 1185–1192.

Earl, D.A. 2009. Structure harvester v0.3. http://users.Soe.Ucsc.Edu/ dearl/software/struct_harvest/.

Evanno, G., Regnaut, S. & Goudet, J. 2005. Detecting the number of clusters of individuals using the software structure: a simulation study. *Mol. Ecol.* 14: 2611–2620.

Excoffier, L., Laval, G., & Schneider, S. 2005. Arlequin (version 3.0): an integrated software package for population genetics data analysis. *Evol. Bioinform. Online* 1: 47–50.

Excoffier, L., Smouse, P.E. & Quattro, J.M. 1992. Analysis of molecular variance inferred from metric distances among DNA haplotypes: application to human mitochondrial DNA restriction data. *Genetics* 131: 479–491.

Falush, D., Stephens, M. & Pritchard, J.K. 2003. Inference of population structure: extensions to linked loci and correlated allele frequencies. *Genetics* 164: 1567–1587.

Faxon, W. 1914. Notes on the crayfishes in the United States National Museum and the Museum of Comparative Zoology with descriptions of new species and subspecies to which is appended a catalogue of the known species and subspecies. *Mem. Mus. Comp. Zool.* 40: 352–427.

Feldmann, R.M., Grande, L., Birkhimer, C.P., Hannibal, J.T. & McCoy, D.L. 1981. Decapod fauna of the Green River Formation (Eocene) of Wyoming. *J. Paleontol.* 55: 788–799.

Fetzner, J.W., Jr. & Crandall, K.A. 2003. Linear habitats and the nested clade analysis: an empirical evaluation of geographic vs. river distances using an Ozark crayfish (Decapoda: Cambaridae). *Evolution* 57: 2101–2118.

Folmer, O., Black, M., Hoeh, W., Lutz, R. & Vrijenhoek, R. 1994. DNA primers for amplification of mitochondrial cytochrome c oxidase subunit I from diverse metazoan invertebrates. *Mol. Mar. Biol. Biotech.* 3: 294–299.

Fu, Y. 1997. Statistical tests of neutrality of mutations against population growth, hitchhiking and background selection. *Genetics* 147: 915–925.

Hobbs, H.H., Jr. 1942. *The Crayfishes of Florida*. Gainesville, FL: University of Florida Press.

Hobbs, H.H., Jr. 1981. The Crayfishes of Georgia. *Smiths. Contr. to Zool.* 318: 1–549.

Horwitz, P.H.J. & Richardson, A.M.M. 1986. An ecological classification of the burrows of Australian freshwater crayfish. *Austr. J. Mar. Freshw. Res.* 37: 237–242.

Imaizumi, R. 1938. Fossil crayfishes from Jehol. *Sci. Rep. Tokyo Imp. Univ., Sendal, Japan, 2nd Ser.* 19: 173–179.

Kass, R. & Raftery, A. 1995. Bayes factors. *J. Amer. Stat. Assoc.* 90: 773–795.

Katoh, K., Kuma, K. & Miyata, T. 2001. Genetic algorithm-based maximum-likelihood analysis for molecular phylogeny. *J. Mol. Evol.* 53: 477–484.

Katoh, K., Kuma, K., Toh, H. & Miyata, T. 2005. MAFFT version 5: improvement in accuracy of multiple sequence alignment. *Nucleic Acids Res.* 33: 511–518.

LeConte, J. 1856. Descriptions of a new species of *Astacus* from Georgia. *Proc. Nat. Acad. Sci. USA* 7: 4000–4402.

Maddison, W.P. & Maddison, D.R. 2009. Mesquite: a modular system for evolutionary analysis. Version 2.6. http://mesquiteproject.org.

Mantel, N. 1967. The detection of disease clustering and a generalized regression approach. *Cancer Res.* 27: 209–220.

Martin, B.A., Saiki, M.K. & Fong, D. 2009. Habitat requirements of the endangered California freshwater shrimp (*Syncaris pacifica*) in Lagunitas and Olema creeks, Marin County, California, USA. *J. Crust. Biol.* 29: 595–604.

McMahon, B.R. 2001. Respiratory and circulatory compensation to hypoxia in crustaceans. *Resp. Physiol.* 128: 349–364.

McMahon, B.R. & Stuart, S.A. 1999. Haemolymph gas exchange and ionic and acid–base regulation during long-term air exposure an aquatic recovery in *Procambarus clarkii*. *Freshw. Crayfish* 12: 134–153.

Nei, M. & Kumar, S. 2000. *Molecular Evolution and Phylogenetics*. New York, NY: Oxford University Press.

Neigel, J.E. 2002. Is F_{ST} obsolete? *Conserv. Genetics* 3: 167–173.

Peakall, R. & Smouse, P.E. 2006. Genalex 6: genetic analysis in Excel. Population genetic software for teaching and research. *Mol. Ecol. Notes* 6: 282–295.

Pearse, D.E. & Crandall, K. 2004. Beyond F_{ST}: analysis of population genetic data for conservation. *Conserv. Genetics* 5: 585–602.

Porter, M.L., Pérez-Losada, M. & Crandall, K.A. 2005. Model based multi-locus estimation of decapod phylogeny and divergence times. *Mol. Phylogenet. Evol.* 37: 355–369.

Posada, D. & Crandall, K.A. 1998. Modeltest: testing the model of DNA substitution. *Bioinformat-*

ics 14: 817–818.

Rambaut, A. & Drummond, A.J. 2007. Tracer v1.4. http://beast.Bio.Ed.Ac.Uk/tracer.

Rand, D.M. & Harrison, R.G. 1989. Ecological genetics of a mosaic hybrid zone: mitochondrial, nuclear, and reproductive differentiation of crickets by soil type. *Evolution* 43: 432–449.

Rozas, J., Sánchez-DelBarrio, J.C., Messegyer, X. & Rozas, R. 2003. DnaSP, DNA polymorphism analyses by the coalescent and other methods. *Bioinformatics* 19: 2496–2497.

Schultz, M.B., Smith, S.A., Horwitz, P., Richardson, A.M.M., Crandall, K.A. & Austin, C.M. 2009. Evolution underground: a molecular phylogenetic investigation of Australian burrowing freshwater crayfish (Decapoda: Parastacidae) with particular focus on *Engaeus erichson. Mol. Phylogenet. Evol.* 50: 580–598.

Song, H., Buhay, J.E., Whiting, M.F. & Crandall, K.A. 2008. Many species in one: DNA barcoding overestimates the number of species when nuclear mitochondrial pseudogenes are coamplified. *Proc. Nat. Acad. Sci. USA* 105: 13486–13491.

Spielman, D., Brook, B.W. & Frankham, R. 2004. Most species are not driven to extinction before genetic factors impact them. *Proc. Nat. Acad. Sci. USA* 101: 15261–15264.

Stadler, T., Haubold, B., Merino, C., Stephan, W. & Pfaffelhuber, P. 2009. The impact of sampling schemes on the site frequency spectrum in nonequilibrium subdivided populations. *Genetics* 182: 205–216.

Stamatakis, A., Ludwig, T. & Meier, H. 2005. RAxML-III: a fast program for maximum likelihood-based inference of large phylogenetic trees. *Bioinformatics* 21: 456–463.

Suchard, M.A., Weiss, R.E. & Sinsheimer, J.S. 2001. Bayesian selection of continuous-time Markov chain evolutionary models. *Mol. Biol. Evol.* 18: 1001–1013.

Tajima, F. 1989. Statistical method for testing the neutral mutation hypothesis by DNA polymorphism. *Genetics* 123: 585–595.

Tamura, K., Dudley, J., Nei, M. & Kumar, S. 2007. MEGA4: Molecular evolutionary genetics analysis (MEGA) software version 4.0. *Mol. Biol. Evol.* 24: 1596–1599.

Templeton, A.R., Crandall, K.A. & Sing, C.F. 1992. A cladistic analysis of phenotypic associations with haplotypes inferred from restriction endonuclease mapping and DNA sequence data. III. Cladogram estimation. *Genetics* 132: 619–633.

Van Straelen, V. 1928. On a fossil freshwater crayfish from eastern Mongolia. *Bull. Geol. Soc. China* 7: 173–178.

Weir, B.S. & Cockerham, C.C. 1984. Estimating F–statistics for the analysis of population structure. *Evolution* 38: 1358–1370.

Welch, S.M. & Eversole, A.G. 2006. The occurrence of primary burrowing crayfish in terrestrial habitat. *Biol. Conserv.* 130: 458–464.

Wolf, J.B.W., Harrod, C., Brunner, S., Salazar, S., Trillmich, F. & Tautz, D. 2008. Tracing early stages of species differentiation: ecological, morphological and genetic divergence of Galápagos sea lion populations. *BMC Evol. Biol.* 8: 150.

Zaccara, S., Stefani, F. & Crosa, G. 2005. Diversity of mitochondrial DNA of the endangered white-clawed crayfish (*Austropotamobius italicus*) in the Po River catchment. *Freshw. Biol.* 50: 1262–1272.

APPENDIX

Table A1 summarizes information about sample size at each site, haplotypes in each population, haplotype and nucleotide diversities for *Procambarus paeninsulanus* and *Procambarus spiculifer*. Table A2 contains locality data, sampling sites, sample size, and GenBank accession numbers for 16S and COI gene fragments.

Table A1. Data for each species: genus, species, site names, sample size, population, haplotypes in population, number of haplotypes from species level alignment (N_h, haplotype diversity (h_d), and nucleotide diversity (π).

Genus	Species	Sample site	Sample Size	Population	Haplotype	N_h	h_d	π
Procambarus	*paeninsulanus*	5	5	PN1	PH1(3), PH2(2)	2	0.6	0.00055
Procambarus	*paeninsulanus*	6	10	PN1	PH2(9), PH3	2	0.2	0.00018
Procambarus	*paeninsulanus*	8	10	PN1	PH4(9), PH9	1	0.2	0.00036
Procambarus	*paeninsulanus*	9	8	PN1	PH5(4), PH6, PH7, PH8(2)	4	0.75	0.00348
Procambarus	*paeninsulanus*	10	1	PN1	PH9	1	0	0
Procambarus	*paeninsulanus*	11	1	PN1	PH10	0	0	0
Procambarus	*paeninsulanus*	12	1	PN1	PH11	0	0	0
Procambarus	*paeninsulanus*	13	1	PN1	PH12	0	0	0
Procambarus	*paeninsulanus*	14	2	PN1	PH9(2)	1	0	0
Procambarus	*paeninsulanus*	16	2	PN1	PH9, PH13	2	1	0.00182
Procambarus	*paeninsulanus*	17	2	PN1	PH9(2)	1	0	0
Procambarus	*paeninsulanus*	21	1	PN2	PH14	1	0	0
Procambarus	*paeninsulanus*	23	1	PN1	PH9	1	0	0
Procambarus	*paeninsulanus*	25	7	PN1	PH15, PH16(2), PH17(3), PH18, PH18	4	0.80952	0.00182
Procambarus	*paeninsulanus*	26	1	PN1	PH9	1	na	na
Procambarus	*paeninsulanus*	27	8	PN4	PH19(5), PH20, PH21, PH22	3	0.46429	0.00137
Procambarus	*paeninsulanus*	28	14	PN1	PH16(3), PH23, PH24(9), PH25	4	0.571	0.00071
Procambarus	*paeninsulanus*	30	3	PN3	PH26(3)	1	0	0
Procambarus	*paeninsulanus*	32	5	PN4	PH27(3), PH28(2)	2	0.6	0.00109
Procambarus	*paeninsulanus*	34	11	PN4	PH29(5), PH30(2), PH31(3), PH32	2	0.18182	0.00017
Procambarus	*paeninsulanus*	35	1	PN4	PH29	1	na	na
Procambarus	*paeninsulanus*	36	6	PN5	PH33, PH34(5)	2	0.33333	0.0003
Procambarus	*paeninsulanus*	37	9	PN5	PH33(4), PH34(5)	2	0.55556	0.00051
Procambarus	*paeninsulanus*	38	16	PN4	PH29(8), PH35(2), PH36, PH37(2), PH38, PH39, PH40	7	0.75	0.00432
Procambarus	*paeninsulanus*	39	7	PN5	PH41, PH42, PH43, PH44, PH45, PH46(2)	6	0.95238	0.00322
Procambarus	*paeninsulanus*	41	4	PN5	PH47(4)	1	0	0
Procambarus	*paeninsulanus*	42	15	PN5	PH48, PH49, PH50(13)	2	0.13333	0.00012
Procambarus	*paeninsulanus*	44	15	PN5	PH51(14), PH52	2	0.13333	0.00012
Procambarus	*paeninsulanus*	45	10	PN5	PH53(2), PH54, PH55(7)	3	0.51111	0.00051
Procambarus	*paeninsulanus*	46	8	PN5	PH56(6), PH57(2)	2	0.42857	0.00039
Procambarus	*paeninsulanus*	47	2	PN5	PH58, PH59	2	1	0.00641

Table A1. Continuation.

Genus	Species	Sample site	Sample Size	Population	Haplotype	N_h	h_d	π
Procambarus	*paeninsulanus*	50	1	PN5	PH60	1	0	0
Procambarus	*paeninsulanus*	51	1	PN5	PH61	1	0	0
Procambarus	*paeninsulanus*	52	1	PN5	PH62	1	0	0
Procambarus	*spiculifer*	1	7	SN1	SH1(3), SH2(3), SH3	3	0.71729	0.00322
Procambarus	*spiculifer*	2	1	SN1	SH4	1	0	0
Procambarus	*spiculifer*	3	2	SN1	SH5(2)	1	0	0
Procambarus	*spiculifer*	4	2	SN1	SH6(2)	1	0	0
Procambarus	*spiculifer*	7	19	SN3	SH7(7), SH8(10), SH17(2)	3	0.60819	0.00315
Procambarus	*spiculifer*	9	11	SN2	SH9, SH10(6), SH11(4)	2	0.54545	0.00102
Procambarus	*spiculifer*	15	3	SN4	SH12(3)	1	0	0
Procambarus	*spiculifer*	18	1	SN3	SH13	1	0	0
Procambarus	*spiculifer*	19	13	SN3	SH14(13)	1	0	0
Procambarus	*spiculifer*	20	5	SN4	SH12(4), SH15	2	0.4	0.00038
Procambarus	*spiculifer*	22	1	SN5	SH16	1	0	0
Procambarus	*spiculifer*	24	11	SN3	SH17(10), SH18	2	0.018182	0.00017
Procambarus	*spiculifer*	26	17	SN5	SH19(12), SH20, SH21, SH22, SH23, SH24	6	0.051471	0.00087
Procambarus	*spiculifer*	28	10	SN5	SH19(4), SH20, SH25, SH26, SH27,SH28(2)	5	0.066667	0.00075
Procambarus	*spiculifer*	31	8	SN5	SH29	1	0	0
Procambarus	*spiculifer*	32	3	SN7	SH30(2), SH31	2	0.66667	0.00063
Procambarus	*spiculifer*	33	14	SN5	SH30, SH32(10), SH33(3)	2	0.36264	0.0034
Procambarus	*spiculifer*	40	17	SN5	SH19(6), SH34, SH35, SH36, SH37(2), SH38, SH39, SH40, SH41(2), SH42	7	0.82353	0.00254
Procambarus	*spiculifer*	43	5	SN6	SH43(4),SH44	1	0	0
Procambarus	*spiculifer*	48	4	SN7	SH45(2), SH46(2)	2	0.66667	0.00063
Procambarus	*spiculifer*	49	8	SN7	SH45(5), SH46(3)	2	0.053571	0.0005

Genus	Species	No. of sample sites	No. of samples	No. of populations	N_h	h_d (total)	π
Procambarus	*paeninsulanus*	34	190	5	51	0.955	0.01476
Procambarus	*spiculifer*	21	162	7	39	0.948	0.02036

Table A2. Data for each species: genus, species, sample site, sample size, state, county, locality, latitude, longitude, 16S GenBank accession number, COI GenBank accession numbers.

Genus	Species	Sample site	State	County	Locality	Latitude	Longitude	16S	COI
Procambarus	*paeninsulanus*	5	FL	Walton	SR 20 0.7 mi W Washington Co. line	30.4536	-85.90598	JF737191–JF737195	JF737546–JF737550
Procambarus	*paeninsulanus*	6	FL	Walton	CR 284A 1.5 mi W 284	30.55744	-85.86892	JF737199–JF737208	JF737554–JF737563
Procambarus	*paeninsulanus*	8	FL	Washington	junction Interstate 10 and CR 279	30.7598055	-85.8125277	JF737209–JF737218	JF737564–JF737573
Procambarus	*paeninsulanus*	9	FL	Holmes	Wrights Creek at CR 177A	30.85795	-85.76279	JF737219–JF737226	JF737574–JF737581
Procambarus	*paeninsulanus*	10	FL	Holmes	Alligator Creek at CR 160	30.924841	-85.682463	JF737037	JF737392
Procambarus	*paeninsulanus*	11	FL	Holmes	US 90 1.0 mi E Bonifay	30.7872	-85.6608	JF737038	JF737393
Procambarus	*paeninsulanus*	12	FL	Washington	CR 280 at Holmes Creek, 5 mi E of SR 79	30.73042	-85.61987	JF737039	JF737394
Procambarus	*paeninsulanus*	13	FL	Washington	Holmes Creek at US 90	30.7785	-85.6156	JF737040	JF737395
Procambarus	*paeninsulanus*	14	FL	Washington	Flat Creek at SR 77, S of Chipley	30.6954	-85.5645	JF737041–JF737042	JF737396–JF737397
Procambarus	*paeninsulanus*	16	FL	Washington	Helms Creek at SR 77 N of Chipley	30.8344	-85.533	JF737043–JF737044	JF737398–JF737399
Procambarus	*paeninsulanus*	17	FL	Holmes	Holmes Creek at SR 2	30.9645	-85.5319	JF737044–JF737045	JF737400–JF737401
Procambarus	*paeninsulanus*	21	FL	Gulf	US 98 9.1 mi W Franklin Co. line	29.7805277	-85.2998	JF737047	JF737402
Procambarus	*paeninsulanus*	23	FL	Jackson	Marshall Creek at SR 2	30.9379	-85.2946	JF737048	JF737403
Procambarus	*paeninsulanus*	25	FL	Jackson	SR 73, 1.6 mi S I-10	30.7089722	-85.231	JF737049–JF737055	JF737404–JF737411
Procambarus	*paeninsulanus*	26	FL	Jackson	Chipola River at Caverns Rd NE of Marianna	30.79286	-85.22233	JF737056	JF737411
Procambarus	*paeninsulanus*	27	FL	Liberty	W floodplain Apalachicola River at SR 20	30.435395	-85.000002	JF737057–JF737064	JF737412–JF737419
Procambarus	*paeninsulanus*	28	FL	Calhoun	N of SR 20 at Chipola River	30.4318611	-85.1718611	JF737065–JF737078	JF737420–JF737433
Procambarus	*paeninsulanus*	30	FL	Jackson	Grand Ridge, US 90 1.5 mi E SR 69	30.7189722	-84.992805	JF737079–JF737081	JF737434–JF737436
Procambarus	*paeninsulanus*	32	FL	Liberty	Telogia Creek at SR 20	30.42641	-84.92779	JF737082–JF737086	JF737437–JF737441
Procambarus	*paeninsulanus*	34	FL	Gadsden	SR 20 W of Ochlockonee River	30.38777	-84.66157	JF737087–JF737097	JF737442–JF737452
Procambarus	*paeninsulanus*	35	FL	Gadsden	floodplain of Little River at CR 268	30.5124	-84.5238	JF737098	JF737453

Table A2. Continuation.

Genus	Species	Sample site	State	County	Locality	Latitude	Longitude	16S	COI
Procambarus	paeninsulanus	36	FL	Wakulla	Apalachicola Nat. Forest Rd 309 3 mi W of CR 272	30.290277	-84.480833	JF737099–JF737104	JF737454–JF737459
Procambarus	paeninsulanus	37	FL	Wakulla	Apalachicola Nat. Forest Rd 309 3.4 mi W of CR 267	30.29267	-84.47603	JF737105–JF737113	JF737460–JF737468
Procambarus	paeninsulanus	38	FL	Gadsden	US 90 W Ochlockonee River	30.482833	-84.419222	JF737114–JF737129	JF737469–JF737484
Procambarus	paeninsulanus	39	FL	Leon	St. Marks River headwaters at US 27	30.421	-84.1017	JF737130–JF737136	JF737485–JF737491
Procambarus	paeninsulanus	41	FL	Jefferson	US 27 0.7 mi W US 19	30.4113	-83.9232	JF737137–JF737140	JF737492–JF737495
Procambarus	paeninsulanus	42	FL	Madison	Hixtown Swamp, I-10 6 mi E of US 221	30.4350277	-83.526222	JF737141–JF737155	JF737496–JF737510
Procambarus	paeninsulanus	44	FL	Union	Lake Butler, culvert at SE 6th street and SE 6th Ave	30.01818	-82.3308	JF737156–JF737170	JF737511–JF737525
Procambarus	paeninsulanus	45	FL	Marion	SR 318 3.4 mi E of U.S. 301	29.43222	-82.035	JF737171–JF737180	JF737526–JF737535
Procambarus	paeninsulanus	46	FL	Clay	US 301 0.8 mi S Duval Co. Line, roadside ditch	30.17515166	-82.0246	JF737181–JF737188	JF737536–JF737543
Procambarus	paeninsulanus	47	FL	Nassau	US 90 0.3 mi W Duval Co. Line, roadside ditch	30.302766	-82.02152	JF737189–JF737190	JF737544–JF737545
Procambarus	paeninsulanus	50	FL	Duval	US 1 0.3 mi N I-95	30.1888	-81.562	JF737196	JF737551
Procambarus	paeninsulanus	51	FL	Saint Johns	CR 210 0.6 mi W US 1	30.0832	-81.4636	JF737197	JF737552
Procambarus	paeninsulanus	52	FL	Saint Johns	SR 207 1.1 mi E US 1	29.7567	-81.2945	JF737198	JF737553
Procambarus	spiculifer	1	FL	Escambia	Pine Barren Creek at US 29	30.77628	-87.33868	JF737228–JF737234	JF737583–JF737589
Procambarus	spiculifer	2	FL	Escambia	Canoe Creek at US 29	30.91931	-87.31414	JF737252	JF737607
Procambarus	spiculifer	3	FL	Santa Rosa	SR 4 5 mi W Berrydale	30.95206	-87.09505	JF737305–JF737306	JF737660–JF737661
Procambarus	spiculifer	4	FL	Okaloosa	Shoal River at SR 85	30.69708333	-86.57111	JF737332–JF737333	JF737687–JF737688
Procambarus	spiculifer	7	FL	Washington	Pinelog Creek SR 20 1.9 mi E Ebro	30.441611	-85.8569722	JF737360–JF737378	JF737715–JF737733
Procambarus	spiculifer	9	FL	Holmes	Wrights Creek at CR 177A	30.85795	-85.76279	JF737379–JF737389	JF737734–JF737744
Procambarus	spiculifer	15	FL	Washington	Econfina WMA, Creek Flowing to Whitewater Lake	30.4769	-85.553	JF737235–JF737237	JF737590–JF737592
Procambarus	spiculifer	18	FI	Bay	SR 231 at Bay Creek	30.29923	-85.48625	JF737238	JF737593
Procambarus	spiculifer	19	FL	Bay	Bear Creek at SR 20, Econfina drainage	30.4355	-85.40996	JF737239–JF737251	JF737594–JF737606

Table A2. Continuation.

Genus	Species	Sample site	State	County	Locality	Latitude	Longitude	16S	COI
Procambarus	*spiculifer*	20	FL	Jackson	Econfina Creek at Freeman Rd	30.56956	-85.36222	JF737253–JF737257	JF737608–JF737612
Procambarus	*spiculifer*	22	FL	Calhoun	Juniper Creek at SR 20	30.42634	-85.29479	JF737258	JF737613
Procambarus	*spiculifer*	24	FL	Gulf	Wetappo Creek at SR 22	30.1395	-85.255305	JF737259–JF737269	JF737614–JF737624
Procambarus	*spiculifer*	26	FL	Jackson	Chipola River at Caverns Rd NE Marianna	30.79286	-85.22233	JF737270–JF737286	JF737625–JF737641
Procambarus	*spiculifer*	28	FL	Calhoun	N of SR 20 Chipola River	30.4318611	-85.1718611	JF737287–JF737296	JF737642–JF737651
Procambarus	*spiculifer*	31	FL	Liberty	Sweetwater Creek at SR 12	30.5251	-84.97	JF737307–JF737314	JF737662–JF737669
Procambarus	*spiculifer*	32	FL	Liberty	Telogia Creek at SR 20	30.42641	-84.9277972	JF737315–JF737317	JF737670–JF737672
Procambarus	*spiculifer*	33	FL	Gadsden	Flat Creek at CR 269	30.6285	-84.8347	JF737318–JF737331	JF737673–JF737686
Procambarus	*spiculifer*	40	FL	Liberty	Big Gully Creek at SR 12	30.24715	-85.00488	JF737297–JF737304, JF737334–JF737342	JF737652–JF737659, JF737689–JF737697
Procambarus	*spiculifer*	43	FL	Alachua	Santa Fe River High Springs at US 441 bridge	29.85208	-82.61145	JF737343–JF737347	JF737698–JF737702
Procambarus	*spiculifer*	48	FL	Clay	SR 16 ca. 2.5 mi W SR 21 (Trib. of Bull Creek)	29.981611	-81.9475	JF737348–JF737351	JF737703–JF737706
Procambarus	*spiculifer*	49	FL	Clay	CR 215 at Bull Creek	30.02111	-81.90277	JF737352–JF737359	JF737707–JF737714
Procambarus	*clarkii*	na	FL	Escambia	US 90 at I-10 exit no. 5.	30.53471	-87.378023	JF737227	JF737582
Procambarus	*clarkii*	na	na	na	na	na	na	AF235990	AY701195
Procambarus	*versutus*	na	FL	Okaloosa	US 90 1 mi W of Milligan in stream	30.744944	-86.644778	JF737390	JF737745
Astacus	*astacus*	na	na	na	na	na	na	AF235983	AF517104
Cambarellus	*shufeldtii*	na	na	na	na	na	na	AF235986	EU921149
Cambaroides	*japonicas*	na	na	na	Korea	na	na	JF737391	JF737747
Cambarus	*maculatus*	na	na	na	na	na	na	AF235988	JF737746
Orconectes	*virilis*	na	na	na	na	na	na	AF235989	AF474365
Pacifastacus	*leniusculus*	na	na	na	na	na	na	AF235985	EU921148

Shallow phylogeographic structure of Puerto Rico freshwater crabs: an evolutionary explanation for low species diversity compared to Jamaica

CHRISTOPH D. SCHUBART[1], NICOLE T. RIVERA[1], KEITH A. CRANDALL[2] & TOBIAS SANTL[1]

[1] *Biology 1, University of Regensburg, 93040 Regensburg, Germany*
[2] *Department of Biology, Bringham Young University, Provo, UT 84602-5255, U. S. A.*

ABSTRACT

Freshwater crabs constitute a common faunal component of tropical and subtropical river systems. They have a worldwide occurrence in these warmer regions, being represented by different taxonomic lineages on different continents or even within continents. Due to their mostly direct development and assumed dependency on fresh water, freshwater crabs are considered reliable model organisms to genetically reconstruct the hydrographical history of a region. However, very few studies have been carried out to directly document within-river dispersal or overland dispersal of these crabs. Thus the questions remain, in how far the restriction to river systems is comparable throughout the different taxa of freshwater crabs, and if all of the taxa can be used similarly well to reconstruct the history of watersheds, orogeny, island formation, and continental drift. In the current study, we analyze the phylogeographic structure of *Epilobocera sinuatifrons* (Decapoda: Brachyura: Pseudothelphusidae), a freshwater crab species endemic to the Caribbean island of Puerto Rico. Results show limited morphometric and genetic (mitochondrial and nuclear DNA) differentiation among metapopulations along a west-east gradient, paralleling the direction of the main mountain chain. The north-south comparison, in turn, does not show any differentiation, suggesting that the crabs must be able to migrate between headwaters of unconnected river systems. These results are compared to recently published ones on phylogeographic structure within species of *Sesarma* (Decapoda: Brachyura: Sesarmidae) from Jamaican rivers. The Jamaican freshwater crabs are endemic to a much smaller geographic area and show a pronounced genetic-geographical structure with restricted gene flow among many of the studied rivers systems. These results are unexpected, because the colonization of Jamaica occurred much more recently according to geological history and because the Jamaican crabs still have an abbreviated larval development (González-Gordillo et al. 2010) which should favor distribution within a drainage system and possibly among rivers, if able to survive in coastal areas. This comparison gives evidence for different distribution potential in freshwater crabs and cautions about the assumption that these crabs do not migrate between rivers and are thus infallible biogeographic model systems.

1 INTRODUCTION

The Caribbean islands (or West Indies) consist of the four Greater Antillean islands, Jamaica, Cuba, Hispaniola and Puerto Rico, the Leeward Antilles and the Lesser Antilles. The arc formed by the Greater and Lesser Antilles delimits the Caribbean Sea. Different scenarios have been put forward to explain the geological history of the Caribbean (Buskirk 1985; Pindell 1994; Hedges 2001). Today there is growing evidence that the Proto-Caribbean Plate formed in the eastern Pacific during the Mid Cretaceous around 100 mya. This newly formed plate then moved northeast towards its present position (Pindell 1994). Subduction from the North American Plate under the lighter Caribbean Plate caused formation of the Proto-Antilles which can be considered the ancestral island of the present-day Greater Antilles. The geological history of this region remains very complex and until today there is uncertainty as to which islands were above the sea level at which time (Hedges 1996, 2006). Iturralde-Vinent & MacPhee (1999) claim that no land areas in the Greater Antilles were constantly above sea level before 45 mya, thus not allowing survival of a possible Proto-Antillean fauna and flora after early contact with North and South America. Instead they postulate a mid-Cenozoic vicariant event via the Aves Ridge to South America that allowed land mammals to colonize the Antilles. On the other hand, Hedges (2006, 2010) postulates that all life on the Caribbean islands originated from dispersal, with most terrestrial vertebrates probably arriving via flotsam and prevalent currents from South America, whereas most of the birds, bats, and freshwater fishes appear to have come from North and Central America. The island of Jamaica became submerged around 20 mya, starting in the late Eocene. The limestone and karst formations covering large parts of Jamaica are a result of these submarine epochs. Starting in the late Miocene, Jamaica was lifted again above the sea level (Lewis & Draper 1990; Robinson 1994). The newly emerged island was then available for new biological colonizations, resulting in plenty of endemic animal and plant species. The Greater Antillean island, Puerto Rico, reached its present position around 35 mya. The island lost its connection with Hispaniola in the Miocene (Graham 2003) and was separated from the Virgin Islands due to sea level changes resulting from glacial events in the Quaternary. These changes in water level also altered the amount and distribution of land mass on Puerto Rico, whereas the central mountain range, Cordillera Central, is the result of Eocene volcanism, uplift, and later deformation followed by erosion.

Possibly in consequence of its geological complexity, the Caribbean is today considered one of the biodiversity hotspots of the world (Mittermeier et al. 2004). Within this Caribbean hotspot, the islands of the Greater Antilles harbour a particularly high degree of endemic flora and fauna. These islands cover more than 90% of the 229,549 square kilometres of terrestrial surface in the Caribbean. They also present the highest elevation with 3071 m above sea level, the Pico Duarte on Hispaniola (Orvis 2003). Very different vegetation occurs on the islands, from cactus shrubs, savannahs, evergreen bushland, to freshwater swamps, mangrove forests or lowland rainforests, which are today mostly deforested. In higher elevation, seasonal forest and mountain cloud forest occur (Beard 1955). Among the vertebrate species, frogs show more than 99% endemism (164 out of 165 species). Most of these are endemic to specific islands. The reptiles also bear a high percentage of endemism with around 94% (Hedges 1996). This includes some interesting species radiations, one of which is the genus *Anolis*, with 150 endemic out of 154 species (Roughgarden 1995). In the islands' freshwater systems, 74 species of fish can be found, of which 71 are endemic, some of them even inhabiting single lakes (Hedges 1996). Also the invertebrate fauna of the Caribbean islands has developed a huge amount of endemic species, even if they are not documented as thoroughly as vertebrates. According to Woods & Sergile (2001), the diversity of invertebrates known from the West Indies is only a small fraction of the undocumented present diversity. These authors also remarked that the known species groups tend to be the result of adaptive radiation. As an example, only thirteen species of ostracods were known from Jamaican ponds, most of them widespread in the neotropics. In 1996, Little & Hebert discovered and described eleven new species of ostracods, all from bromeliads, ten of which are endemic to Jamaica. Similarly, the number of endemic milli-pedes of the genus *Anadenobolus* from Jamaica increased from one to three after the genetic study by Bond & Sierwald (2002). The terrestrial mollusc fauna from Jamaica also has a high percent-age of endemic species. Nearly 90%, that is 505 species out of 562, are only found on this island

(Rosenberg & Muratov 2006). In the present study, we will focus our attention on the freshwater crab diversity of two Caribbean islands, the potential of finding undescribed diversity and the mechanisms generating current diversity. In order to do so, we present new data on the Puerto Rican freshwater crab *Epilobocera sinuatifrons* and compare it with already published data on selected Jamaican freshwater crabs (Schubart et al. 2010).

The freshwater crab *Epilobocera sinuatifrons* (A. Milne Edwards 1866) belongs to the family Pseudothelphusidae and is the only freshwater crab of Puerto Rico with a complete freshwater life cycle. It is endemic to the Caribbean islands Puerto Rico and Saint Croix (Chace & Hobbs 1969; Villalobos-Figueroa 1982; Covich & McDowell 1996). Its closest relative is assumed to be the endemic freshwater crab of Hispaniola *Epilobocera haytensis* Rathbun 1893 (see Pretzmann 1974). *Epilobocera sinuatifrons* has a trapezoidal carapace with one anterolateral tooth. Adult individuals can grow to a carapace width of up to 150 mm, maturity is reached with a size of around 30 mm carapace width (Zimmerman & Covich 2003). The species undergoes direct development and females carry relatively large eggs, from which juveniles hatch while the eggs are still carried by the mother. After hatching, the juveniles stay with the mother for some time before they are released into suitable habitats, but do not moult during that time. These habitats can vary greatly. Crabs can be found in rivers of varying structure, from small headwater creeks to large lowland streams, from riverbeds with mainly boulder and rocky composition to sandy and silty ones. According to Zimmermann & Covich (2003), the average flow velocity has an influence on the abundance of juvenile crabs, which tend to prefer higher velocities. Juveniles are often found hiding under rocks, wooden debris or in leaf litter, whereas large adults prefer burrows in sandy or muddy river banks. *Epilobocera sinuatifrons* is omnivorous, with a high percentage of the normal diet comprising palm seeds and fruits, other freshwater invertebrates, and terrestrial snails (Covich & McDowell 1996; March & Pringle 2003). The regular diet of juvenile crabs is unknown (Henry et al. 2000). Unlike its Hispaniolan relative, *Epilobocera haytensis*, *E. sinuatifrons* is no longer a regular component of local human diet, but is more endangered by commercial land use through deforestation and river regulation. The role of humans in translocating animals between rivers is unknown. The phylogeography of the species was described for the first time by Cook et al. (2008) based on representatives of nine rivers. The goal of this study is to describe connectivity in *E. sinuatifrons* among different river systems and to understand the mechanisms of dispersal by comparing morphometric data, mitochondrial, and nuclear DNA of freshwater crabs from 40 localities from throughout Puerto Rico.

2   MATERIALS AND METHODS

Freshwater crabs of the species *Epilobocera sinuatifrons* (Decapoda: Brachyura: Pseudothelphusidae) were collected from 40 localities, corresponding to 23 river systems from throughout Puerto Rico, during four sampling trips between 1997 and 2008 (Figure 1; Appendix, Table A1). For later comparative analyses, these localities were clustered into six (mitochondrial DNA, see Figure 1) or eight (nuclear DNA) geographic regions. These arbitrary regions follow a general west-to-east direction parallel to the central mountain range Cordillera Central, which stretches in the same direction in the southern half of Puerto Rico. In addition, river systems from northern slopes were always distinguished from southern slopes to test the influence of this hydrographic divide. Intentionally, streams with geographically close headwaters, but belonging to either southern or northern drainage systems were sampled. Under the assumption, that each locality should be considered a distinct population, especially if belonging to independent watersheds, as is often the case (see Figure 1), our somewhat artificially created geographic regions will here be regarded as metapopulations. This term was first coined in Levins' (1969) pioneering work, with his own words "a population of populations." It is often applied to species in fragmented habitats and will here be used subsequently for all our geographic clusters.

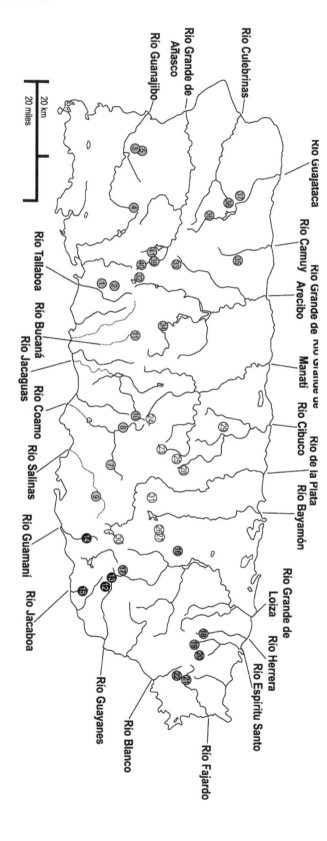

Figure 1. Map of Puerto Rico with 40 (population 6=4) collection points, 23 rivers systems with corresponding names, and the color coding for metapopulations of the freshwater crab *Epilobocera sinuatifrons* as used in the mitochondrial DNA phylogeographic comparisons based on mitochondrial DNA. Green: North-West, blue: South-West, yellow: North-Center, orange: South-West, red: North-East, dark red: South-East (see Figure 7 in Color insert).

buffer, 2.5 μl of 1.25 mM dNTPs, 0.5 μl of both primers (20 mM), 2 μl of 25 mM MgCl$_2$, 1 μl of 0.5u/μl TAQ and 15 μl of double-distilled water in addition of 1 μl DNA. 40 cycles were run at an annealing temperature of 48 °C for the COI and ND1 primers and 50 °C for the ITS primers. COI and ND1 PCR products were cleaned using QuickClean (GenScript, Piscataway NJ) and sequenced with an ABI-PRISM 310 (Applied Biosystems, Carlsbad CA) or outsourced for sequencing.

Cloning of the ITS genes was carried out at Brigham Young University. Initially, PCR products were treated with an A-Addition kit from Quiagen (Qiagen GmbH, Dsseldorf) to add an A-overhang. Cloning itself was performed using the TOPO-TA cloning kit from Invitrogen (Invitrogen Corporation, Carlsbad, CA). 2 μl PCR product was added to a mix of 2 μlddH$_2$O, 1 μl salt solution and 1 μl TOPO vector. This mix was incubated for 30 minutes at room temperature. Chemical competent TOP10 One Shot® *Escherichia coli* cells were thawed on ice, and 2 μl of the TOPO cloning reaction was added. The cells were incubated on ice for 30 minutes, subsequently heat-shocked for 30 seconds at 42 °C, and immediately placed back on ice. 250 μl SOC medium was added and reaction tubes were shaken horizontally (200 rpm) at 37 °C. After 1 hour, 25 μl of the cells were spread evenly on pre-warmed LB plates containing 50 μg/ml ampicillin. Plates were incubated overnight at 37 °C. Colonies that had successfully included the vector with the PCR product were picked and transferred to 50 μl of ddH$_2$O. This solution was denatured for 10 min at 96 °C, and 1 μl was used for a PCR with 35 cycles and 55 °C as an annealing temperature to check if the correct fragment had been cloned. This PCR product was cleaned with PCR Cleanup Millipore plates (Millipore Corporation, Billerica MA) and thereafter cycle-sequenced in both directions using 1/16th of Big Dye v3.0 reaction and standard protocols. The sequencing was performed on an automated ABI 3730 machine. All sequences obtained were proofread for possible errors made by the sequencer software analyses. We used ChromasLite (http://www.technelysium.com) to read chromatograms and edit possible errors. DNA sequences are deposited at Genbank under accession numbers FR871245–FR871285 (ND1 mtDNA) and FN395370–FN395607 (ITS nDNA).

Due to the lack of indels, the corrected sequences for COI and ND1 could be aligned completely by eye using BioEdit (Hall 1999). The ND1 alignment was exported as a Phylip file to construct a statistical parsimony network using the algorithm outlined in Templeton et al. (1992) and implemented in the TCS software package version 1.21 (Clement et al. 2000). Based on the obtained haplotype network of the ND1 data, a nested clade analysis (NCA) was performed (Templeton et al. 1995; Templeton 2004) to test the null hypothesis of no association between the geographic distributions of the haplotypes. The haplotype network was converted into a nested statistical design using the instructions given in Templeton & Sing (1993) and in Crandall & Templeton (1996). To test for an association between the genetic composition and the geographic distribution of the haplotypes, two distances were calculated. First, the clade distance D_c, which estimates how geographically widespread a clade is and second, the nested clade distance D_n, which measures the relative distribution of a clade compared to the other clades in the same higher clade level (see Posada et al. 2006 for details). All calculations were carried out with the application GEODIS 2.5 (Posada et al. 2000), using 1,000,000 permutations and direct distances. The direct-distances option was favored over river distances as all species in this study are freshwater species without marine forms, which would theoretically be able to maintain gene flow among independent watersheds (Fetzner & Crandall 2003). The direct distances between single sample locations were measured with GoogleEarth. We used the most recent inference key from Templeton (http://darwin.uvigo.es/software/geodis.html) to infer the historical events that caused the observed genetic population structure. To measure the genetic differentiation between populations, Φ_{ST} values were calculated using an Analysis of Molecular Variance (AMOVA) in ARLEQUIN ver. 3.0 (Excoffier et al. 2005).

The alignment of ITS sequences was created with the ClustalW plugin of BioEdit. After this initial alignment, we manually checked the microsatellite regions as they are not always correctly recognized by the automated alignment. To be able to analyze the ITS dataset, some pre-processing was necessary. An important part of the information provided by ITS sequences is the high number of indels (Simmons & Ochoterena 2000). These indels are necessary to align microsatellite-like positions in the ITS1/ITS2 region in the noncoding region. The simple indel coding method (Simmons et al. 2001) was applied and calculated with the program GapCoder (Young & Healy 2003) to render the indels phylogenetic information for tree search methods. For further analysis, all alignment files were converted to the Nexus file format. The great amount of variation within the ITS datset did not allow the use of the statistical parsimony algorithm of the TCS software package for network calculation. Therefore, the software Splitstree version 4 was used (Huson & Bryant 2006) to construct minimum spanning networks of the gap-coded ITS sequence data.

3   RESULTS

3.1   Ecology

The collection of *Epilobocera sinuatifrons* during four field seasons between 1997 and 2008 enables us to provide some ecological observations. This species, like other species of *Epilobocera* on Hispaniola and Cuba, is inactive during the daytime, except for cave populations (personal observations). The best way to obtain specimens during the daytime is thus to actively search the river bed and banks of small- to medium-sized rivers. Larger fast-flowing rivers, which probably have a considerable degree of bed-load shift during high water periods and exclusively consist of large boulders, are normally devoid of crabs. On the other hand, it is difficult or impossible to find these crabs in slow-flowing or standing waters (e.g., lakes) with lack of bottom structure. Smaller individuals were easiest to collect from under rocks in the shallow part of rivers by hand, or in deeper parts by placing a hand net downstream of the rock to be turned. In deeper pools of mountainous regions, crabs can be seen and collected by snorkelling and exploring cracks in the rocky walls, which they inhabit together with species of *Macrobrachium* sp. Largest individuals (up to 9.8 cm carapace width) were invariably collected from within caves or relatively deep burrows in the sediment of the water banks with entrances above the water level, but often reaching down to the water table and having more than one burrow opening. This differs from observations by Covich & McDowell (1996) and Zimmerman & Covich (2003), who reported groups of adult crabs under rocks in saturated sections of stream banks. Their findings are from streams of the El Yunque National Forest that are relatively steep, with many pools and often lack muddy banks. This may account for the differences among adult habitats in most of the rivers sampled by us from other regions of Puerto Rico. Our observations on the terrestrial movements of these crabs will be summarized in the Discussion.

Table 1. Classification percentage based on the morphometric classification function for three geographically defined metapopulations of *Epilobocera sinuatifrons* from Puerto Rico. Correct classifications are indicated in bold. The mean correct classification corresponds to 64.4%.

Metapopulation	East	Center	West
East	**66.7**	22.9	10.4
Center	10.5	**84.2**	5.3
West	34.8	21.7	**43.5**

3.2   Morphometry

According to the Kolmogorov-Smirnov test, 10 out of the 15 measured characters in the morphometric dataset of *Epilobocera sinuatifrons* showed a normal distribution; the measurements of the interorbital distance, the carapace height and all measurements from the larger chelae (sexually dimorphic) were not normally distributed. Analyses were continued exclusively with the normal-distributed data: nineteen different populations, from which morphometric data were available, were differently pooled into geographic groups and tested for the highest signal of differentiation. We compared a west-to-east differentiation with a north-to-south differentiation. In addition, we compared the influence of subdividing the dataset into more (five) or less (three) geographic sub-groupings. The clearest signal of differentiation resulted from a subdivision into three groups, namely West, Center, and East. These three groups showed significant differences (Wilk's Lambda 0.614; $P < 0.005$) with an overall correct classification of 64.4% (from 43.5 to 84.2%, see Table 1, Figure 2). When the sampling points were pooled into northern, central and southern groups, no significant differences were found (Wilk's Lambda 0.779; $P = 0.557$) and the corresponding classification only revealed no more than 52.2% (from 48.7 to 56.2%) overall correct placement (see Santl 2009). The morphometric data thus show that there are subtle phenotypic differences in *Epilobocera sinuatifrons* following a west-east direction, compared to less and not significant differences in a north-south direction. However, all these morphometric differences are not very pronounced and do not allow consistent distinction of morphotypes.

3.3   Genetics

Amplification of COI resulted in fragments of 658 bp, of which 624 were compared in an alignment. Many of the sequences showed double peaks in several positions in addition to a surprising homogeneity. We thus suspected the presence of pseudogenes and refrained from continuing with this gene (which was furthermore already used in the study by Cook et al. 2008. Instead, we concentrated on the second mitochondrial gene, the ND1. In total, sequences of ND1 were obtained from 103 individuals, which had to be cropped to a length of 572 bp in order to be used for network and AMOVA analyses. The final alignment included 89 variable positions of which 27 were parsimony-informative, resulting in 41 different haplotypes that are distributed over the six defined metapopulations (Table 2). Relatively high haplotype diversities (h), but moderate nucleotide diversities (π) are noteworthy, indicating that at least some populations are out of equilibrium (Grant & Bowen 1998): all six metapopulations have h values of at least 0.83 and the number of haplotypes is always larger than the 50% value of the sample size, suggesting that additional sampling would uncover many more haplotypes. The dataset was furthermore condensed into three metapopulations in a west-east direction or two metapopulations in a north-south direction. While there is a noticeable decrease from west to east in haplotype diversity, such a gradient is not visible from north to south.

We constructed a TCS haplotype network on these haplotypes and documented their occurrence across the six metapopulations (Figure 3). It can be seen that the distribution is not random and thus geographic influence can be determined. However, there is also no clean genetic separation between the six metapopulations. In the upper half of the network plus cluster 1–5, blue and green colors (with the exception of a dark red coded individual comprised in haplotype 24, which needs to be confirmed) indicate that these haplotypes are restricted to the western part of Puerto Rico as delimited by our metapopulations North-West and South-West. This, however, does not mean that western animals are not represented in the rest of the network. They are still very abundant in the most common haplotype 1 (ht1), which they share with animals from North-Center and South-Center. Eastern animals, as defined by metapopulations North-East and South-East, are not represented in the com-

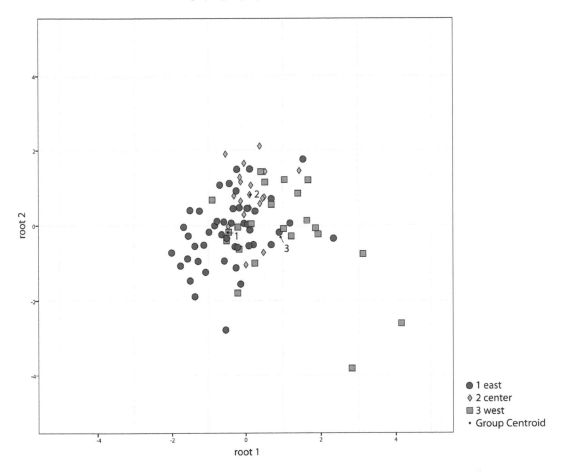

Figure 2. Morphometric analysis of *Epilobocera sinuatifrons* throughout Puerto Rico. Canonical analysis showing plot of the first two discriminant functions. Discrimination based on 10 normally distributed measurements among three metapopulations: West, Center, East. Coloration explained in legend (see Figure 8 in Color insert).

mon haplotype, but they form rare satellite haplotypes 9 and 11 in close vicinity to ht1. Otherwise the lower third of the network, separated by at least five mutational steps from all other haplotypes, is restricted to animals from the central (mainly in haplotype 37) and from the eastern (mainly in haplotype 38) part of Puerto Rico. Table 2 also shows negative and significant Fu's F_S values for all the western populations (North-West -5.096, South-West -7.825, lumped West -12.034), indicating departure from neutrality and population expansion in the west, as also concluded from the increased haplotype diversities in western Puerto Rico. These values, in addition to the structure of the network, suggest that the colonization of Puerto Rican rivers probably took place from west to east.

The estimation of gene flow among the six metapopulations by means of an AMOVA revealed low Φ_{ST} values that nevertheless resulted in restricted gene flow at various significance levels (Table 3). Strongest level of significance ($P < 0.001$) and highest Φ_{ST} values (up to 0.127) were detectable between all pairwise combinations of the three northern populations, between North-Central and South-East and South-West respectively and between North-East and South-

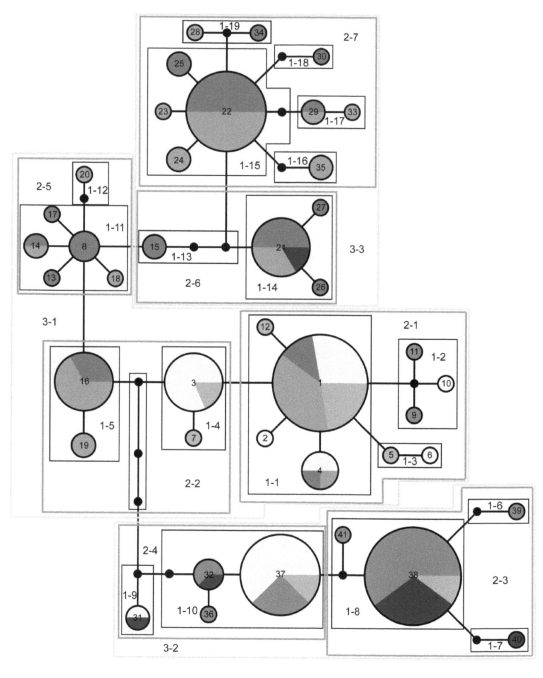

Figure 3. Statistical parsimony network of ND1 haplotypes of *Epilobocera sinuatifrons* ($N = 103$) from Puerto Rico constructed with TCS and the corresponding nesting design for the Nested Clade Analysis. Each black line represents one substitution, dots on lines indicate additional substitutions between haplotypes. The size of a circle represents the frequency of the corresponding haplotype. Coloration according to Figure 1 (see Figure 9 in Color insert).

Table 2. Sample size, number of haplotypes, molecular diversities and population demographic statistics of the ND1 gene (572 bp) of different metapopulations of *Epilobocera sinuatifrons* with a sample size $N \geq 7$. Significant values for demographic parameters are shown in bold.

Metapopulation	Sample Size	Haplotypes	h	π	Tajima's D	Fu's F_S
North-West	28	15	0.942	0.0070	−0.7667	**−5.096**
South-West	26	16	0.954	0.0063	−0.6543	**−7.825**
North-Center	20	9	0.874	0.0068	−0.2634	−0.839
South-Center	7	6	0.952	0.0071	−0.4753	−1.540
North-East	15	8	0.838	0.0061	−0.7373	−1.214
South-East	7	5	0.857	0.0072	−0.7865	−0.047
West	54	24	0.948	0.0067	−0.8797	**−12.034**
Center	27	12	0.883	0.0067	−0.5992	−2.402
East	22	11	0.827	0.0064	−1.0905	−2.534
North	63	29	0.956	0.0098	−0.7036	**−11.693**
South	40	24	0.965	0.0092	−0.4433	**−11.393**
Total	103	41	0.960	0.0097	−0.5262	−2.366

West. Three other pairwise relationships show significant limitation of gene flow, but at a much weaker level ($0.01 < P < 0.05$, Φ_{ST} values from 0.036 to 0.085). It is noteworthy that all three comparisons of local gene flow across the Cordillera Central were nonsignificant and thus imply unrestricted gene flow: North-West vs. South-West: $\Phi_{ST} = 0.001$; North-Center vs. South-Center: $\Phi_{ST} = 0$; North-East vs. South-East: $\Phi_{ST} = 0$. These are by far the lowest values in all pairwise comparisons. This phenomenon is also reflected in Table 4 where condensing the dataset in either West-Central-East or North-South metapopulations reveals that highly significant reduction of gene flow is only detectable in an west-east, but not in a north-south direction.

The Nested Clade Analysis (NCA), which is based on the clades as drawn on top of the haplotype network in Figure 3, proposes significant conclusions for clades 1-14, 1-15 (both restricted gene flow with isolation by distance = ibd) and 1-11 (contiguous range expansion = crg) at the first clade level, for 2-2 (crg) and 2-7 (inconclusive outcome) at the second clade level, and for 3-1 and 3-3 (ibd) at the third clade level. The total network is interpreted by NCA to be shaped by long distance colonization or past fragmentation followed by range expansion.

In order to obtain comparable results from the nuclear genome, we amplified the ITS1-5.8S-ITS2 complex with an average length of around 1620 bp in 40 individuals of *Epilobocera sinuatifrons*. These amplicons were then cloned and treated as described in Material and Methods. In total, we obtained 238 clones and the number of clones per individuals varied between one and seventeen. After incorporation of all indel positions in the dataset, the number of aligned sites increased to 1765. In total, we found 236 alleles resulting in 729 variable sites, of which 258 were parsimonious informative. The fact that 236 alleles were obtained from 40 specimens gives clear evidence for the existence of more than two alleles per animal. Furthermore only two alleles were found twice, suggesting that the dataset is by far not exhaustive and many more alleles can be expected within these 40 animals (see Discussion on nonconcerted evolution of ITS genes in arthropods in Schubart et al. 2010). In Figure 4, the minimum spanning network constructed from this data is displayed. Overall, the ITS dataset does not provide a clear picture of geographic distribution of alleles, but the trends already observed in the morphometric and mitochondrial data can be

Table 3. Φ_{ST} (lower left) and P values (upper right) of six metapopulations of *Epilobocera sinuatifrons*; *: $0.01 < P < 0.05$; **: $0.001 < P < 0.01$; ***: $P < 0.001$ (in bold); –: $P > 0.05$ (not significant).

Φ_{ST} \ P	North-West	South-West	North-Center	South-Center	North-East	South-East
North-West		–	***	–	***	*
South-West	0.001		***	*	***	*
North-Center	0.068	0.077		–	***	***
South-Center	0.013	0.036	0		***	***
North-East	0.107	0.101	0.114	0.037		–
South-East	0.085	0.072	0.127	0.036	0	

Table 4. Φ_{ST} (lower left) and P values (upper right) of three west-eastern or two north-southern metapopulations of *Epilobocera sinuatifrons*; *: $0.01 < P < 0.05$; **: $0.001 < P < 0.01$; ***: $P < 0.001$ (in bold); –: $P > 0.05$ (not significant).

Φ_{ST} \ P	West	Center	East	North	South
West		***	***		
Center	0.064		***		
East	0.104	0.112			
North					–
South				0.001	

confirmed. Three major groupings can be recognized: a western, an eastern, and a central group, respectively, including alleles from the neighboring regions, and standing in a triangular relation to each other. There is no pattern that would allow separating northern from southern metapopulations.

4   DISCUSSION

In the present study, we present morphometric and genetic data describing intraspecific geographic structure of the Puerto Rican freshwater crab *Epilobocera sinuatifrons*. Three datasets (morphometry, mitochondrial DNA, and nuclear DNA) agree in proposing a west-eastern orientation of differentiation processes, alongside the central mountain range (Cordillera Central). On the other hand, in three or four subjectively defined geographic regions of Puerto Rico (e.g., West, Center, East), no restriction of gene flow across the Cordillera Central (northern versus southern drainages) could be determined, thereby giving evidence for faunal exchange across the water divide, not only in one case, but in several regions independently. This finding supports previous observations that this species of freshwater crabs has well-developed capacities for overland dispersal.

Overland movement of *Epilobocera sinuatifrons* has been previously documented by Covich & McDowell (1996) and March & Pringle (2003), with observations that adult crabs often feed on leaf-based detritus, forest fruits, and terrestrial invertebrates on the rainforest floor. Several locals in Puerto Rico also report that during the rainy season, crabs can be found crossing roads or even wandering through human settlements, if adjacent to forest with streams (pers. comm. to authors).

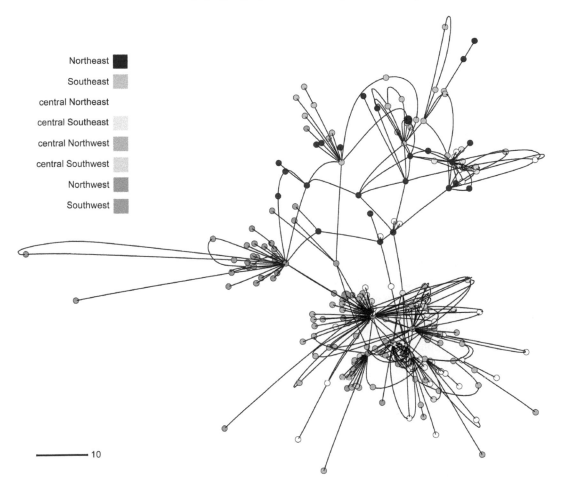

Figure 4. Minimum spanning network based on 238 clones of 1765 gap-coded basepairs of the ITS1-5.8S-ITS2 nuclear DNA region of *Epilobocera sinuatifrons* ($N = 40$) from Puerto Rico. Coloration explained in legend (see Figure 10 in Color insert).

In the Bosque Estatal de Guajataca, we found crabs inhabiting rock rubble in karst sinkholes, thriving in natural crevices and burrows that are probably connected with subterranean water; they are also abundant in several cave systems, which are only partially flooded (personal observations). Unpublished trapping experiments by Alan Covich, and Nicole Rivera and colleagues (pers. comm.) give further evidence that crabs venture out of the streams and thrive on the forest floor at night or during periods of rainfall. This behavior explains how crabs may reach headwaters of one river system after having left their home waters. Forest floor is probably more predominant in the upper rainforests of the Cordillera Central than in the lower mountainous areas or coastal plains, providing an explanation, why exchange in a north-south direction seems to be more frequent than in a west-eastern one.

The population genetic structure of *Epilobocera sinuatifrons* has previously been described by Cook et al. (2008) based on the mitochondrial cytochrome c oxidase subunit I gene (COI). They compare nine river systems from throughout Puerto Rico with a total of 126 individuals and 548 basepairs of aligned sequences. Direct comparison allows concluding that either our sampling scheme of less individuals from more rivers allowed to detect more genetic variability or alter-

Table 5. Comparison of variability in mitochondrial markers in two independent studies on population geographic structure within *Epilobocera sinuatifrons* from throughout Puerto Rico. COI = cytochrome c oxidase subunit I; ND1 = NADH dehydrogenase sununit 1.

Mitochondrial marker	COI (Cook et al. 2008)	ND1 (This study)
Number of individuals	126	103
Number of catchments	9	23
Alignment length	548	572
Variable positions	25	89
Percent variable positions	4.56	15.37
Haplotypes	26	41
Maximal haplotype distance	7	19

natively that the ND1 gene is more variable than the COI gene used by Cook and co-workers (see Table 5). A forthcoming study by Rivera et al. (unpublished data), comparing fewer populations and more individuals of *E. sinuatifrons* with the ND1 gene, and without lumping river systems into metapopulations, will allow a more direct comparison of haplotype and nucleotide diversities of the two genes and determine the influence of sampling strategy or the selected gene. Since the number of mutational steps between haplotypes is clearly higher in the ND1 gene (Table 5), despite including the same geographic extremes, we assume that ND1 in this case is more variable and thus has a higher resolution power.

Cook et al. (2008) recorded an absence of genetic population subdivision among the western rivers, whereas they found restricted gene flow among most other populations, with sample sizes ranging from 4 to 22. They conclude that the western region experienced relatively recent geographic dispersal. What they do not mention in the text is that their pairwise comparisons fail to show differences between Río Guayanés and Río Espíritu Santo in the east of the island and between Río Coamo and Río Grande de Manatí in the central part of the island. Interestingly these rivers, as well as the three western ones, are all north-south counterparts of each other. Only one of Cook et al.'s (2008) north-south river pairs (Río Guayanés versus Río Grande de Arecibo) has restricted gene flow, with a low Φ_{ST} value of 0.189 (P values are not specified in this study, except for being below 0.05). This seems to reflect the same phenomenon as recorded in our paper: limited differentiation in a north-south direction as opposed to much higher differences in an east-west direction. Cook et al. (2008: 160–161) doubt that walking ability alone is likely to facilitate among-river dispersal in this species and propose that "recent gene flow between rivers at the western end of the island was facilitated by recent (Holocene) drainage rearrangements associated with faulting and erosion processes." A conclusion in the same direction is later made (p. 162): "The geologically dynamic history of Puerto Rico appears to have sporadically and repeatedly facilitated continuity in riverine habitat over evolutionary time, thereby prohibiting strong divergence among populations of *E. sinuatifrons* in different rivers." We challenge this explanation and do not think that extrinsic factors are necessary to explain the population structure of *E. sinuatifrons* in Puerto Rico. We recognize in our data as well as in those of Cook et al. (2008) more instances of genetic homogeneity than the ones reported by them in the west of the island. Since most of these cases are in a north-south direction, there seems to be a pattern that can be explained by overland motility alone. This would be more parsimonious than proposing several independent drainage rearrangements via erosion and fault processes, especially, if they are claimed to have occurred in geologically recent times (Pleistocene: 0.6 mya). Clearly, the Eocene volcanism in Puerto Rico cannot have played a role in these postulated hydrographic changes.

One of the rationales to study population genetic structure within *Epilobocera sinuatifrons* from Puerto Rico and *E. haytensis* from Hispaniola (latter data in Rivera & Schubart, in preparation) was to compare and understand the apparent lack of diversity (at least at the species level) of the pseudothelphusid freshwater crabs from these islands (1–2 species/island) with species diversity of sesarmid freshwater crabs from Jamaica (10 species, see Schubart & Koller 2005). Two underlying questions were: 1) is the lack of diversity the consequence of morphological stasis in the genus *Epilobocera*, and is genetic diversity possibly much higher, revealing the existence of old cryptic lineages, and 2) is the different potential for diversification of Jamaican versus Puerto Rican crabs already expressed at the population level?

The results of the present study on *Epilobocera sinuatifrons* from Puerto Rico, and the ones by Rivera & Schubart (in preparation) on *E. haytensis* from Hispaniola, show that there is no deep intraspecific structure within the two species and thus the first question finds an easy and straight-forward response: the apparent lack of diversity in these species is not only based on morphological stasis, but is real and reflected in the genomes. The second question shall be answered by comparing the here presented results of *E. sinuatifrons* with those of a previous study by Schubart et al. (2010) on intraspecific divergence within three freshwater crab species from western and central Jamaica, *Sesarma dolphinum*, *S. meridies*, and *S. windsor*.

Sesarma dolphinum was described as a distinct species from rivers in westernmost Jamaica by Reimer et al. (1998) in order to separate it from the freshwater crab species *S. fossarum* from the neighboring western Cockpit Country. Likewise, *S. meridies* was described by Schubart & Koller (2005) after obtaining evidence from mtDNA and, to a lesser extent, from morphology of its distinctness from *S. windsor* from the eastern Cockpit Country. Preliminary genetic analyses showed that mtDNA in *S. meridies* and *S. windsor* (12S and 16S rRNA genes: Schubart & Koller 2005), *S. dolphinum* (ND1: Santl 2009) and all other Jamaican endemic species (Schubart group, unpublished) often allows to distinguish intraspecific geographically separated populations by diagnostic differences in their DNA. That means that some sequence positions allow to unmistakeably assigning sequences to specific populations, thus resulting in reciprocally monophyletic groups, if represented as phylogenetic trees or networks. This is not really the case in *Epilobocera sinuatifrons* (Figure 3). Even if the distribution is not even and shows clear geographic trends, there is not a single branch that would separate one specific metapopulation from the others and at the same unite all individuals from this metapopulation. The present pattern is thus paraphyletic and reflects recent separation with incomplete lineage sorting (Neigel & Avise 1986) or incomplete separation due to ongoing mixing among neighboring populations.

Nuclear DNA sequence data also confirm higher differentiation potential in Jamaican crabs from freshwater streams (*Sesarma dolphinum*, *S. meridies*, and *S. windsor*) when compared to *Epilobocera sinuatifrons*. The best evidence for this is obtained when comparing the separation of populations in the networks based on the ITS1-5.8S-ITS2 gene regions from *S. dolphinum* (see Schubart et al. 2010: Figure 4) and the one based on the homologous genetic region in *E. sinuatifrons* (see present study: Figure 4). The corresponding F_{ST} values in *S. dolphinum* among single rivers lie between 0.04 and 0.63 (see Schubart et al. 2010: Table 3), whereas between metapopulations (lumping nearby river systems) of *E. sinuatifrons*, these values lie between 0.03 and 0.36 (see Santl 2009). This is even more striking, when considering that *S. dolphinum* only inhabits the western tip of Jamaica and thus an at least five-fold smaller region than *Epilobocera sinuatifrons*, which is distributed throughout Puerto Rico.

Referring back to the second question, stated earlier in this Discussion, whether the different potential for diversification of Jamaican versus Puerto Rican crabs is already expressed at the population level, we can confirm that this is really the case. Pseudothelphusid crabs have probably been present on Puerto Rico, Hispaniola, and Cuba much longer than endemic sesarmid crabs on Jamaica, the latter becoming independent from their marine relatives approximately 4.5 mya (see Schubart et al. 1998). This means that the Pseudothelphusidae did not "use" the available evolutionary time for differentiation and adaptive radiation in the same way as the Jamaican Sesarmidae. The Jamaican endemic crab species still have an abbreviated larval development that should favor distribution within a drainage system and even among rivers, if able to withstand higher salinities in coastal areas. Nevertheless, results from Santl (2009) and Schubart et al. (2010) show that in different species of *Sesarma*, genetic differentiation takes place within single river systems and thus suggest a high local retention of adult as well as larval stages.

The most likely explanation for the observed differences in genetic differentiation potential according to the current findings is that species of *Epilobocera* migrate more regularly between river systems, thereby crossing watersheds, and thus experience less allopatric differentiation and less local specializations due to regular genetic mixing. This differs from distribution models for other freshwater crab families from Africa (Potamonautidae) and Asia (Gecarcinucidae and Potamidae), for which published evidence exists that strong genetic differentiation regularly occurs even among neighboring river populations (Daniels et al. 2001, 2006; Shih et al. 2006; Klaus, Koller & Schubart, unpublished data). However, genetic similarity has been found among widely separated populations of the European freshwater crab *Potamon fluviatile* (see Jesse et al. 2009). These comparisons give evidence for different dispersal potential in different freshwater crab lineages and cautions about the assumption that these crabs do not migrate between rivers and are thus infallible biogeographic model systems.

ACKNOWLEDGEMENTS

The authors are grateful to Silke Reuschel for her participation in one of the collecting trips. Owen McMillan hosted Tobias Santl during several weeks laboratory work at the University of Puerto Rico, campus of Rios Piedras and Nikolaos V. Schizas hosted Nicole T. Rivera during several months laboratory work (Feb.–Aug. 2008) at the University of Puerto Rico, Department of Marine Sciences, Isla Magüeyes Laboratories (funding provided by NSF-Epscor Puerto Rico to NVS; 31 DNA sequences run at the sequencing facility of UPR, Río Piedras, supported in part by NCRR AABRE grant no. P20 RR16470, NIH-SCORE grant no. S06GM08102, University of Puerto Rico Biology Department and NSFCREST grant no. 0206200). Collection permits were issued by the Puerto Rico Department of Natural and Environmental Resources. In Puerto Rico we also enjoyed the hospitality of Florentina Cuevas Rivera, Yogani Govender and Dani Dávila. Special thanks are also due to the team of the Crandall lab in Provo (Utah), who supported T. Santl, while cloning a maximum of ITS alleles in relatively short time intervals and associated support from the U.S. National Science Foundation (EF-0531762). Lapo Ragionieri commented on the methods. This study was financially supported by a six year research project to Christoph D. Schubart (Schu 1460/3) through the Deutsche Forschungsgemeinschaft within the priority program 1127: "Adaptive Radiation—Origin of Biological Diversity." We acknowledge constructive criticism by Sebastian Klaus, one anonymous reviewer, and Stefan Koenemann. Marco T. Neiber was of invaluable help with the formatting.

REFERENCES

Beard, J.S. 1955. The classification of tropical American vegetation-types. *Ecology* 36: 89–100.

Bond, J.E. & Sierwald, P. 2002. Cryptic speciation in the *Anadenobolus excisus* millipede species complex on the island of Jamaica. *Evolution* 56: 1123–1135.

Buskirk, R. 1985. Zoogeographic patterns and tectonic history of Jamaica and the northern Caribbean. *J. Biogeogr.* 12: 445–461.

Chace Jr., F.A. & Hobbs Jr., H.H. 1969 The freshwater and terrestrial decapod crustaceans of the West Indies with special reference to Dominica. *Bull. U.S. Natl. Mus.* 292: 1–258.

Clement, M., Posada, D. & Crandall, K.A. 2000. TCS: a computer program to estimate gene genealogies. *Mol. Ecol.* 9: 1657–1659.

Cook, B.D., Pringle, C.M. & Hughes, J.M. 2008. Phylogeography of an island endemic, the Puerto

Rican freshwater crab (*Epilobocera sinuatifrons*). *J. Hered.* 99: 157–164.

Covich, A. & McDowell, W. 1996. The stream community. In: Reagan, D.P. & Waide, R.B. (eds.), *The food web of a tropical rainforest*: 433–459. Chicago, IL: University of Chicago Press.

Crandall, K.A. & Templeton, A.R. 1996. Applications of intraspecific phylogenetics. In: Harvey, P.H., Brown, A.J.L., Smith, J.M. & Nee, S. (eds.), *New Uses for New Phylogenies*: 81–99. New York, Oxford: Oxford University Press.

Daniels, S.R., Gouws, G. & Crandall, K.A. 2006. Phylogeographic patterning in a freshwater crab species (Decapoda: Potamonautidae: *Potamonautes*) reveals the signature of historical climatic oscillations. *J. Biogeogr.* 33: 1538–1549.

Daniels, S.R., Stewart, B.A. & Burmeister, L. 2001. Geographic patterns of genetic and morphological divergence amongst populations of a river crab (Decapoda: Potamonautidae) with the description of a new species from mountain streams in the Western Cape, South Africa. *Zool. Scripta* 30: 181–197.

Excoffier, L., Laval., G. & Schneider, S. 2005. Arlequin (version 3.0): an integrated software package for population genetics data analysis. *Evol. Bioinform. Online* 1: 47–50.

Fetzner Jr., J.W. & Crandall, K.A., 2003. Linear habitats and the nested clade analysis: an empirical evaluation of geographic versus river distances using an Ozark crayfish (Decapoda: Cambaridae). *Evolution* 57: 2101–2118.

González-Gordillo, J.I., Anger, K., Schubart, C.D. 2010. Morphology of the larval and first juvenile stages of two Jamaican endemic crab species with abbreviated development, *Sesarma windsor* and *Metopaulias depressus* (Decapoda: Brachyura: Sesarmidae). *J. Crust. Biol.* 30: 101–121.

Graham, A. 2003. Geohistory models and Ceonozoic paleoenvironments of the Caribbean region. *Syst. Bot.* 28: 378–386.

Grant, W.A.S. & Bowen, B.W. 1998. Shallow population histories in deep evolutionary lineages of marine fishes: insights from sardines and anchovies and lessons for conservation. *J. Hered.* 89: 415–426.

Hall, T.A. 1999. BioEdit: a user-friendly biological sequence alignment editor and analysis program for Windows 95/98/NT. *Nucleic Acids Symp. Ser.* 41: 95–98.

Harris, D.J. & Crandall, K.A. 2000. Intragenomic variation within ITS1 and ITS2 of freshwater crayfishes (Decapoda: Cambaridae): implications for phylogenetic and microsatellite studies. *Mol. Biol. Evol.* 17: 284–291.

Hedges, S.B. 1996. Historical biogeography of West Indian vertebrates. *Annu. Rev. Ecol. Syst.* 27: 163–196.

Hedges, S.B. 2001. Caribbean biogeography: an overview. In: Woods, C.A. & Sergile, F.E. (eds.), *Biogeography of the West Indies: Patterns and Perspectives*: 15–33. Boca Raton, FL: CRC Press.

Hedges, S.B. 2006. Paleogeography of the Antilles and origin of West Indian Terrestrial Vertebrates. *Ann. Missouri Bot. Gard.* 93: 231–244.

Hedges, S.B. 2010. Molecular clocks, flotsam, and Caribbean islands. In: Cox, C.B. & P.D. Moore (eds.), *Biogeography: An Ecological and Evolutionary Approach. 8th Edition*: 353–354. Hoboken, NJ: John Wiley & Sons, Inc.

Henry, J.K., Covich, A.P., Bowden, T.S. & Crowl, T.A. 2000. Mayfly predation by juvenile freshwater crabs: implications for crab habitat selection. *Bull. North Amer. Benthol. Soc.* 17: 123.

Huson, D.H. & Bryant, D. 2006. Application of phylogenetic networks in evolutionary studies. *Mol. Biol. Evol.* 23: 254–267.

Iturralde-Vinent, M.A. & MacPhee, R.D.E. 1999. Paleogeography of the Caribbean region: implications for Cenozoic biogeography. *Bull. Amer. Mus. Nat. Hist.* 238: 1–95.

Jesse, R., Pfenninger, M., Fratini, S., Scalici, M., Streit, B. & Schubart, C.D. 2009. Disjunct distribution of the freshwater crab *Potamon fluviatile*—natural expansion or human introduction?

Biol. Invas. 11: 2209–2221.

Levins, R. 1969. Some demographic and genetic consequences of environmental heterogeneity for biological control. *Bull. Entomol. Soc. America* 15: 237–240.

Lewis, J.F. & Draper, G. 1990. Geological and tectonic evolution of the northern Caribbean margin. In: Dengo, G. & Case, J.E. (eds.), *Decade of North American Geology. The Caribbean. Volume H*: 77–140. Boulder, CO: Geological Society of America.

Little, T. & Hebert, P. 1996. Endemism and ecological islands: the ostracods from Jamaican bromeliads. *Freshw. Biol.* 36: 327–338.

March, J.G. & Pringle, C.M. 2003. Food web structure and basal resource utilization along a tropical island stream continuum, Puerto Rico. *Biotropica* 35: 84–93.

Milne Edwards, A. 1866. Description de trois nouvelles especes du genre *Boscia*, Crustaces Brachyures de la tribu des Telpheusiens. *Ann. Soc. Entomol. France* (4)6: 203–205.

Mittermeier, R.A., Gil, P.R., Hoffman, M., Pilgrim, J., Brooks, T., Goettsch Mittermeier, C. & Lamoreux, J. 2004. *Hotspots Revisited: Earth's Biologically Richest and Most Endangered Terrestrial Ecoregions.* Mexico City: CEMEX.

Neigel, J.E. & Avise, J.C. 1986. Phylogenetic relationships of mitochondrial DNA under various models of speciation. In: Nevo, E. & Karlin, S. (eds.), *Evolutionary Processes and Theory*: 515–534. New York, NY: Academic Press.

Orvis, K.H. 2003. The highest mountain in the Caribbean: controversy and resolution via GPS. *Carib. J. Sci.* 39: 378–380.

Pindell, J.L. 1994. Evolution of the Gulf of Mexico and the Caribbean. In: Donovan, S.K. & Jackson, T.A. (eds.), *Caribbean Geology: An Introduction*: 13–39. Kingston, Jamaica: The University of the West Indies Publishers Association.

Posada, D., Crandall, K.A. & Templeton, A.R. 2000. GeoDis: a program for the cladistic nested analysis of the geographical distribution of genetic haplotypes. *Mol. Ecol.* 9: 487–488.

Posada, D., Crandall, K.A. & Templeton, A.R. 2006. Nested clade analysis statistics. *Mol. Ecol. Notes* 6: 590–593.

Pretzmann, G. 1974. Zur Systematik der Pseudothelphusidae (Decapoda, Brachyura). *Crustaceana* 27: 294–304.

Rathbun, M.J. 1893. Descriptions of new species of American fresh-water crabs. *Proc. U. S. Natl. Mus.* 16: 649–661, pls. LXXIII–LXXVII.

Reimer, J., Schubart, C.D. & Diesel, R. 1998. Description of a new freshwater crab of the genus *Sesarma* Say, 1817 (Brachyura: Grapsidae: Sesarminae) from western Jamaica. *Crustaceana* 71: 186–196.

Reuschel, S. & Schubart C.D. 2006. Phylogeny and geographic differentiation of two Atlanto-Mediterranean species of the genus *Xantho* (Crustacea: Brachyura: Xanthidae) based on genetic and morphometric analyses. *Mar. Biol.* 148: 853–866.

Rivera, N.T. & Schubart C.D. (in preparation). Phylogeography of the freshwater crab *Epilobocera haytensis* (Brachyura: Pseudothelphusidae) from Hispaniola reveals partly restricted gene flow among different river systems.

Robinson, E.J. 1994. Jamaica. In: Donovan, S.K. & Jackson, T.A. (eds.), *Caribbean Geology: An Introduction*: 111–127. Kingston, Jamaica: The University of the West Indies Publishers Association.

Rosenberg, G. & Muratov, I.V. 2006. Status report on the terrestrial Mollusca of Jamaica. *Proc. Acad. Nat. Sci. Philadelphia* 155: 117–161.

Roughgarden, J. 1995. *Anolis Lizards of the Caribbean: Ecology, Evolution and Plate Tectonics.* New York, Oxford: Oxford University Press.

Santl, T. 2009. Comparative diversification potential of an old and a young lineage of freshwater crabs on two Caribbean islands explained at the population level. Electronically published Ph.D. dissertation, Universität Regensburg, http://epub.uni-regensburg.de/13391/.

Schubart, C.D. 2009. Mitochondrial DNA and decapod phylogenies; the importance of pseudogenes and primer optimization. In: Martin, J.W., Crandall, K.A. & Felder, D.L. (eds.), *Crustacean Issues 18: Decapod Crustacean Phylogenetics*: 47–65. Boca Raton, FL: Taylor & Francis/CRC Press.

Schubart, C.D., Diesel, R. & Hedges, S.B. 1998. Rapid evolution to terrestrial life in Jamaican crabs. *Nature* 393: 363–365.

Schubart, C.D. & Huber, M.G.J. 2006. Genetic comparisons of German populations of the stone crayfish, *Austropotamobius torrentium* (Crustacea: Astacidae). *Bull. Franç. Pêche Piscic.* 380–381: 1019–1028.

Schubart, C.D., Koller, P. 2005. Genetic diversity of freshwater crabs (Brachyura: Sesarmidae) from central Jamaica with description of a new species. *J. Nat. Hist.* 39: 469–481.

Schubart, C.D., Weil, T., Stenderup, J.T., Crandall, K.A. & Santl, T. 2010. Ongoing phenotypic and genotypic diversification in adaptively radiated freshwater crabs from Jamaica. In: Glaubrecht, M. (ed.), *Evolution in Action*: 323–349. Berlin, Heidelberg: Springer-Verlag.

Schulenburg, J.H.G.v.d., Hancock, J.M., Pagnamenta, A., Sloggett, J.J., Majerus, M.E.N. & Hurst, G.D.D. 2001. Extreme length and length variation in the first ribosomal internal transcribed spacer of ladybird beetles (Coleoptera: Coccinellidae). *Mol. Biol. Evol.* 18: 648–660.

Shih, H.-T., Hung, H.-C., Schubart, C.D., Chen, C.A. & Chang, H.-W. 2006. Intraspecific genetic diversity of the endemic freshwater crab *Candidiopotamon rathbunae* (Crustacea: Decapoda, Brachyura, Potamidae) reflects five million years of geological history of Taiwan. *J. Biogeogr.* 33: 980–989.

Simmons, M.P. & Ochoterena, H. 2000. Gaps as characters in sequence-based phylogenetic analyses. *Syst. Biol.* 49: 369–381.

Simmons, M.P., Ochoterena, H. & Carr, T.G. 2001. Incorporation, relative homoplasy, and effect of gap characters in sequence-based phylogenetic analysis. *Syst. Biol.* 50: 454–462.

Tang, B., Zhou, K., Song, D., Yang, G. & Dai, A., 2003. Molecular systematics of the Asian mitten crabs, genus *Eriocheir* (Crustacea: Brachyura). *Mol. Pylogenet. Evol.* 29: 309–316.

Templeton, A.R. 2004. Statistical phylogeography: methods of evaluating and minimizing inference errors. *Mol. Ecol.* 13: 789–809.

Templeton, A.R., Crandall, K.A. & Sing, C.F. 1992. A cladistic analysis of phenotypic associations with haplotypes inferred from restriction endonuclease mapping and DNA sequence data. III. Cladogram estimation. *Genetics* 132: 619–633.

Templeton, A.R., Routman, E. & Phillips, C.A. 1995. Separating population structure from population history: a cladistic analysis of the geographical distribution of mitochondrial DNA haplotypes in the tiger salamander, *Ambystoma tigrinum. Genetics* 140: 767–782.

Templeton, A.R. & Sing, C.F. 1993. A cladistic analysis of phenotypic associations with haplotypes inferred from restriction endonuclease mapping. IV. Nested analyses with cladogram uncertainty and recombination. *Genetics* 134: 659–669.

Vogler, A.P. & DeSalle, R. 1994. Evolution and phylogenetic information content of the ITS-1 region in the tiger beetle *Cicindela dorsalis. Mol. Biol. Evol.* 11: 393-405.

Villalobos-Figueroa, A. 1982. Decapoda. In: Hurlbert, S.H. & Villalobos-Figueroa, A. (eds.), *Aquatic Biota of Mexico, Central America and the West Indies*: 215–239. San Diego, CA: San Diego State University Press.

Woods, C.A. & Sergile, F.E. 2001. *Biogeography of the West Indies: Patterns and Perspectives.* Boca Raton, FL: CRC Press.

Young, N.D. & Healy, J. 2003. GapCoder automates the use of indel characters in phylogenetic analysis. *BMC Bioinform.* 4: 6.

Zimmerman, J.K.H. & Covich, A.P. 2003. Distribution of juvenile crabs (*Epilobocera sinuatifrons*) in two Puerto Rican headwater streams: effects of pool morphology and past land-use legacies. *Archiv Hydrobiol.* 158: 343–357.

APPENDIX

Table A1. List of localities and specimens of *Epilobocera sinuatifrons* used for genetic and morphometric comparisons (sample numbers according to map in Figure 1).

Assignment	Sample No.	Drainage System	Locality	Sampling Date	Coordinates	N (Morphometrics)	N (Genetics)	Haplotype No. (see Figure 3)	Museum No.
Southwest (Blue)	1	Río Tallaboa	Convento Cave	07.03.2008	18°02.631'N 66°44.903'W	0	3	8, 14, 22	ZRC 2011-0204
	2	Río Tallaboa	Guayanés 2	18.10.2004	18°05.773'N 66°44.196'W	5	5	1, 4, 16, 22, 27	SMF 38907
	3	Río Guanajibo	Nueve Pasos	17.10.2004	18°09.467'N 67°04.534'W	14	11	8 (2×), 13, 15 (2×), 21 (3×), 22, 25, 29	SMF 38905
	4	Río Guanajibo	Pico Fraile	17.10.2004	18°07.524'N 66°54.975'W	6	4	17, 25, 26, 34	BMNH
	5	Río Guanajibo	Quebrada Flora	05.03.2008	18°10.045'N 67°04.135'W	0	3	16, 22, 30	
South-Center (Orange)	7	Río Coamo	Cuyon	09.05.2006	18°05.253'N 66°16.266'W	0	2	5, 37	
	8	Río Coamo	Coamo	10.05.2006	18°07.001'N 66°21.901'W	0	1	3	RMNH Crust.D.53425
	9	Río Salinas	Jajome	09.05.2006	18°02.856'N 66°11.950'W	0	2	12, 38	SMF 38909
	10	Río Jacaguas	Toa Vaca	10.05.2006	18°09.320'N 66°23.621'W	0	1	1	
	11	Río Bucaná	Cerillo	15.05.2006	18°08.753'N 66°36.472'W	0	1	1	
Southeast (Darkred)	12	Río Guayanés	Arenas	02.03.2008	18°03.581'N 65°57.896'W	0	3	31, 38 (2×)	MNHN U-2011-861
	13	Río Guayanés	Guayanés 1	12.10.2004	18°17.717'N 65°71.117'W	4	0	3	SMF 38911
	14	Río Guamaní	Guamaní	13.10.2004	18°02.310'N 66°06.110'W	4	2	21, 32	
	15	Río Jacaboa	Jacaboa	13.10.2004	18°01.149'N 65°57.704'W	4	2	38, 40	ZSM A 20110102
Northeast (Red)	16	Río Grande de Loiza	Cuevas Aguas Buenas	03.03.2008	18°13.950'N 66°06.490'W	0	3	11, 32, 38	SMF 38908
	17	Río Grande de Loiza	Grande de Loiza	13.10.2004	18°05.072'N 65°59.936'W	4	2	38 (2×)	ZRC 2011-0206
	18	Río Herrera	Herrera	10.03.2008	18°19.507'N 65°51.546'W	0	3	32, 39, 41	
	19	Río Espíritu Santo	El Verde	09.03.2008	18°19.264'N 65°49.185'W	0	1	38	RMNH Crust.D.53426
	20	Río Espíritu Santo	Espíritu Santo	23.10.2004	18°19.464'N 65°49.140'W	4	1	37	RMNH Crust.D.53426
	21	Río Fajardo	Fajardo	16.10.2004	18°16.904'N 65°43.896'W	8	4	9, 36, 38 (2×)	NHMW 25234
	22	Río Blanco	Blanco	16.10.2004	18°14.407'N 65°45.329'W	8	1	37	SMF 38910

Table A1. Continuation.

Assignment	Sample No.	Drainage System	Locality	Sampling Date	Coordinates	N (Morpho-metrics)	N (Genetics)	Haplotype No. (see Figure 3)	Museum No.
North-Center (Yellow)	23	Río Manatí	Manatí	09.03.2008	18°15.640'N 66°17.880'W	0	3	1, 2, 7	BMNH
	24	Río Manatí	Canabon	09.05.2006	18°13.685'N 66°20.488'W	0	2	3, 4	
	25	Río Manatí	Bauta	10.05.2006	18°10.444'N 66°24.439'W	0	2	3 (2×)	NHMW 25233
	26	Río Bayamón	Bayamón 1	04.03.2008	18°12.323'N 66°08.365'W	0	3	1, 31, 37	
	27	Río Bayamón	Bayamón 2	09.05.2006	18°12.328'N 66°08.352'W	0	1	37	
	28	Río Cibuco	Mavilla	09.03.2008	18°16.106'N 66°16.424'W	0	3	3, 6, 10	ZSM A 20110101
	29	Río Cibuco	Cueva Buruquena	25.07.2008	18°21.761'N 66°22.649'W	0	3	1, 3, 4	
	30	Río de la Plata	Plata	13.10.2004	18°05.710'N 66°04.854'W	4	2	37 (2×)	ZRC 2011-0205
	31	Río de la Plata	Arroyata	09.05.2006	18°11.966'N 66°12.528'W	0	1	37	MNHN U-2011-860
Northwest (Green)	32	Río Grande de Arecibo	Río Vacas	17.10.2004	18°08.624'N 66°44.952'W	2	2	19, 28	SMF 38906
	33	Río Grande de Arecibo	Río Tanama	21.10.2004	18°13.150'N 66°45.471'W	3	2	29, 33	MNHN U-2011-859
	34	Río Grande de Arecibo	Río Jauca	15.05.2006	18°11.158'N 66°38.365'W	0	2	4, 21	
	35	Río Camuy	Cueva Represa	08.03.2008	18°24.031'N 66°47.609'W	0	3	14, 20, 21	RMNH Crust.D.53424
	36	Río Guajataca	Guajataca	20.10.2004	18°19.822'N 66°54.955'W	7	5	1, 16, 18, 22, 35	SMF 38917
	37	Río Guajataca	Busque Estatal	20.10.2004	18°24.791'N 66°57.980'W	7	1	35	SMF 38904
	38	Río Culebrinas	Culebrinas	12.10.2004	18°22.105'N 66°57.185'W	6	1	22	
	39	Río Grande de Añasco	Guilarte 11	17.10.2004	18°10.300'N 66°46.292'W	2	2	1, 24	NHMW 25232
	40	Río Grande de Añasco	Guilarte 12	15.05.2006	18°08.550'N 66°45.951'W	3	8	1 (2×), 16 (3×), 19, 22, 23	
	41	Río Grande de Añasco	Limani	17.10.2004	18°10.392'N 66°48.217'W	5	2	22, 24	ZSM A 20110100

Total: 100   Total: 103

Contributors

Abatzopoulos, Theodore J.: Department of Genetics, Development & Molecular Biology, School of Biology, Aristotle University of Thessaloniki, Thessaloniki, Greece
E-mail: abatzop@bio.auth.gr (corresponding author)

Alexandrino, Paulo: CIBIO, Centro de Investigacão em Biodiversidade e Recursos Genéticos, Universidade do Porto, Campus Agrário de Vairão, 4485-661 Vairo, Portugal
E-mail: palexan@mail.icav.up.pt

Ambariyanto: Faculty of Fisheries and Marine Sciences, Diponegoro University, Semarang, Indonesia
E-mail: ambariyanto@undip.ac.id

Barber, Paul H.: Department of Ecology and Evolutionary Biology, University of California at Los Angeles, Los Angeles, U. S. A.
E-mail: paulbarber@ucla.edu

Baxevanis, Athanasios D.: Department of Genetics, Development & Molecular Biology, School of Biology, Aristotle University of Thessaloniki, Thessaloniki, Greece
E-mail: tbaxevan@bio.auth.gr

Bird, Christopher E.: Hawai'i Institute of Marine Biology, School of Ocean and Earth Sciences and Technology, University of Hawai'i at Mānoa, Kāne'ohe, HI 96744, U. S. A.
E-mail: cbird@hawaii.edu

Bracken-Grissom, Heather D.: Department of Biology, Brigham Young University, Provo, UT 84602, U. S. A.
E-mail: heather.bracken@gmail.com

Breinholt, Jesse W.: Department of Biology, Brigham Young University, Provo, UT, U. S. A.
E-mail: jessebreinholt@gmail.com; jessewayne34@hotmail.com

Cabezas, Patricia: MNCN, Museo Nacional de Ciencias Naturales (CSIC), Biodiversidad y Biología Evolutiva, José Gutiérrez Abascal 2, 28006 Madrid, Spain
E-mail: pcabezaspadilla@gmail.com

Chan, Benny K. K.: Biodiversity Research Center, Academia Sinica, Taipei, Taiwan
E-mail: chankk@gate.sinica.edu

Cheng, Samantha H.: Department of Ecology and Evolutionary Biology, University of California at Los Angeles, Los Angeles, U. S. A.; Joint Science Department, Scripps College, Claremont, U. S. A.
E-mail: scheng87@gmail.com

Chu, Ka Chou: Simon F. S. Li Marine Science Laboratory, School of Life Sciences, The Chinese University of Hong Kong, Hong Kong
E-mail: kahouchu@cuhk.edu.hk (corresponding author)

Cook, Benjamin D.: Australian Rivers Institute, Griffith University, Nathan, Queensland, Australia, 4111; Tropical Rivers and Coastal Knowledge Commonwealth Environmental Research Facility
E-mail: ben.cook@griffith.edu.au

Crandall, Keith A.: Department of Biology & Monte L. Bean Life Science Museum, Brigham Young University, Provo, UT 84602, U. S. A.
E-mail: keith_crandall@byu.edu

Dufresne, France: Département de Biologie, Centre d'Études Nordiques, Université du Québec à Rimouski, Rimouski, Canada
E-mail: france_dufresne@uqar.qc.ca

Enders, Tiffany: Department of Biology, Brigham Young University, Provo, UT 84602, U. S. A.
E-mail: tenders3@gmail.com

Erdmann, Mark V.: Conservation International, Renon Denpasar, Indonesia
E-mail: m.erdmann@conservation.org

Fratini, Sara: Department of Evolutionary Biology, University of Florence, Florence, Italy
E-mail: sarafratini@unifi.it

Froufe, Elsa: CIIMAR, Centro Interdisciplinar de Investigação Marinha e Ambiental, Rua dos Bragas 289, 4050-123 Porto, Portugal
E-mail: elsafroufe@gmail.com

Grosberg, Richard K.: College of Biological Sciences, Center for Biology, University of California at Davis, Davis, U. S. A.
E-mail: rkgrosberg@ucdavis.edu

Held, Christoph: Alfred Wegener Institute for Polar and Marine Research, Functional Ecology, Am alten Hafen 26, D-27568 Bremerhaven, Germany
E-mail: christoph.held@awi.de

Hughes, Jane M.: Australian Rivers Institute, Griffith University, Nathan, Queensland, Australia, 4111; Tropical Rivers and Coastal Knowledge Commonwealth Environmental Research Facility
E-mail: jane.hughes@griffith.edu.au

Jara, Carlos G.: Instituto de Zoología, Casilla 567, Universidad Austral de Chile, Valdivia, Chile
E-mail: cjara@uach.cl

Kappas, Ilias: Department of Genetics, Development & Molecular Biology, School of Biology, Aristotle University of Thessaloniki, Thessaloniki, Greece
E-mail: ikappas@bio.auth.gr

Karl, Stephen A.: Hawai'i Institute of Marine Biology, School of Ocean and Earth Sciences and Technology, University of Hawai'i at Mānoa, Kāne'ohe, HI 96744, U. S. A.
E-mail: skarl@hawaii.edu

Leese, Florian: Ruhr University of Bochum, Department of Animal Ecology, Evolution and Bio-diversity, Universitätsstraße 150, D-44801 Bochum, Germany; British Antarctic Survey, High Cross, Madingley Road, Cambridge CB3 0ET, United Kingdom
E-mail: florian.leese@rub.de

Moler, Paul E.: Florida Fish and Wildlife Conservation Commission, Gainesville, FL, U. S. A.
E-mail: paul.moler@fwc.state.fl.us

Neigel, Joseph E.: Department of Biology, University of Louisiana at Lafayette, Lafayette, U. S. A.
E-mail: jneigel@louisiana.edu

Ng, Wai Chuen: The Swire Institute of Marine Science and The School of Biological Sciences, The University of Hong Kong, Hong Kong
E-mail: chueneugene@yahoo.com

Page, Timothy J.: Australian Rivers Institute, Griffith University, Nathan, Queensland, Australia, 4111
E-mail: t.page@griffith.edu.au

Pérez-Losada, Marcos: CIBIO, Centro de Investigacão em Biodiversidade e Recursos Genéticos, Universidade do Porto, Campus Agrário de Vairão, 4485-661 Vairão, Portugal
E-mail: mlosada@genoma-llc.com

Phongdara, Amornrat: Center for Genomics and Bioinformatics Research, Faculty of Science, Prince of Songkla University, Hat-Yai, Songkhla, Thailand
E-mail: pamornra@yahoo.com

Ragionieri, Lapo: Department of Evolutionary Biology, University of Florence, Florence, Italy
E-mail: lapo.ragionieri@gmail.com

Rivera, Nicole T.: Biology 1, University of Regensburg, 93040 Regensburg, Germany
E-mail: Nicole.Rivera@biologie.uni-regensburg.de

Santl, Tobias: Biology 1, University of Regensburg, 93040 Regensburg, Germany
E-mail: tobias.santl@googlemail.com

Schubart, Christoph D.: Biology 1, University of Regensburg, 93040 Regensburg, Germany
E-mail: christoph.schubart@biologie.uni-regensburg.de

Smouse, Peter E.: Ecology, Evolution and Natural Resources, School of Environmental and Biological Sciences, Rutgers, The State University of New Jersey, New Brunswick, NJ 08901, U. S. A.
E-mail: smouse@aesop.rutgers.edu

Tenggardjaja, Kimberly: Department of Ecology and Evolutionary Biology, University of California at Santa Cruz, Santa Cruz, U. S. A.
E-mail: kimberly.tenggardjaja@gmail.com

Toonen, Robert J.: Hawai'i Institute of Marine Biology, School of Ocean and Earth Sciences and Technology, University of Hawai'i at Mānoa, Kāne'ohe, HI 96744, U. S. A.
E-mail: toonen@hawaii.edu

Tsang, Ling Ming: Simon F. S. Li Marine Science Laboratory, School of Life Sciences, The Chinese University of Hong Kong, Hong Kong
E-mail: kiryusky@gmail.com

von Rintelen, Kristina: Museum für Naturkunde, Leibniz-Institut für Evolutions- und Biodiversitätsforschung an der Humboldt-Universität zu Berlin, Invalidenstr. 43, 10115 Berlin, Germany
E-mail: kristina.rintelen@mfn-berlin.de

Wanna, Warapond: Center for Genomics and Bioinformatics Research, Faculty of Science, Prince of Songkla University, Hat-Yai, Songkhla, Thailand
E-mail: w.wanna@yahoo.com

Williams, Gray A.: The Swire Institute of Marine Science and The School of Biological Sciences, The University of Hong Kong, Hong Kong
E-mail: hrsbwga@hkucc.hku.hk

Wu, Tsz Huen: Simon F. S. Li Marine Science Laboratory, School of Life Sciences, The Chinese University of Hong Kong, Hong Kong
E-mail: sharontszv@yahoo.com.hk

Xu, Jiawu: Department of Biology & Monte L. Bean Life Science Museum, Brigham Young University, Provo, UT 84602, U. S. A.
E-mail: jiawuxu@gmail.com

Yednock, Bree K.: Department of Biology, University of Louisiana at Lafayette, Lafayette, U. S. A.
E-mail: bky8151@louisiana.edu

Index

Color inserts

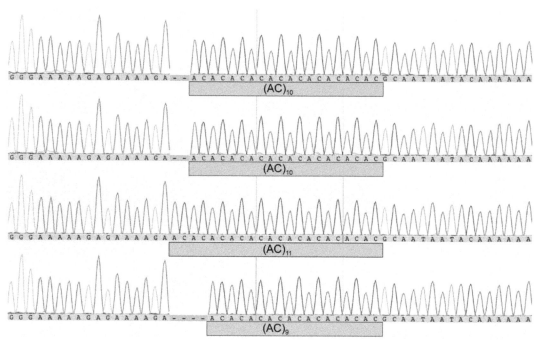

Figure 1 (Figure 4 in Leese & Held). Using a DNA mix of different individuals allows testing for variation at individual loci after shotgun 454 sequencing. In this case, three different alleles for the microsatellite locus, i.e., $(CA)_9$, $(CA)_{10}$ and $(CA)_{11}$, are clearly distinguishable (own data from an enrichment according to Leese et al. 2008)

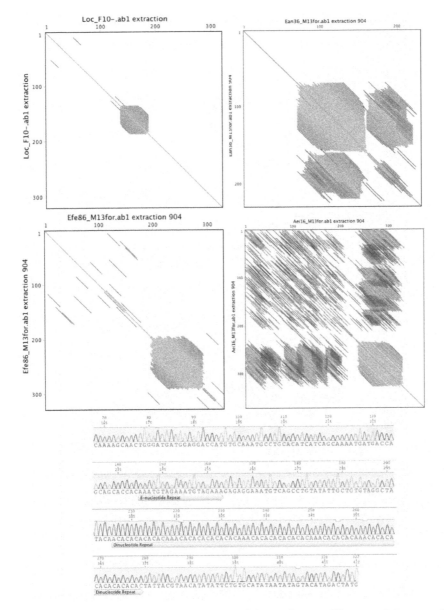

Figure 2 (Figure 5 in Leese & Held). Dotplots of four microsatellite-containing contigs against themselves (Geneious Pro version 5.3). The main diagonal indicates the perfect identity of the sequence (X axis) against itself (Y axis). Off-diagonal parallels to the main diagonal indicate that sequence motifs occur elsewhere in the sequence, albeit slightly modified and thus not easily detected using bioinformatic approaches or inspection by eye. Candidate sequences with enough room to place primers in unique parts of the flanking sequence (indicated by lack of parallels, upper left panel) are to be preferred over candidates with cryptically repetitive flanking regions. Note that the Phobos search for repeats with commonly used parameters only identifies one additional 8 bp repeat (annotations in lower panel; Geneious Pro). The remaining repeat structures are camouflaged by their lower degree of sequence conservation in the sequence view (lower panel), but clearly visible in the dotplot of the same sequence (lower left panel).

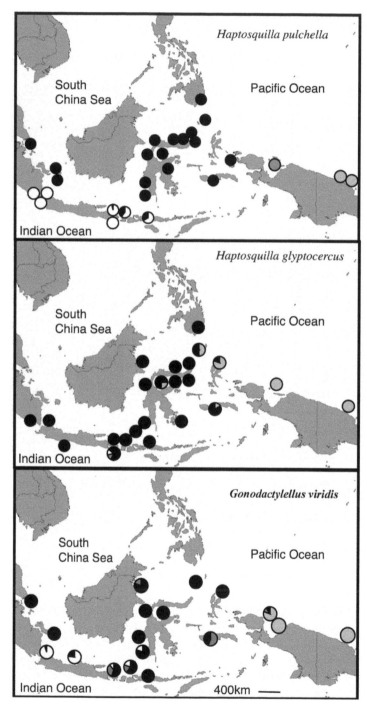

Figure 3 (Figure 3 in Barber et al.). Phylogeographic structure of *Haptosquilla pulchella*, *H. glyptocercus* and *Gonodactylellus viridis* across the Coral Triangle (after Barber et al. 2006) showing both divergence among Pacific and Indian Ocean populations as well as divergence between Eastern and Central Indonesia.

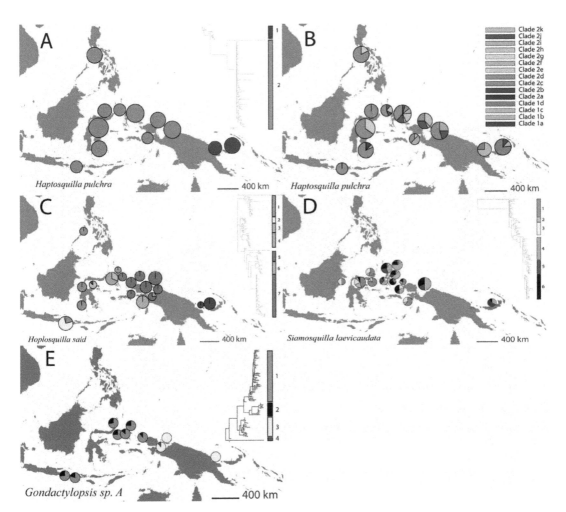

Figure 4 (Figure 4 in Barber et al.). Phylogeographic structure of A) *Haptosquilla pulchra* (broad scale), B) *H. pulchra* (fine scale), C) *Hoplosquilla said*, D) *Siamosquilla laevicaudata* and E) *Gonodactylopsis* sp. A. across the Coral Triangle based on unique mtDNA COI haplotypes.

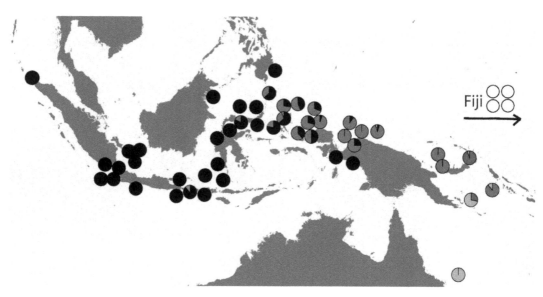

Figure 5 (Figure 6 in Barber etal.). Distribution of five mtDNA clades in *Haptosquilla glyptocercus* across the Coral Triangle and Central Pacific. Relationship of clades is shown in Figure 7 in Barber et al.

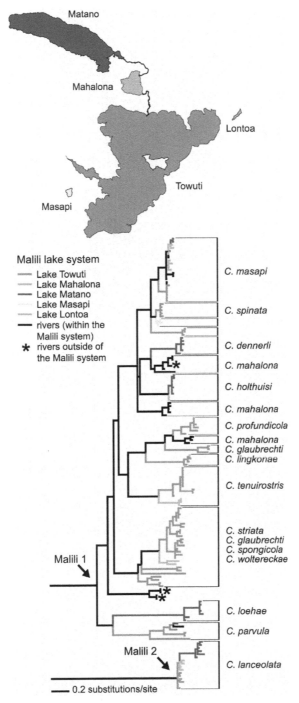

Figure 6 (Figure 2 in von Rintelen) Bayesian Inference phylogram (1332 basepairs of combined 16S and COI mtDNA) of *Caridina* from the Malili lake system. Detail topology of the two Malili clades with its 15 species (for the entire topology, compare Figure 1D in von Rintelen). The oc-currence of sequenced specimens in single lakes and surrounding rivers are colour-coded (modified from von Rintelen et al. 2010).

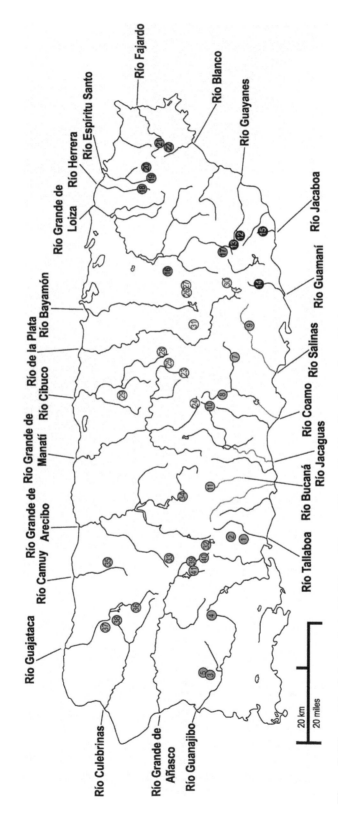

Figure 7 (Figure 1 in Schubart et al.). Map of Puerto Rico with 40 collection points, 23 rivers systems with corresponding names, and the colour coding for metapopulations of the freshwater crab *Epilobocera sinuatifrons* as used in the mitochondrial DNA phylogeographic comparisons based on mitochondrial DNA. Green: North-West, blue: South-West, yellow: North-Center, orange: South-Center, red: North-East, dark red: South-East.

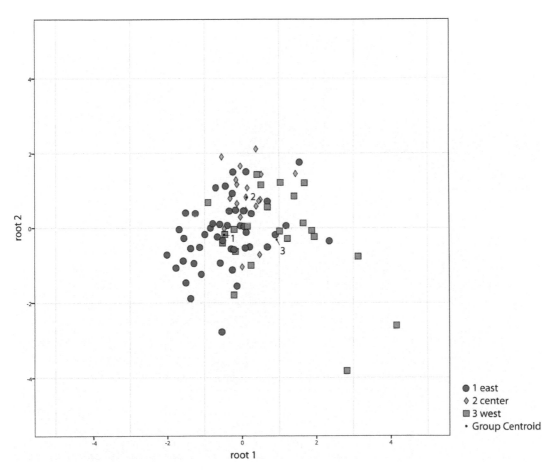

Figure 8 (Figure 2 in Schubart et al.). Morphometric analysis of *Epilobocera sinuatifrons* throughout Puerto Rico. Canonical analysis showing plot of the first two discriminant functions. Discrimination based on 10 normally distributed measurements among three metapopulations: West, Center, East. Coloration explained in legend.

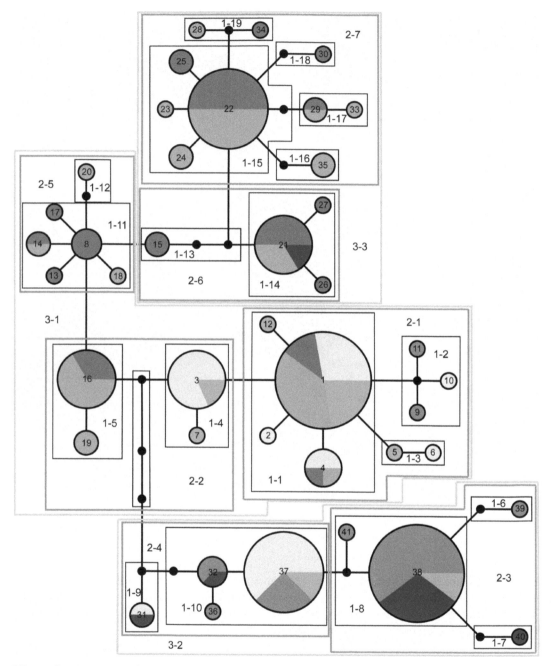

Figure 9 (Figure 3 in Schubart et al.). Statistical parsimony network of ND1 haplotypes of *Epilobocera sinuatifrons* ($N = 103$) from Puerto Rico constructed with TCS and the corresponding nesting design for the Nested Clade Analysis. Each black line represents one substitution, dots on lines indicate additional substitutions between haplotypes. The size of a circle represents the frequency of the corresponding haplotype. Coloration according to Figure 1 in Schubart et al.

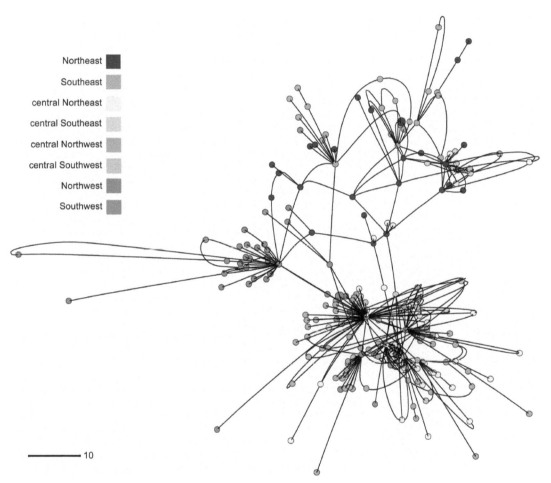

Figure 10 (Figure 4 in Schubart et al.). Minimum spanning network based on 238 clones of 1765 gap-coded basepairs of the ITS1-5.8S-ITS2 nuclear DNA region of *Epilobocera sinuatifrons* (N = 40) from Puerto Rico. Coloration explained in legend.